MICRODROPS AND DIGITAL MICROFLUIDICS

MICRO & NANO TECHNOLOGIES

Series Editor: Jeremy Ramsden
Professor of Nanotechnology
Microsystems and Nanotechnology Centre, Department of Materials
Cranfield University, United Kingdom

The aim of this book series is to disseminate the latest developments in small scale technologies with a particular emphasis for accessible and practical content. These books will appeal to engineers from industry, academia and government sectors.

For more information about the book series and new book proposals please contact the Publisher, Dr. Nigel Hollingworth at nhollingworth@williamandrew.com.

http://www.williamandrew.com/MNT

MICRODROPS AND DIGITAL MICROFLUIDICS

Jean Berthier
CEA-LETI, Grenoble, France

William Andrew
Norwich, NY, USA

ISBN: 978-0-8155-1544-9

Library of Congress Cataloging-in-Publication Data

Berthier, Jean, 1952-
Microdrops and digital microfluidics / Jean Berthier.
p. ; cm. -- (Micro & nano technology)
Includes bibliographical references.
ISBN 978-0-8155-1544-9 (alk. paper)
1. Atomizers. 2. Microfluidics--Equipment and supplies. 3. Spraying--Equipment and supplies. 4. Electrostatic atomization--Methodology. 5. Microelectromechanical systems. I. Title. II. Series.
[DNLM: 1. Microfluidics--methods. 2. Microchemistry--methods. QT 36.5 B541m 2008]
TP159.A85B48 2008
629.8'042--dc22

2007047532

Printed in the United States of America

This book is printed on acid-free paper.

10 9 8 7 6 5 4 3 2 1

Published by:
William Andrew Inc.
13 Eaton Avenue
Norwich, NY 13815
1-800-932-7045
www.williamandrew.com

NOTICE

To Erwin, Linda, Rosanne, and Susanne

Contents

Series Editor's Preface

From the classical histories of micro- and nanotechnologies, whether referring to the ideas of Richard Feynman and others concerning machines that make smaller machines, and so on down to the nanometre scale, that were aired around 1960, or to the relentless improvement in machining accuracy in both precision and ultraprecision machining (and, indeed, in normal machining) clearly traced by Norio Taniguchi, who invented the term nanotechnology in 1974, these technologies are very firmly associated with mechanical devices such as miniature motors and accelerometers. As the feature size of components on electronic 'chips' was progressively reduced, very large scale integrated circuits also came to be seen as part of micro- and nanotechnologies—indeed Moore's 'law', according to which the number of components on a chip doubles approximately every 18 months, has come to be seen as epitomizing these technologies—a link that has been further strengthened as semiconductor processing technologies used to fabricate the chips are also used to fabricate microelectromechanical devices (MEMS).

Fluids only later appeared on this stage, in the 1980s, through the necessity of delivering microscopic quantities of fuel and oxidizer to the miniature rocket motors used for correcting the courses of spacecraft, and through the inkjet printers that were attached to digital computers that were by then already associated with micro- and nanotechnologies. There was also, in the background, the ideas associated with the biological "proof-of-principle" of the possibility of genuine nanotechnology, emphasized by Eric Drexler and others, which naturally implicated fluids as the medium in which biology operates at the cellular level. More recently, the development of miniature, MEMS-based medical sensing devices has been powerfully boosted by the advantage of needing only microliter-sized samples of blood and other clinical fluids for analysis.

In some ways, the apotheosis of the assimilation of fluidics into micro- and nanotechnologies has been the emergence of discrete fluidics, which is fluidics based on the manipulation of discrete droplets, now usually called, with an obvious allusion to digital electronics, "digital microfluidics".

These developments have only been possible through the careful and consequential application of a great deal of knowledge carefully accumulated for quite different purposes, and involving several different disciplines. Until now, the necessary combination of the different areas of knowledge was the prerogative of the relatively small number of pioneers in the field, but as it now enters the mainstream, it is very necessary for a much larger community of practising engineers, scientists and technologists to become acquainted with areas of knowledge that, in the crowded curricula of modern university physics and engineering undergraduate courses, are practically relegated to footnotes.

This book is unique in combining in one volume all the knowledge necessary for launching into cutting-edge research and development in the field of digital microfluidics. Formerly esoteric subjects such as the physics of droplets and electrowetting, and the relevant parts of acoustics, are thoroughly treated, so that the reader will have an excellent grounding in the fundamental theory underpinning the technology. At the same time, applications are never far from the mind of the author, who is himself an active practitioner in the field, and enrich the entire book, not only the chapters specifically devoted to applications in, for example, biomedicine and chemistry. Several chapters benefit from collaboration with distinguished experts in those topics.

This is a book that will lead the reader from some possibly hazy recollections of the classical results of scientists such as Laplace and Young, right up to the rapidly increasing and astonishingly imaginative current developments in microdroplets and digital microfluidics.

Jeremy Ramsden
Cranfield University, United Kingdom
December 2007

Preface

This book deals with the physics of micro-drops and its application to micro-electro-mechanical systems (MEMS), especially digital microfluidics systems, and to some extent to droplet microfluidics. Many of the physical principles on which this book is based have not been recently discovered; they were stated between the end of the 18th century and the beginning of the 20th century. Interface and capillarity laws are based on the successive theories of Laplace, Young, Dupré, Wenzel, Cassie and Baxter, Jurin, Concus and Finn between the years 1780 and 1940; the electrowetting effect was discovered by Lippmann at the end of the 19th century, and dielectrophoresis by Pohl in the middle of the 20th century.

However, all these theories have been recently subject to new attention and increasing interest with the development of microfluidics. This development has been first promoted by the generalization of inkjet printing and, to some extent, by the space industry where droplets are used to feed micro-motors; and more recently by the formidable expansion of biotechnology. This expansion follows the development of microfabrication techniques pioneered by the microelectronic industry. Microsystems like MEMS and labs-on-chips (LOC) have been built to address the many goals of biotechnology, from bio-analysis and recognition, to DNA, protein and cell handling, and to screening and drug discovery. All these systems make extensive use of microfluidic principles, like microflows, digital, or droplet microfluidics.

Hence, the early interface and capillarity laws have been revisited and further developed to conform to their new environment of microsystems. Israelachvili has worked on molecular interactions and improved the notion of interface, de Gennes has clarified the notion of surface tension and developed an understanding of the mechanical behavior of colloids; Lippmann's law has been revisited by Berge to give birth to digital microfluidic systems. Pohl's dielectrophoresis has been re-interpreted by Jones. Coupling between hydrodynamic and electric forces has given birth to electro-osmotic actuation. Interactions between a force field (electric, acoustic) and a liquid-fluid interface have been investigated, and the modeling of the shape and deformation of interfaces has seen considerable progress with the energy minimization approach for static problems and the volume of fluid (VOF) and level set approaches for dynamic problems. A huge volume of new information has been produced covering all these domains.

Because they constitute the first microfluidic approach, microflows—continuous flows in micro-channels—have been the subject of hundreds of publications. Many books have been published concerning the different applications of microflows for micro-electro-mechanical systems MEMS and labs-on-chips LOC. The situation is very different for the techniques more recently developed using micro-drops instead of continuous microflows. Although many publications and some reviews in journals are addressing these techniques, a comprehensive book is still lacking. This book is intended to fill this gap.

In an introductory chapter, the different types of microfluidics are presented and categorized, specifying the place of digital and droplet microfluidics in the global picture of today's microfluidics.

In the second chapter, the notions of interface and surface tension are exposed, and the theory of capillarity is presented. Laplace's, Young's, and Jurin's laws are derived and it is shown how they are used for the determination of the shapes of interfaces in complex geometries. It is shown how the principle of energy minimization governs the shape of a static interface.

These prerequisites constitute the basis of the third chapter, which deals with the physics of micro-drops and their behavior in microsystems with either lyophilic, lyophobic—or both—surfaces, textured and patterned surfaces comporting micro-pillars, arêtes, and grooves. Especially, the importance of the pinning of droplets by sharp edges is pointed out.

The fourth chapter is devoted to the theory of electrowetting, with a focus on electrowetting on dielectrics (EWOD). In spite of its apparent simplicity, it is shown that the Lippmann law hides complex physical phenomena, some being still debated. The most important constraints for the functioning of EWOD actuation are analyzed: first, the saturation limit of the electrowetting effect at large voltages; this phenomenon is still poorly understood and the different explanations that have been proposed are discussed. Second, it is shown how contact angle hysteresis hinders electrowetting at low voltages. Finally, the limitations imposed by the breakdown limit of the dielectric are discussed.

Global applications of electrowetting theory are presented in Chapter 5. The mechanisms enabling the basic operations on an electrowetting on dielectrics (EWOD)-based chip are investigated: these operations consist of moving, merging, splitting, mixing, and diluting droplets on EWOD-based microsystems. Based on these results, fabrication and design constraints are discussed.

Applications of EWOD-based microchips to biotechnology and biology are the subject of Chapter 6. Some typical applications in the different domains of sample collection, DNA analysis, DNA repair, protein analysis, and cell handling have been chosen amongst all the possible applications. Because it is a good test for comparison with other systems (conventional and continuous microfluidics), a special focus is made on DNA analysis using the polymerase chain reaction (PCR).

Chapter 7 is devoted to the study of the behavior of chemical components on EWOD-based chips, especially of organic solvents and ionic liquids. The maneuverability of these different liquids on the substrate of a microsystem is analyzed, and it is shown how these carrier fluids can be used to realize controlled chemical reactions in micro-drops.

Acoustic wave actuation of droplets constitutes another mode of actuation, and Chapter 8 presents its principle. It is shown how it can be used to design a digital droplet microsystem. Applications to biology, biotechnology and electronics are discussed. Chapter 9 is an introduction to another approach to the use of micro-drops, based on the merging of microflows in T-junctions or flow-focusing devices (FFD) to form controlled micro-droplets transported by the continuous phase. First, the physical behavior of two-phase flow droplets and plugs is investigated, with a focus on plugs moving in capillary tubes, and on immiscible fluids flowing in parallel. Pressure loss is determined in both cases. Finally, we analyze how droplets are produced in T-junctions or FFD and what parameters control the size and reproducibility of these droplets.

The Epilog (Chapter 10) reflects on the potential of digital microfluidics, and future trends and perspectives are discussed.

This book has been conceived and written to enable the reader to acquire a solid understanding of the physical behavior of droplets in complex microsystems, and a profound knowledge of the functioning of digital and droplet microfluidic systems.

Jean Berthier
Grenoble, France
December 2007

Acknowledgments

It is a very common statement to say that writing a scientific book is a long and difficult task. I would like to add my contribution to this statement. I had the chance however not to be alone in this enterprise and, first of all, I express here my gratitude to Ken Brakke, Hsueh-Chia Chang and Achim Wixforth.

Although I do not know Ken personally—we always worked together through the Internet—I admire his capacity to tackle difficult problems and solve them in a very short time. I have not counted the times he has sent me the answers to my questions late at night. I thank him for his help in using Surface Evolver on tricky cases. Chapters 2 and 3 would not have been the same without him.

Many thanks also to Hsueh-Chia Chang. I had the chance to be welcomed by Chia at the University of Notre-Dame, and to visit his laboratory. I was extremely impressed by the high standard of the research and developments going on there. So when I looked for someone to back me up in writing Chapters 4 and 5 dedicated to electrowetting theory and applications, I immediately thought of him. I express to him here all my gratitude for his help.

I am also very much indebted to Achim Wixforth and his colleagues Thomas Frommelt and Matthias Schneider from Augsburg University and Advalytix for writing the chapter dedicated to acoustic actuation of droplets. This chapter brings another view of digital microfluidics, and I was keen to have it in this book. Many thanks to all three.

I also thank the colleagues of my lab who helped me in this task: Christine Peponnet—my laboratory director—for helping me write Chapter 6 (concerning the biological applications of EWOD). Many thanks also to Philippe Dubois, presently a PhD student in the labs of the LETI, for his writing of Chapter 7 dedicated to chemical applications of EWOD, together with his tutors, Gilles Marchand and Michel Vaultier. I am also grateful to all the persons who have contributed photographs, David Hu, Thomas Frommelt, Anthony Mescher, Michael Bestehorn, and my colleagues Olivier Raccurt, Philippe Clementz, Yves Fouillet, Dorothé Jary, Roland Bavière, Anne-Gaël Brachet, and Nicolas Chiaruttini.

I express my gratitude to Pierre Puget, scientific advisor in the Biotechnology and Health Department of the LETI, for his careful reviewing of the manuscript, to Jean Chabbal, our department leader, and Patrice Caillat, my group leader, for their support to this project.

Finally, and not least, I thank Nigel Hollingworth from William Andrew for his help and guidance during the long process of writing this book.

1 Digital Microfluidics in Today's Microfluidics

1.1 The Development of Microfluidics

The development of microfluidics has been remarkable since the 1980s. Historically, in the 1980s, the first applications of microfluidics were triggered by the generalization of ink jet printing, and also to some extent by space applications, where droplets are used to feed micro-motors. Today, the developments are boosted by biotechnology. Indeed, microfluidics is the basis of most biotechnological developments, simply because biological samples are in a liquid state and biological organisms live in an aqueous environment. Compared to conventional techniques, the use of microfluidic devices has many advantages [1]: the volumes of sample and reagents are significantly reduced—which is practical and diminishes the cost of the process; the results (diagnostic) or the products (biochemistry) are obtained in a much shorter time—which also contributes to reducing costs, or can be necessary for medical emergency reasons; the level of sensitivity is increased due to higher precision and selectivity of the process; the risks with the manipulation of toxic or dangerous products are decreased. Besides, miniaturization allows for an increase in automation and parallelization, which opens the way to screening and systematic testing in the domain of drug discovery. Finally, portability is a unique feature that will permit diagnostic and bio-analysis on the field.

However, the revolution that fluidic microsystems brings to biotechnology depends on the response to some challenges, the most important being the integration of all the components of the system on the same chip. Hence, pumps, valves, mixers, etc. must be miniaturized in order to obtain an integrated system, introducing a complicated choice between active methods—efficient but difficult to miniaturize and requiring energy sources—and passive methods—easier to integrate, but less efficient. Another challenge is the growing complexity of the microfabrication required to support all the functions of the system: it is foreseen that composite materials will have to be assembled and connected. In biotechnology, the volume reduction from the macroscopic environment where the targets are located to the microscopic environment of the microsystem is also a huge challenge. Molecules and/or particles existing in macroscopic volumes have to be concentrated in micro-volumes in order to be introduced in a biochip. Miniaturization of the whole chain of treatment is still to be done. Finally, the very large surface/volume ratio of miniaturized systems modifies the physical behavior of the system and some new problems arise, like, for instance, the adherence of target molecules to the solid walls, or the effect of capillary forces that may prevent the fluid from entering microchannels.

1.2 The Advantages of Digital and Droplet Microfluidics Compared to Conventional Microflows

1.2.1 From Microflows to Microdrops

The most common—and historically first—microsystems, such as micro-electrical-mechanical systems (MEMS) or lab on a chip (LOC), are using microflows; conceptually, they are more or less derived from macroscopic devices, taking into account the particularities brought by the reduction of the dimensions and the physical modifications due to the influence of the solid walls. In such systems, microfluidics is often combined with electrics, magnetic, optics, acoustics, and chemistry in order to perform adequate functions. These systems have brought considerable progress to biotechnology. A striking example is that of screening which consists in searching for a specific component in a sample. Conventionally, screening is performed in microplates comprising tenth of 10 μl wells. With this format, when screening for 100,000 components, the process can last for about 4 months and require up to 10 liters of reagent. Using microflow systems such as the one described in Fig. 1.1 [2] the duration of the same screening is reduced to a week and the quantities of sample are reduced 100 times. However, a week is still a long time; nowadays the use of droplet microfluidics reduces the time to about 20 minutes. We see here the considerable advantage brought by digital or droplet microfluidics which is closely linked to the dramatic reduction of volume. The working volumes can be reduced down to 50 nl and probably less than that in the near future.

The second advantage brought by the use of microdroplets is the reduction of the contact with the solid walls: a large part of the contact surface of the liquid with the solid wall being replaced by a liquid–liquid or liquid–gas interface, the solid contact surface to liquid volume ratio is reduced compared to that of microflow systems; hence, adherence and adhesion problems are reduced (but not suppressed).

1.2.2 Microdrops: Comparison Between Digital and Droplet Microfluidics

Digital and droplet microfluidics have in common the use of microdrops and the associated reduction of volume (Fig. 1.2). However they differ considerably. Droplet microfluidics is a particular example of two or multi-phase microflows, where droplets are transported by a carrier flow, whereas digital microfluidics is based on a totally different approach, inspired by micro-plate systems where the manipulation robot would be incorporated; in such systems the micro-drops are moved and treated individually as digital entities on a planar surface (hence

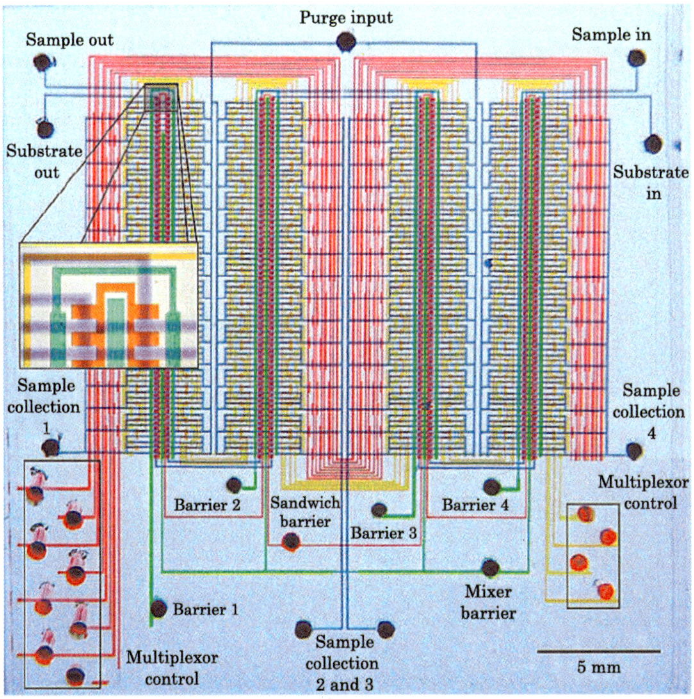

Figure 1.1 Scheme of a complex microflow system governed by pressure actuated microvalves. From Thorsen et al., Science, Vol. 298, pp. 580–584, 2002. Reprinted with permission from AAAS.

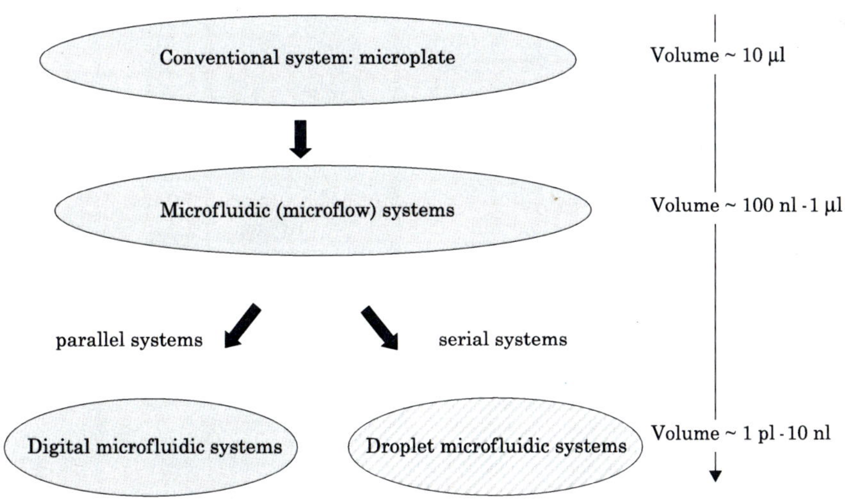

Figure 1.2 Scheme of the different scales of fluidic microsystems.

the other name of digital microfluidics is planar microfluidics), allowing for a very reconfigurable system [3].

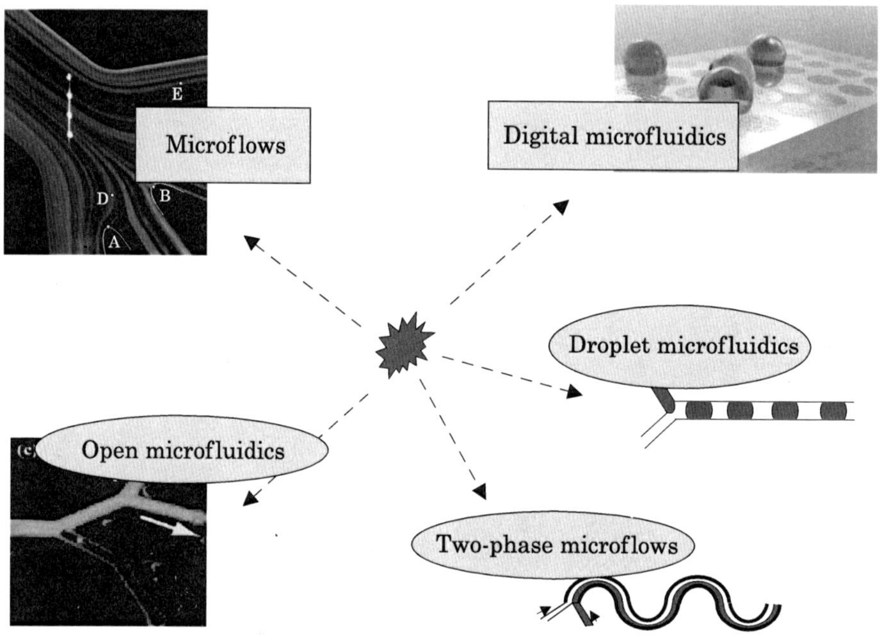

Figure 1.3 The different domains (toolbox) of microfluidics. Photo bottom left reprinted with permission from [5]. © 2005 American Chemical Society. Photo top right courtesy Advalytix.

Contrary to microflow and droplet microfluidic systems, where pumps and valves are actuated by mechanical, electrical, magnetic, or acoustic forces, pumping and valving functions are inherent in the digital microfluidic construction.: A unique actuation force (electric or acoustic) performs all the different tasks like pumping, valving, and motioning the liquid samples [4].

In the conclusions of Chapter 9, it is shown that digital and droplet microfluidic systems are in fact very complementary. Digital microfluidics can handle extremely small volumes of liquids and realize operations and manipulation in parallel, and with great accuracy, whereas droplet microfluidics is well adapted to perform operations in series, like screening or encapsulation.

1.3 The Respective Place of Digital and Droplet Microfluidics in Today's Microfluidics

It seems interesting to position the place of digital microfluidics in the global panel of state-of-the-art microfluidics. The different "toolboxes" proposed by today's microfluidics are shown in Fig. 1.3. The first and most important toolbox is the "microflow" toolbox (top left in Fig. 1.3). It is now mature and has already been the object of a few books and numerous publications.

Digital microfluidics (top right) is a new toolbox that is the subject of this book. It is still at a developing stage, and its possibilities have not all been determined. More recently, new concepts have been developed to complement the possibilities offered by microfluidics: on the one hand, the two toolboxes constituted by two-phase microflows and droplet microfluidics—that will be the object of Chapter 9—are schematized on the bottom right of Fig. 1.3; on the other hand, "open" microfluidics [5], i.e., microflows in partially opened channels, where the flow is stabilized by capillarity forces in the absence of a solid boundary, is at the crossroads between single-phase and two-phase microflows (Fig. 1.3, bottom left). Open microfluidics is probably the less advanced type of microfluidics today.

1.4 Summary

The aim of this book is to provide the reader with the background for the physical behavior of micro-drops in general, with a focus on digital microfluidics. A profound understanding of the physics of micro-drops—subjected only to surface tension and capillary forces, and to a lesser extent to gravity—is required to be able to tackle the study of digital microfluidics—in which electric or acoustic forces are involved. In Chapter 9 an introduction on droplet microfluidics has been inserted in order to give a more complete view of the use of droplets in microsystems.

References

1. Petra S. Dittrich and Andreas Manz, "Lab-on-a-chip: microfluidics in drug discovery," *Nature Reviews Drug Discovery*, Vol. 5, pp. 210–218, 2006.
2. Todd Thorsen, Sebastian J. Maerkl and Stephen R. Quake, "Microfluidic Large-Scale Integration," *Science*, Vol. 298, pp. 580–584, 2002.
3. Krishnendhu Chakrabarty and Fei Su, *Digital Microfluidic Biochips: Synthesis, Testing, and Reconfiguration Techniques*, Taylor and Francis, 2007.
4. Richard B. Fair, "Digital microfluidics: is a true lab-on-a-chip possible?," *Microfluidics and Nanofluidics*, Vol. 3, pp. 245–281, 2007.
5. W. Satoh, H. Hosono and H. Suzuki, "On-chip microfluidic transport and mixing using electrowetting and incorporation of sensing functions," *Anal. Chem.*, Vol. 77, pp. 6857–6863, 2005.

2 Theory of Wetting☆

2.1 Introduction

The understanding, development, and design of fluidic microsystems require dealing with forces that are usually overlooked and neglected at the macroscopic scale. In the case of digital microfluidics, interfaces are omnipresent and capillary and surface tension forces have a tremendous influence. These forces are of utmost importance at the microscopic level, whereas at a macroscopic scale, they are usually neglected in front of forces like pressure or gravity. Typical ranges of pressure in macroscopic applications are 1–100 bars, i.e., 10^5 to 10^7 Pa, whereas pressure levels seldom reaches 0.1 bar (10^4 Pa) above atmospheric pressure in microfluidic systems. Most of the time gravity has no or little consequence in the design of fluidic microdevices.

Numerous examples of the importance of capillarity can be found in existing microsystems for biotechnology, or microchemistry, and in space applications. In order to point out the importance of capillary forces we give in Figs. 2.1–2.4 some examples illustrating the importance of capillary forces in microsystems.

A starting point for many analyses of biological substances is the deposition, i.e., dispensing or spotting, of liquids in microwells (Fig. 2.1 (a)). In each of these microwells biological reactions take place simultaneously, leading to a high efficiency parallelization of the reactions. Spotting of liquids in microwells requires a careful control of the wettability of the surfaces as shown in Fig. 2.1 (b). The amount of liquid deposited on the plate is both a function of the wettability of the pipette and the plate, and a function of the applied pressure in the pipette.

A second example is that of electrowetting based microsystems. Electrowetting is a combination of capillary and electric forces, where electric forces modify the apparent capillarity; such a technique has been found to be a promising way to realize digital microfluidic devices. Such microsystems are being developed for biological applications and microchemistry. This technique will be detailed in Chapters 4 and 5. A schematic view of a digital microfluidic system is shown in Fig. 2.2 [3]. Liquid droplets are displaced on a digital grid—or along a path—paved with electrodes and displaced by electric actuation of the electrodes.

Another illustration of the use of capillarity in microfluidic systems is the use of T-junctions to produce droplets immersed in a carrier fluid; such systems are principally devoted to performing chemical reactions or the encapsulation of biological objects. The principle is to use a contrast of wettability of the

☆This Chapter was written with the collaboration of **Kenneth A. Brakke** (Mathematics Department, Susquehanna University, Selinsgrove, PA).

Figure 2.1 (a) View of a microplate for Biotechnology applications, from [1]. © 2004 JALA. Reprinted with permission. (b) Capillarity is a determining force when spotting a droplet on a microplate, from [2].

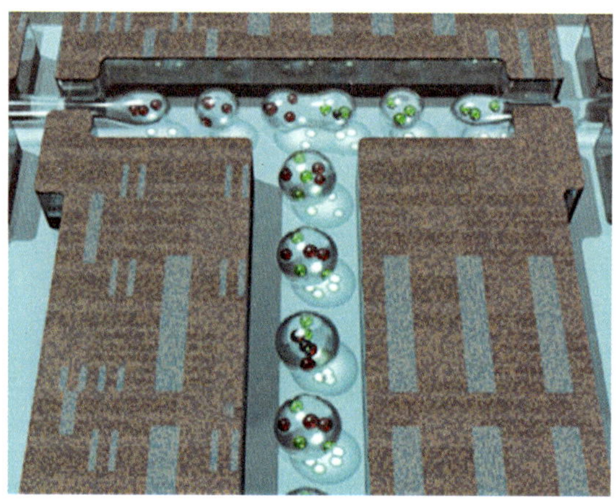

Figure 2.2 Schematic view of a digital microfluidic system based on electrowetting. Courtesy Rain Dance Technology Inc. [3].

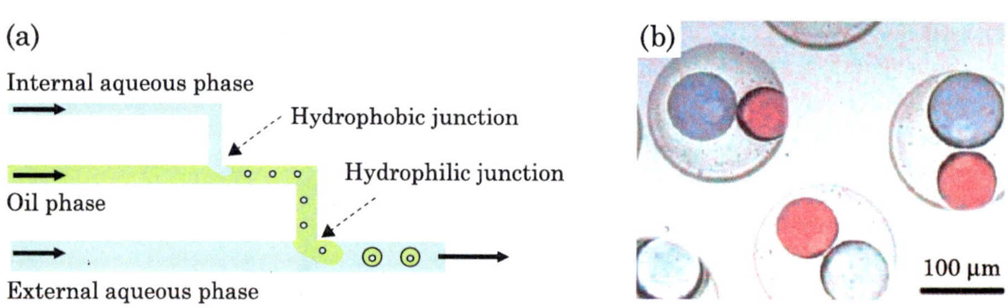

Figure 2.3 (a) Successive use of T-junctions with wetting and non-wetting walls results in the encapsulation of two liquids in a droplet [5]; (b) clever use of T-junctions can be used to obtain simultaneous encapsulation; reprinted with permission from [5]. © 2004 American Chemical Society.

solid walls at microfluidic T-junctions to create micro-drops (Fig. 2.3). At each step, the wettability of the walls of the carrier liquid should be chosen so that the carrier liquid wets the walls, forcing the incoming immiscible liquid

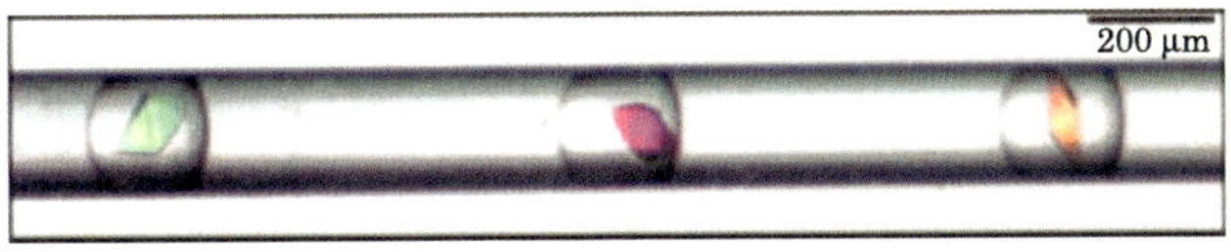

Figure 2.4 Droplets in a glass capillary: crystals of proteins were grown under microbatch condition in these droplets, from [6]. © 2004 Wiley, reprinted with permission.

to form droplets (Fig. 2.3 (a)). By a clever combination of multiple junctions, encapsulation of two aqueous liquids in an oil droplet can be achieved [4–6].

Encapsulation of biomolecules in their liquid environment is now foreseen as a tool for targeted drug delivery [4]; it is also a tool for protein crystallization. Fig. 2.4 shows how proteins can be continuously crystallized in micro-drops, using the preceding technique.

All these examples clearly show the essential role of surface tension and capillary forces in microsystems.

In this chapter, the fundamentals of capillarity are presented. Surface tension of an interface is the fundamental notion in capillarity theory; this notion leads naturally to that of wetting and to the introduction of capillarity and capillary forces. Next, different applications of capillary forces are shown, and the keystone problem of the measurement of surface tensions is presented. The consequences of surface tension and capillary forces for micro-drop shape and behavior are presented in the Chapter 3.

2.2 Interfaces and Surface Tension

2.2.1 The Notion of Interface

Mathematically speaking, an interface is the geometrical surface that delimits two fluid domains. This definition implies that an interface has no thickness and is smooth (i.e., has no roughness). As practical as it is, this definition is in reality a schematic concept. The reality is more complex and the separation of two immiscible fluids (water/air, water/oil, etc.) depends on molecular interactions between the molecules of each fluid [7] and on Brownian diffusion (thermal agitation). A microscopic view of the interface between two fluids looks more like the scheme of Fig. 2.5.

However, in engineering applications, it is the macroscopic behavior of the interface that is the focus of attention, and the mathematical concept regains its utility. At a macroscopic size, the picture of Fig. 2.5 can be replaced by that of Fig. 2.6, where the interface is a mathematical surface without thickness and the contact angle θ is uniquely defined by the tangent to the surface at the contact line.

In a condensed state, molecules attract each other. Molecules located in the bulk of a liquid have interactions with all neighboring molecules; these

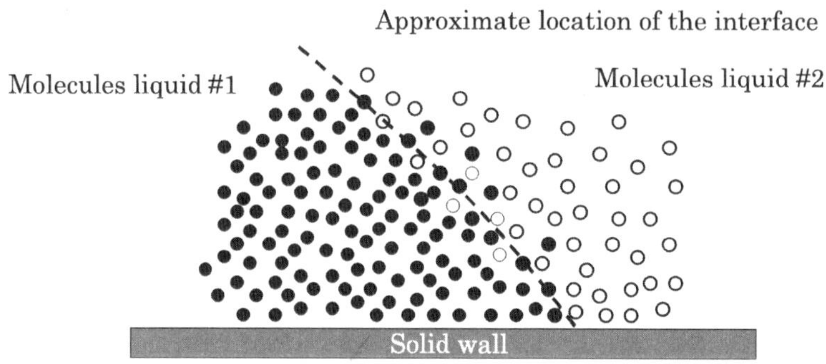

Figure 2.5 Schematic view of an interface at the molecular size.

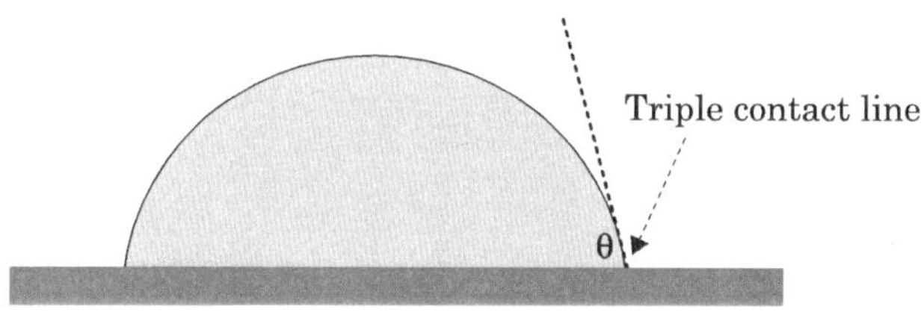

Figure 2.6 Macroscopic view of the interface of a drop.

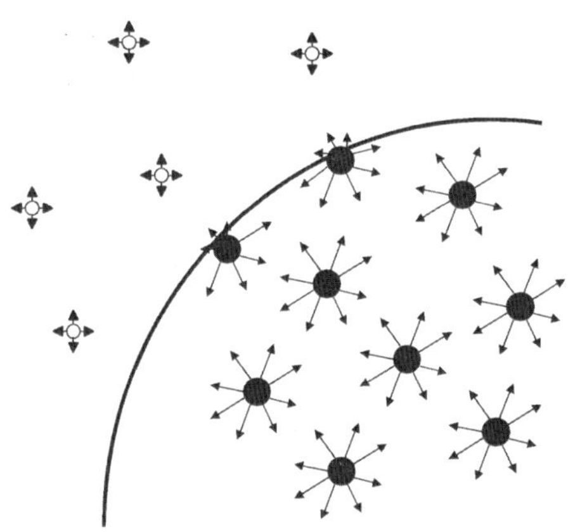

Figure 2.7 Simplified scheme of molecules near an air/water interface. In the bulk, molecules have interaction forces with all the neighboring molecules. At the interface, half of the interactions have disappeared.

interactions are mostly van der Waals attractive interactions for organic liquids and hydrogen bonds for polar liquids like water [7]. On the other hand, molecules at an interface have interactions in a half space with molecules of the same liquid, and in the other half space interactions with the molecules of the other fluid or gas (Fig. 2.7).

Consider an interface between a liquid and a gas. In the bulk of the liquid, a molecule is in contact with 4 to 12 other molecules depending on the liquid (4 for water and 12 for simple molecules); at the interface this number is divided by 2. Of course, a molecule is also in contact with gas molecules, but, due to the low densities of gases, there are less interactions and less interaction energy than in the liquid side. The result is that there is locally a dissymmetry in the interactions, which results in a defect of surface energy. At the macroscopic scale, a physical quantity called "surface tension" has been introduced in order to take into account this molecular effect. The surface tension has the dimension of energy per unit surface and in the International System it is expressed in J/m^2 or N/m (sometimes, it is more practical to use mN/m as a unit for surface tension.) An estimate of the surface tension can be found by considering the molecules' cohesive energy. If U is the total cohesive energy per molecule, a rough estimate of the energy loss of a molecule at the interface is $U/2$. Surface tension is a direct measure of this energy loss, and if δ is a characteristic molecular dimension and δ^2 the associated molecular surface, then the surface tension is approximately

$$\gamma \approx \frac{U}{2\delta^2}. \tag{2.1}$$

This relation shows that surface tension is important for liquids with large cohesive energy and small molecular dimension. This is why mercury has a large surface tension whereas oil and organic liquids have small surface tensions.

Another consequence of this analysis is the fact that a fluid system will always act to minimize surface areas: the larger the surface area, the larger the number of molecules at the interface and the larger the cohesive energy imbalance. Molecules at the interface always look for other molecules to equilibrate their interactions. As a result, in the absence of other forces, interfaces tend to adopt a flat profile, and when it is not possible due to capillary constraints at the contact of solids, they take a convex rounded shape, as close as possible to that of a sphere. Another consequence is that it is energetically costly to expend or create an interface: we will come back on this problem in Chapter 5 when dividing a droplet into two "daughter" droplets by electrowetting actuation.

The same reasoning applies to the interface between two liquids, except that the interactions with the other liquid will usually be more energetic than a gas and the resulting dissymmetry will be less. For example, the contact energy (surface tension) between water and air is 72 mN/m whereas it is only 50 mN/m between water and oil (Table 2.1). Interfacial tension between two liquids may be zero: fluids with zero interfacial tension are said to be miscible. For example, there is no surface tension between fresh and salt water. Salt molecules will diffuse freely across a boundary between fresh and salt water.

The principle applies for a liquid at the contact of a solid. The interface is just the solid surface at the contact of the liquid. Molecules in the liquid

Table 2.1 Values of Surface Tension of Different Liquids at the Contact with Air at a Temperature of 20 °C(Middle Column) and Thermal Coefficient α (Right Column)

Liquid	γ_0	α
Acetone	25.2	−0.112
Benzene	28.9	−0.129
Ethanol	22.1	−0.0832
Ethylene-Glycol	47.7	−0.089
Glycerol	64.0	−0.060
Methanol	22.7	−0.077
Mercury	425.4	−0.205
Perfluorooctane	14.0	−0.090
Polydimethylsiloxane	19.0	−0.036
Pyrrol	36.0	−0.110
Toluene	28.4	−0.119
Water	72.8	−01514

are attracted towards the interface by van der Waals forces. But usually these molecules do not "stick" to the wall because of Brownian motion. However, impurities contained in the fluid, like particles of dust or biological polymers like proteins, may well adhere permanently to the solid surface because, at the contact with the solid interface, they experience more attractive interactions. The reason is that the size of polymers is much larger than water molecules and van der Waals forces are proportional to the number of contacts.

Usually surface tension is denoted by the Greek letter γ with subscripts referring to the two components on each side of the interface, for example γ_{LG} at a liquid/gas interface. Sometimes, if the contact is with air, or if no confusion can be made, the subscripts can be omitted. It is common to speak of "surface tension" for a liquid at the contact with a gas, and "interfacial tension" for a liquid at the contact with another liquid.

According to the definition of surface tension, for a homogeneous interface (same molecules at the interface all along the interface), the total energy of a surface is

$$E = \gamma S \tag{2.2}$$

where S is the interfacial surface area.

2.2.2 Surface Tension

In the literature or on the Internet there exist tables for surface tension values [8,9]. Typical values of surface tensions are given in Table 2.1. Note that surface tension increases as the intermolecular attraction increases and the molecular size decreases. For most oils, the value of the surface tension is in the range $\gamma \sim 20$–30 mN/m, while for water, $\gamma \sim 70$ mN/m. The highest surface tensions are for liquid metals; for example, liquid mercury has a surface tension $\gamma \sim 500$ mN/m.

2.2.2.1 The Effect of Temperature on Surface Tension

The value of the surface tension depends on the temperature. The first empirical equation for the surface tension dependence on temperature was given by Eötvös in 1886 [10]. Observing that the surface tension goes to zero when the temperature tends to the critical temperature T_C, Eötvös proposed the semi-empirical relation

$$\gamma = \left(\frac{1}{\nu_L}\right)^{\frac{2}{3}} (T - T_C)$$

where ν_L is the molar volume. Katayama (1915) and later Guggenheim (1945) [11] improved Eötvös's relation to obtain

$$\gamma = \gamma^* \left(1 - \frac{T}{T_C}\right)^n \tag{2.3}$$

where γ^* is a constant for each liquid and n is an empirical factor, whose value is 11/9 for organic liquids. Equation (2.3) produces very good results for organic liquids. If temperature variation is not very important, and taking into account that the exponent n is close to 1, a good approximation of the Guggenheim–Katayama formula is the linear approximation

$$\gamma = \gamma^* \left(1 + \alpha T\right). \tag{2.4}$$

It is often easier and more practical to use a measured reference value (γ_0, T_0) and consider a linear change of the surface tension with the temperature

$$\gamma = \gamma_0 \left(1 + \beta \left(T - T_0\right)\right). \tag{2.5}$$

Comparison between (2.4) and (2.5) requires

$$\beta = -\frac{1}{T_C - T_0}. \tag{2.6}$$

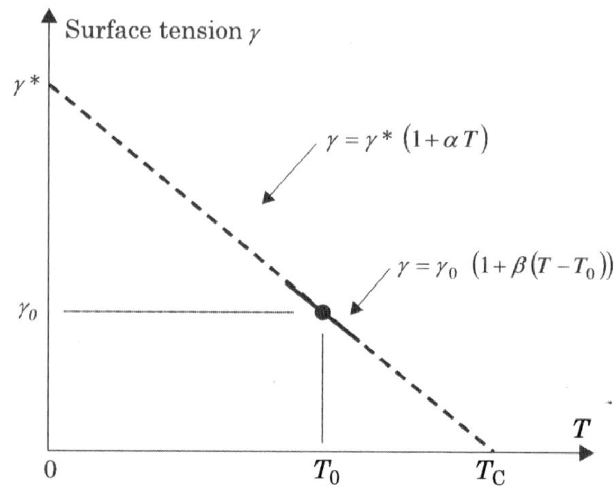

Figure 2.8 Representation of the relations (2.4) and (2.5).

Relations (2.4) and (2.5) are shown in Fig. 2.8. The value of the reference surface tension γ_0 is linked to γ^* by the relation

$$\gamma_0 = \gamma^* \frac{T_C - T_0}{T_C}.$$

Typical values of surface tensions and their temperature coefficients a are given in Table 2.1.

The coefficient α being always negative, the value of the surface tension decreases with temperature. This property is at the origin of a phenomenon which is called either Marangoni convection or thermocapillary instabilities (Fig. 2.9). If an interface is locally heated by any heat source (like radiation, convection or conduction), the surface tension is reduced on the heated area according to equations (2.4) or (2.5). A gradient of surface tension is then induced at the interface between the cooler interface and the warmer interface. We will show in Section 2.7 that surface tensions can be viewed as forces; as a consequence, there is an imbalance of tangential forces on the interface, creating a fluid motion starting from the warm region (smaller value of the surface tension) towards the cooler region (larger value of the surface tension). This surface motion propagates to the bulk under the influence of viscosity.

If the temperature source is temporary, the motion of the fluid tends to homogenize the temperature and the motion progressively stops. If a difference of temperature is maintained on the interface, the motion of the fluid is permanent; this is the case of a film of liquid spread on a warm solid. Depending on the contrast of temperature between the solid surface and the liquid surface, the motion of the liquid in the film has the morphology of convective rolls, hexagons, or squares. Fig. 2.10 shows hexagonal patterns of Marangoni convection in a film of liquid heated from below [12]. The white segments show

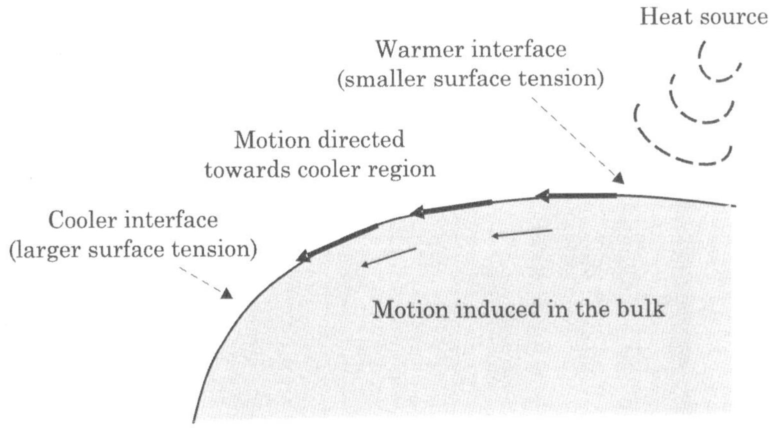

Figure 2.9 Sketch of interface motion induced by a thermal gradient between two regions of the surface. The motion of the interface propagates into the bulk under the action of the viscous forces.

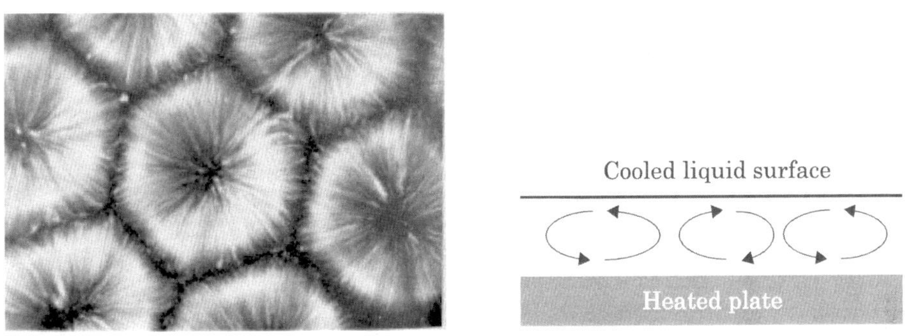

Figure 2.10 Marangoni flow (thermocapillary instabilities) patterns in a thin film of liquid; reused with permission from [12]. © 2005 American Institute of Physics.

the trajectories of the liquid molecules. In Chapter 3, we will look in more details at Marangoni convection inside droplets.

2.2.2.2 The Effect of Surfactants

Surfactant is the short form for surface active agent. Surfactants are long molecules characterized by a hydrophilic head and a hydrophobic tail and are for this reason called amphiphilic molecules. Very often surfactants are added to biological samples in order to prevent the formation of aggregates and to prevent target molecules from adhering to the solid walls of the microsystem (remember that microsystems have extremely large ratios between the wall surfaces and the liquid volumes). Surfactants diffuse in the liquid and when reaching the interface they are captured because their amphiphilic nature prevents them

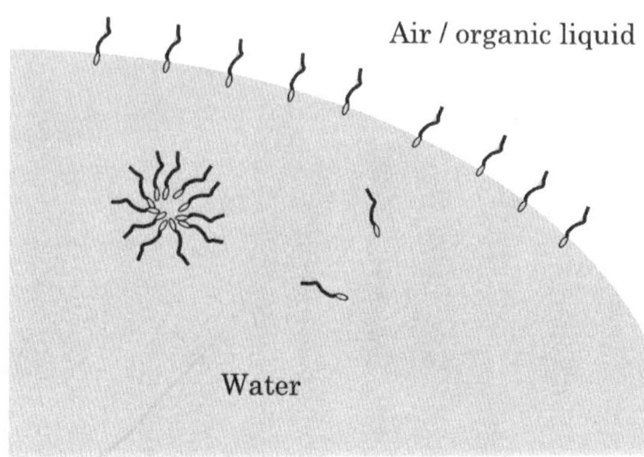

Figure 2.11 Schematic view of surfactants in a liquid drop.

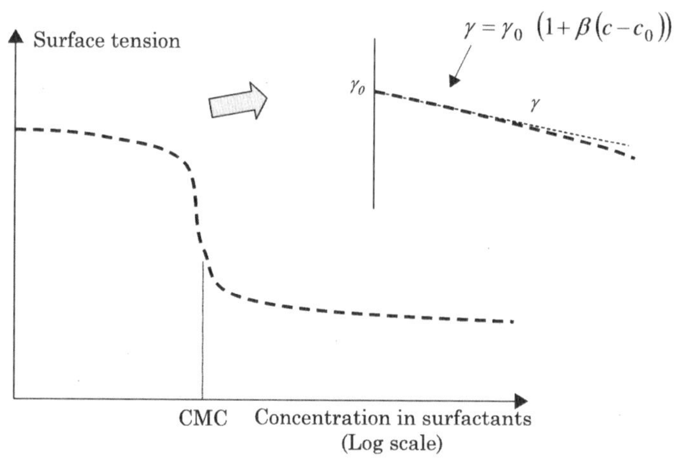

Figure 2.12 Evolution of the value of the surface tension as a function of the surfactant concentration.

from escaping easily from the interface. As a consequence, they gather on the interface as is sketched in Fig. 2.11, lowering the surface tension of the liquid.

As the concentration in surfactants increases, the surface concentration increases too. Above a critical value of the concentration, called CMC for critical micelle concentration, the interface is saturated with surfactants and surfactant molecules in the bulk of the fluid group together to form micelles.

The evolution of the value of the surface tension as a function of the concentration in surfactants is shown in Fig. 2.12. At very low concentration, the slope is nearly linear. W;en concentration approaches the CMC, the value of the surface tension drops sharply; above CMC, the value of the surface tension is nearly constant [13]. For example, pure water has a surface tension 72 mN/m and water with Tween 10 at a concentration above the CMC has a surface tension of only 30 mN/m.

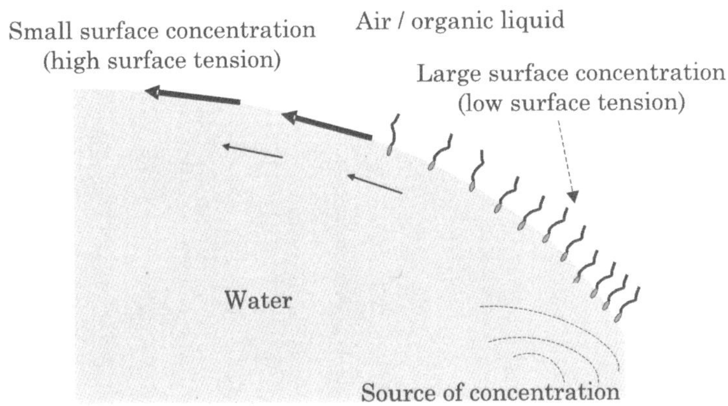

Figure 2.13 Schematic of Marangoni convection induced by a gradient of concentration.

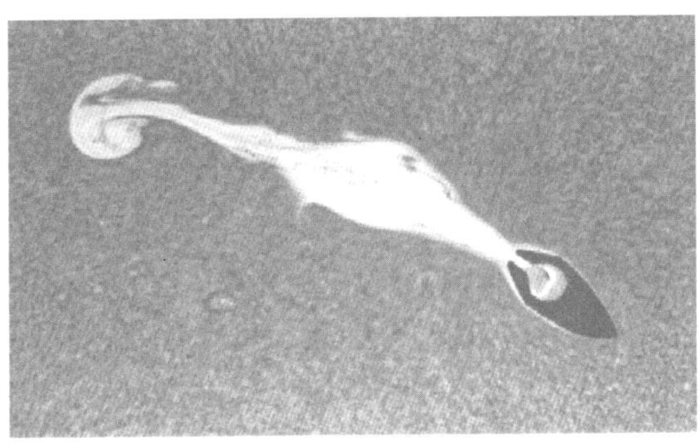

Figure 2.14 Soap boat: a floating body contains a small volume of soap. At first, the soap exits the rear of the boat under Marangoni stress. Hence a low surface tension region is created behind the boat, whereas the unsoaped region in front of the boat has a larger surface tension. This difference of surface tension pulls the boat forward [14].

In the limit of small surfactant concentration ($c \ll$ CMC), the surface tension can be expressed as a linear function of the concentration

$$\gamma = \gamma_0 \left(1 + \beta \left(c - c_0\right)\right). \tag{2.7}$$

Equation (2.7) is similar to equation (2.5). We have seen how a temperature gradient results in a gradient of surface tension leading to Marangoni type of convection. Similarly, a concentration gradient results in a gradient of surface tension, and consequently to a Marangoni convection. Note that the direction of the motion is always towards the largest value of surface tension (Fig. 2.13).

Spreading of surfactant molecules on an interface can be easily seen experimentally: an instructive example is that of a thin paper boat with a cavity at the rear (Fig. 2.14). When the boat is placed gently on the surface of water, it rests on the surface of water suspended by surface tension forces.

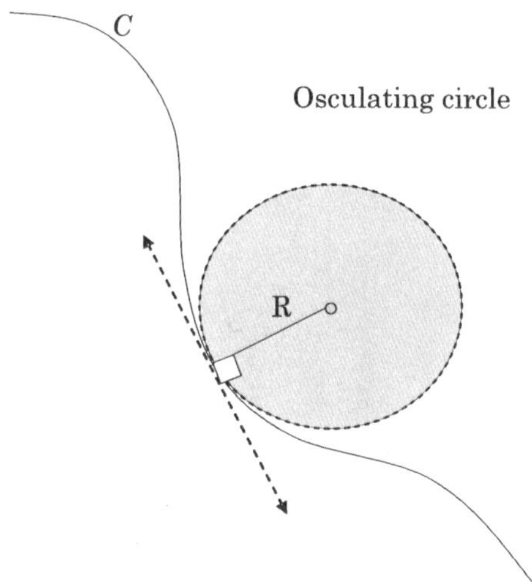

Figure 2.15 Radius of curvature and osculating circle.

When a drop of soap solution/detergent is put on the notch, the boat accelerates rapidly. Soap molecules try to spread over the surface of water. Since they are confined in the cavity of the boat with only one way out, they jet from the rear end creating a reaction force which drives the boat forward.

2.3 Laplace Law and Applications

2.3.1 Laplace's Law

Laplace's law is fundamental when dealing with interfaces and micro-drops. It relates the pressure inside a droplet to the curvature of the droplet. Let us first describe the notion of curvature.

2.3.1.1 Radius of Curvature and Curvature

For a planar curve the radius of curvature is the radius of the osculating circle—the circle which is the closest to the curve at the contact point (Fig. 2.15). The curvature of the curve is defined by

$$\kappa = \frac{1}{R}. \tag{2.8}$$

Note that the curvatures as well as the curvature radii are signed quantities. Curvature radius can be positive or negative depending on the orientation (convex or concave) of the curve.

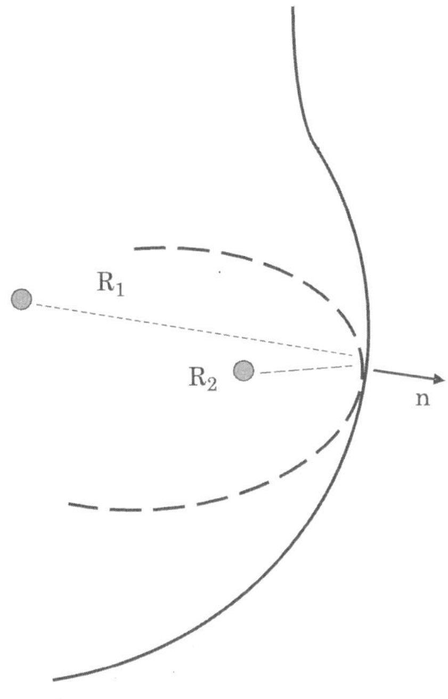

Figure 2.16 Schematic view of the radius of curvature of a surface.

In the case of a parametric curve $c(t) = (x(t), y(t))$, the curvature is given by the relation [15]

$$\kappa = \frac{\dot{x}\ddot{y} - \dot{y}\ddot{x}}{(\dot{x}^2 + \dot{y}^2)^{\frac{3}{2}}} \tag{2.9}$$

where the dot denotes a differentiation with respect to t. For a plane curve given implicitly as $f(x, y) = 0$, the curvature is

$$\kappa = \nabla \cdot \left(\frac{\nabla f}{\|\nabla f\|} \right) \tag{2.10}$$

that is, the divergence of the direction of the gradient of f. And for an explicit function $y = f(x)$, the curvature is defined by

$$\kappa = \frac{\frac{d^2y}{dx^2}}{\left(1 + \left(\frac{dy}{dx}\right)^2\right)^{\frac{3}{2}}}. \tag{2.11}$$

The situation is more complex for a surface. Any plane containing the vector normal to the surface intersects the surface along a curve. Each of these curves has its own curvature. The mean curvature of the surface is defined using the principal (maximum and minimum) curvaturesκ_1 and κ_2 (Fig. 2.16) in the whole set of curvatures

$$H = \frac{1}{2}(\kappa_1 + \kappa_2). \tag{2.12}$$

It can be shown that the principal curvatures κ_1 and κ_2 are located in two perpendicular planes. Introducing the curvature radii in (2.12) leads to

$$H = \frac{1}{2}(\kappa_1 + \kappa_2) = \frac{1}{2}\left(\frac{1}{R_1} + \frac{1}{R_2}\right). \tag{2.13}$$

For a sphere of radius R, the two curvatures are equal to $1/R$ and the mean curvature is $H = 1/R$. For a cylinder of base radius R, the maximum curvature is R and the minimum curvature zero, hence $H = 1/2R$. For a plane, the two curvatures are zero and $H = 0$: a plane has no curvature.

At a saddle point of a surface (Fig. 2.17), one of the curvature radii is positive because it corresponds to a convex arc, whereas the other one is negative, because it corresponds to a concave arc. If $|R_1| = |R_2|$ the mean curvature H is zero: $H = \frac{1}{2}\left(\frac{1}{R_1} + \frac{1}{R_2}\right) = \frac{1}{2}\left(\frac{1}{|R_1|} - \frac{1}{|R_2|}\right) = 0$.

2.3.1.2 Derivation of the Laplace Law

Suppose there is a spherical droplet of liquid surrounded by a fluid. Let us calculate the work necessary to increase its volume from the radius R to the radius $R + dR$ (Fig. 2.18). The part of the work due to the internal volume increase is

$$\delta W_i = -P_0 dV_0$$

where dV_0 is the increase of the volume of the droplet

$$dV_0 = 4\pi R^2 dR.$$

The work to pull out the external fluid is

$$\delta W_e = -P_1 dV_1$$

where dV_1 is the decrease of the external volume, equal to—dV_0. The work corresponding to the increase of interfacial area is

$$\delta W_s = \gamma dA$$

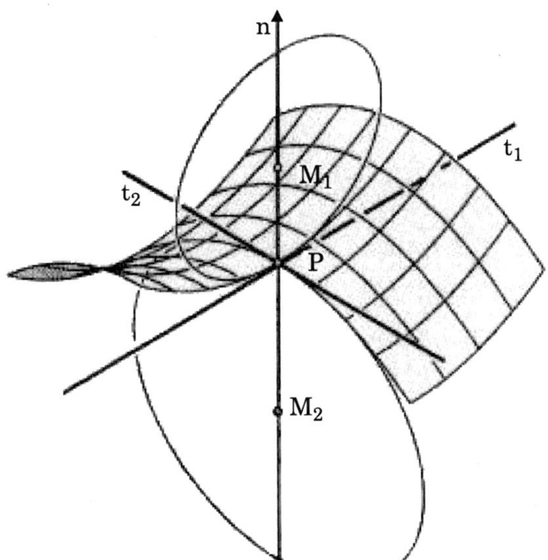

Figure 2.17 Mean curvature at a saddle point is zero if $|R_1| = |R_2|$.

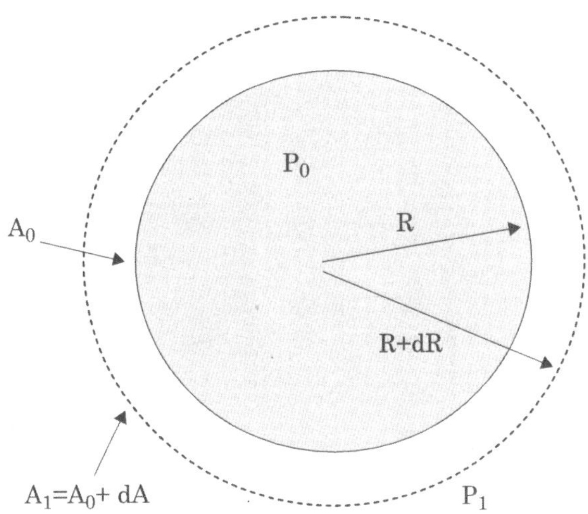

Figure 2.18 Schematic of a liquid drop immersed in a fluid; initially, the droplet radius is R and its surface area A_0. An increase of its radius by a quantity dR corresponds to the new surface area A_1 and the pressure P_1.

where dA is the increase of the surface area $dA = 8\pi R dR$. The mechanical equilibrium condition is then

$$\delta W = \delta W_i + \delta W_e + \delta W_s = 0. \tag{2.14}$$

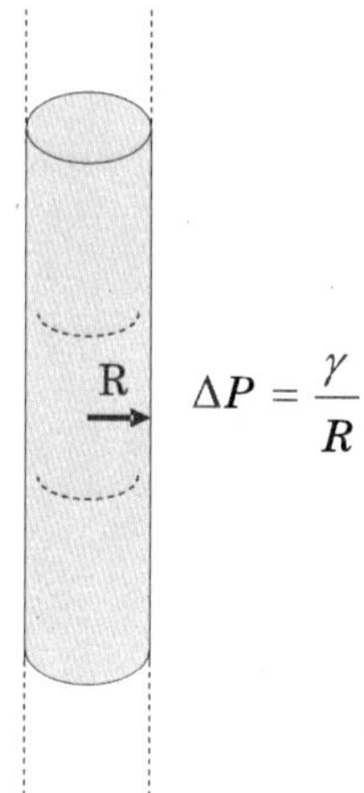

Figure 2.19 Laplace's law for cylindrical interface.

Substituting the values of the work found previously, it follows that

$$\Delta P = P_0 - P_1 = \frac{2\gamma}{R}. \tag{2.15}$$

Equation (2.15) is the Laplace equation for a sphere.
The reasoning we have used to obtain equation (2.15) can be generalized

$$\Delta P = \gamma \frac{dA}{dV}. \tag{2.16}$$

For simplicity, we have derived Laplace's equation for the case of a sphere, but we can use (2.16) for an interface locally defined by two (principal) radii of curvature R_1 and R_2; the result would then be

$$\Delta P = \gamma \left(\frac{1}{R_1} + \frac{1}{R_2} \right). \tag{2.17}$$

For a cylindrical interface, as sketched in Fig. 2.19, one of the two radii of curvature is infinite, and the Laplace law reduces to

$$\Delta P = \frac{\gamma}{R}. \tag{2.18}$$

Equation (2.17) is called the Laplace law. Keep in mind that it is closely linked to the minimization of the energy. Laplace's law is fundamental when dealing with interfaces, micro-drops and in digital microfluidics. In the following section, we give some examples of application of Laplace's law.

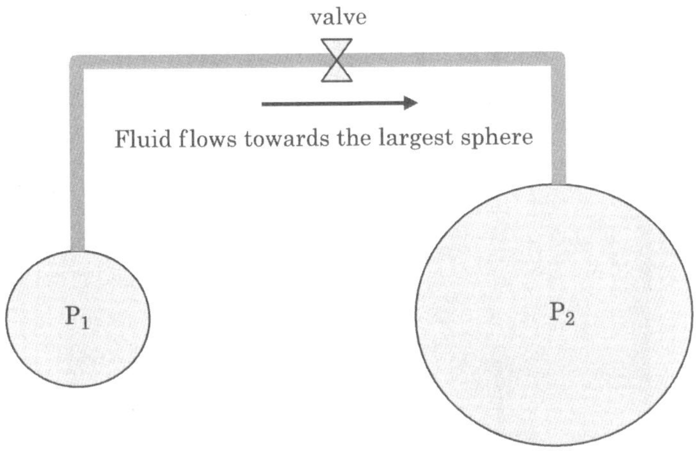

Figure 2.20 Fluid flow from the smaller bubble/droplet to the larger is a direct application of Laplace's law.

2.3.2 Examples of Application of Laplace's Law

Amongst other things, Laplace's law explains many phenomena occurring during electrowetting actuation. The use of Laplace's law for electrowetting will be detailed in Chapter 5. In the following section, applications of Laplace's law outside the electrowetting domain are presented.

2.3.2.1 Liquid Transfer From a Smaller Drop to a Bigger Drop

It has been observed that when two bubbles or droplets are connected together, there is a fluid flow from the small bubble/droplet to the larger one (Fig. 2.20). This is a direct application of Laplace's law: the pressure inside the small bubble/droplet is larger than that of the larger bubble/droplet, inducing a flow from the former towards the latter. This flow continues until the smaller bubble/droplet disappears to the profit of the larger one.

2.3.2.2 Precursor Film and Coarsening

The concept of infinitely thin interface and unique contact angle presented at the beginning of this chapter is a mathematical simplification of the reality. When a partially wetting droplet is deposited on a flat solid surface, a very thin film of a few nanometers spreads before the contact line, and the contact between the liquid and the solid resembles the sketch of Fig. 2.21. The precursor film can be explained by thermodynamic considerations; because a jump between the chemical potential of the gas and of the solid is not physical, liquid molecules intercalate between the gas and the solid [16]. Molecules of liquids

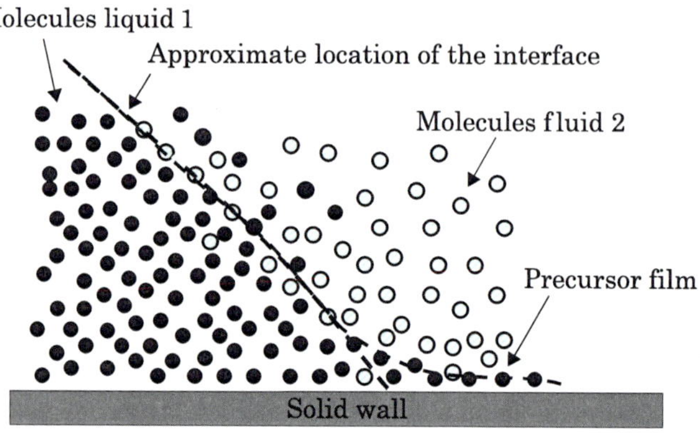

Figure 2.21 Interface with precursor film.

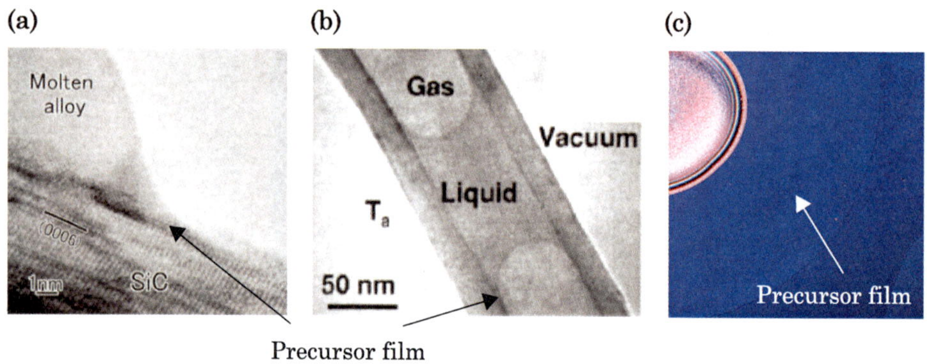

Figure 2.22 (a) Precursor film of spreading molten alloy; (b) precursor film of a liquid plug inside a carbon nanotube [19]; (c) scan of a liquid crystal precursor film; the precursor film spreads on a distance of 1 mm; (a) from [18]. © 2002 Elsevier, reprinted with permission. (b) reused with permission from [19]. © 2005 American Institute of Physics. (c) from [20] (Sarfus Image: Courtesy Nanolane).

progressively spread under the action of the "disjoining pressure" caused by the van der Waals interactions between the liquid and solid molecules [17].

Such precursor films have been observed for different wetting situations, as shown in Fig. 2.22. Note the extreme thinness of the film in the photographs.

When droplets of different size are deposited on a flat solid surface, if the droplets are sufficiently close to one another, it is observed that the smaller droplets disappear to the benefit of the large droplets. This phenomenon is called "coarsening". Experimental evidence of coarsening is shown in Fig. 2.23.

The explanation of the phenomenon requires two steps: first, the existence of a precursor film (extremely thin film on the solid surface spreading around each droplet) that links the droplets together. The second step is identical to that of the previous section: the pressure is larger in a small droplet according to the Laplace law, and there is a liquid flow towards the largest droplet (Fig. 2.24).

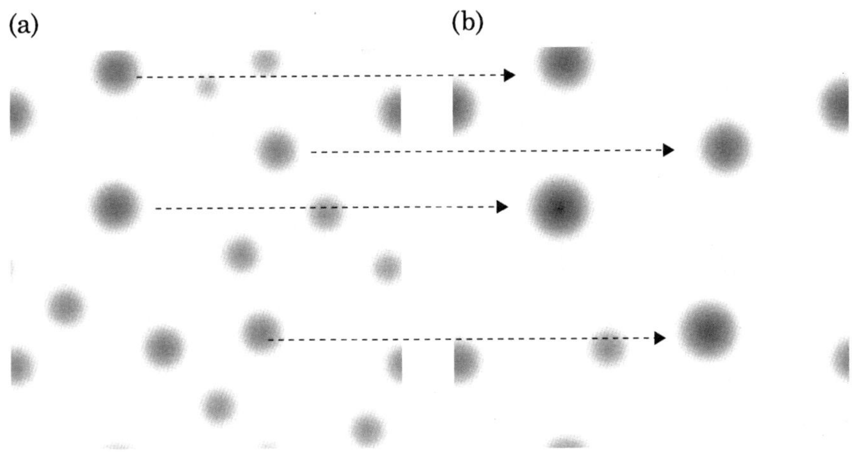

Figure 2.23 Experimental observation of coarsening: the number of droplets diminishes; only the largest droplets remain: (a) beginning of the observation, (b) increase in size of the large droplets and vanishing of the smaller droplets [21] (Courtesy Michael Bestehorn).

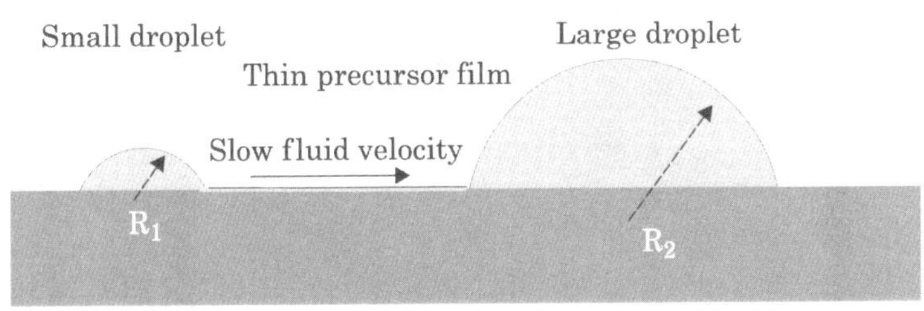

Figure 2.24 When two droplets are linked by a precursor film, a fluid flow is established from the smaller droplet to the larger droplet. The smaller droplet progressively disappears.

The precursor film is very thin, thus the flow rate between droplets is very small and mass transfer is extremely slow. Hence, experimental conditions require that the droplets do not evaporate.

2.3.2.3 Pressure in Droplets Constrained Between Two Parallel Plates

When using Laplace's law, one should be careful of the orientation of the curvature. A convex surface has two positive radii of curvature. A "saddle" surface has one positive and one negative curvature radius. Take the example of a water droplet flattened between two horizontal plates (we will see in Chapter 4 that this situation is frequent in EWOD-based microsystems [22]). Suppose that the droplet is placed at the intersection of a hydrophobic band and a hydrophilic band. As a result, the droplet is squashed by the hydrophobic band and elongated on the hydrophilic band (Fig. 2.25). The pressure in the droplet

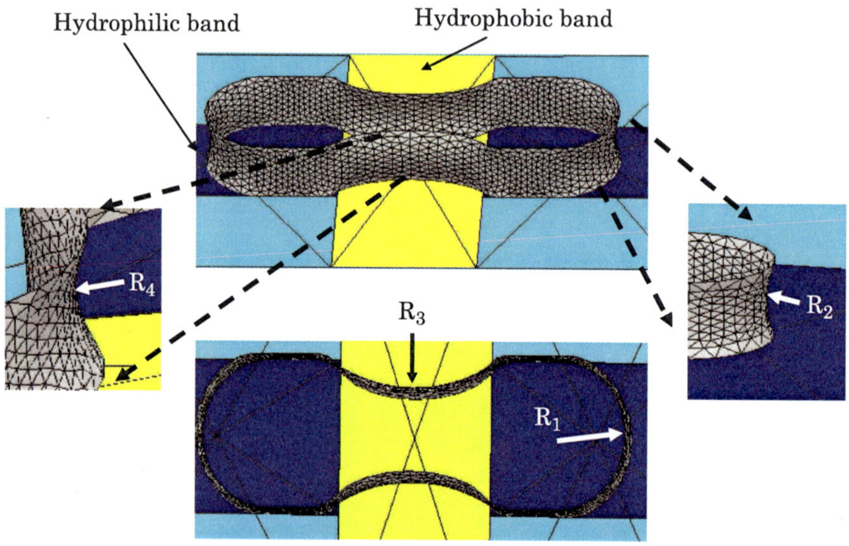

Figure 2.25 Sketch of a water droplet pinched by a hydrophobic surface. Case of a droplet constrained between two parallel planes (the upper plane has been dematerialized for visualization). Simulation performed with the Surface Evolver numerical software [23].

is given by the Laplace law

$$P_{\text{droplet}} - P_0 = \gamma\left(\frac{1}{R_1} + \frac{1}{R_2}\right) = \gamma\left(\frac{1}{R_3} + \frac{1}{R_4}\right) \tag{2.19}$$

where R_1, R_2, R_3, and R_4 are respectively the horizontal curvature radius in the hydrophilic region, the vertical curvature radius in the hydrophilic region, the horizontal curvature radius in the hydrophobic region, and the vertical curvature radius in the hydrophobic region. Taking into account the sign of the curvatures, we obtain

$$P_{\text{droplet}} - P_0 = \gamma\left(\frac{1}{|R_1|} - \frac{1}{|R_2|}\right) = \gamma\left(\frac{1}{|R_4|} - \frac{1}{|R_3|}\right). \tag{2.20}$$

The pressure in the drop being larger than the exterior pressure, the following relations must be satisfied

$$|R_4| < |R_3| \tag{2.21}$$

and

$$|R_2| > |R_1|. \tag{2.22}$$

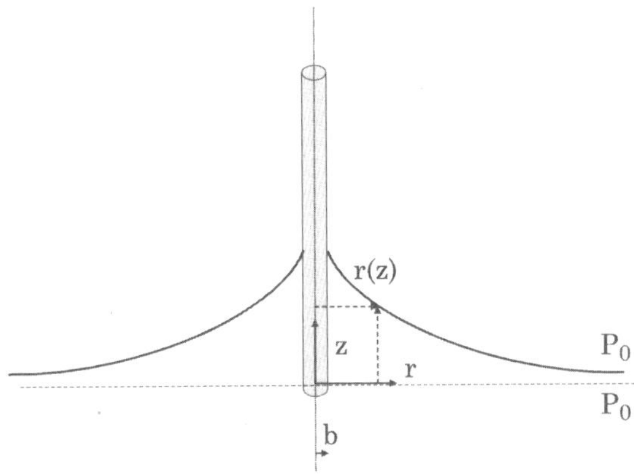

Figure 2.26 Sketch of wire dipped into a liquid (wetting case).

The vertical curvature radius R_4 in the hydrophobic region is smaller than the concave horizontal radius R_3 and the vertical curvature radius in the hydrophilic region R_2 is larger than the convex horizontal radius R_1. We shall see in Chapter 5 the use of a hydrophobic band to "cut" the droplet into two daughter droplets. For that to happen, the curvature radius R_3 must be sufficiently small, so that the two concave contact lines contact each other. The inequality (2.21) then produces a condition on the level of hydrophobicity required to obtain droplet division.

2.3.2.4 Zero Pressure Surfaces: Example of a Meniscus on a Wire

Laplace's law is often seen as determining a pressure difference on the two sides of the interface from the observation of curvature. But it is interesting to look at it the other way: knowing the pressure difference, what conclusion may be reached on the curvature of the interface?

We consider an example in which the pressure difference is zero: this is the case of a cylindrical wire dipped in a wetting liquid. We suppose that the angle between the surface and the wire (contact angle) is $\theta = 0$. We shall develop the notion of contact angle in Section 2.5. The liquid rises along the wire, deforming the free surface (Fig. 2.26). What is the shape of the surface? Laplace's law combined with the hydrostatic pressure yields

$$P_0 + \gamma \left(\frac{1}{R_1} + \frac{1}{R_2} \right) = P_0 - \rho g z \tag{2.23}$$

where R_1 and R_2 are the two (signed) principal curvature radii. $R_1(z)$ is the horizontal (positive) radius of the circle delimited by the intersection of a horizontal plane and the surface, while $R_2(z)$ is the (negative) curvature radius of the vertical profile at the elevation z. Assuming that the system is small

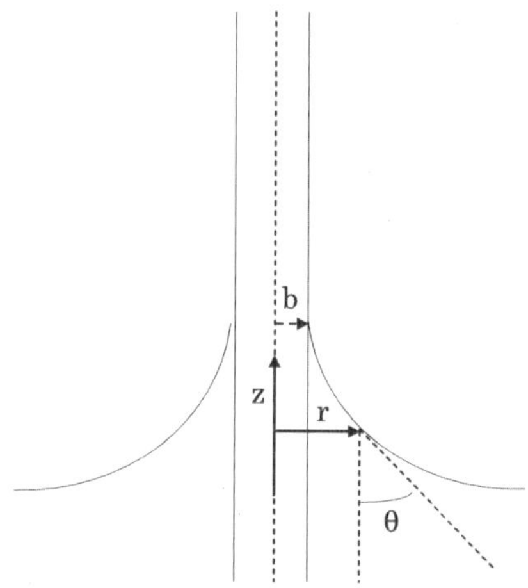

Figure 2.27 Vertical profile of liquid surface at the vicinity of a vertical rod.

enough that the gravity term can be neglected compared to the pressure P_0, we are left with

$$\frac{1}{R_1} + \frac{1}{R_2} = 0. \tag{2.24}$$

This is the equation of a zero curvature surface, also called a minimal surface. The equation of the surface can be obtained by writing that the vertical projection of the surface tension force is constant [23]. Using the notations of Fig. 2.27, we find

$$2\pi r \gamma \cos\theta = 2\pi b \gamma. \tag{2.25}$$

Substituting the relation $\tan\theta = \frac{dr}{dz} = \dot{r}$, we are left with

$$\frac{r}{\sqrt{1+\dot{r}^2}} = b. \tag{2.26}$$

Integration of (2.26) yields the equation of the vertical profile

$$r = b^* \cosh\frac{z-h^*}{b^*} \tag{2.27}$$

where b^* and h^* are constants depending on the contact angle with the vertical cylinder. In the case of a wetting contact ($\theta = 0$) the exact solution is

$$r = b\cosh\frac{z-h_0}{b} \tag{2.28}$$

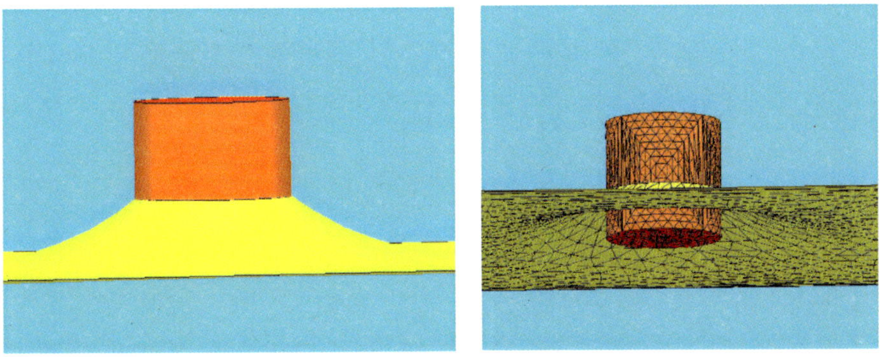

Figure 2.28 Vertical wire dipped into a fluid. The surface is deformed depending on the surface tension of the liquid and on the contact angle (Surface Evolver calculation). Left: view of the surface in the vicinity of the wire. Right: view from below showing the gain in elevation of the surface at the contact of the wire.

where b is the wire radius and h_0 is the height of the interface along the wire. We note that equation (2.28) is the equation of a catenoid [23]. However, this equation describes the surface close to the wire (for $r = b$, $z = h_0$) but not far from the wire. Far from the wire, the value of z increases again in equation (2.28). This is not physical, and we should take into account gravity that flattens the surface. We shall see in the following chapter that the expression

$$\kappa^{-1} = \sqrt{\frac{\gamma}{\rho g}} \tag{2.29}$$

is a characteristic length of the capillarity, called the capillary length. Scaling the horizontal extent of the surface deformation with the capillary length $r_{\max} \approx \kappa^{-1}$, equation (2.28) yields $\kappa^{-1} = b \cosh(-h_0/b)$. Assuming that $b \ll h_0$, we obtain an approximate value for the maximum height h_0 [24]

$$h_0 \approx b \ln\left(\frac{2\kappa^{-1}}{b}\right). \tag{2.30}$$

Fig. 2.28 shows the deformed surface obtained by a numerical simulation. Relation (2.30) shows that the elevation h_0 along the wire increases with the surface tension γ. At first sight, this may seem a paradox because the surface is pulled tighter when the surface tension increases. However, we show later in this chapter that the capillary force exerted by the wire is proportional to the surface tension. The force pulling the surface is thus larger for high surface tension liquids.

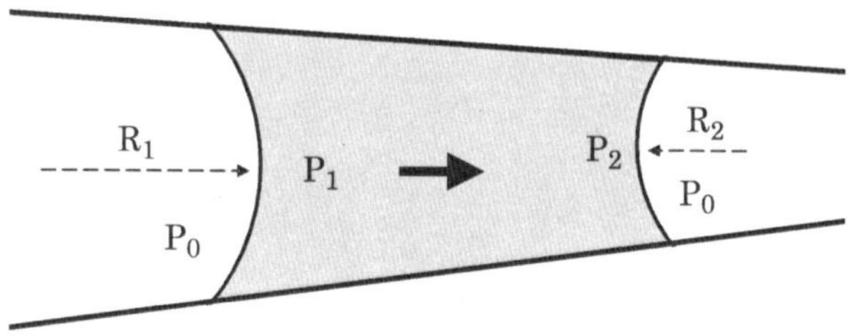

Figure 2.29 Sketch of a liquid plug moving under capillary forces between two plates. The contact angle is $\theta < 90°$.

2.3.2.5 Self Motion of a Liquid Plug Between Two Non-parallel Wetting Plates

It was first observed by Hauksbee [25] that a liquid plug limited by two non-parallel wetting plates moves towards the narrow gap region. A sketch of the plug is shown in Fig. 2.29. Laplace's law furnishes a very clear explanation of this phenomenon. Supposing that Fig. 2.29 is a wedge (2D situation), let us write Laplace's law for the left side interface

$$P_0 - P_1 = \frac{\gamma}{R_1} \tag{2.31}$$

and, for the right side interface

$$P_0 - P_2 = \frac{\gamma}{R_2}. \tag{2.32}$$

Subtraction of the two relations leads to

$$P_1 - P_2 = \gamma \left(\frac{1}{R_2} - \frac{1}{R_1} \right). \tag{2.33}$$

Next, we show that $R_2 < R_1$. Looking at Fig. 2.30, we have

$$R_2 \sin \beta = d_2$$

where d_2 is the half-distance between the plates.

The angle β is linked to θ and α by the relation

$$\beta = \frac{\pi}{2} + \alpha - \theta.$$

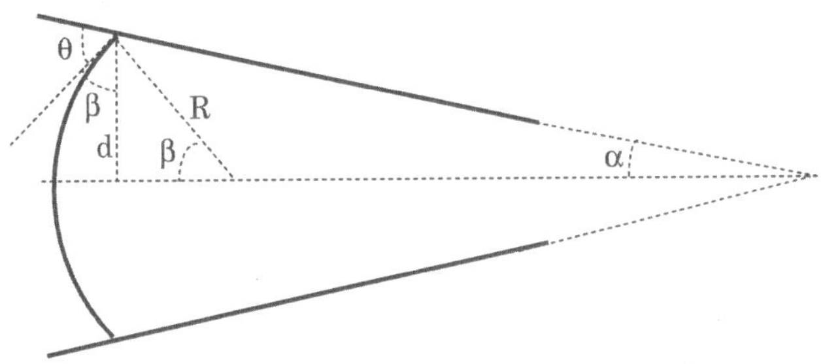

Figure 2.30 Curvature of the interface in a dihedral.

Finally we obtain

$$R_2 = \frac{d_2}{\cos(\alpha - \theta)}. \tag{2.34}$$

Using the same reasoning with a meniscus oriented in the opposite direction, we obtain the expression of R_1

$$R_1 = \frac{d_1}{\cos(\alpha + \theta)}. \tag{2.35}$$

Comparing relations (2.34) and (2.35), remarking that $d_2 < d_1$ and $\cos(\alpha - \theta) > \cos(\alpha + \theta)$, we deduce that R_2 is then smaller than R_1, and $P_1 > P_2$. The situation is not stable. Liquid moves from the high pressure region to the low pressure region, and the plug moves towards the narrow gap region. It has also been observed that the plug accelerates; it is due to the fact that the difference of the curvatures in equation (2.33) is increasing when the plug moves to a narrower region. Bouasse [25] has remarked that the same type of motion applies for a cone, where the plug moves towards the tip of the cone. In reality, Bouasse used a conical frustum (slice of cone) in order to let the gas escape during plug motion.

2.3.2.6 Laplace's Law in Medicine: Normal and Shear Stress in Blood Vessels

2.3.2.6.1 Shear stress

A human body—or any mammalian organism—respects the rules of physics. Take the example of blood vessels. The arrangement of blood vessel networks very often satisfies Murray's law (Fig. 2.31). In 1926, Murray observed the morphology of the blood system and found a very general relation between the dimensions of a "parent" branch and of a "daughter" branch, and he found that the same relation applies at any level of bifurcation.

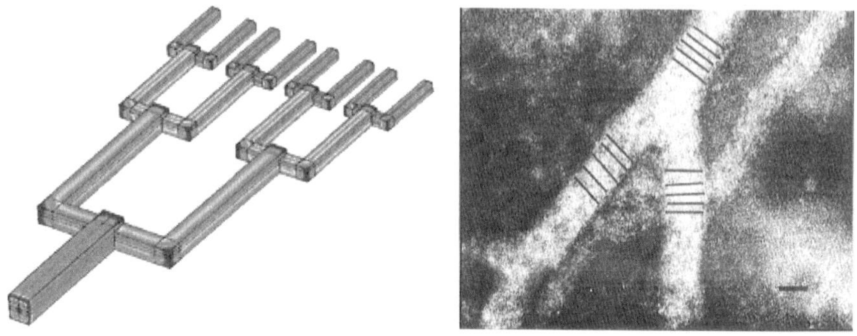

Figure 2.31 Left: schematic of blood vessel system. Right: photograph of blood vessel division in chickens. From [27]. © 2001 Elsevier; reprinted with permission.

Soon after, he published this discovery [26]. Since that time, this relation is known as Murray's law and can be written as

$$d_0^3 = d_1^3 + d_2^3 \tag{2.36}$$

where d_0, d_1, and d_2 are the "parent" and "daughter" channel diameters. Usually, daughter branches have the same dimension $d_1 = d_2$. A recurrence relation can be deduced from (2.36); it shows that the diameter, flow rates, and average velocities at the nth generation, i.e., at a bifurcation rank n (Fig. 2.31), are related to the "origin" diameter, flow rate, and average velocity by the relations

$$\begin{aligned} d_n &= \frac{d_0}{2^{n/3}} \\ Q_n &= \frac{Q_0}{2^n} \\ V_n &= \frac{V_0}{2^{n/3}}. \end{aligned} \tag{2.37}$$

Relation (2.37) can be developed further to show that the wall friction is the same at each level [27,28]. This property simply stems from the expression of the shear stress of a cylindrical duct

$$\tau_n = \frac{8\eta V_n}{d_n} = \frac{8\eta V_0}{d_0} = \tau_0. \tag{2.38}$$

The equality of the shear stress still stands if we take into account the non-Newtonian character of blood. Murray showed that, from a physiological point of view, such a relation minimizes the work of the blood circulation. The important thing here is that the shear stress is constant in most blood vessel networks. Now what about the normal stress?

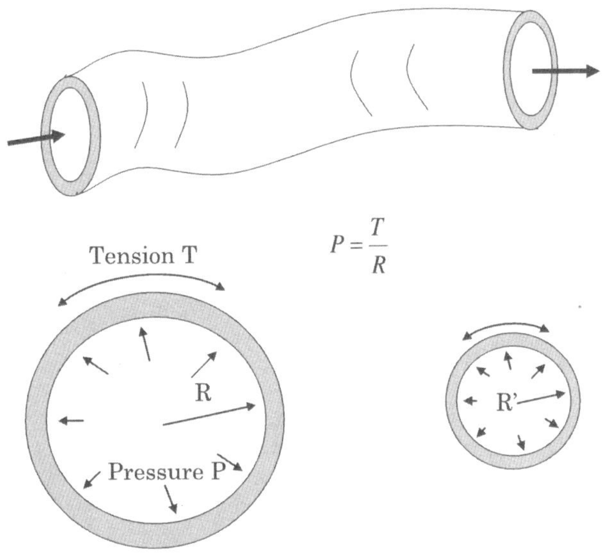

Figure 2.32 Schematic view of a blood vessel: if the internal pressure is P, the tension in the wall is $T = PR$. Small blood vessels have smaller walls than large blood vessels.

2.3.2.6.2 Normal stress

In the particular case of human or mammalian blood systems, the normal stress is simply the internal pressure, because, in a first approximation, the flow is purely axial and there is no radial component of the velocity. It has been observed that the thickness of the walls of blood vessels satisfies Laplace's law (Fig. 2.32). In this particular case, the surface tension is replaced by the wall tension T, and Laplace's law becomes

$$P = \frac{T}{R}. \tag{2.39}$$

At a given distance from the heart, the pressure is approximately the same and Laplace's equation (2.39) has the consequence that the wall tension increases together with the curvature radius. As a consequence, arteries have larger wall thickness than veins, and similarly veins compared to capillaries.

2.3.2.7 Laplace Law in Medicine: The Example of Lung Alveoli

It is very tempting to refer systematically to the Laplace law because of its simplicity. But one should refrain from doing that systematically. A striking example is that of lung ventilation.

Ventilation of lungs has been widely studied for medical purposes. It was usual to consider the alveoli to be like spherical balloons inflated during lung

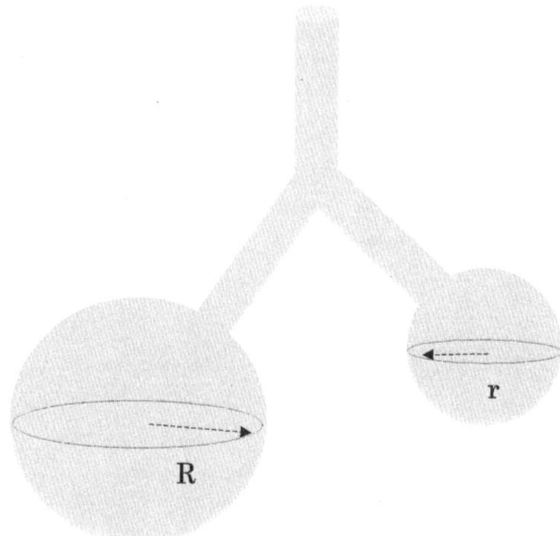

Figure 2.33 Wrong sketch for the alveoli leading to a biased Laplace's law application.

ventilation (Fig. 2.33). The problem is that the alveoli are connected, and when applying Laplace's law, the air in the smaller alveoli should be driven to the larger alveoli and a general collapse of the lungs would occur. Because the collapse of the alveoli does not—luckily—correspond to the reality, it has been suggested that the concentration of surfactant in the alveoli is not uniform and compensates for the different Laplace pressures.

Recently, a different, more realistic analysis has been made [29]: the alveoli are not "free" spheres but are packed together (Fig. 2.34) and there are pores in the alveoli walls. Alveoli cannot expand freely, and they are limited in their inflation. As a result, smallest alveoli do not collapse during ventilation, because large alveoli cannot grow indefinitely and Laplace's law is not the answer in this kind of problem.

2.4 Wetting—Partial or Total Wetting

So far, we have dealt with interfaces between two fluids. Triple contact lines are the intersections of three interfaces involving three different materials: for example, a droplet of water on a solid substrate has a triple contact line.

Liquids spread differently on a horizontal plate according to the nature of the solid surface and that of the liquid. In reality, it depends also on the third constituent, which is the gas or the fluid surrounding the drop. Two different situations are possible: either the liquid forms a droplet and the wetting is said to be partial, or the liquid forms a thin film wetting the solid surface, with the horizontal dimension of the film depending on the initial volume of liquid (Fig. 2.35). For example, in the environment of a room, water spreads like a

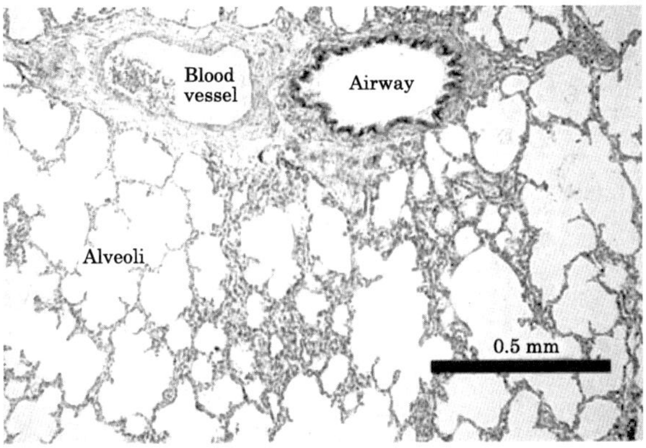

Figure 2.34 Image of lung alveoli; detail from an original slide generously provided by A. Mescher.

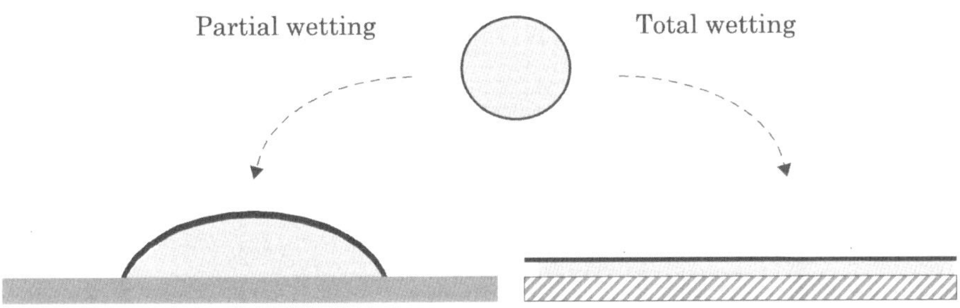

Figure 2.35 Wetting is said to be total when the liquid spreads like a film on the solid surface.

film on a very clean and smooth glass substrate, whereas it forms a droplet on a plastic substrate. In the case of partial wetting, there is a line where all three phases come together. This line is called the *contact line* or the *triple line*.

A liquid spreads on a substrate in a film if the energy of the system is lowered by the presence of the liquid film (Fig. 2.36). The surface energy per unit surface of the dry solid surface is γ_{SG}; the surface energy of the wetted solid is $\gamma_{\mathrm{SL}}+\gamma_{\mathrm{LG}}$. The spreading parameter S determines the type of spreading (total or partial)

$$S = \gamma_{\mathrm{SG}} - (\gamma_{\mathrm{SL}} + \gamma_{\mathrm{LG}}) . \tag{2.40}$$

If $S > 0$, the liquid spreads on the solid surface; if $S < 0$ the liquid forms a droplet.

When a liquid does not totally wet the solid, it forms a droplet on the surface. Two situations can occur: if the contact angle with the solid is less than 90°, the contact is said to be "hydrophilic" if the liquid has a water base, or more generally "wetting" or "lyophilic." In the opposite case of a contact angle larger

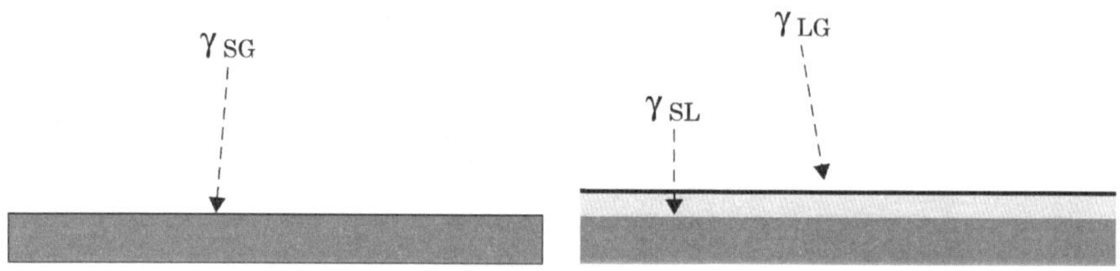

Figure 2.36 Comparison of the energies between the dry solid and the wetted solid.

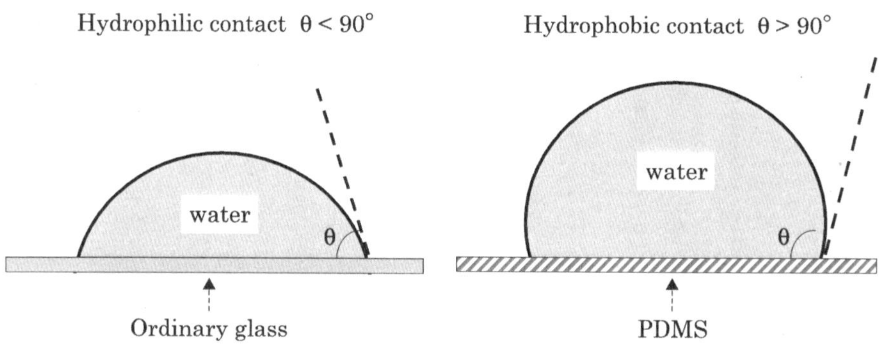

Figure 2.37 Water spreads differently on different substrate.

than 90°, the contact is said to be "hydrophobic" with reference to water or more generally "not wetting" or "lyophobic" (Figs. 2.37 and 2.38).

2.5 Contact Angle—Young's Law

2.5.1 Young's Law

Surface tension is not exactly a force; its unit is N/m. However, it represents a force that is exerted tangentially to the interface. Surface tension can be looked at as a force per unit length. This can be directly seen from its unit. But it may be interesting to give a more physical feeling by making a very simple experiment (Fig. 2.39) [24]. Take a solid frame and a solid tube that can roll on this frame. If we form a liquid film of soap between the frame and the tube—by plunging one side of the structure in a water-soap solution, the tube starts to move towards the region where there is a liquid film. The surface tension of the liquid film exerts a force on its free boundary.

On the other hand, we can increase the film surface by exerting a force on the tube. The work of this force is given by the relation

$$\delta W = F dx = 2\gamma L dx. \tag{2.41}$$

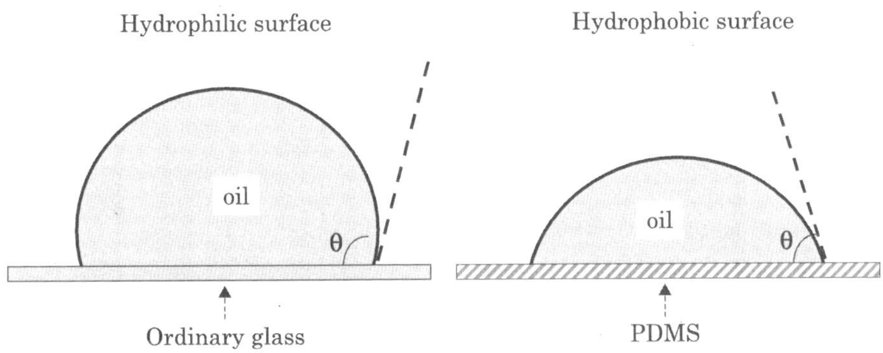

Figure 2.38 Silicone oil has an opposite wetting behavior than water.

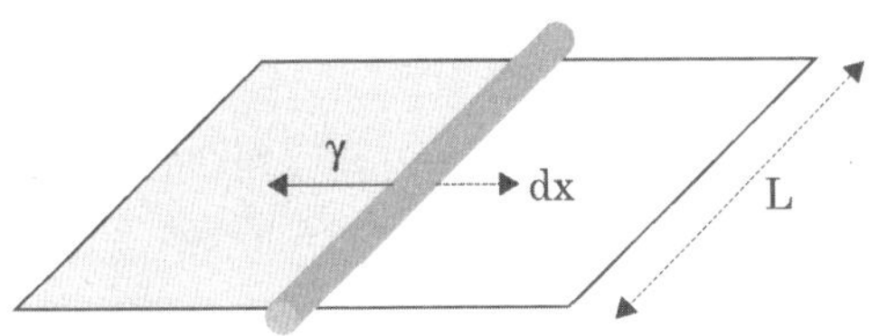

Figure 2.39 A tube placed on a rigid frame whose the left part is occupied by a soap film requires a force to be displaced towards the right; this force opposed the surface tension that tends to bring the tube to the left.

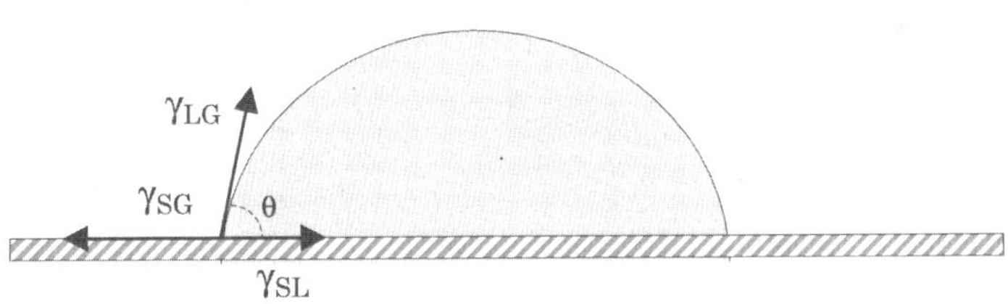

Figure 2.40 Schematic of the forces at the triple contact line.

The coefficient 2 stems from the fact that there are two interfaces between the liquid and the air. This relation shows that the surface tension γ is a force per unit length, perpendicular to the tube, in the plane of the liquid and directed towards the liquid.

We can then draw the different forces that are exerted by the presence of a fluid on the triple line (Fig. 2.40).

At equilibrium, the resultant of the forces must be zero. We use a coordinate system where the x-axis is the tangent to the solid surface at the contact line (horizontal) and the y-axis is the perpendicular (vertical). At equilibrium, the projection of the resultant on the x-axis is zero and we obtain the relation

$$\gamma_{LG}\cos\theta = \gamma_{SG} - \gamma_{SL}. \tag{2.42}$$

This relation is called Young's law and is very useful for understanding the behavior of a drop. Especially, it shows that the contact angle is determined by the surface tensions of the three constituents. For a micro-drop on a solid, the contact angle is given by the relation

$$\theta = \arccos\frac{\gamma_{SG} - \gamma_{SL}}{\gamma_{LG}}. \tag{2.43}$$

Note that sometimes it happens that, in real experimental situations when we deal with real biological liquids, one observes unexpected changes in the contact angle with time. This is just because biological liquids are inhomogeneous and can deposit a layer of chemical molecules on the solid wall, thus progressively changing the value of the tension γ_{SL}, and consequently the value of θ, as stated by Young's law.

Young's law can be more rigorously derived from free energy minimization. Consider a sessile droplet large enough for the effect of the triple line to be neglected. The change of free energy due to a change in droplet size can be written as [30]

$$\begin{aligned} dF &= \gamma_{SL} dA_{SL} + \gamma_{SG} dA_{SG} + \gamma_{LG} dA_{LG} \\ &= (\gamma_{SL} - \gamma_{SG} + \gamma_{LG}\cos\theta)\, dA_{SL} \end{aligned} \tag{2.44}$$

where θ is the contact angle. At mechanical equilibrium $dF = 0$ and

$$\gamma_{SL} - \gamma_{SG} + \gamma_{LG}\cos\theta = 0. \tag{2.45}$$

Equation (2.45) is the Young law, identical to (2.42).

2.5.2 Nano-bubbles on Hydrophobic Walls, Line Tension and the Modified Young's Law

It has been observed that bubbles often form along hydrophobic walls, even when the surface is smooth. The size of these bubbles is in the mesoscopic range—between the microscopic and nanoscopic scales: bubble dimension is usually less than 200 nm. A paradox arises when calculating the internal pressure. Using the Laplace law with a curvature radius of the order of the observed contact at the wall, one finds that the internal pressure should be of

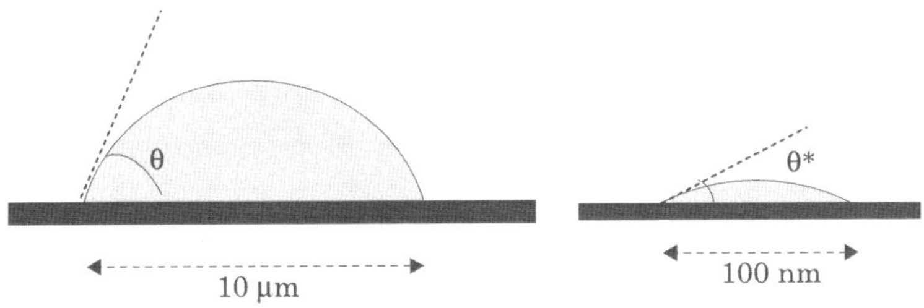

Figure 2.41 Comparative sketches of micro-bubble and nano-bubble (not to scale).

the order of

$$P \approx \frac{\gamma}{R} \approx \frac{70 \times 10^{-3}}{200 \times 10^{-9}} \approx 5 \times 10^5 \text{ Pa}.$$

At this level of pressure the gas should dissolve, and the bubble would disappear rapidly in the liquid. So why are these bubbles stable? From a Laplace's law point of view, either their surface tension should be smaller than that of a "macroscopic" bubble, or their curvature radius should be larger. It is easy to see that a reduced surface tension is not sufficient to find a sustainable internal pressure. On the other hand, at this scale, contact angle measurements are very tricky. However, recent measurements [31] have shown that nano-bubbles have very flat profiles—the base radius is 5–20 times larger than the height—because the contact angle of the bubble is much smaller than the macroscopic contact angle of the bubble on the same substrate (Fig. 2.41).

The problem has now shifted from Laplace's law to Young's law. What has changed in Young's law for the contact angle between the nanometric scale and the microscopic/macroscopic scale? The answer to this question is not yet clear. A plausible answer is connected to the notion of "line tension" and to the so-called modified Young's law. Young's law has been derived for a triple line without consideration of the interactions near the triple contact line. A sketch of the interactions leading to surface tension and line tension is shown in Fig. 2.42. The molecules close to the triple line experience a different set of interactions than at the interface. To take into account this effect, a line tension term has been introduced in Young's law [32]

$$\gamma_{SG} = \gamma_{SL} + \gamma_{LG} \cos\theta^* + \frac{\gamma_{SLG}}{r} \tag{2.46}$$

where r is the contact radius, γ_{SLG} the line tension (unit N), and θ^* the real contact angle. The contact angle is then changed by the line tension according to

$$\cos\theta^* = \cos\theta - \frac{\gamma_{SLG}}{r\gamma_{LG}}. \tag{2.47}$$

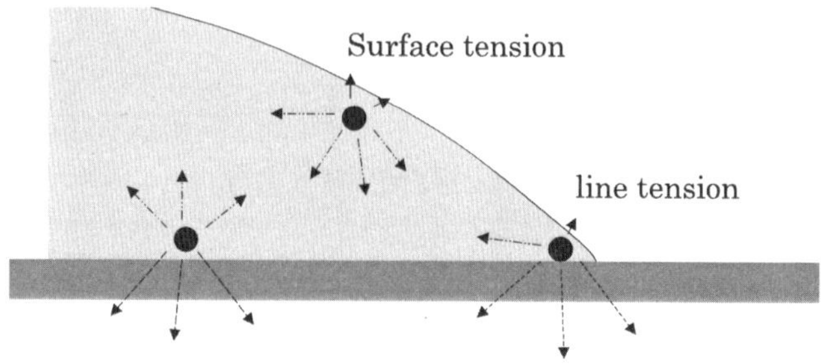

Figure 2.42 Sketch of interactions leading to surface tension and line tension.

For droplet contact radius larger than 10 μm, the effect of the line tension is negligible; the value of the term $\tau/r = \gamma_{SLG}/r\gamma_{LG}$ is of the order of 10^{-4} [24]. But it is not the case for nano-drops and nano-bubbles.

2.6 Work of Adhesion, Work of Cohesion, and the Young–Dupré Equation

In this section, we introduce the notions of work of adhesion and work of cohesion. These notions are valid for solids or immiscible liquids. When applied to a solid and a liquid, the concept of work of adhesion combined with Young's law produces the Young–Dupré equation. Work of adhesion and Young–Dupré's equation have been widely used to determine the surface tension of solids (Section 2.9).

2.6.1 Work of Adhesion

Imagine a body contacting another body on a surface S (Fig. 2.43). The surface energy of S when there is contact is

$$E_{12} = \gamma_{12} S.$$

The work of adhesion is the work required to separate the two bodies. After separation, the surface energies are

$$E = E_1 + E_2 = (\gamma_1 + \gamma_2) S.$$

The work of adhesion is then

$$W_a = \gamma_1 + \gamma_2 - \gamma_{12}. \tag{2.48}$$

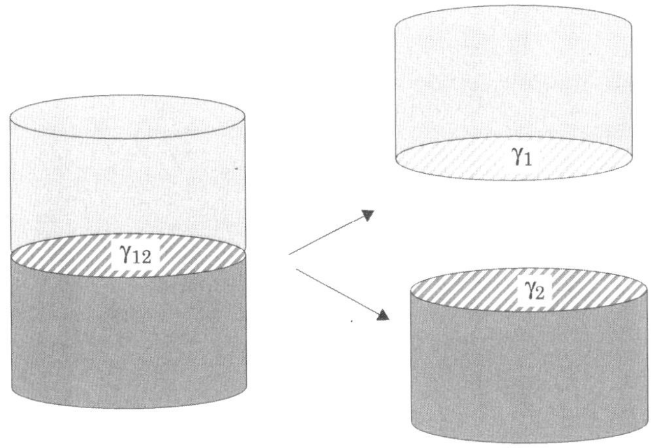

Figure 2.43 Work of adhesion is the work done to separate 2 surfaces of incompatible substance.

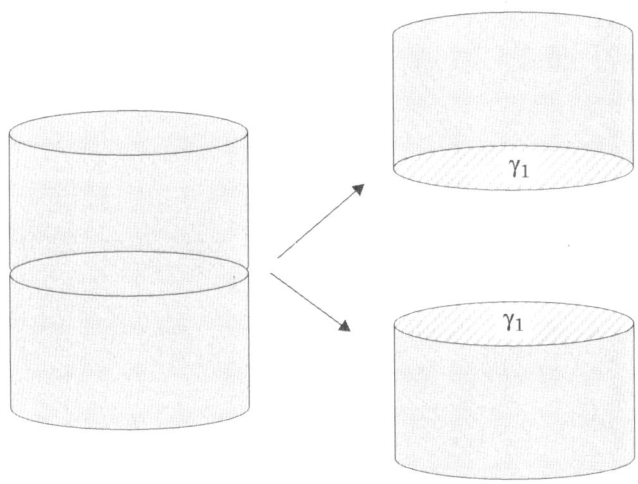

Figure 2.44 Work of cohesion is the work done to separate 2 surfaces of the same substance.

2.6.2 Work of Cohesion

The work of cohesion is obtained similarly, but this time the body being split is homogeneous (Fig. 2.44). The same reasoning yields

$$W_c = 2\gamma_1. \tag{2.49}$$

In other words, the surface energy is half the work of cohesion.

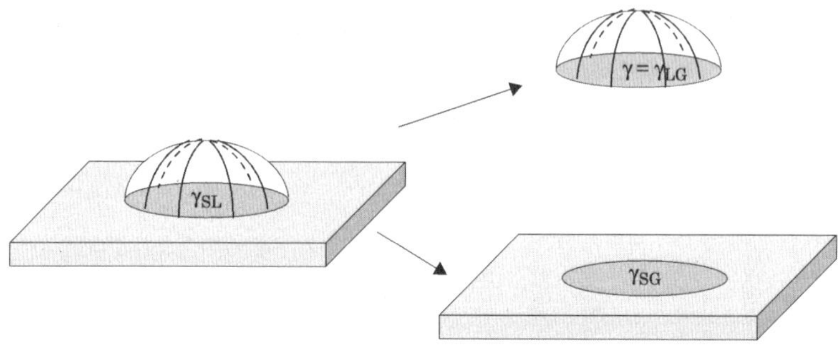

Figure 2.45 Sketch of the Young–Dupré equation for the work of adhesion of a liquid and a solid: $W_a = \gamma(1 + \cos\theta)$.

2.6.3 Young–Dupré Equation

Let us express the work of adhesion for a liquid and a solid (Fig. 2.45). Using (2.48) with the surface tensions $\gamma_1 = \gamma_{LG} = \gamma$, $\gamma_2 = \gamma_{SG}$, and $\gamma_{12} = \gamma_{SL}$, we obtain

$$W_a = \gamma + \gamma_{SG} - \gamma_{SL}. \tag{2.50}$$

Upon substitution of Young's law, we derive the Young–Dupré equation

$$W_a = \gamma(1 + \cos\theta). \tag{2.51}$$

For a super-hydrophobic contact, $\theta = \pi$ and $\cos\theta = -1$; we deduce that $W_a = 0$: there is no work to furnish to separate a super-hydrophobic liquid from a solid. Concretely, a droplet of water rolls freely over a super-hydrophobic surface. The Young–Dupré equation indicates that the more hydrophobic (non-wetting) is the contact between a liquid and a solid, the smaller is the work of adhesion.

2.7 Capillary Force, Force on a Triple Line

2.7.1 Introduction

We have seen in Section 2.5 the correspondence between surface tension and capillary forces. Capillary forces are extremely important at a micro-scale. We have all seen insects "walking" on the surface of a water pond (Fig. 2.46). Their hydrophobic legs do not penetrate the water surface and their weight is balanced by the surface tension force. More than that, it is observed that some insects can walk up meniscus, i.e., can walk inclined water surface. The

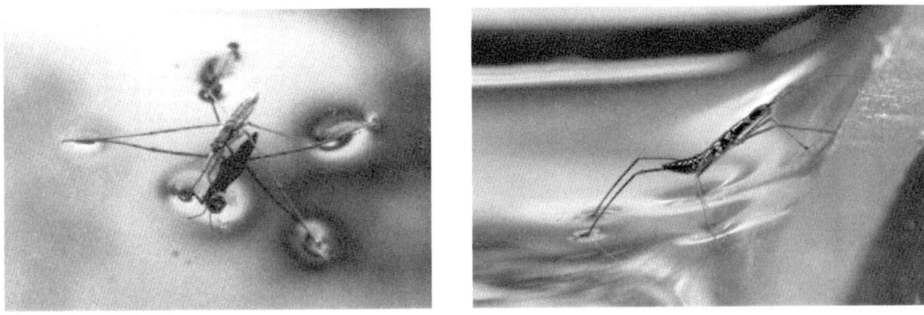

Figure 2.46 Left: capillary forces make the water surface resist the weight of an insect. Right: an insect walking up a meniscus [33] (photo courtesy David Hu).

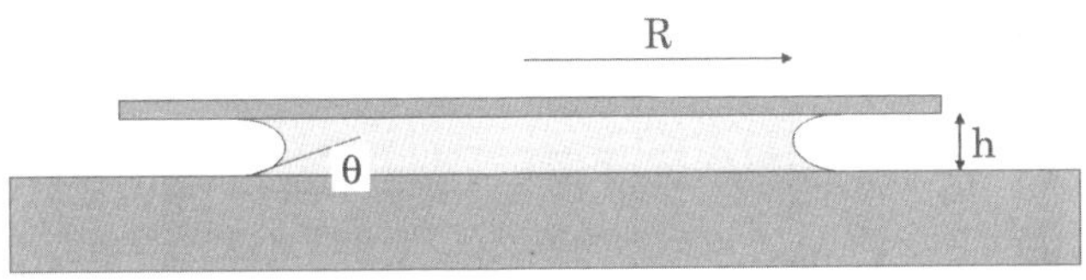

Figure 2.47 Film of water between two glass plates.

explanation of this phenomenon was recently given by Hu et al. [33] and refers to complex interface deformation under capillary forces.

In the domain of microfluidics, capillary forces are predominant; some examples of the action of capillary forces are given in the following sections.

2.7.2 Capillary Force Between Two Parallel Plates

A liquid film placed between two parallel plates makes the plates very adhesive. For instance, when using a microscope to observe objects in a small volume of liquid deposited on a plate and maintained by a secondary glass plate, it is very difficult to separate the plates. This situation is schematized in Fig. 2.47. In the first place, it is observed that the meniscus has a round shape (in order to minimize the free energy).

Let us write Laplace's law at the free interface. The first (horizontal) radius of curvature is approximately R. The second (vertical) radius of curvature, shown in Fig. 2.48, is calculated by

$$R_2 \sin\left(\frac{\pi}{2} - \theta\right) = \frac{h}{2}$$

or

$$R_2 = \frac{h}{2\cos\theta}. \tag{2.52}$$

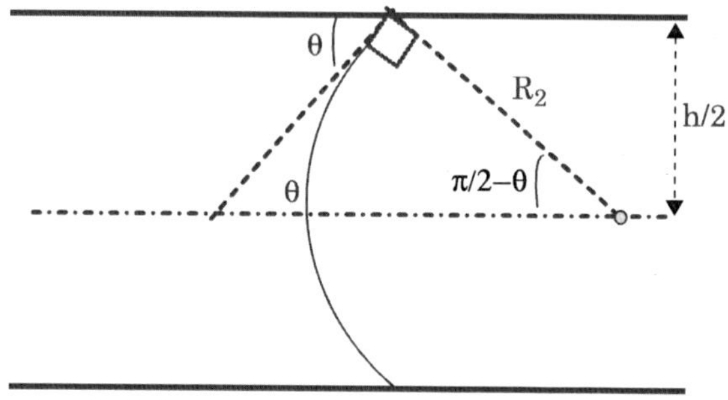

Figure 2.48 Calculation of the vertical curvature.

Laplace's law states that

$$\Delta P = \gamma \left(\frac{1}{R} - \frac{2\cos\theta}{h} \right). \tag{2.53}$$

In (2.53) the minus sign derives from the concavity of the interface. Because the vertical gap h is much less than the horizontal dimension R, we have the approximation

$$\Delta P \approx -\frac{2\gamma\cos\theta}{h}.$$

And the capillary force that links the plates together is

$$F \approx \frac{2\gamma\cos\theta}{h}\pi R^2. \tag{2.54}$$

This capillary force can be quite important; for $h = 10\ \mu m$ and $R = 1$ cm, the force F is of the order of 2.5 N.

2.7.3 Capillary Rise in a Tube

When a capillary tube is plunged into a volume of wetting liquid, the liquid rises inside the tube under the effect of capillary forces (Figs. 2.49 and 2.50). It is observed that the height reached by the liquid is inversely proportional to the radius of the tube.

Historically, many scientists have investigated this phenomenon, from Leonardo da Vinci, to Hauksbee, and Jurin. This property is now referred to as Jurin's law.

Using the principle of minimum energy, one can conclude that the liquid goes up in the tube if the surface energy of the dry wall is larger than that of the

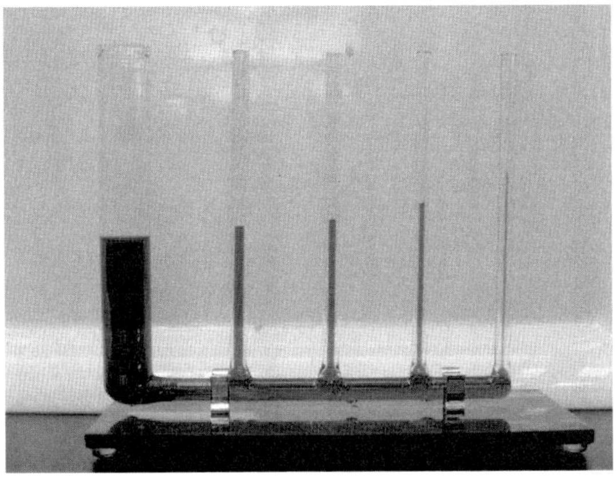

Figure 2.49 Capillary rise in tubes of different internal cross section.

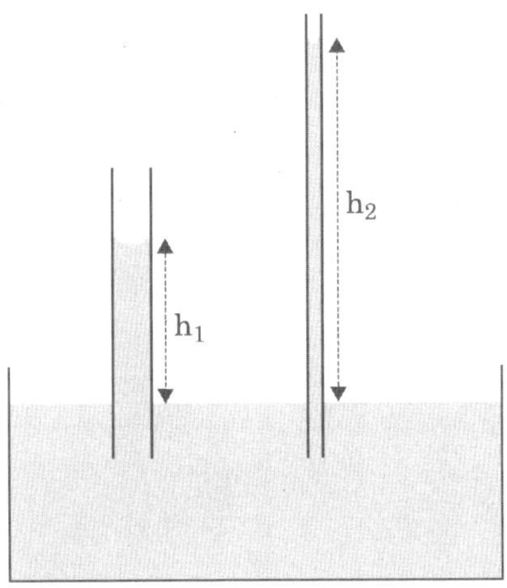

Figure 2.50 Capillary rise is inversely proportional to the capillary diameter.

wetted wall. If we define the impregnation criterion I by

$$I = \gamma_{SG} - \gamma_{SL}. \tag{2.55}$$

The liquid rises in the tube if $I > 0$; Otherwise the liquid descends in the tube. Using Young's law, the impregnation criterion can be written under the form

$$I = \gamma \cos\theta. \tag{2.56}$$

When the liquid rises in the tube, the system gains potential energy—because of the elevation of a volume of liquid—and loses capillary energy—due to the reduction of the surface energy. The balance is [24]

$$\begin{aligned} E &= \frac{1}{2}\rho g h V_{\text{liquid}} - S_{\text{contact}} I = \frac{1}{2}\rho g h \left(\pi R^2 h\right) - 2\pi R h I \\ &= \frac{1}{2}\rho g \pi R^2 h^2 - 2\pi R h \gamma \cos\theta. \end{aligned} \tag{2.57}$$

Note that we have not taken into account in (2.57) the detailed shape of the meniscus for the height h. The equilibrium elevation of the liquid is given by

$$\frac{\partial E}{\partial h} = 0$$

which results in

$$h = \frac{2\gamma\cos\theta}{\rho g R}. \tag{2.58}$$

Equation (2.58) is called Jurin's law. The capillary rise is inversely proportional to the tube radius. It can be also applied to the case where the liquid level in the tube decreases below the outer liquid surface; this situation happens when $\theta = 90$ degrees.

The maximum possible height that a liquid can reach corresponds to $\theta = 0$ and is $h = 2\gamma/\rho g R$.

In microfluidics, capillary tubes of 100 μm diameter are currently used; if the liquid is water ($\gamma = 72$ mN/m), and using the approximate value $\cos\theta \sim 1/2$, the capillary rise is of the order of 14 cm, which is quite important at the scale of a microcomponent.

Equation (2.58) characterizes the capillary rise; note that the capillary rise can be important. What is the corresponding capillary force? The capillary force balances the weight of the liquid in the tube. This weight is given by

$$F = \rho g \pi R^2 h.$$

Replacing h by its value from equation (2.54), we find the capillary force

$$F = 2\pi R \gamma \cos\theta. \tag{2.59}$$

The capillary force is the product of the length of the contact line $2\pi R$ times the line force $f = \gamma\cos\theta$. This line force is sketched in Fig. 2.51.

Note that the capillary force per unit length f is identical to the impregnation criterion

$$f = \gamma\cos\theta = I. \tag{2.60}$$

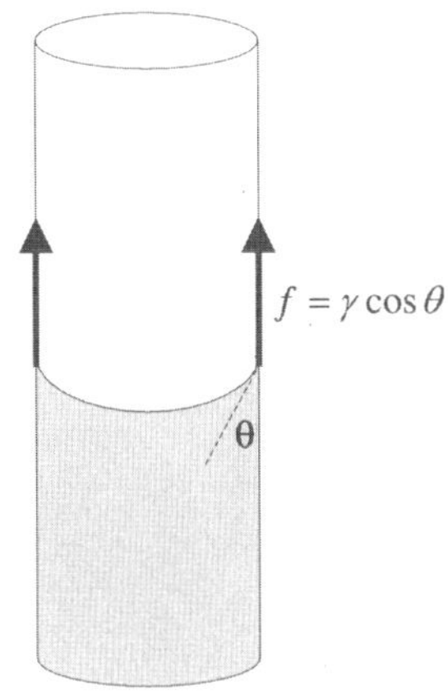

Figure 2.51 Sketch of the capillary force of a liquid inside a tube.

For $f > 0$ the liquid goes up in the tube and for $f < 0$ the liquid goes down. Note that Fig. 2.50 is not quite exact. There are also capillary forces on the outside of the tube, as shown in Fig. 2.52.

To derive the expression of the capillary rise inside the tube, a control volume corresponding to the liquid volume inside the tube was first considered. Let us consider now a control volume defined by the pipette (Fig. 2.53). The force to maintain the pipette is

$$F = P - P_A + P_{c,e} + P_{c,i} \tag{2.61}$$

where P is the weight of the tube, P_A the floatation force, and $P_{c,i}$ and $P_{c,e}$ are respectively the interior and exterior capillary forces *exerted on the solid.*

$$F = P - P_A + 2\pi R_{\text{int}}\gamma\cos\theta + 2\pi R_{\text{ext}}\gamma\cos\theta. \tag{2.62}$$

This force is a function of the surface tension γ. In Section 2.8, we will see that the measure of such a force constitutes a way to determine the surface tension.

2.7.4 Capillary Rise Between Two Parallel Vertical Plates

The same reasoning can be made for a meniscus between two parallel plates (Fig. 2.54) separated by a distance $d = 2R$. It is easy to show that in this case

$$h = \frac{\gamma\cos\theta}{\rho g R}. \tag{2.63}$$

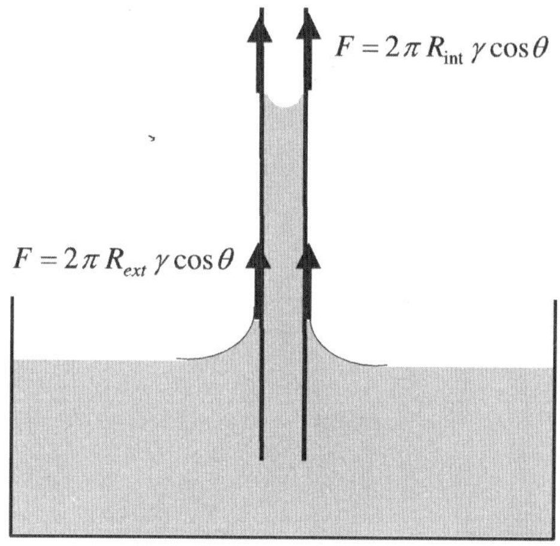

Figure 2.52 Capillary forces act also on the tube exterior, raising the level of the liquid around the tube (if the liquid wets the solid it would be the opposite if the liquid were not wetting).

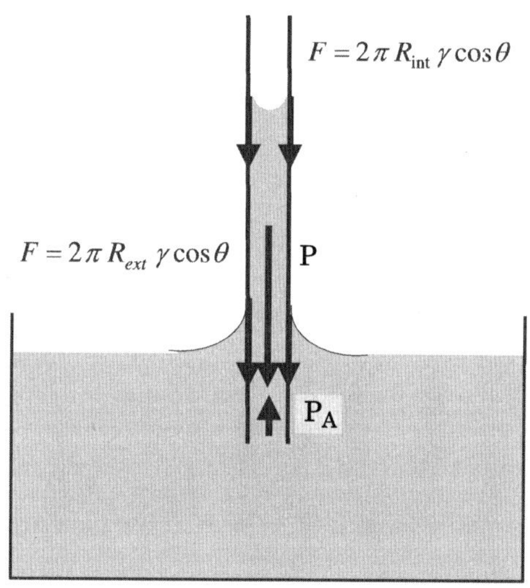

Figure 2.53 Forces acting on the pipette.

If we introduce the capillary length defined by

$$\kappa^{-1} = \sqrt{\frac{\gamma}{\rho g}} \tag{2.64}$$

we can rewrite (2.63) under the form

$$h = \kappa^{-2} \frac{\cos\theta}{R}. \tag{2.65}$$

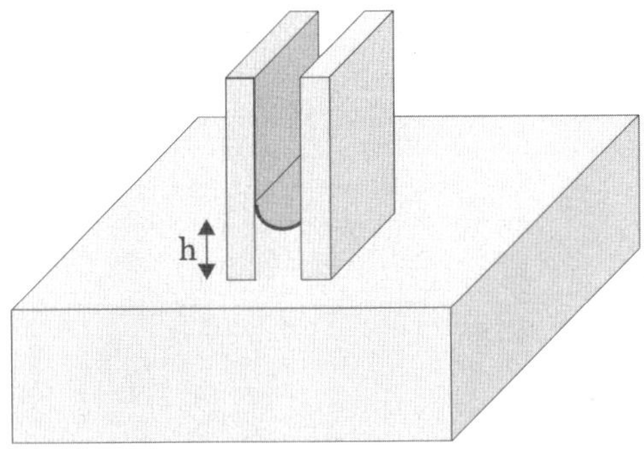

Figure 2.54 Capillary rise between two parallel vertical plates.

Note: The expressions for the two geometries (cylinder and two parallel plates) are similar. If we use the coefficient c, with $c = 2$ for a cylinder and $c = 1$ for parallel plates [34], we have

$$h = c\kappa^{-2}\frac{\cos\theta}{R} \tag{2.66}$$

where R is either the radius of the cylinder or the half-distance between the plates.

2.7.5 Capillary Rise in a Pipette

The analysis of the capillary rise—or descent—in cylindrical tubes or between two parallel plates has been recently extended by Tsori [34] to the case where the walls are not parallel, as, for instance, for a conical pipette (Fig. 2.55).

The mechanic equilibrium states that the Laplace pressure is balanced by the hydrostatic pressure

$$P_0 + \frac{c\gamma}{r} = P_0 - \rho g h \tag{2.67}$$

where c is the index defined previously, $c = 2$ for cones and $c = 1$ for wedges, and r is the curvature radius of the meniscus. Note that the depth h is counted negatively from the surface. Using the same approach as that of Section 2.3.2.5, the curvature radius is expressed by

$$r(h) = -\frac{R(h)}{\cos(\theta + \alpha)} \tag{2.68}$$

with

$$R(h) = R_0 + h\tan\alpha \tag{2.69}$$

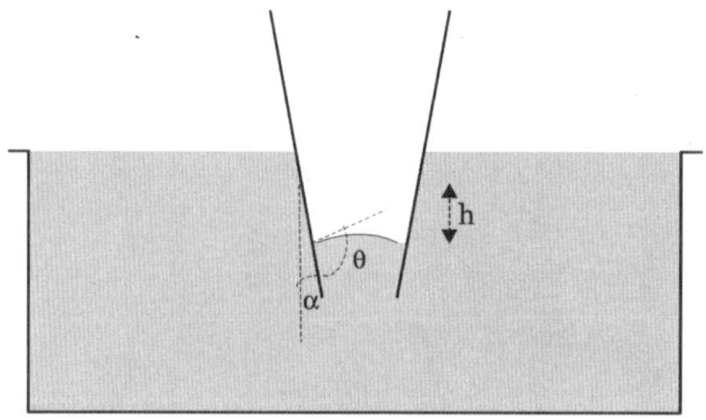

Figure 2.55 Capillary descent inside a hydrophobic pipette.

where R_0 is the internal radius of the pipette at $h = 0$. Substituting (2.69) in (2.68) and then in (2.67), we obtain

$$\cos(\theta + \alpha) = \frac{1}{c} h \left(R_0 + h \tan \alpha\right) \frac{\rho g}{\gamma}. \tag{2.70}$$

Using capillary length κ^{-1} to scale the variables, we obtain the dimensionless variables $\bar{R} = \kappa R$ and $\bar{h} = \kappa h$. Equation (2.70) then becomes

$$\tan \alpha \bar{h}^2 + \bar{R}_0 \bar{h} - c \cos(\theta + \alpha) = 0. \tag{2.71}$$

This is a quadratic equation in $\bar{h}$. The discussion of this equation is somewhat complex. Depending on the values of α and θ, the meniscus may be stable or not stable; in this latter case, the meniscus jumps to the top or the bottom of the pipette where it gets stabilized by pinning (anchoring to an angle). The diagram of Fig. 2.56 summarizes the meniscus behavior.

The important information here is that there are two domains where the meniscus "jumps" inside the pipette until it finds a pinning edge. The first case is that of a cone/wedge angle α larger than a critical value α^* (and contact angle θ sufficiently large); the meniscus stays pinned at the bottom of the pipette, and no liquid penetrates the pipette, unless a negative pressure is established. On the other hand, when the angles α and θ are sufficiently small (α smaller than a negative critical value $-\alpha^*$), the liquid jumps to the top of the cone/wedge. The critical values depend on the internal radius R_0 of the pipette at $h = 0$.

In conclusion, a cone-shaped micro-pipette dipped into a liquid does not always have the expected behavior, i.e., there might not be the expected capillary rise.

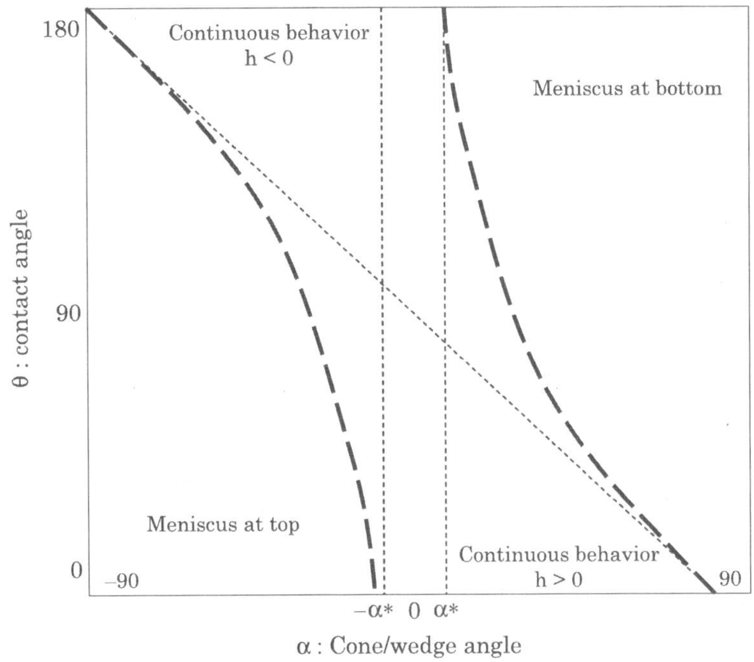

Figure 2.56 Diagram of meniscus behavior in the (θ, α) plane. The angle α^* is defined by $\sin \alpha^* = \bar{R}_0^2/4c$. Depending on the contact angle, if the angle of a pipette is sufficiently large, whether it is hydrophobic or hydrophilic, the meniscus will stay at the bottom. No liquid will penetrate the pipette unless a negative pressure is established in the pipette [34].

2.7.6 Force on a Triple Line

The analysis of the capillary rise in tubes has shown the expression of the capillary force on the triple contact line [35]. This expression can be generalized to any triple contact line [36]. For a triple contact line Ω—as sketched in Fig. 2.57—the capillary force is

$$\vec{F} = \int_\Omega \vec{f} dl = \int_\Omega \gamma \cos \theta \vec{n} dl. \tag{2.72}$$

Suppose that we want to find the value of the resultant of the capillary forces in a particular direction, say x-direction. Using the notations of Fig. 2.58, the projection along the x-direction of equation (2.68) can be written as

$$F_x = \int_\Omega \vec{f} \cdot \vec{i} dl = \int_\Omega \gamma \cos \theta \vec{n} \cdot \vec{i} dl. \tag{2.73}$$

Equation (2.73) can be simplified and cast under the form

$$F_x = \int_\Omega \gamma \cos \theta \vec{n} \cdot \vec{i} dl = \gamma \cos \theta \int_\Omega \vec{n} \cdot \vec{i} dl = \gamma \cos \theta \int_\Omega \cos \alpha dl = \gamma \cos \theta \int_0^e dl'.$$

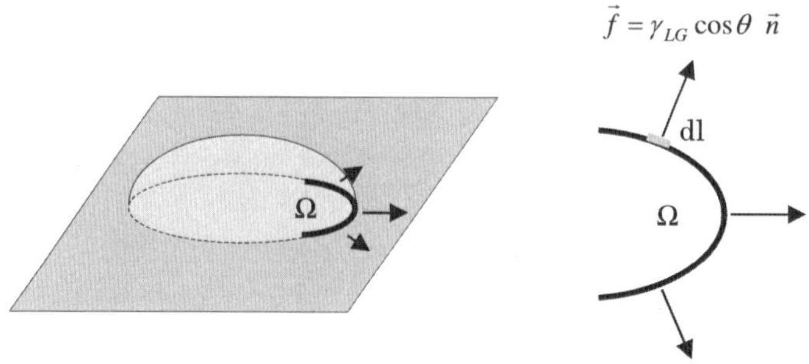

Figure 2.57 Schematic of the capillary force on a triple line.

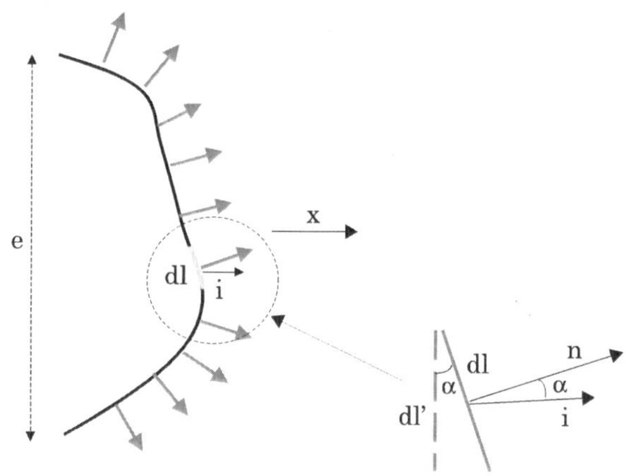

Figure 2.58 Capillary force on a triple contact line in the x-direction.

Finally we obtain the expression

$$F_x = e\gamma \cos\theta. \tag{2.74}$$

Equation (2.74) shows that the resulting force on a triple contact line in any direction does not depend on the shape of the interface [36]; it depends only on the distance between the two ends of the triple line normal to the selected direction.

2.7.7 Examples of Capillary Forces in Microsystems

It is very common in biotechnology to use plates comprising thousands of micro-holes or cusps. The position of the free surface of the liquid in the cusps is of utmost importance. In particular, the liquid must not exit the holes under the action of capillary forces.

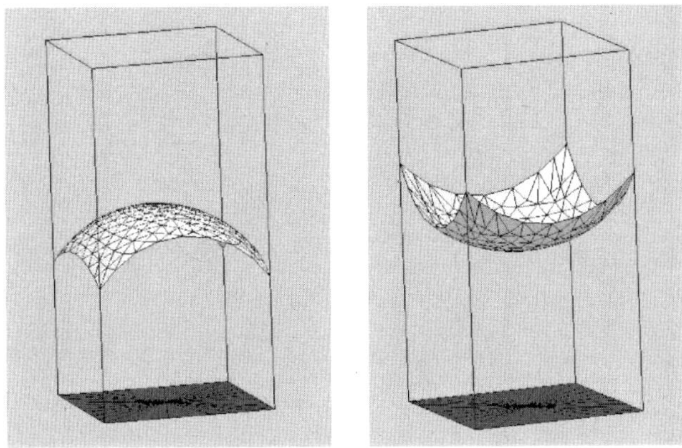

Figure 2.59 The surface of a liquid in a micro-well is not flat due to capillary forces. The figure is a simulation with the Surface Evolver software. Left: case of water in a hydrophilic well (contact angles of 140°); right: case of water in a hydrophilic well (contact angles 60°). The "free" surface is tilted downwards or upwards depending on the contact angle. The walls have been dematerialized for clarity.

As an example, Fig. 2.59 shows a free liquid interface in a square hole, calculated with the Surface Evolver numerical software [23]. In the following chapter, we will study in more detail the position of an interface in a hole, as a function of the wetting characteristics of the solid surfaces and the shape of the hole.

2.8 Measuring Surface Tension of Liquids

It has already been mentioned that surface tension can be seen as a force. Surface tension can then be measured by comparison with another known and calibrated force. This other force can be pressure as in the "bubble pressure method," also known as the Schrödinger method, can be a capillary force as in the Wilhelmy plate method, or can be gravity as in the pendant drop method, and more recently it can be fluid stress as in the drop deformation method. In the following, we analyze these methods. Devices that are used to measure the surface tension are called tensiometers.

2.8.1 Using Pressure (Bubble Pressure Method)

The idea of using pressure to balance surface tension was first proposed by Erwin Schrödinger. The principle is based on the maximum pressure that an interface can support. A sketch of the experimental set up is shown in Fig. 2.60.

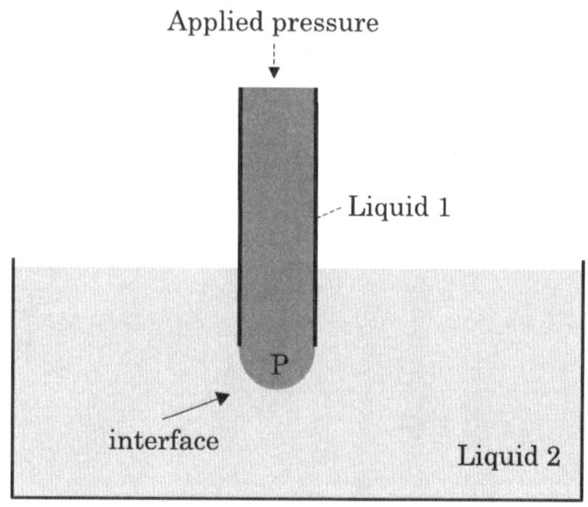

Figure 2.60 Principle of surface tension measurement using pressure: the maximum pressure which a liquid interface can support is when the interface has the shape of a half-sphere. This pressure is related to the surface tension by the Laplace law.

A tube filled with liquid 1 is plunged into a beaker containing liquid 2. As pressure in liquid 1 is increased, the interface at the tube outlet deforms until it reaches the form of a half-sphere. Above this maximum pressure, the interface blows out and liquid 1 breaks down into droplets flowing through liquid 2. The maximum pressure is related to the surface tension between liquid 1 and liquid 2 by the Laplace relation

$$P = 2\frac{\gamma_{L1L2}}{R_{\text{tube}}}. \tag{2.75}$$

This value can be found numerically by using the Surface Evolver software (Fig. 2.61). We have used a capillary tube of 100 μm radius so that the gravity force does not introduce a bias in the result. Using a surface tension value of 72 mN/m (water in air), we find a maximum critical pressure of 1440 Pa, which is the value expected from (2.75).

Remark: Theoretically, when the pressure is increased above the value defined by equation (2.71), the interface is no longer stable. It expands suddenly (Fig. 2.62) and breaks down. Physically, it corresponds to the instability of the interface and the formation of droplets.

Experimentally the maximum pressure before the interface breaks is difficult to determine very precisely. This is the reason why other methods of measuring surface tension have been developed.

Figure 2.61 Result of a calculation with Surface Evolver: the surface deforms with increasing pressures, till it reaches a half-sphere.

2.8.2 Using the Capillary Rise on a Plate—Wilhelmy Plate

Surface tension of a liquid can be determined by measuring the capillary force exerted by the liquid on a solid plate. The principle is that already described in Section 2.7.3: the surface of the liquid rises along a vertical plate and the height reached by the liquid on the plate is proportional to the surface tension as shown by equation (2.58).

In the standard method, a thin plate (perimeter about 40 mm) is lowered to the surface of a liquid and the downward force directed to the plate is measured. Surface tension is directly the force divided by the perimeter of the plate. A couple of very important points with this method must be noted. First, the plate must be completely wetted before the measurement to ensure that the contact angle between the plate and the liquid is zero. If this is not true, the Wilhelmy method is not valid. Secondly, one must be sure that the position of the plate is correct, meaning that the lower end of the plate is exactly on the same level as the surface of the liquid. Otherwise the buoyancy effect must be calculated separately. Fig. 2.63 shows a numerical simulation of the Wilhelmy method.

2.8.3 Using Gravity: The Pendant Drop Method

By definition, a pendant drop is a drop suspended from a fixed solid, as shown in Fig. 2.64.

The two forces acting on the drop are gravitation and surface tension. The shape of such a drop is then a function of the surface tension (Fig. 2.65). The pendant drop method consists of extracting the surface tension from an image of the drop shape.

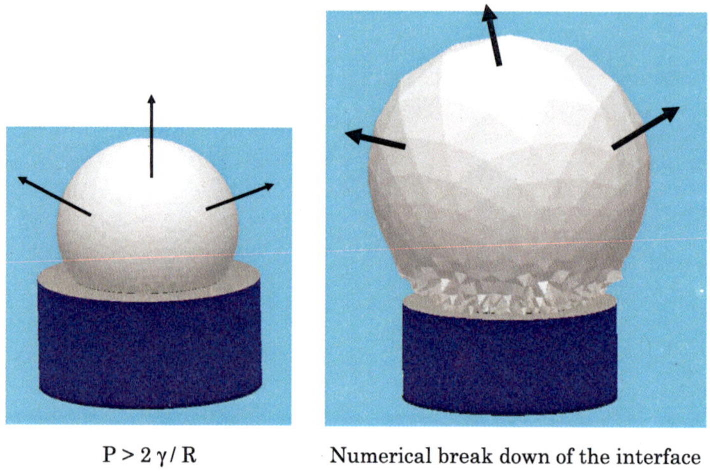

Figure 2.62 Above the threshold value $P = 2\gamma_{L1L2}/R_{\text{tube}}$ the interface is no longer stable; it starts expanding and breaks down.

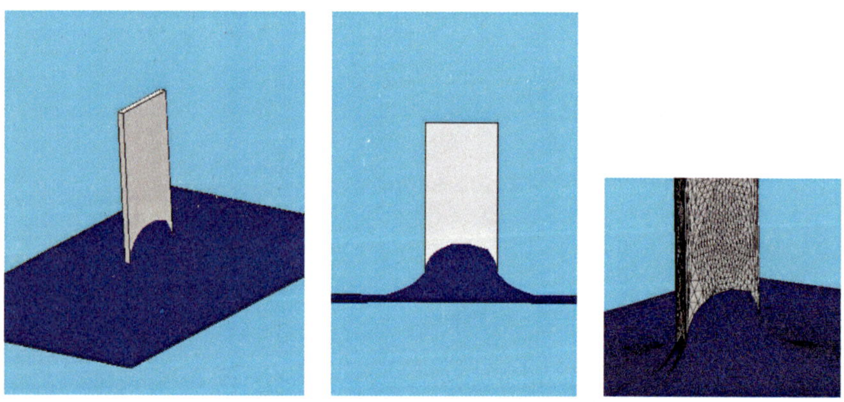

Figure 2.63 Simulation of the capillary rise along a Wilhelmy plate using the Surface Evolver software: left, perspective view; middle, front view; right close up on the contact line in the angle.

2.8.3.1 Bond Number

For a pendant drop, the ratio of the gravitational and surface tension forces is scaled by a dimensionless number, the Bond number, defined by

$$Bo = \frac{\Delta \rho g R^2}{\gamma} \tag{2.76}$$

where $\Delta\rho$ is the difference of the density of the liquid and the surrounding fluid, g the gravitational constant, γ the surface tension, and R a typical dimension of the droplet. Here we choose R to be the maximum horizontal radius of the

Figure 2.64 Image of a pendant drop.

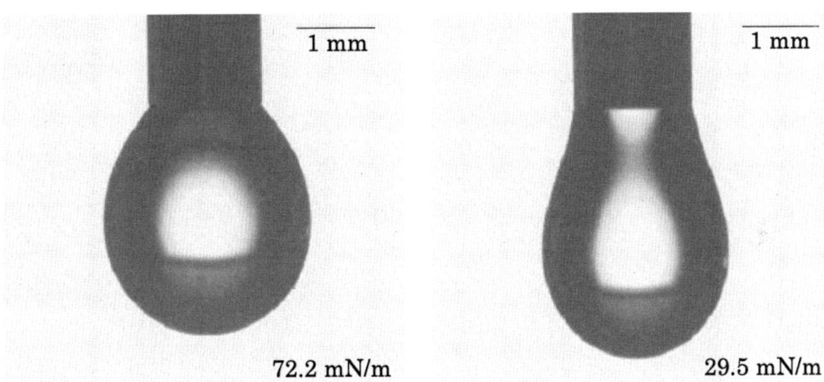

Figure 2.65 Experimental view of two drops: left: water droplet in air; right: oil droplet in air.

pendant drop. The shape of the pendant drop is shown in Fig. 2.66 for different Bond numbers.

Fig. 2.66 shows that, for a well chosen Bond number, the competition between the gravitational force and the surface tension determines the shape of the droplet. On one hand, the gravitational force, i.e., the weight of the drop, tends to elongate the droplet vertically; on the other hand, the surface tension tends to minimize the interface by making it spherical. From an image analysis, the volume—and consequently the weight—of the drop can be determined, and also a vertical profile which is a function of the surface tension and the weight of the drop (Fig. 2.67) can be found. In the following section, the numerical approach used to determine the surface tension is detailed.

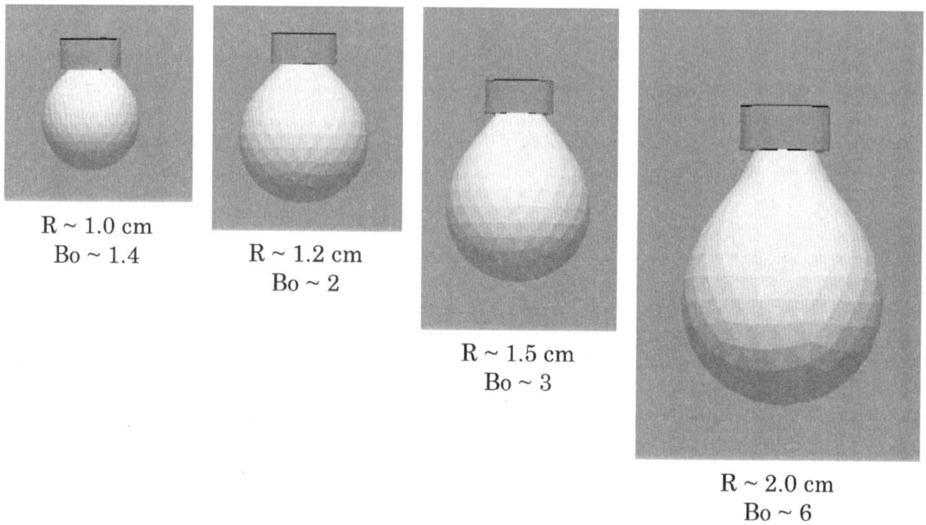

Figure 2.66 Shape of a pendant drop vs. Bond number: the drop shape departs from that of a sphere when its volume increases, until it detaches from the solid support. The liquid of the drop is water and is immersed in silicone oil; the surface tension is $\gamma = 33$ mN/m. Calculation performed with Surface Evolver software.

2.8.3.2 Method

There exist different numerical schemes to extract the surface tension from the pendant drop shape. A well known software is that pioneered by del Rio et al. [37] called ADSA. We do not give here the details of the method—which would be long—but just the main lines.

The drop pressure is determined by writing the Laplace equation at the bottom of the droplet. Due to axisymmetry

$$\Delta P = \gamma \left(\frac{1}{R} + \frac{1}{R'} \right) = \frac{2\gamma}{R}. \tag{2.77}$$

The curvature radius being derived from an image analysis, equation (2.77) produces the internal pressure of the drop.

Second, we write the equilibrium equation of the drop in any horizontal section. Because of axisymmetry, this equilibrium imposes a zero vertical resultant

$$2\pi R\gamma \sin\theta = V\left(\rho_h - \rho_l\right) g + \pi R^2 P. \tag{2.78}$$

In (2.78) R is the horizontal radius in the considered section, θ is the angle of the tangent at M to the contour of the image of the drop, V is the volume of the fluid under the plane of altitude z, ρ_h and ρ_l are the densities of the two

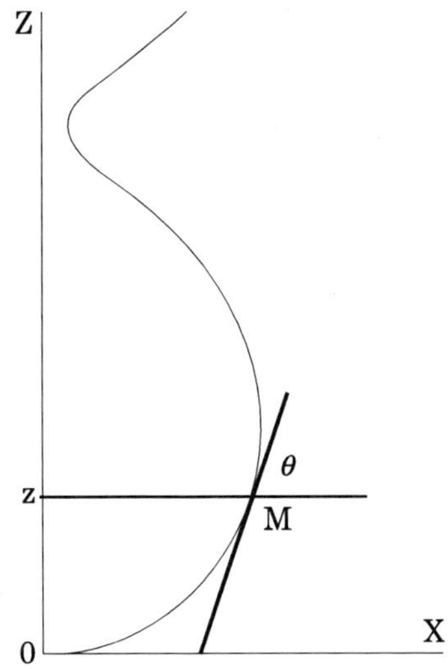

Figure 2.67 Typical drop contour in the pending drop method.

fluids, and g the gravitational acceleration. The left term in equation (2.78) corresponds to the surface tension force; the first term to the right hand side, to the weight of the liquid below the considered section; and the second term to the right hand side to the pressure force. For each section, (2.78) produces a value for the surface tension γ. An averaging of all the values of γ determines precisely the real value of the surface tension.

Because the pendant drop method reacts very quickly, it is even possible to determine the surface tension as a function of the surface concentration in surfactants [38]. When CMC is reached, the shape of the drop does not evolve any more.

In conclusion, the advantages of the pendant drop method are

- Small volume of liquid
- Easy to spread a known amount of surfactant at the surface
- Rapid rates of surface area change compared to Langmuir–Wilhelmy balance

However, one must be very careful that the capillary is very smooth and cylindrical to preserve axisymmetry.

2.8.4 Using Shear Stress in a Microflow

Hydrodynamic forces are usually difficult to control and to monitor accurately. However, the situation has changed drastically with the use of microflows. Such flows are completely laminar and their velocity is largely predictable [39]. With this in mind, and observing that a droplet/bubble

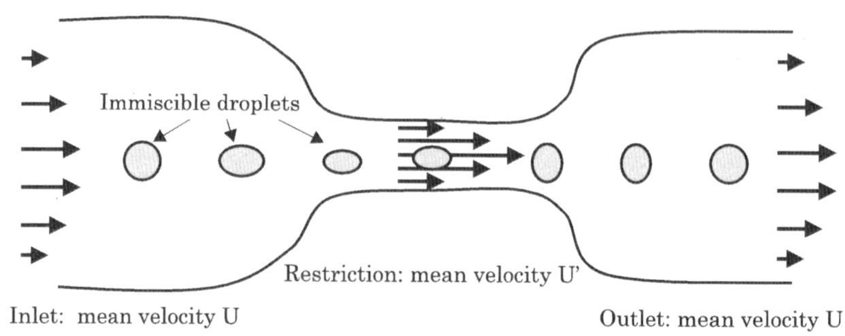

Figure 2.68 Schematic of droplet stretching in a contracted section for surface tension measurement.

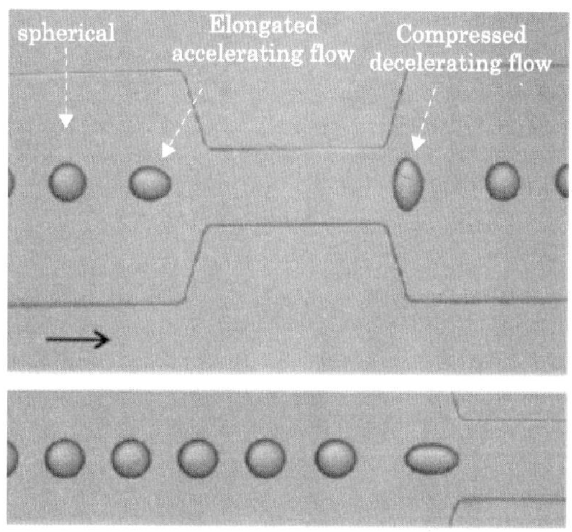

Figure 2.69 Experimental image of droplets in accelerated and decelerated fluid flows, from [41], reproduced by permission of The Royal Society of Chemistry.

changes its shape according to the change in the flow mean velocity, a method for measuring surface tensions between two immiscible liquids or between a liquid and a gas has been proposed [40,41]. The principle is sketched in Fig. 2.68.

An experimental view of the phenomenon is shown in Fig. 2.69. A droplet traveling in the middle of the tube experiments a very small shear stress. However, when it reaches a constricted section, the flow accelerates and the shear stress increases. As a result the droplet takes an ellipsoidal shape oriented along the axis of the tube. At the entrance of the divergent section, the shear stress acts oppositely and the droplet takes an ellipsoidal shape oriented perpendicularly to the tube axis. After a while, the spherical shape is regained.

Let us define the drop deformation by

$$D = \frac{a-b}{a+b} \tag{2.79}$$

where a and b are the two vertical dimensions of the droplet. When the droplet is spherical $D = 0$. Let $\dot{\varepsilon}$ be the extension rate given by the change of axial velocity u

$$\dot{\varepsilon} = \frac{du}{dx}.$$

Then, it can be shown that D is given by the solution of the differential equation

$$\frac{dD}{dt^*} = \frac{5}{2\hat{\eta} + 3}\tau\dot{\varepsilon} - D \tag{2.80}$$

where $t*$ is a nondimensional time defined by $t^* = t/\tau$, and $\hat{\eta} = \eta_{\mathrm{drop}}/\eta_{\mathrm{carrier}}$ where η denotes the dynamic viscosity. Finally τ is defined by

$$\tau = \frac{\alpha\eta_c a_0}{\gamma}$$

where a_0 is the undistorted radius and α a rational function of $\hat{\eta}$. Instead of solving (2.80), it is convenient to re-write it under the form

$$\alpha\eta_{\mathrm{carrier}}\left[\frac{5}{2\hat{\eta} + 3}\dot{\varepsilon} - u\frac{\partial D}{\partial x}\right] = \gamma\frac{D}{a_0}. \tag{2.81}$$

D and its axial evolution $\partial D/\partial x$ are estimated by image analysis. Then, a plot of the quantity $\alpha\eta_{\mathrm{carrier}}\,[5/2\hat{\eta} + 3\dot{\varepsilon} - u\partial D/\partial x]$ as a function of D/a_0 is a linear curve with slope γ. Fig. 2.70 shows experimental results of the method.

This microfluidic tensiometer is particularly interesting because it produces the instantaneous surface tension on line.

2.9 Surface Tension of Solids

2.9.1 Introduction

In the preceding sections, the concept of surface tension of solids has been extensively used in Young's law. However, surface tension of solids is a complex notion; on one hand, it is very difficult to measure intrinsically and on the other hand, different analytical relations have been proposed, but their results are still not completely satisfactory [42]. In our microfluidic approach, we will avoid as much as possible having to use surface tension of solids. We shall use Young's law as soon as it is possible to eliminate the solid surface tension, and make our approaches depend on only two values: γ_{LG} and θ. Still we present the notion of *surface free energy* and in particular the notion of *critical surface tension* which is important for wetting analysis.

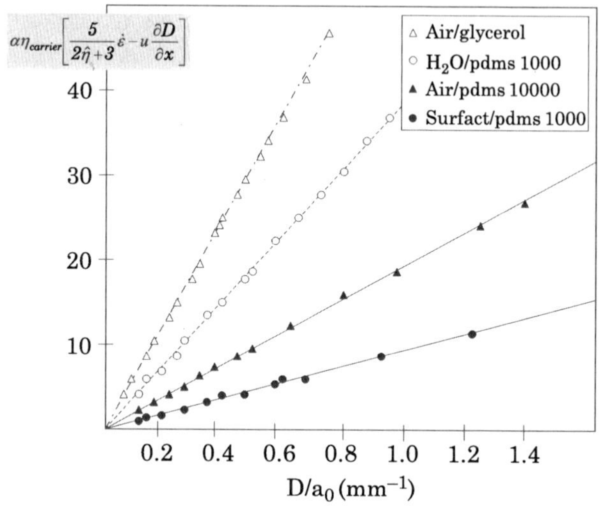

Figure 2.70 Plot of the flow extension rate versus the droplet deformation. The slope is the surface tension; from [41], reproduced by permission of The Royal Society of Chemistry.

2.9.2 Surface Free Energy

2.9.2.1 Method of Good–Girifalco

The approach of Girifalco and Good [43] consists in expressing the work of adhesion in terms of γ_{LG} and γ_S

$$W_a = 2\Phi\left(\gamma_{\mathrm{LG}}\gamma_S\right)^{\frac{1}{2}} \tag{2.82}$$

where Φ is a function of the molar volumes of the liquid and the solid. Inserting equation (2.82) in the Young–Dupré relation (2.51) yields

$$W_a = 2\Phi\left(\gamma_{\mathrm{LG}}\gamma_S\right)^{\frac{1}{2}} = \gamma_{\mathrm{LG}}\left(1+\cos\theta\right) \tag{2.83}$$

leading to

$$\gamma_S = \gamma_{\mathrm{LG}}\frac{\left(1+\cos\theta\right)^2}{4\,\Phi^2}. \tag{2.84}$$

The remaining problem is to estimate Φ. Kwok and Neumann [44] made a systematic study of the contact between organic fluids and low energy surfaces. They based their approach on a dependency of the type

$$\Phi = e^{-\beta\left(\gamma_{\mathrm{LG}}-\gamma_S\right)^2} \approx 1-\beta\left(\gamma_{\mathrm{LG}}-\gamma_S\right)^2 \tag{2.85}$$

where β is a constant slightly depending on the substrate. Substitution of (2.85) in (2.83) and using Young's law yields

$$\cos\theta = -1 + 2\sqrt{\frac{\gamma_S}{\gamma_{LG}}}e^{-\beta(\gamma_{LG}-\gamma_S)^2}.$$

For any given substrate, γ_S is constant, and by using many different liquids and measuring γ_{LG} and θ, a least-squares analysis leads to the values of β and γ_S.The parameter β was found to be of the order of 0.000100 to 0.000130 $(m^2/mJ)^2$ for low energy substrate such as polystyrene and polymethacrylate).

2.9.2.2 Fowkes Method

The preceding approach does not take into consideration the polar or apolar nature of the solid. The different components of the surface energy can be taken into account by writing [45]

$$\gamma_S = \gamma_S^D + \gamma_S^P \tag{2.86}$$

where the superscripts D and P respectively stand for "diffusive" (non-polar) and polar. Neumann's approach [46] consists in subdividing the polar component into a Lewis acid and a Lewis base, so that the work of adhesion can be cast under the form

$$W_a = \gamma_{LG}(1+\cos\theta) = 2\left(\gamma_{LG}^D\gamma_S^D\right)^{\frac{1}{2}} + 2\left(\gamma_{LG}^-\gamma_S^-\right)^{\frac{1}{2}} + 2\left(\gamma_{LG}^+\gamma_S^+\right)^{\frac{1}{2}}. \tag{2.87}$$

Equation (2.87) contains eight parameters. The contact angle θ can be measured, and the properties of the liquid are also known (they can be measured). Then we are left with three unknowns describing the solid. If we first use a non-polar liquid ($\gamma_{LG}^+ = \gamma_{LG}^- = \gamma_{LG}^P = 0$) then (2.87) simplifies to

$$\gamma_{LG}(1+\cos\theta) = 2\left(\gamma_{LG}^D\gamma_S^D\right)^{\frac{1}{2}}$$

and we deduce γ_S^D. More generally, if we use three liquids—including a non-polar one—we find a system of three equations with the three unknowns $(\gamma_S^D, \gamma_S^-, \gamma_S^+)$. The total solid surface tension is then given by the relation

$$\gamma_S = \gamma_S^D + \gamma_S^P = \gamma_S^D + 2\left(\gamma_S^+\gamma_S^-\right)^{\frac{1}{2}}. \tag{2.88}$$

A very instructive example has been given by Combe et al. for measuring the surface tension of the enamel of human teeth [47].

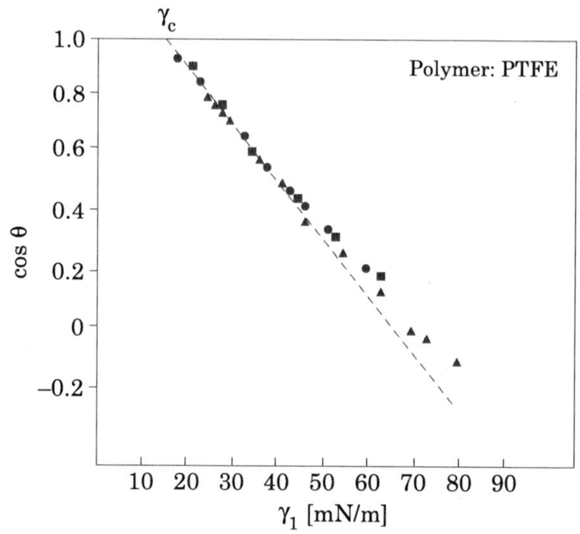

Figure 2.71 Zisman plot of the contact angle of different liquids on a PTFE surface.

2.9.3 Critical Surface Tension and Surface Free Energy

A very general question in capillary studies is: would a given liquid completely wet a given solid or not? In other words, what is the condition for spreading? Zisman and coworkers pioneered this problem in the 1950s, with the introduction of the notion of "critical surface tension," denoted here CST for simplicity, and defined by the proposition: a solid surface cannot be completely wetted by a liquid if the value of the liquid surface tension is above the critical surface tension.

2.9.3.1 Zisman Plot

Zisman and coworkers established an empirical connection between the cosine of the contact angle $\cos\theta$ and the liquid/air interfacial tension γ [48]. For a given low energy solid surface, they measured the contact angle θ for many different liquids. The plot of $\cos\theta$ as a function of γ is approximately to a straight line; it has the shape shown in Fig. 2.71.

When the contact angle is such that $\cos\theta = 1$, i.e., $\theta = 0$, the liquid wets the solid surface. If we note γ_C the point where $\cos\theta = 1$, obtained by a linear extrapolation of Zisman's curves, then a liquid with a surface tension smaller than γ_C wets the surface. As a matter of fact, this nearly linear behavior is valid not only for pure liquids, but also for aqueous solutions.

Zisman's approach was later refined by Bargeman, et al. [49,13]. Observing that the Zisman's curves were not exactly linear, especially for aqueous solutions, they have shown that linearity could be achieved by considering

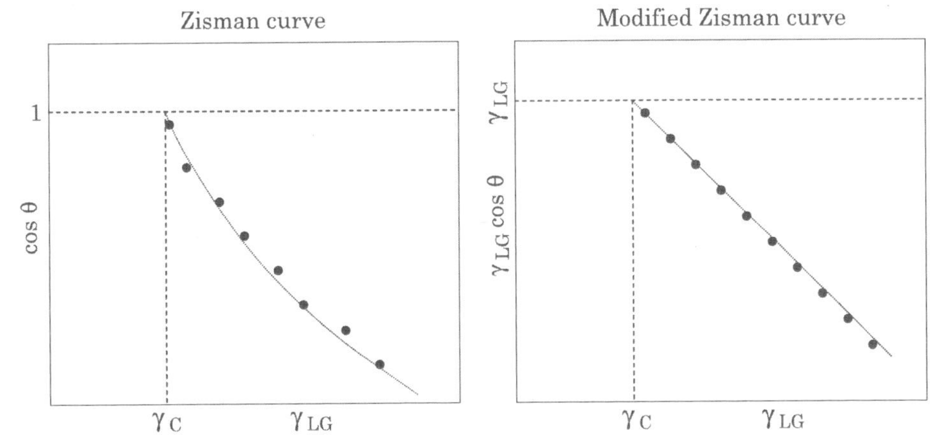

Figure 2.72 Modified Zisman's plot for PMMA (polymethyl methacrylate) and PTFE (polytetrafluoroethylene).

the relation between $\gamma \cos\theta$ and γ instead of the relation between $\cos\theta$ and γ (Fig. 2.72).

In summary there is a linear relation between adhesion tension and surface tension of aqueous solutions

$$\gamma_{\rm LG} \cos\theta = a\gamma_{\rm LG} + b \tag{2.89}$$

where a and b are constants depending on the solid surface.

Zisman's approach is very practical; it easily produces the value of the CST. It has been observed that the value of the CST is often close to that of the surface tension of the solid, i.e., the surface free energy (SFE). For a liquid whose surface tension approaches the critical surface tension, Young's law can be written

$$\gamma_{\rm LG} = \gamma_C = \gamma_{\rm SG} - \gamma_{\rm SL}. \tag{2.90}$$

In partial wetting conditions, there is a precursor film on the solid substrate; this is evidently the case at the onset of total wetting. This precursor film has a film pressure P_e^* (we discuss the film pressure in the next section) so that the surface tension γ_{SG} is given by

$$\gamma_S - P_e^* = \gamma_{\rm SG}. \tag{2.91}$$

And (2.90) can be cast under the form

$$\gamma_S = \gamma_C + (\gamma_{\rm SL} + P_e^*)\,. \tag{2.92}$$

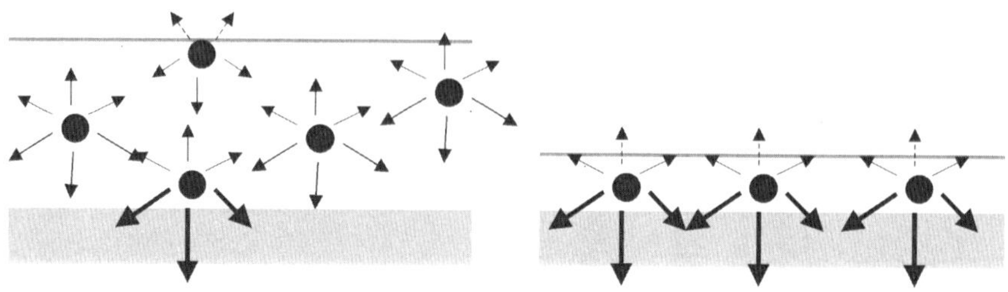

Figure 2.73 In a thin film all the molecules in the liquid interact with molecules of the solid wall.

For a liquid having a surface tension γ_C, the term inside the parentheses on the right hand side of (2.92) is often small [13]. This is in particular the case when the solid surface is apolar, i.e., the polar component of the surface free energy is negligible. For example, PMMA has a CST of 33.1 mN/m and a SFE of 39 mN/m; PTFE has a CST of 20 mN/m and a SFE of 23.4 mN/m. In such cases the CST is a good approximation of the SFE. The advantage is that the critical surface tension can easily be obtained by Zisman plots.

2.9.3.2 Disjoining Pressure

2.9.3.2.1 Definition
When a liquid film on a solid is very thin (like the precursor film), all the liquid molecules have an interaction with the solid wall, as sketched in Fig. 2.73. We are then at a very small scale (less than 1 nm). The energy of the film takes a special form. For a thick film, the energy is $\gamma + \gamma_{SL}$. For a thin film of thickness e the film energy is [23]

$$E = \gamma + \gamma_{SL} + P_e(e) \tag{2.93}$$

where $P_e(e)$ is a function of e such that $P_e(\infty) = 0$ and $P_e(0) = \gamma_S - (\gamma + \gamma_{SL})$.
From a thermodynamical point of view, Derjaguin has introduced the notion of "disjoining pressure" in the chemical potential

$$\mu = \mu_0 + \frac{dP_e}{de} v_0 = \mu_0 - \Pi_e v_0 \tag{2.94}$$

where v_0 is the molar volume in the liquid phase. The quantity Π_e in (2.94) defined by

$$\Pi_e = -\frac{dP_e}{de} \tag{2.95}$$

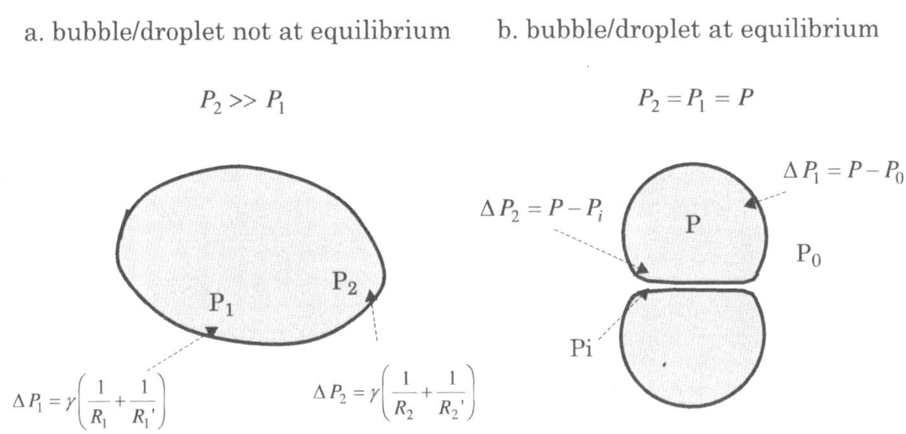

Figure 2.74 Schematic of a droplet outside equilibrium (left) and an assembly of two droplets/bubbles at equilibrium (right).

has the dimension of a pressure. Π_e is called the "disjoining pressure". With this definition, and using (2.92), de Gennes [23] shows that the total surface tension of a thin film is given by

$$\gamma_{\text{film}} = \gamma_{\text{LG}} + \gamma_{\text{SL}} + P_e + e\Pi_e. \tag{2.96}$$

2.9.3.2.2 Example

Bubbles and droplets at equilibrium tend to take a spherical shape. If this were not the case, the internal pressure would not be uniform since the Laplace law applied in regions of different curvatures would produce different internal pressures. In consequence, there is liquid motion inside the bubble/droplet in order to equilibrate the internal pressure (Fig. 2.74).

But what about adjacent droplets or bubbles? The sketch of Fig. 2.74 shows that a droplet is deformed by the presence of the neighboring droplet and its curvature is not constant. A first analysis using the Laplace law would conclude that the internal pressure is not uniform. In reality, the pressure in the thin film between the drops/bubbles is not the external pressure P_0 but a different pressure related to the "disjoining pressure." According to the Laplace equation, the film pressure equilibrates the pressure jump across the interface, and we have the relations

$$\begin{aligned}\Delta P_1 &= \gamma\left(\frac{1}{R_1}+\frac{1}{R_1'}\right) = P - P_0\\ \Delta P_2 &= \gamma\left(\frac{1}{R_2}+\frac{1}{R_2'}\right) = P - P_i\end{aligned} \tag{2.97}$$

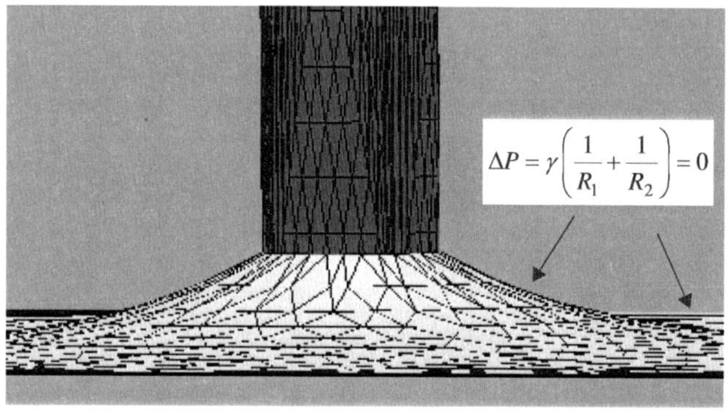

Figure 2.75 Minimal surface formed by the interface of a liquid with a vertical rod.

where indices 1 and 2 refer to two locations, a first one far from the deformed interface and the other close to the deformed interface. In (2.97) P_i is the film pressure.

2.10 Minimization of the Surface Energy and Minimal Energy Surfaces

2.10.1 Minimal Surfaces

Minimal surfaces are a particular type of surface. From a mathematical standpoint, a minimal surface is a surface with zero mean curvature. This definition is very restrictive: there are not many surfaces that are minimal surfaces. One can find a description of these surfaces in reference [50]. These surfaces locally minimize their area. In the domain of physics of liquid interfaces, we have already seen an example of a minimal surface in Section 2.3.2.4 with the surface of a liquid rising along a cylindrical vertical tube (Fig. 2.75).

Another example of minimal surface is that of the catenoid (Fig. 2.76). Physically, a catenoid is the surface formed by a liquid film maintained by two solid circles.

It is a general thermodynamic principle in physics that systems evolve to their minimal energy level. In particular, this is the case of interfaces. Surfaces of droplets in static equilibrium have a minimal free energy. Thus, in the domain of liquid interfaces, we are tempted to give a less restrictive definition for minimal surfaces. A minimal surface is a surface that minimizes its area under some constraints. The goal is to minimize the energy of the system (surface, gravitational, etc.) under some constraints imposed by external conditions, like walls, wires, fixed volume or fixed pressure. With this definition, minimal surfaces comprise a much larger set of surfaces and describe liquid films and

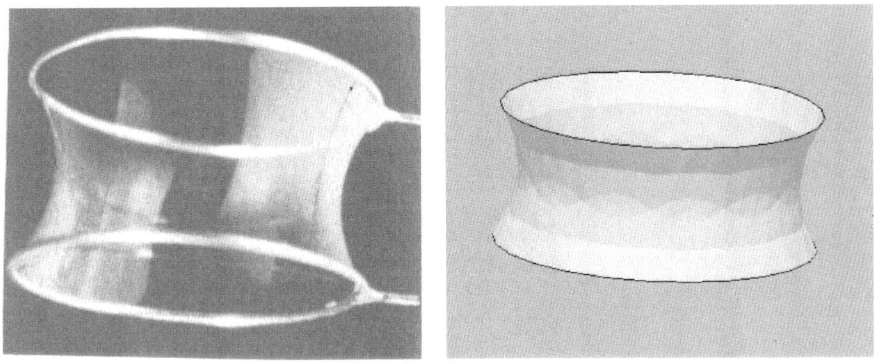

Figure 2.76 Left: image of a soap film between two solid circles; right: result of the calculation with the Surface Evolver software.

interfaces [36]; when gravity is negligible, these surfaces have a constant mean curvature [51].

Note that "minimal surfaces" present a general interest, not only in the domain of microfluidics. For example, they have a great appeal in architecture and arts. The "Cloud Gate" sculpture of Anish Kapoor in downtown Chicago is having a huge success and has been adopted by the public as the new symbol of Chicago (Fig. 2.77). It can be shown that it is a minimal surface—in the extended sense, corresponding to the magnification of a droplet sitting on a solid cylinder.

2.10.2 Minimization of the Surface Energy

According to the extended definition of a minimal surface, the droplet interface can be calculated by minimization of its surface energy. In the absence of a droplet on the surface of the solid substrate, the surface energy is

$$E_{SG,0} = \gamma_{\mathrm{SG}} S_{SG,0}. \tag{2.98}$$

After deposition of the droplet, the surface energy is the sum of the three surface energies

$$E = E_{\mathrm{LG}} + E_{\mathrm{SL}} + E_{SG,1} \tag{2.99}$$

where $E_{SG,1}$ is the surface energy of the solid surface in contact with the gas. Then, we have

$$E = \gamma_{\mathrm{LG}} S_{\mathrm{LG}} + \int_S \int_{\mathrm{SL}} (\gamma_{\mathrm{SL}} - \gamma_{\mathrm{SG}})\, dA + E_{SG,0}. \tag{2.100}$$

Figure 2.77 Left: striking example of minimal surfaces in Architecture: the "Cloud Gate" sculpture by Anish Kapoor in the Millennium Park of Chicago. Right: calculation of a similar surface with the Surface Evolver numerical software (right).

The last term to the right hand side of (2.96) does not depend on the drop shape. Thus, we have to minimize

$$E = \gamma_{\rm LG} S_{\rm LG} + \int_S \int_{\rm SL} (\gamma_{\rm SL} - \gamma_{\rm SG})\, dA. \tag{2.101}$$

Taking into account Young's law, the energy to be minimized is [52,22]

$$E = \gamma_{\rm LG} S_{\rm LG} - \gamma_{\rm LG} \int_S \int_{\rm SL} \cos\theta dA. \tag{2.102}$$

As mentioned earlier, the parameters intervening in equation (2.102) are θ and $\gamma_{\rm LG}$. Thanks to Young's equation, we don't need the surface tension of the solid with the liquid or the gas. This is a real simplification since we have shown that θ and γ_{LG} can be measured relatively easily.

Equation (2.102) constitutes the basis for the calculation of droplet shapes that we develop in the next chapter.

2.11 Summary

This chapter was devoted to the study of surface tensions and capillary forces. The main notions presented in this chapter are the relation between curvature and pressure expressed by Laplace's law, and the relation between the different surface tensions and contact angle at the triple line expressed by Young's law. From these two relations, an expression for the capillary force on a triple line has been deduced. Such an expression has a key role in determining the behavior of droplets on different substrates and geometry in microsystems that we shall consider in the next chapter.

This chapter has shown the essential role of surface tension and capillarity at the microscale. These forces often screen out forces like gravity or inertia which are predominant at the macroscopic scale. Although we have taken the stance of presenting capillarity and surface tension from an engineering point of view by considering global effects, one has to keep in mind that interactions at the nanoscopic scale are the real underlying causes of these global effects.

References

1. R. Steger, B. Bohl, R. Zengerle and P. Koltay, "The dispensing well plate: a novel device for nanoliter liquid handling in ultra high-throughput screening," *JALA*, Vol. 9, pp. 291–299, 2004.
2. Jun Zeng, Manish Deshpande, Heng-Chuan Kan and John R. Gilbert, A dynamic spotting method for SPLIT-PIN based microarrays, http://www.memcad.com/media/papers/mtas4'01.pdf.
3. Rain Dance Technology: http://www.raindancetechnologies.com/tech.html.
4. J. Atencia and D.J. Beebe, "Controlled microfluidic interfaces," *Nature*, Vol. 437, pp. 648–654, 2005.
5. Shingo Okushima, Takasi Nisisako, Toru Torii and Toshiro Higuchi, "Controlled Production of Monodisperse Double Emulsions by Two-Step Droplet Breakup in Microfluidic Devices," *Langmuir*, Vol. 20, pp. 9905–9908, 2004.
6. B. Zheng, J.D. Tice, L.S. Roach and R.F. Ismagilov, "A droplet-based, composite PDMS/glass capillary microfluidic system for evaluating protein crystallization conditions by micro-batch and vapor-diffusion methods with on-chip X-ray diffraction," *Angew. Chem., International Edition English*, Vol. 43, pp. 2508–2511, 2004.
7. J. Israelachvili, *J. Intermolecular and Surface Forces*, Academic Press, 1992.
8. Table of surface tension for chemical fluids: http://www.surface-tension.de/.
9. G. Navascués, "Liquid surfaces: theory of surface tension," *Rep. Prog. Phys.*, Vol. 42, pp. 1133–1183, 1979.
10. R. Eötvös, *Wied. Ann. Phys.*, Vol. 27, pp. 445–459, 1886.
11. E.A. Guggenheim, "The principle of corresponding states," *J. Chem. Phys.*, Vol. 13, pp. 253–261, 1945.
12. Alexander Alexeev, Tatiana Gambaryan-Roisman and Peter Stephan, "Marangoni convection and heat transfer in thin liquid films on heated walls with topography: Experiments and numerical study," *Phys. Fluids*, Vol. 17, p. 062106, 2005.
13. K. Szymczyk, A. Zdiennicka, B. Janczuk and W. Wocik, "The wettability of polytetrafluoroethylene and polymethyl methacrylate by aqueous solution of two cationic surfactants mixture," *J. Colloid. Interface Science*, Vol. 293, pp. 172–180, 2006.
14. MIT, Lecture 4, Marangoni flows, http://web.mit.edu/1.63/www/Lec-notes/Surfacetension/Lecture4.pdf.
15. E. Weisstein, http://mathworld.wolfram.com/Curvature.html.
16. L.M. Pismen, B.Y. Rubinstein and I. Bazhlekov, "Spreading of a wetting film under the action of van der Waals forces," *Physics of Fluids*, Vol. 12, no. 3, 2000.
17. P.G. deGennes, "Wetting: statistics and dynamics," *Rev. Mod. Phys.*, Vol 57, p. 827, 1985.
18. C. Iwamoto and S. Tanaka, "Atomic morphology and chemical reactions of the reactive wetting front," *Acta Materiala*, Vol. 50, p. 749, 2002.
19. A.L. Yarin, A.G. Yazicioglu and C.M. Megaridis, "Thermal stimulation of aqueous volumes contained in carbon nanotubes: experiment and modelling," *Appl. Phys. Lett.*, Vol. 86, p. 013109, 2005.
20. Nanolane: http://www.nano-lane.com/wetting-afm.php.
21. L. Pismen, Dewetting patterns, moving contact lines, and dynamic diffuse interfaces, CISM lecture notes, Udine, Italy, 16–20 October, 2006.

22. J. Berthier, Ph. Clementz, O. Raccurt, P. Claustre, C. Peponnet and Y. Fouillet, "Computer aided design of an EWOD microdevice," *Sensors and Actuators, A*, Vol. 127, pp. 283–294, 2006.
23. K. Brakke, "The Surface Evolver," *Exp. Math.*, Vol. 1, p. 141, 1992.
24. P.G. deGennes, F. Brochart-Wyart and D. Quéré, *Drops, Bubbles, Pearls and Waves*, Springer, 2005.
25. A.A. Darhuber and S.M. Troian, "Principles of microfluidic actuation by modulation of surface stresses," *Annu. Rev. Fluid Mech.*, Vol. 37, pp. 425–455, 2005.
26. C.D. Murray, "The physiological principle of minimum work. 1. The vascular system and the cost of blood volume," *Proc. Natl. Acad. Sci. USA*, Vol. 12, pp. 207–214, 1926.
27. L.A. Taber, S. Ng, A.M. Quesnel, J. Whatman and C.J. Carmen, "Investigating Murray's law in the chick embryo," *J. Biomech*, Vol. 34, no. 1, pp. 121–124, 2001.
28. K.A. Mc Culloh, J.S. Sperry and F.R. Adler, "Water transport in plants obeys Murray's law," *Nature*, Vol. 421, pp. 939–942, 2003.
29. H.D. Prange, "Laplace's law and the alveolus: a misconception of anatomy and a misapplication of physics," *Adv. Physiol. Educ.*, Vol. 27, pp. 34–40, 2003.
30. J.Y. Wang, S. Betelu and B.M. Law, "Line tension approaching a first-order wetting transition: Experimental results from contact angle measurements," *Physical Review E*, Vol. 63, pp. 031601-1–031601-10, 2001.
31. X.H. Zhang, X.D. Zhang, S.T. Lou, Z.X. Zhang, J.L. Sun and J. Hu, "Degassing and temperature effects on the formation of nanobubbles at the mica/water interface," *Langmuir*, Vol. 20, no. 9, pp. 3813–3815, 2004.
32. J. Drehlich, "The significance and magnitude of the line tension in three-phase (solid–liquid–fluid) systems," *Colloids and Surfaces A: Physicochemical and Engineering Aspects*, Vol. 116, pp. 43–54, 1996.
33. David L. Hu and John W.M. Bush, "Meniscus-climbing insects," *Nature*, Vol. 437, pp. 733–736, 2005.
34. Y. Tsori, "Discontinuous liquid rise in capillaries with varying cross-sections," *Langmuir*, Vol. 22, pp. 8860–8863, 2006.
35. Jun Zeng and T. Korsmeyer, "Principles of droplet electrohydrodynamics for lab-on-a-chip," *Lab Chip*, Vol. 4, pp. 265–277, 2004.
36. J. Berthier, Ph. Dubois, Ph. Clementz, P. Claustre, C. Peponnet and Y. Fouillet, "Actuation potentials and capillary forces in electrowetting based microsystems," *Sensors and Actuators, A: Physical*, Vol. 134, no. 2, pp. 471–479, 2007.
37. O.I. del Rio and A.W. Neumann, "Axisymmetric drop shape analysis: computational methods for the measurement of interfacial properties from the shape and dimensions of pendant and sessile drops," *J. Coll. Int. Sci.*, Vol. 196, pp. 134–147, 1997.
38. H.A. Wege, J.A. Holgado-Terriza, M.J. Galvez-Ruiz and M.A. Cabrerizo-Vilchez, "Development of a new Langmuir-type pendant drop film balance," *Colloids and Surfaces B: Biointerfaces*, Vol. 12, pp. 339–349, 1999.
39. J. Berthier and P. Silberzan, *Microfluidics for Biotechnology*, Artech House, 2005.
40. S.D. Hudson, J.T. Cabral, W.J. Goodrum Jr., K.L. Beers and E.J. Amis, "Microfluidic interfacial tensiometry," *Applied Physics Letters*, Vol. 87, p. 081905, 2005.
41. J.T. Cabral and S.D. Hudson, "Microfluidic approach for rapid multicomponent interfacial tensiometry," *Lab on a Chip*, Vol. 6, pp. 427–436, 2006.
42. L. Makkonen, "On the methods to determine surface energies," *Langmuir*, Vol. 16, pp. 7669–7672, 2000.
43. R.J. Good and L.A. Girifalco, "A theory for estimation of surface and interfacial energies. 3. Estimation of surface energies of solids from contact angle data," *J. Phys. Chem.*, Vol. 64, pp. 561–565, 1960.
44. D.Y. Kwok and A.W. Neumann, "Contact angle interpretation in terms of solid surface tension," *Colloids and Surfaces A: Physicochemical and Engineering Aspects*, Vol. 161, nos. 1, 15, pp. 31–48, 2000.

45. F.M. Fowkes, "Additivity of intermolecular forces at interfaces.1. Determination of contribution to surface and interfacial tensions of dispersion forces in various liquids," *J. Phys. Chem.*, Vol. 67, no. 12, p. 2538, 1963.
46. A.W. Neumann, R.J. Good, C.J. Hope and M. Seipal, "Equation of state to determine surface tensions of low energy solids from contact angles," *J. Colloid Interf. Sci.*, Vol. 49, pp. 291–304, 1974.
47. E.C. Combe, B.A. Owen and J.S. Hodges, "A protocol for determining the surface energy of dental materials," *Dental Materials*, Vol. 20, pp. 262–268, 2004.
48. W.A. Zisman, "Contact angle, wettability and adhesion," Advances in Chemistry Series, Vol. 43, Am. Chem. Soc., Washington, DC, p. 1, 1964.
49. D. Bargeman and F. van Voorst Vader, "Tensile strength of water," *J. Coll. Sci.*, Vol. 42, pp. 467–472, 1973.
50. Brandeis University: http://rsp.math.brandeis.edu/3D-XplorMath/Surface/gallery_m.html.
51. D.E. Hewgill, "Computing surfaces of constant mean curvature with singularities," *Computing*, Vol. 32, pp. 81–92, 1984.
52. N.A. Patankar and Y. Chen, Numerical simulation of droplets shapes on rough surfaces, Nanotech 2002. Technical Proceedings of the 5th International Conference on Modeling and Simulation of Microsystems, pp. 116–119, 2002.

3 The Physics of Droplets☆

3.1 Introduction

The importance of droplets in microfluidics has been pointed out in the introduction of the preceding chapter, where we presented the principles of capillarity and surface tension effects. This chapter is particularly devoted to the behavior of droplets and micro-drops.

In this chapter, the shape of micro-drops in different situations typical of microsystems is investigated, assuming that these micro-drops are in an equilibrium state, i.e., at rest, or moving at a sufficiently low velocity that the inertial forces can be neglected. Different situations will be examined: sessile droplets deposited on a plate, droplets constrained between two horizontal planes, pendant droplets, droplets on lyophilic strips, in corners and dihedrals, in wells and cusps, etc.

Because microfabricated substrates are not ideally smooth, the microscopic state of the solid substrate which the droplets contact has to be taken into account. Its effect is described by the Wenzel and Cassie laws and we describe applications to the case of micro-patterned substrates.

In a third section, the effect of capillarity forces on micro-drops is investigated and it is shown how such forces can displace micro-drops on smooth surfaces. However, it is shown that such capillarity induced motions can be hindered by hysteresis of contact angle (difference between the advancing and receding contact angles) and by pinning on the solid surface.

Finally, a focus is made on two very common phenomena occurring when working with micro-drops: the Marangoni convection induced by a gradient of surface tension, and the evaporation process which applies to volatile liquids such as water in air or organic solvent—octane, benzene—in air.

3.2 The Shape of Micro-drops

3.2.1 Sessile Droplets

It is easily observed that large droplets on horizontal surfaces have a flattened shape, whereas small droplets have a spherical shape (Fig. 3.1). This observation is linked to the balance between gravity and surface tension. A microscopic drop is governed solely by surface tension, whereas the shape of a larger droplet results from a balance between the two forces. The scale length of this transition

☆This Chapter was written with the collaboration of **Kenneth A. Brakke** (Mathematics Department, Susquehanna University, Selinsgrove, PA).

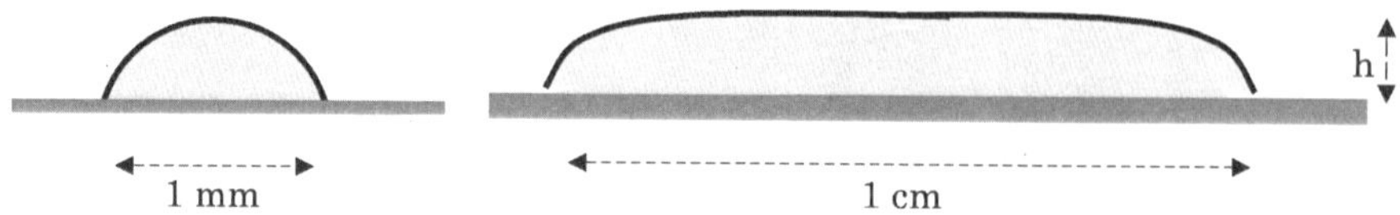

Figure 3.1 Comparison of the shape between micro-drops and macro-drops (not to scale): micro-drops have the shape of spherical caps whereas larger drops are flattened by the action of gravity and their height is related to the capillary length.

ℓ is the "capillary length". This length is defined by the ratio of the Laplace pressure—characterizing the size of a microscopic drop—to the hydrostatic pressure [1]. If we compare the two pressures for a drop, we obtain

$$\frac{\Delta P_{\text{Laplace}}}{\Delta P_{\text{hydrostatic}}} \approx \frac{\frac{\gamma}{\ell}}{\rho g \ell} \tag{3.1}$$

where γ is the surface tension, ρ the density, and g the gravitational constant. The two pressures are of the same order when

$$\ell \approx \sqrt{\frac{\gamma}{\rho g}} \tag{3.2}$$

ℓ is called the capillary length. A drop of dimension smaller than the capillary length has a shape resembling that of a spherical cap. A drop larger than the capillary length is flattened by gravity. A dimensionless number—the Bond number—can be derived from (3.2) yielding a similar meaning. The Bond number is expressed by

$$Bo = \frac{\rho g R^2}{\gamma} \tag{3.3}$$

where R is of the order of the drop radius. If $Bo < 1$, the drop is spherical, or else the gravitational force flattens the drop on the solid surface. A numerical simulation of the two shapes of droplets obtained with the numerical software Surface Evolver [2] is shown in Fig. 3.2.

The capillary length is of the order of 2 mm for most liquids, even for mercury. In the following sections we analyze successively the characteristics of drops having respectively large and small Bond number.

3.2.1.1 Height of a "Large Droplet": $Bo \gg 1$

According to the observation of the preceding section, a "large" droplet has a flat upper surface and its shape is shown in Fig. 3.3.

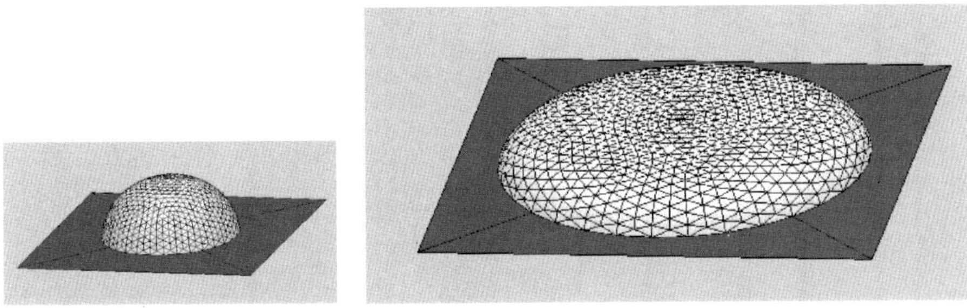

Figure 3.2 Numerical simulations of a micro-drop ($Bo \ll 1$) and a larger drop ($Bo \gg 1$) obtained with Surface Evolver software (not to scale).

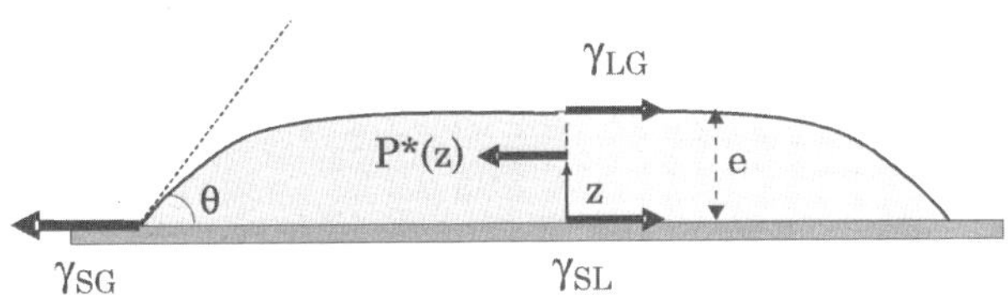

Figure 3.3 Equilibrium of the forces (per unit length) on a control volume of the drop.

Let us calculate the height of such a droplet as a function of contact angle and surface tension. Take the control volume shown in Fig. 3.3 and write the balance of the forces that act on this volume [1]. The surface tension contribution is

$$S = \gamma_{SG} - (\gamma_{\mathrm{SL}} + \gamma_{\mathrm{LG}}). \tag{3.4}$$

And the hydrostatic pressure contribution is

$$P^* = \int_0^e \rho g\,(e - z)\,dz = \frac{1}{2}\rho g e^2. \tag{3.5}$$

The equilibrium condition yields $P^* + S = 0$ which results in the relation

$$\frac{1}{2}\rho g e^2 + \gamma_{SG} - (\gamma_{\mathrm{SL}} + \gamma_{\mathrm{LG}}) = 0. \tag{3.6}$$

Recall that Young's law imposes a relation between the surface tensions

$$\gamma_{SG} - \gamma_{\mathrm{SL}} = \gamma_{\mathrm{LG}} \cos\theta. \tag{3.7}$$

Upon substitution of (3.7) into (3.6), we obtain

$$\gamma_{\mathrm{LG}}\,(1 - \cos\theta) = \frac{1}{2}\rho g e^2.$$

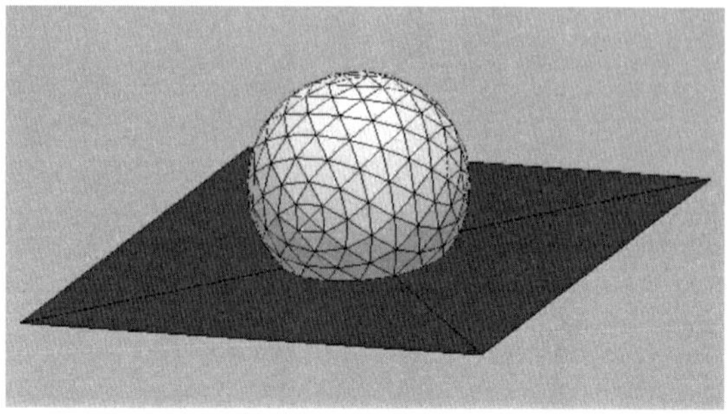

Figure 3.4 Shape of a micro-drop calculated with the software Surface Evolver.

Using the trigonometric expression

$$1 - \cos\theta = 2\sin^2\frac{\theta}{2}$$

we finally find

$$e = 2\sqrt{\frac{\gamma_{\mathrm{LG}}}{\rho g}}\sin\frac{\theta}{2} = 2\ell\sin\frac{\theta}{2}. \tag{3.8}$$

Relation (3.8) shows that the height of a "large" droplet is proportional to the capillary length. The capillary length is of the order of 2 mm, and the height of large droplets is less than 4 mm.

3.2.1.2 Microscopic Drops: $Bo \ll 1$

3.2.1.2.1 Shape of the droplet

As was mentioned earlier, a micro-drop has the form of a spherical cap. A spherical cap is a surface of minimum energy if only surface tension is taken into account. This can be checked by using the Surface Evolver software. The result is shown in Fig. 3.4 for a contact angle of 110°.

3.2.1.2.2 Volume of a sessile droplet

Fig. 3.5 shows a cross section of the droplet, on non-wetting and wetting surfaces. The volume V of such a droplet is a function of two parameters in the set of the four parameters $\{\theta, a, R, h\}$ where θ is the contact angle, a the contact radius (i.e., the radius of the circular base), R the curvature radius (i.e., the sphere radius), and h the height of the droplet. First, we observe that the height of the droplet is expressed by the same relation in the wetting and

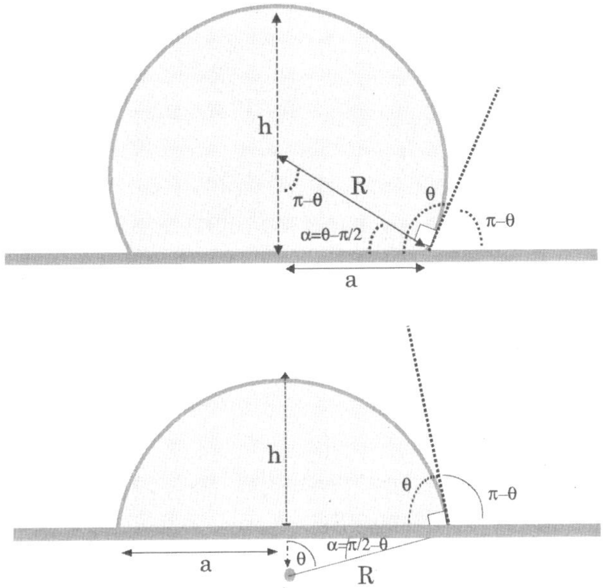

Figure 3.5 Cross section of a micro-drop sufficiently small to be a spherical cap; top: non-wetting droplet; bottom: wetting droplet. Notice that $\alpha = \theta - \pi/2$ in the first (non-wetting) case and $\alpha = \pi/2 - \theta$ in the second (wetting) case.

non-wetting case

$$\begin{aligned} h &= R + R\cos(\pi - \theta) = R(1 - \cos\theta) \\ h &= R - R\cos\theta = R(1 - \cos\theta). \end{aligned} \tag{3.9}$$

The contact radius a—radius of the circular base—is in both cases

$$a = R\sin(\pi - \theta) = R\sin\theta. \tag{3.10}$$

Thus in the set $\{\theta, a, R, h\}$ h and a are functions of R and θ.
Using the Pythagorean theorem

$$|h - R| = \sqrt{R^2 - a^2}. \tag{3.11}$$

Another useful relation can be obtained by combining (3.9) and (3.10)

$$h = a\frac{1 - \cos\theta}{\sin\theta} = a\tan\left(\frac{\theta}{2}\right). \tag{3.12}$$

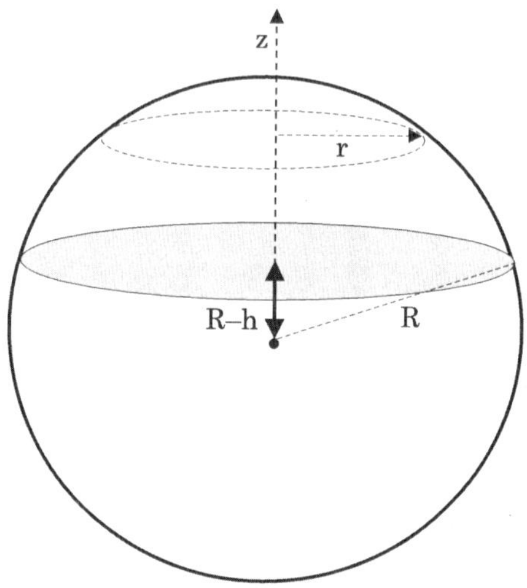

Figure 3.6 Schematic of the volume for the integration of relation (3.14).

The volume of the spherical cap is calculated by the formula applied to the sketch of Fig. 3.6 [3,4]

$$V = \int_{R-h}^{R} \pi r^2 dz. \tag{3.13}$$

Integration of (3.13) leads to

$$V = \int_{R-h}^{R} \pi r^2 dz = \int_{R-h}^{R} \pi \left(R^2 - z^2\right) dz = \frac{\pi h^2}{3} (3R - h). \tag{3.14}$$

Using relation (3.9), one finds an expression for the volume as a function of R and θ

$$V(R, \theta) = \frac{\pi}{3} R^3 \left(2 - 3\cos\theta + \cos^3\theta\right). \tag{3.15}$$

Besides expression (3.14) and (3.15), other expressions for the volume of the spherical cap are

$$V(a, h) = \frac{\pi}{6} h \left(3a^2 + h^2\right) \tag{3.16}$$

and

$$V(a, R) = \frac{\pi}{6} \left(R \pm \sqrt{R^2 - a^2}\right) \left(3a^2 + \left(R \pm \sqrt{R^2 - a^2}\right)^2\right). \tag{3.17}$$

Note that in (3.17) the + sign corresponds to a non-wetting case (lyophobic) and the − sign corresponds to the wetting case (lyophilic). For $\theta = 90°$, $a = R$ and (3.17) reduces to the half sphere volume $V = \frac{2\pi}{3}R^3$. Other expressions of the volume are [4]

$$V(a, \theta) = \frac{\pi}{3}a^3\frac{\left(2 - 3\cos\theta + \cos^3\theta\right)}{\sin^3\theta} \tag{3.18}$$

and

$$V(h, \theta) = \frac{\pi}{3}h^3\frac{\left(2 - 3\cos\theta + \cos^3\theta\right)}{(1 - \cos\theta)^3}. \tag{3.19}$$

3.2.1.2.3 Surface area of the spherical cap

The spherical cap is a surface of revolution obtained by rotating a segment of a circle. More generally, any surface obtained by rotating the curve $y = f(x)$ about the x-axis has the following expression for area [4]

$$S = 2\pi \int f(x)\sqrt{1 + [f'(x)]^2}dx. \tag{3.20}$$

In our case, with the same notations as Fig. 3.6, the curve is defined by $r = f(z)$ and is rotated about the z axis, so that the preceding formula becomes

$$S = 2\pi \int r\sqrt{1 + r'^2}dz.$$

Upon integration between the two limits $R - h$ and R, one obtains

$$S(R, h) = 2\pi Rh. \tag{3.21}$$

Other forms of the expression of the surface of the spherical cap are useful. Using (3.9), we find

$$S(\theta, h) = \frac{2\pi h^2}{1 - \cos\theta}. \tag{3.22}$$

Then using (3.11), we obtain

$$S(a, h) = \pi\left(a^2 + h^2\right). \tag{3.23}$$

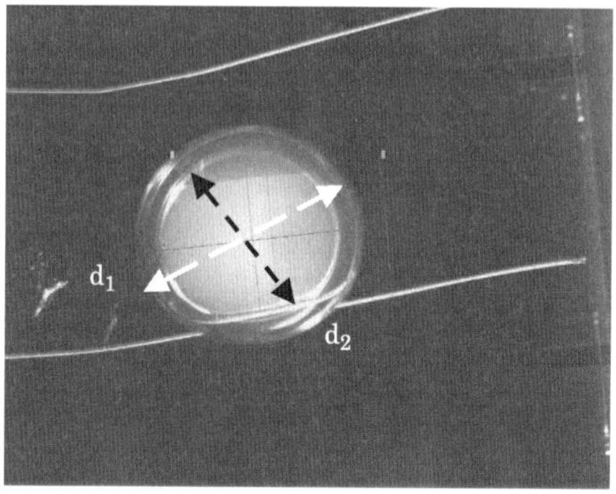

Figure 3.7 Microscope image of a sessile droplet on a hydrophobic surface: the larger diameter d_1 is the diameter of the sphere and d_2 is the diameter of the contact circle of the drop with the substrate (photo Ph. Clementz, CEA/LETI).

After eliminating h in the preceding relation

$$S(a,\theta) = \frac{2\pi a^2}{1+\cos\theta}. \tag{3.24}$$

The surface energy of the spherical cap with surface tension γ is

$$E_{\mathrm{surf}} = \gamma S \tag{3.25}$$

where S is given by any of the preceding relations.

3.2.1.2.4 Measurement of the volume of a droplet from a single image (hydrophobic case)

When observing a non-wetting droplet vertically downwards with a microscope, two circles appear (Fig. 3.7).

A simulation with Evolver leads us to think that the large circle corresponds to the horizontal cross section of the droplet at its equator, and the smaller circle to the contact circle (Fig. 3.8).

The radius of the spherical cap is

$$R = \frac{d_1}{2}$$

and the contact radius

$$a = \frac{d_2}{2}.$$

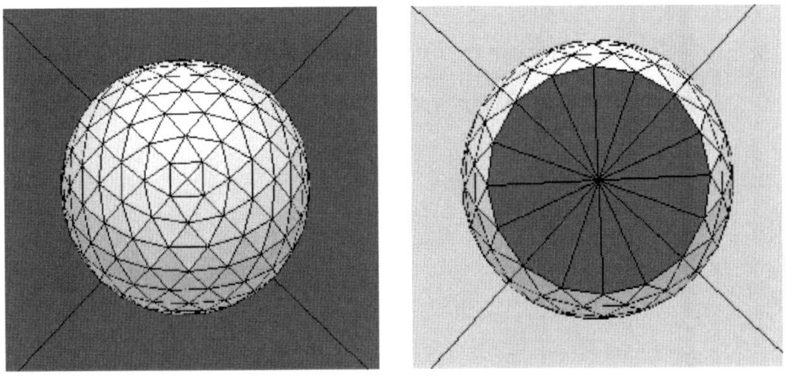

Figure 3.8 Surface Evolver numerical simulation of a sessile droplet: the two diameters are clearly seen on an image taken from below (right).

The information contained in these two relations is enough to determine completely the characteristics of the drop. Using the relation

$$h = R + \sqrt{R^2 - a^2}$$

we find the height of the drop and consequently the drop volume is given by

$$V = \frac{\pi}{48}\left(d_1 + \sqrt{d_1^2 - d_2^2}\right)\left[3d_2^2 + \left(d_1 + \sqrt{d_1^2 - d_2^2}\right)^2\right]. \tag{3.26}$$

Thus the contact angle is

$$\theta = \arcsin\left(\frac{a}{R}\right) = \arcsin\left(\frac{d_2}{d_1}\right). \tag{3.27}$$

In conclusion, a single image (from above) of the drop produces all the characteristics of the droplet. The method can be quite accurate depending on the quality of the image. For example, an estimate of the diameters in the image of Fig. 3.7 is $d_1 = 1332$ μm, and $d_2 = 1146$ μm, leading to a drop volume of $V = 1.05$ μl and a contact angle $\theta = 120.6$ degrees. The expected values were $V = 1.0$ μl and $\theta = 120$ degrees.

3.2.2 Droplets Constrained Between Two Plates

It happens very often in biotechnology that digital microfluidic devices have a planar cover, and the droplets are constrained between two horizontal solid surfaces. Such droplets have a relatively smaller free energy than sessile droplets

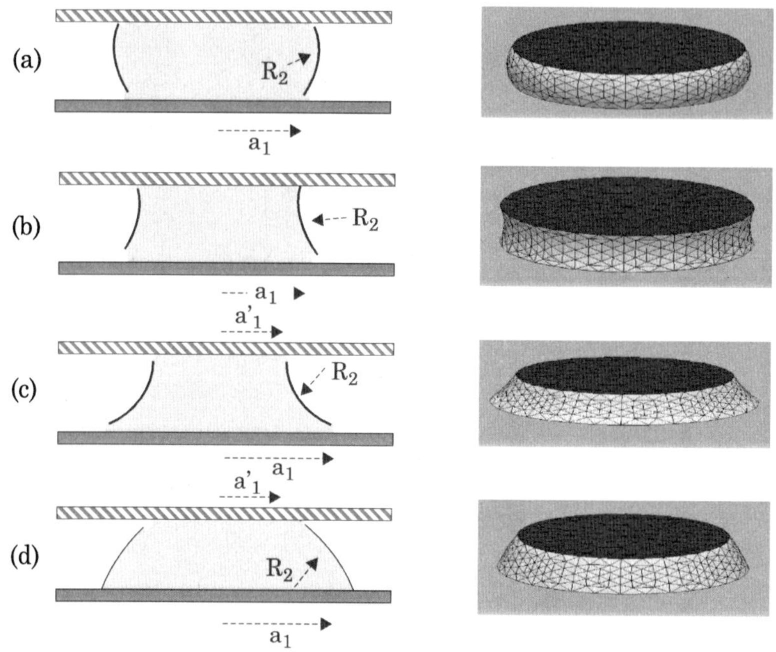

Figure 3.9 Sketch of the shape of a drop between two horizontal plates. (a) non-wetting droplet—hydrophobic wetting for a water drop; (b) partial wetting—hydrophilic wetting for a water drop; (c) drop not wetting the top plate and wetting the bottom plate showing a concave interface (negative curvature radius), corresponding to the relation $\theta_1 < \pi/2, \theta_2 > \pi/2$ and $\theta_1 + \theta_2 < \pi$; (d) same situation as (c), but the interface is convex because $\theta_1 < \pi/2, \theta_2 > \pi/2$ and $\theta_1 + \theta_2 > \pi$.

and are easier to handle. This is particularly the case for electrowetting. We consider here only the case of microsystems where the vertical gap δ is small (usually 50 to 500 μm); their Bond number, given by

$$Bo = \frac{\rho g \delta^2}{\gamma}$$

is less than 0.1, and the free interfaces have circular cross sections.

3.2.2.1 Shape of a droplet constrained between two horizontal planes

A droplet squashed between two parallel plates has two types of contact, one with the bottom plate and another with the top plate; these contacts can be independently either wetting or non-wetting depending on the material of the substrates. Fig. 3.9 shows different possibilities and the corresponding shapes calculated using the Surface Evolver numerical software.

In all cases—due to a small Bond number and the flat upper and lower surfaces—the profile of the droplet in a vertical cross section is nearly a circular arc. Taking advantage of this observation, we can derive a formulation for the calculation of the volume of the droplet.

3.2.2.2 Curvature radius of the free interface

Here, "free interface" denotes the liquid/gas interface as opposed to the liquid/solid interfaces. Referring to the Laplace law, the pressure in a constrained droplet is deduced from the two curvature radii. The expression of the curvature radius in the vertical plane is derived in this section. This curvature radius is shown in Fig. 3.10. Remarking that angles with perpendicular sides are equal, we can write

$$R \cos \theta_1 = H - \delta$$

where H is the distance between the curvature center and the upper plate and δ the gap between the plates; similarly

$$R \cos (\pi - \theta_2) = -R \cos \theta_2 = H.$$

We deduce the value of the curvature radius R

$$R = -\frac{\delta}{\cos \theta_1 + \cos \theta_2}. \tag{3.28}$$

In the case of a concave interface corresponding to $\theta_1 < \pi/2, \theta_2 > \pi/2$ and $\theta_1 + \theta_2 < \pi$ (case (b) in Fig. 3.10), similar reasoning leads to the negative curvature radius

$$R = \frac{\delta}{\cos \theta_1 + \cos \theta_2}.$$

In the particular case where $\theta_1 + \theta_2 = \pi$ (case (c) in Fig. 3.10), the vertical profile of the interface is flat (the interface has a conical shape) and the curvature radius is infinite.

3.2.2.3 Volume of a droplet constrained between two parallel planes and having the same contact angle with both planes

In this section we calculate the volume of the droplet sketched in Fig. 3.11. We will see that we obtain a relation of the type

$$V = f(a, \theta, \delta) = g(R, \theta, \delta) \tag{3.29}$$

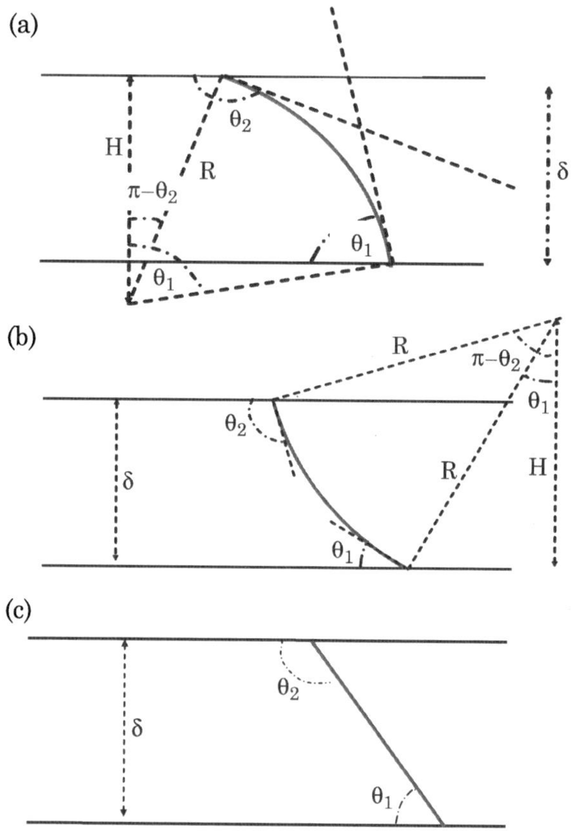

Figure 3.10 Schematic of the geometry of a droplet constrained between two parallel planes; (a) case of a convex interface $\theta_1 < \pi/2, \theta_2 > \pi/2$ and $\theta_1 + \theta_2 > \pi$; (b) case of a concave interface $\theta_1 < \pi/2, \theta_2 > \pi/2$ and $\theta_1 + \theta_2 < \pi$; (c) case of a flat interface $\theta_1 + \theta_2 = \pi$.

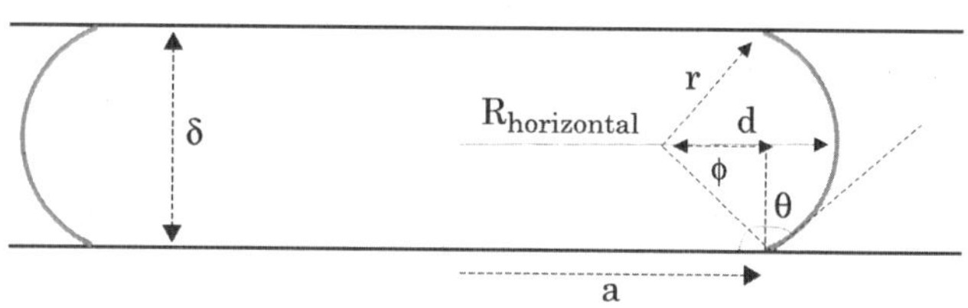

Figure 3.11 Cross section of a droplet constrained by two parallel plates and having the same contact angle with both planes.

where a is the contact radius, θ the contact angle, R the maximum horizontal radius, i.e., the horizontal curvature radius, and δ the vertical gap between the plates. Knowing the volume V, the horizontal curvature radius R can be deduced from (3.29). Using (3.28), we deduce the internal pressure

$$\Delta P = \gamma \left(\frac{1}{R_{\text{vertical}}} + \frac{1}{R_{\text{horizontal}}} \right). \tag{3.30}$$

First, let us introduce the angle ϕ defined by

$$\phi = \theta - \frac{\pi}{2}.$$

The curvature radius r of the interface is given by

$$\cos\theta = -\sin\phi = -\frac{\delta/2}{r}. \tag{3.31}$$

On the other hand, we have

$$\sin\theta = \cos\phi = \frac{d}{r}$$

and we obtain a relation between d and δ:

$$d = -\frac{\delta}{2}\tan\theta.$$

The contact radius a is then

$$a = R - r + d = R + \frac{\delta}{2}\frac{(1-\sin\theta)}{\cos\theta}. \tag{3.32}$$

It can be verified that if $\theta = 90°$ the radius r is infinite and $a = R$. At first glance, the relation (3.32) is indeterminate but using a Taylor development, we find the limit

$$\lim_{\theta\to\frac{\pi}{2}} \frac{(1-\sin\theta)}{\cos\theta} \approx \frac{-\left(\frac{\pi}{2}-\theta\right)^2}{(2!)\left(\frac{\pi}{2}-\theta\right)} = 0.$$

The droplet volume is obtained through the following integral

$$V = 2\int_0^{\frac{\delta}{2}} \pi\tilde{R}^2(y)dy \tag{3.33}$$

where $\tilde{R}(y)$ is the horizontal droplet radius at the vertical coordinate y. Using the equation for a circular segment, one finds that $\tilde{R}(y)$ is given by

$$\tilde{R}(y) = (R-r) + \sqrt{r^2 - y^2}$$

and the volume is then

$$V = 2\int_0^{\frac{\delta}{2}} \pi\left[(R-r) + \sqrt{r^2-y^2}\right]^2 dy. \tag{3.34}$$

Integration of (3.34) is somewhat lengthy. We finally find

$$V = 2\pi \left\{ \left(R^2 - 2rR + 2r^2\right) \frac{\delta}{2} - \frac{1}{3}\left(\frac{\delta}{2}\right)^3 \right.$$
$$\left. + \left(R - r\right) r^2 \left[\theta - \frac{\pi}{2} + \frac{\sin\left(2\theta - \pi\right)}{2}\right] \right\}. \tag{3.35}$$

We notice that, for a droplet constrained by two parallel planes, the droplet volume depends on three independent parameters R, δ and θ.

A similar relation can be found by replacing the equatorial radius R by the contact radius a

$$V = \pi a^2 \delta + 2\pi \left(\frac{\delta}{2\cos\theta}\right)^2 \left\{ a f_1\left(\theta\right) + \frac{\delta}{2} f_2\left(\theta\right) \right\} \tag{3.36}$$

where

$$f_1\left(\theta\right) = \theta - \frac{\pi}{2} + \frac{\sin\left(2\theta - \pi\right)}{2} + 2\sin\theta\cos\theta$$

and

$$f_2\left(\theta\right) = \tan\theta\left(\theta - \frac{\pi}{2} + \frac{\sin\left(2\theta - \pi\right)}{2} + 2\cos\theta\right) - \frac{1}{3}\cos^2\theta + \left(1 - \sin\theta\right)^2.$$

3.2.2.4 Free surface of a droplet constrained by two horizontal planes with different contact angles with the two planes

This case is schematized in Fig. 3.12. The volume V of the droplet now depends on four parameters: a, δ, θ_1 and θ_2

$$V = V\left(a, \delta, \theta_1, \theta_2\right).$$

Completing the figure to obtain a symmetrical droplet, the droplet volume is given by

$$V\left(a, \delta, \theta_1, \theta_2\right) = \frac{V_{\text{sym}}\left(R, \delta_2, \theta_2\right) - V_{\text{sym}}\left(R, \delta_1, \pi - \theta_1\right)}{2} \tag{3.37}$$

where V_{sym} is the volume of a symmetrical droplet, relation (3.36). We now have to express the parameters R, δ_1, δ_2 as functions of a, θ_1, θ_2 and δ. Using equation (3.28), we can write

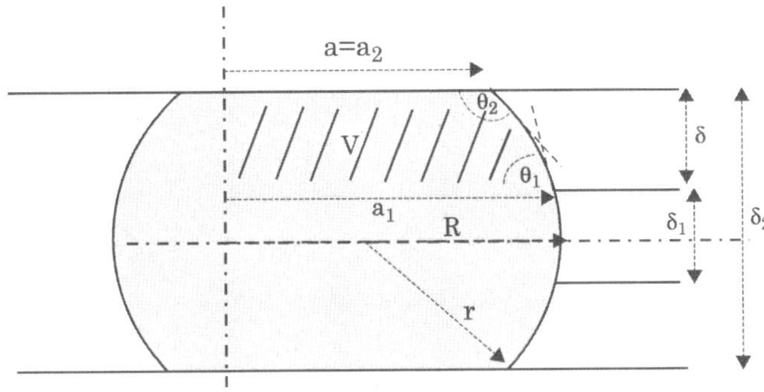

Figure 3.12 Sketch of the droplet (hatched area). The volume V of the droplet is obtained by considering the volume of symmetrical droplet.

$$\begin{aligned} \delta_1 &= -\frac{\delta \cos\theta_1}{\cos\theta_1 + \cos\theta_2} \\ \delta_2 &= \frac{\delta \cos\theta_2}{\cos\theta_1 + \cos\theta_2}. \end{aligned} \tag{3.38}$$

Using (3.32), and substituting (3.38), we find

$$R = a - \frac{\delta_2}{2}\frac{(1-\sin\theta_2)}{\cos\theta_2} = a - \frac{\delta}{2}\frac{(1-\sin\theta_2)}{(\cos\theta_1 + \cos\theta_2)}. \tag{3.39}$$

So, starting from the knowledge of the vector $\{a, \delta, \theta_1, \theta_2\}$, the volume of the droplet is obtained by solving (3.38) and (3.39) to obtain $\{R, \delta_1, \delta_2\}$, and substituting these values in (3.37).

3.2.2.5 Droplet constrained by two planes and bounded by a hydrophilic/hydrophobic boundary

As will be seen in the next chapters, in digital microfluidics applications, droplets are often placed on a hydrophilic surface bounded by hydrophobic regions. In such a case, depending on its volume, either the droplet is small enough and stays far from the hydrophilic/hydrophobic boundary, or it mimics the shape of the hydrophilic substrate (Fig. 3.13). If the volume of liquid increases, the droplet will finally spread over onto the hydrophobic region.

This phenomenon is related to contact line pinning and will be discussed in Section 3.6.

3.2.3 Droplet At a Liquid Interface—Neumann's Construction

In Chapter 2, it was shown that Young's law is obtained by projection of the surface tension forces at the triple line on the plane defined by the solid

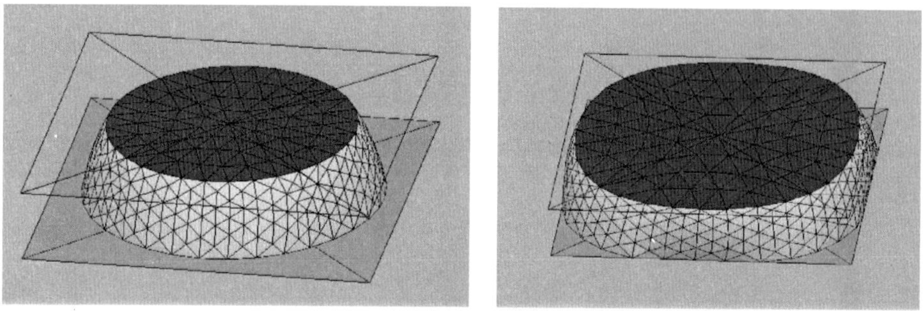

Figure 3.13 Droplet on a bounded wetting region. Left: the droplet has a volume small enough not to touch the boundary; right: the droplet is deformed by pinning on the boundary.

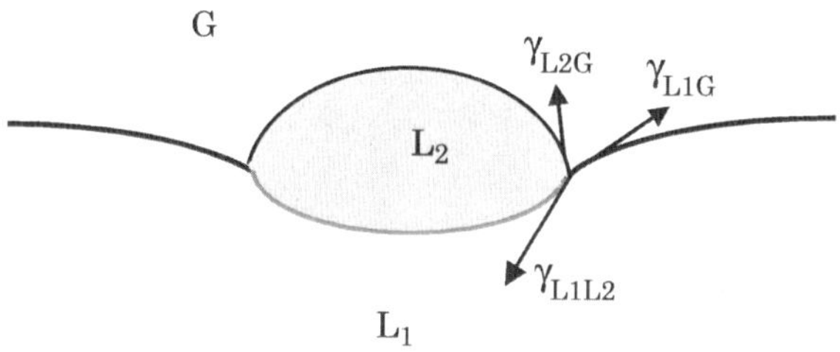

Figure 3.14 Droplet on a liquid surface.

surface. However, no mention is made for the projection on the normal to the solid surface. In reality there is a normal constraint in the solid at its surface to balance the normal component of the surface tensions [1]. But, there is no visible effect if the substrate is a solid. This is not the case for a droplet at a liquid surface, and the resultant of the surface tensions should vanish.

Take two immiscible liquids, denoted 1 and 2, with the droplet of liquid 2 deposited on the interface between liquid 1 and a gas. Even if the density of liquid 2 is somewhat larger than that of liquid 1, the droplet may "float" on the surface, as shown in Fig. 3.14.

The situation is comparable to that of Young's law with the difference that the situation is now two-dimensional. It is called Neumann's construction, and the following equality holds

$$\vec{\gamma}_{L1L2} + \vec{\gamma}_{L1G} + \vec{\gamma}_{L2G} = 0. \tag{3.40}$$

The position of a floating droplet depends on the three interfacial tensions and on the density of each liquid.

In Fig. 3.15 we show some pictures of floating droplets obtained by numerical simulation (Surface Evolver software).

Figure 3.15 Numerical simulations of different positions of a droplet (1 mm) on a liquid surface depending on the three surface tensions. Top: the surface tension of the droplet is very large; bottom: the surface tension of the droplet with the other liquid has been reduced; the drop is at equilibrium due to the balance of buoyancy and surface tensions.

In microfluidic applications for biotechnology, two categories of liquids are used most of the time: the first category consists of aqueous liquids like DIW (de-ionized water), biological solutions (water with a few tenths of molar concentration of salt), and biological buffers with surfactants; the second category comprises organic liquids like silicone oil. It is interesting to know how a microscopic droplet of water behaves at an oil/air interface and a droplet of oil behaves at a water/air interface.

Fig. 3.16 shows the behavior of a droplet of water at an oil/air interface, and Fig. 3.17 shows that oil spreads at the surface of water.

3.2.4 Droplet in a Corner

Microfluidic channels and chambers are etched in silicon, glass, or plastic. Let us investigate first the effect of a corner—or a wedge—on the droplet interface. Take the case of a 90° wedge. The shape of the droplet is shown in Fig. 3.18 depending on the Bond number $Bo = \rho g\ R^2/\gamma$ where R is a characteristic dimension of the droplet, which can be scaled as the 1/3 power of the value of the volume of liquid: $R^3 \approx 3\mathrm{Vol}/\pi$.

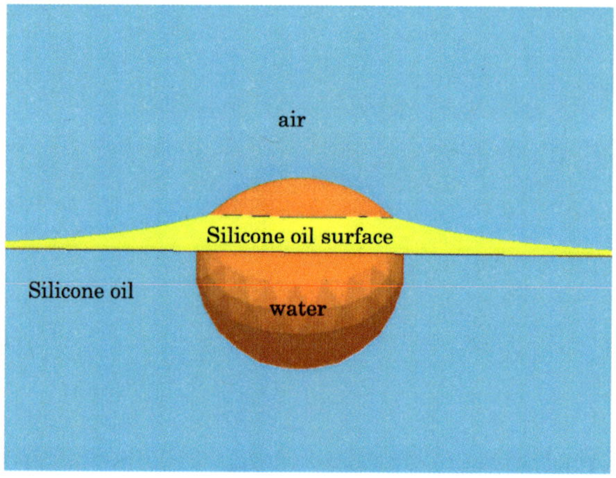

Figure 3.16 Droplet of water at the surface of silicone oil (from Surface Evolver).

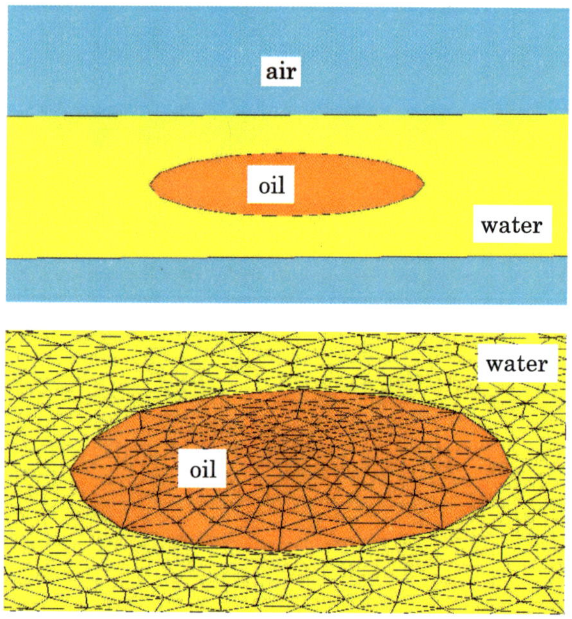

Figure 3.17 A droplet of oil spreads out at a water interface.

3.2.4.1 Concus–Finn Relation

It has been observed that liquid interfaces in contact with highly wetting solid walls forming a wedge tend to spread in the corner (Fig. 3.19). This motion results from the fact that the interface curvature is strongly reduced in the corner (in the case of Fig. 3.19, the vertical curvature radius is small): the Laplace pressure is low in the corner and liquid tends to spread in the corner.

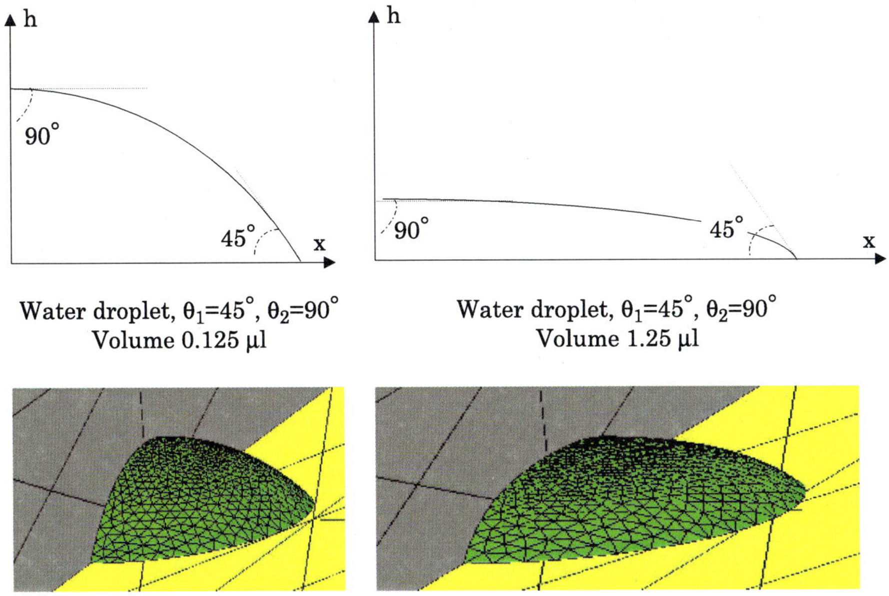

Figure 3.18 Shape of a liquid drop in a 90° wedge. Top: a small volume droplet of 0.125 μl tends to take the form of a sphere despite the different contact angles on the two planes, Bond number of the order of 0.04; a larger droplet of 1.25 μl—Bond number 4—is flattened by gravity.

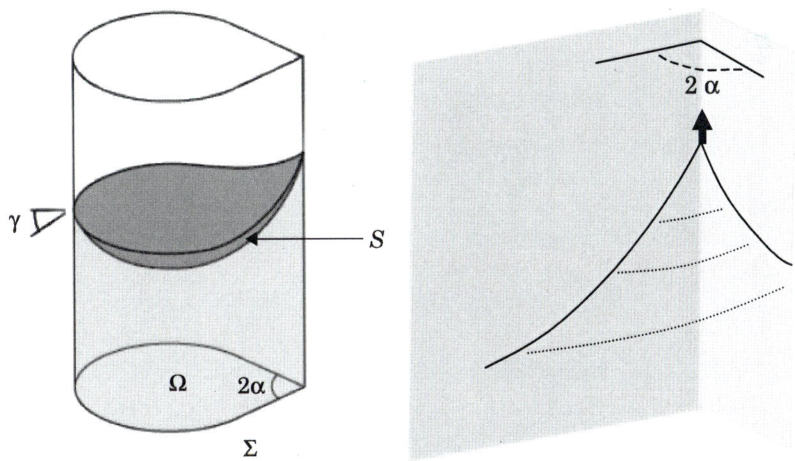

Figure 3.19 A liquid interface is deformed in the corner of a wedge made of two wetting plates. This phenomenon is due to a decrease of curvature at the edge.

Concus and Finn (1969 & 1974) [5] have investigated this phenomenon and they have derived a criterion for capillary motion in the corner of the wedge. If θ is the Young contact angle on both planes and α the wedge half-angle, the

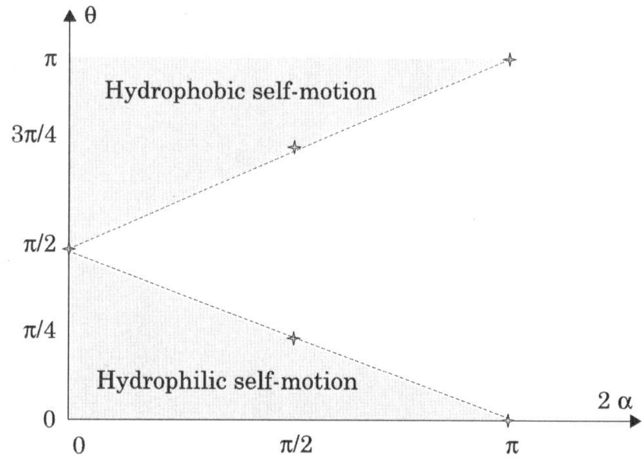

Figure 3.20 Plot of the domains of self motion in a corner, according to the Concus–Finn relations.

condition for capillary self-motion is

$$\theta < \frac{\pi}{2} - \alpha. \tag{3.41}$$

This case corresponds to wetting walls. Conversely, when the walls are non-wetting the condition for de-wetting of the corner is

$$\theta > \frac{\pi}{2} + \alpha. \tag{3.42}$$

In Fig. 3.20, the Concus–Finn relations have been plotted in a $(\theta,\ \alpha)$ coordinate system. One verifies that, for a flat angle, the Concus–Finn relations reduce to the usual capillary analysis.

The Concus–Finn relations can be numerically verified using the Surface Evolver software. Fig. 3.21 shows the spreading of the liquid in the corner when condition (3.41) is met.

In microtechnology, wedges and corners most of the time form a 90° angle, so that a droplet disappears in the form of filaments if the wetting angle on both planes is smaller than 45°. One must be wary that, when coating the interior of microsystems with a strongly wetting layer, in order to have a very hydrophilic (wetting) surface, droplets may disappear; they are transformed into filaments in the corners.

The converse can also be verified: for a rectangular channel, if the coating is strongly hydrophobic, and the contact angles on both planes larger than 135°, the drop detaches from the corner, and does not wet the corner any more (Fig. 3.22).

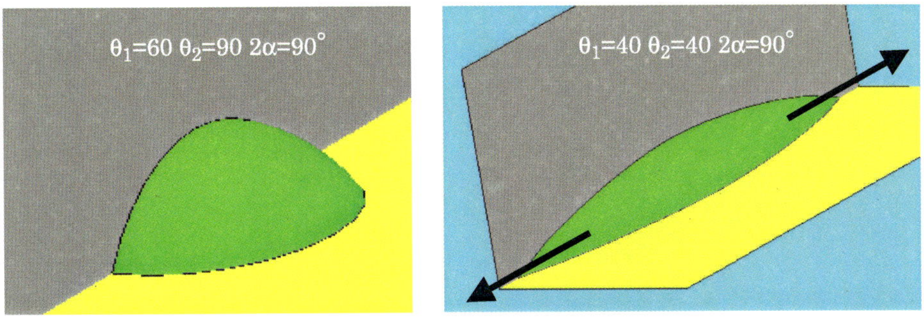

Figure 3.21 A droplet spreads in a corner when the contact angles verify the Concus–Finn condition.

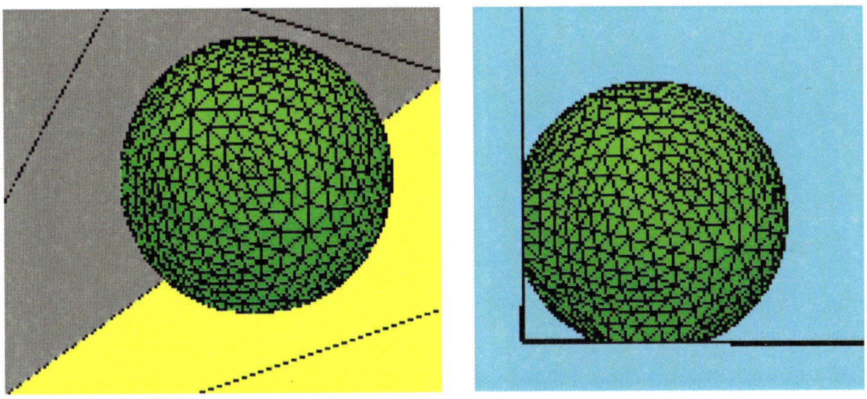

Figure 3.22 Droplet in a strongly hydrophobic corner: the droplet does not wet the tip of the corner, in accordance to the Concus–Finn relation.

3.2.4.2 Extended Concus–Finn Relation

A generalization of the Concus–Finn relation can be derived from the work of K. Brakke [6]. When the two planes do not have the same wettability (contact angles θ_1 and θ_2), the relation (3.41) becomes

$$\frac{\theta_1 + \theta_2}{2} < \frac{\pi}{2} - \alpha. \tag{3.43}$$

Remember that α is the wedge half-angle. An important consequence of relation (3.43) applies to trapezoidal microchannels, a form easily obtained by microfabrication (Fig. 3.23).

When a glass cover is sealed on top, the upper corners form 45° angles and the extended Concus–Finn condition indicates the following limit

$$\frac{\theta_1 + \theta_2}{2} < \frac{\pi}{2} - \alpha = 67.5°.$$

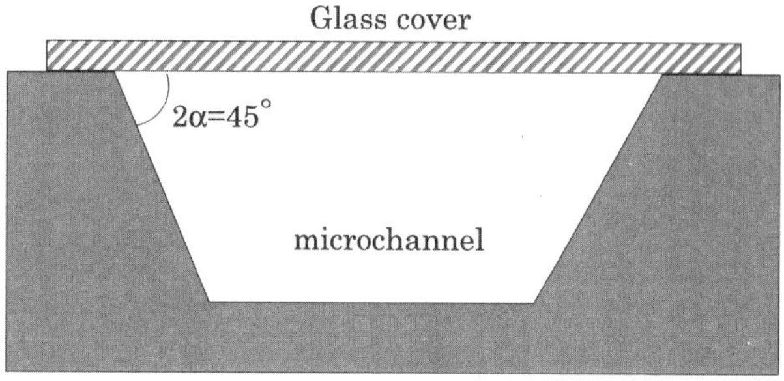

Figure 3.23 Cross section of a trapezoidal microchannel.

The glass cover may be quite hydrophilic, say $\theta_1 \sim 60°$, and if the channel is also hydrophilic, say $\theta_2 \sim 70°$, then $(\theta_1 + \theta_2)/2 \sim 65°$ and the liquid spreads in the upper corners, leading to unwanted leakage.

The same extension applies to the non-wetting case. The extended Brakke–Concus–Finn de-wetting condition for a corner is

$$\frac{\theta_1 + \theta_2}{2} > \frac{\pi}{2} + \alpha. \tag{3.44}$$

In a general way, the surface tension does not appear in the Brakke–Concus–Finn relations. Thus, these relations also apply for two-phase liquids. If we consider water and oil, the contact angles with hydrophilic, respectively hydrophobic, surfaces are of the order of $\theta_1 \sim 60°$, respectively $\theta_2 \sim 130°$. Relation (3.41) shows that a droplet of oil surrounded by water is likely to form in hydrophilic channels, because the water spreads on the solid wall, and relation (3.42) shows the opposite for hydrophobic channels, i.e., oil spreads and a water droplet is formed. This point will be dealt with in more detail in Chapter 9, where solid walls of T-junctions should be hydrophilic or hydrophobic depending on the liquid that must be encapsulated in a droplet.

3.2.4.3 Droplet in a Wetting/Non-wetting Corner

In this section, the behavior of a droplet in a wedge is investigated, in the case where one side of the wedge is wetting and the other non-wetting. It is expected that, if the contact angle on the wetting side is small and that on the non-wetting side is large, the drop will be positioned on the wetting side. It is intuitive to think that this will happen if the difference between the non-wetting contact angle and the wetting contact angle is large.

By referring to [6], a criterion for the drop to be positioned on the wetting side only is

$$\theta_2 - \theta_1 > \pi - 2\alpha \tag{3.45}$$

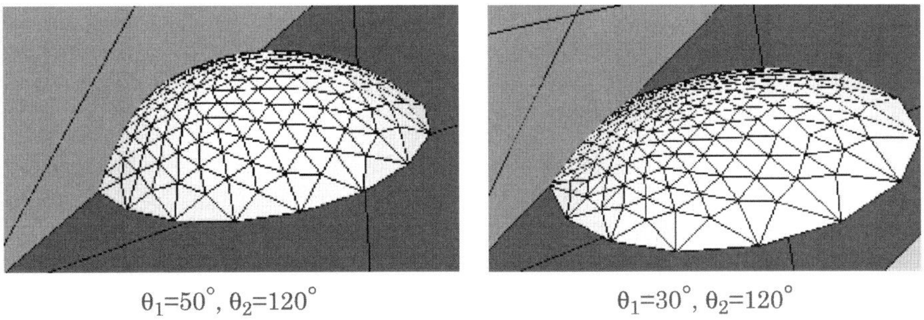

Figure 3.24 Droplet in a corner with wetting and non-wetting sides. Left: the droplet stays attached to the corner; right: the droplet is at equilibrium on the wetting side.

α being the wedge half-angle. Relation (3.45) can be verified by numerical simulation as shown in Fig. 3.24.

The next logical step is to investigate the behavior of droplets in grooves.

3.2.5 Droplet in a Groove

"Open" microfluidics is a relatively new concept which has gained momentum recently. The idea is to use either planar textured surfaces with different wettabilities, or non-planar substrates fabricated by available photo-lithography methods. In a general way, grooves and cusps present the advantages of being easily accessible and easily washable, and they confine the liquid in small volumes, due to pinning of the upper edges. An example is the electrocapillary grooves designed by Baret, et al. [7], shown in Fig. 3.25, where a liquid flows in microgrooves under the actuation of an electric field.

In this section we investigate the shapes of droplets in microgrooves. Different types of morphology have been experimentally observed in rectangular grooves (Fig. 3.26) [8,9]. Seemann et al. and Lipowsky et al. have observed that two parameters govern the morphology of the liquid in a groove: (1) the aspect ratio X of the groove geometry, i.e., the ratio of the groove depth to the groove width; and (2) the contact angle θ of the liquid with the solid substrate. Basically the three morphologies are filaments, wedges, and droplets. Filaments correspond to the case where the liquid spreads in the groove, either in corners if the volume of liquid is small, or in the whole groove if the volume of liquid is sufficient (Fig. 3.27); filaments are obtained for contact angles smaller than 45°, according to the Concus–Finn relation

$$\theta \leq \frac{\pi}{2} - \alpha$$

where α is the corner half-angle. The liquid remains in the form of droplets or stretched droplets if the contact angle is larger than 45°. If the volume of

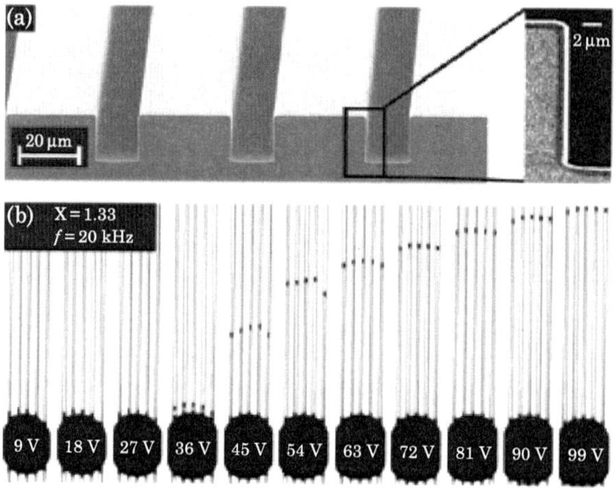

Figure 3.25 Liquid flowing in microgrooves under the action of electro-capillary forces. Top: shape of the channels; bottom: liquid filaments progress in the grooves depending on the applied voltage. Reprinted with permission from [7]. © 2005 American Chemical Society.

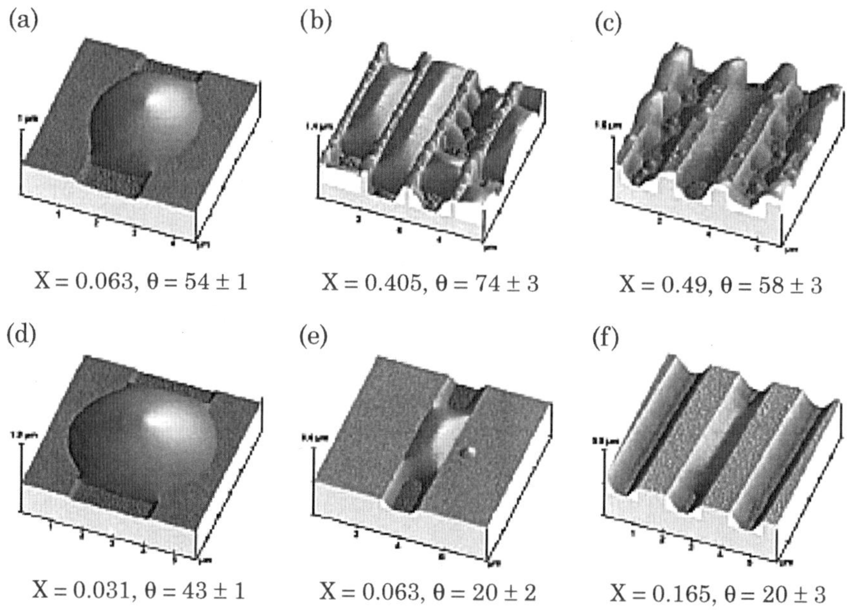

Figure 3.26 AFM observation of a droplet in a groove [8,9]. © 2005 National Academy of Sciences, U.S.A.

liquid is small compared to the dimensions of the groove, the liquid goes to the corners of the groove, forming wedges. For more details on these morphologies, a complete diagram can be found in Seemann et al. [8].

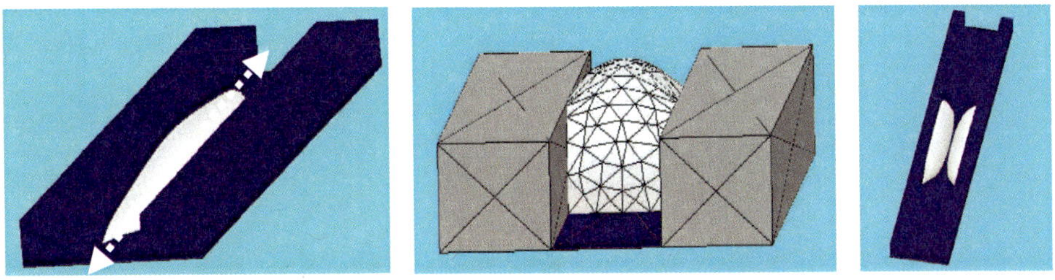

Figure 3.27 The different morphologies of liquid in grooves: filament (left), droplet (middle), wedges (right).

3.2.6 Droplet in a Micro-well

In biotechnology, lab-on-chips and immunoassays are micro-devices aimed at the recognition of DNA and proteins, and for the culture of cells (Fig. 3.28). These devices are constituted of thousands of micro-wells or micro-cusps. Biological liquids—called buffers—are deposited in microscopic wells and biological recognition reactions take place—or not—in each of these wells [10] (Fig. 3.29). It is of importance to be sure that no liquid of a particular well leaks to a neighboring well. The shape of the droplet in a well is then of great importance.

Due to capillary forces at the walls of the well, the free surface of a droplet is never horizontal, but bent depending on the chemical nature of the walls. The convexity/concavity of the surface depends on the wetting/ non-wetting nature of the walls (Fig. 3.30). Non-wetting walls produce convex liquid surfaces; wetting walls produce concave liquid surfaces.

Attention should be given to the fact that the liquid should not overflow on the plate out of the cusp. This can happen when the contact angle in the corners is smaller than the Concus–Finn limit.

Another case for over-spilling can occur when the liquid in the cusp is heated. This is particularly the case of DNA microsystems. DNA arrays are designed for the recognition of DNA segments. The principle is simple: it is based on the matching between a target DNA and its complementary sequence. When the target DNA finds its complementary sequence, there is a binding between the two DNA segments due to hydrogen bonds. Each well is grafted with a predetermined DNA sequence so that each well aims at a specific target. The target can be identified by fluorescence when it binds in a well. Grafting of complementary sequences in the bottom of a cusp requires successive operations of deposition of liquid drops and heating for evaporation. As the heating of the drop is not uniform, surface tension is lower at the walls and the liquid rises along the walls due to increased capillarity (Fig. 3.31). Cusps should then be designed in a way that the liquid cannot exit by capillarity and overflow in the neighboring cusps [11].

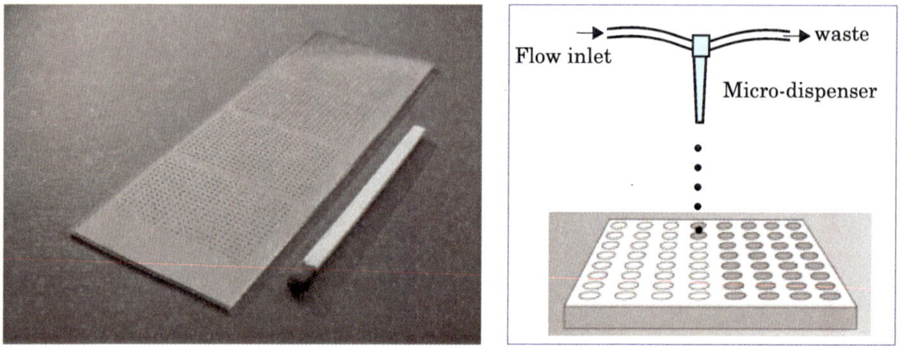

Figure 3.28 Left: DNA microarray made by SCIENION. Right: the wells are filled with biological buffers by droplet spotting.

Figure 3.29 Microwells for spatial confinement of avidin used to the immobilization and housing of single cells from [10], reproduced by permission of The Royal Society of Chemistry.

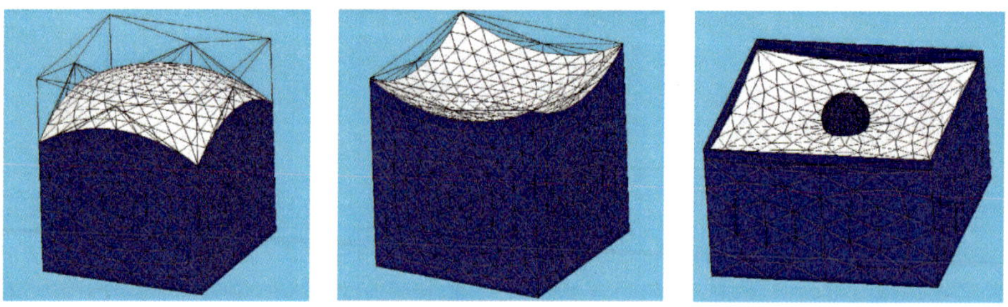

Figure 3.30 Numerical simulation of a droplet in a micro-well; left: case of water in a hydrophobic well; middle: case of water in a hydrophilic well; right: case of a water droplet at the surface of silicone oil in a hydrophobic well (Surface Evolver software).

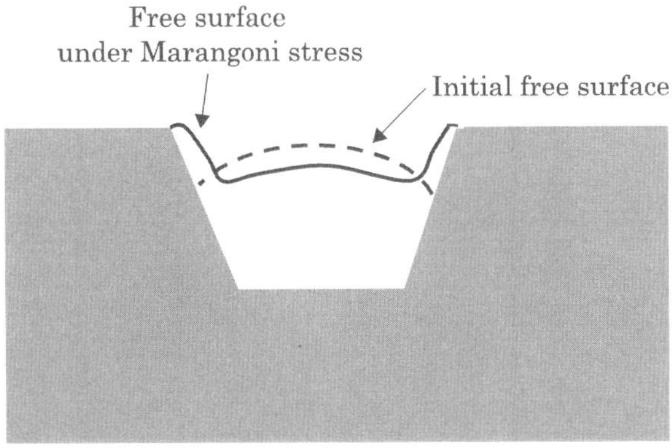

Figure 3.31 Marangoni effect in a micro-well of a DNA array. Remark that in the case of a uniform temperature, the surface tension is constant and the interface is a spherical cap with contact angles defined by the Young relation, whereas in the case where the drop is heated, the interface has a more complicated shape determined by the surface tension distribution.

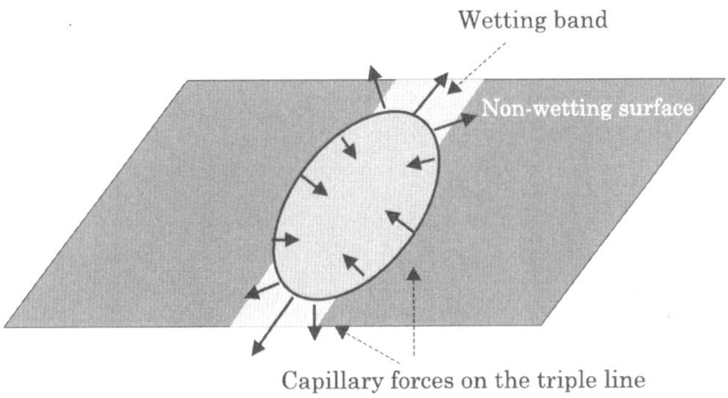

Figure 3.32 Sketch of the capillary forces on the triple contact line.

3.2.7 Droplet on Striped Surface Domains

In this section the behavior of a droplet located on a wetting (lyophilic) band on the surface of an otherwise non-wetting (lyophobic) horizontal plate is investigated. In Fig. 3.32 we have schematized the forces on the contact line and we observe that the contribution of the wetting region (band) tends to elongate the droplet; this effect is reinforced by the forces exerted by the non-wetting region that want to pinch the droplet. However, there is a resisting force to this phenomenon: it is the droplet/air surface tension, whose contribution is to bring back the surface towards that of a sphere.

In Chapter 5, we will see that stretching droplets on lyophilic bands is a very important issue. In particular it will be shown that this is one way to cut a droplet into two "daughter" droplets.

Then the question is: can a droplet be stretched by capillary forces to a point where it is completely resting on the wetting band? This question has been

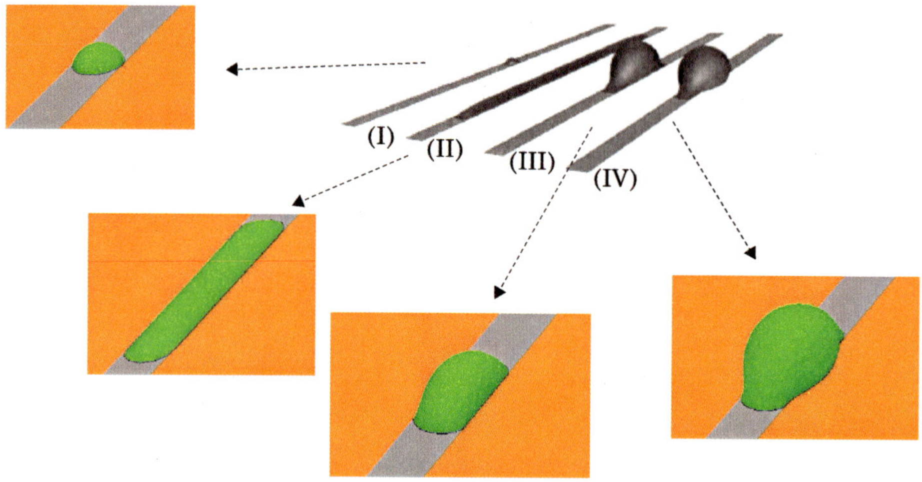

Figure 3.33 Different shapes of a droplet on a lyophilic band: results of a numerical simulation [14].

answered in a series of articles by Gau et al., Lenz et al., Brinkman et al., Klingner et al. [12–16]. They have shown that four morphologies are possible depending mostly on the lyophilic contact angle and the volume of the drop. These four morphologies are shown in Fig. 3.33: (I) when the liquid volume is small, the droplet has a spherical shape and is totally located on the lyophilic band; for larger volumes, the morphology depends on the lyophilic contact angle: if the contact angle is smaller than a threshold value $\theta_{lyo} < \theta_{lim}(V)$ the droplet spreads like a *channel* on the lyophilic surface without overflowing on the lyophobic surface (case II). If $\theta_{lyo} > \theta_{lim}(V)$ the droplet stays localized in a *bulge* state, i.e., does not spread, and two morphologies are possible depending on the volume of the droplet: (III) the volume is sufficiently small and the droplet is constrained by the lyophilic surface limits; (IV) the volume is sufficiently large and the droplet spreads over the transition line onto the lyophobic surface.

The cross sectional profiles of the droplet in the different types of morphologies are shown in Fig. 3.34. These profiles are circle arcs in all cases, with different curvatures. The curvature (and the internal pressure according to Laplace's law) is maximal in the bulge morphology III, where the ratio of the height of the drop to the base width is maximal.

In microfluidic systems, it is interesting to know if a volume of liquid deposited on a striped surface will spread on the lyophilic stripe or will remain in a bulge form. An investigation of the bifurcation between morphologies II and III has been done by Brinkmann et al. [14] and the result is shown in Fig. 3.35. The two main parameters are the lyophilic contact angle θ_{lyo} and the dimensionless volume V/L^3 where L is the width of the band. It is easy to show that when

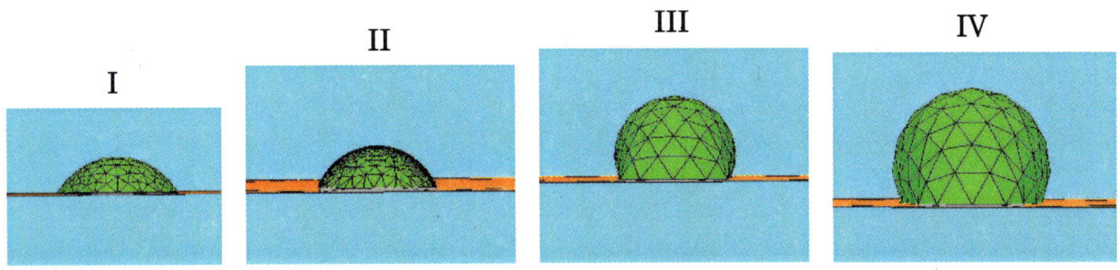

Figure 3.34 Transverse shape of the droplet in the different morphologies: (I) the drop is a spherical cap with Young contact angle; (II) the drop has a circular transverse shape with $h \sim L/2$; (III) the droplet bulge over the hydrophobic band, but its contact surface lies within the hydrophilic band; (IV) curvature is reduced by liquid overflowing onto the lyophobic surfaces.

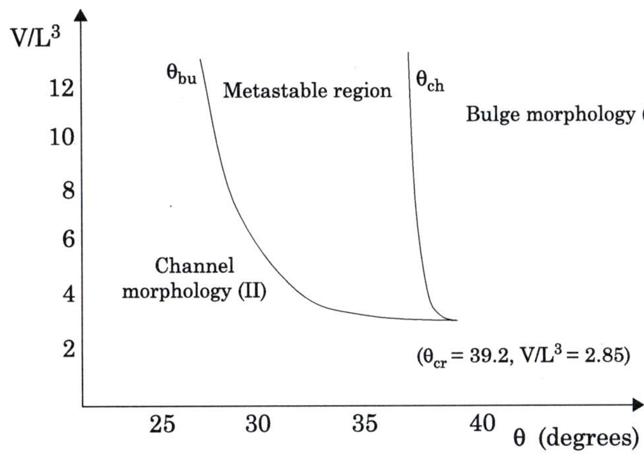

Figure 3.35 Bifurcation diagram for morphologies II and III.

$V/L^3 < \pi/6$ the morphology of the droplet is a spherical cap (number I in Fig. 3.33). For volumes in the range

$$\frac{\pi}{6} \leq \frac{V}{L^3} \leq 2.85$$

there is a continuous smooth change from "channel" droplet to "bulge" droplet. Above the critical volume $V/L^3 = 3.85$ the transition between morphologics II and III is abrupt. It appears that there is a metastable region where the droplet can be in either one of the two morphologies II and III; this metastable region is comprised between the two curves denoted θ_{bu} and θ_{ch} in Fig. 3.35. What is particularly interesting is that the curve $\theta_{ch}(V)$ is nearly vertical, meaning a large volume droplet deposited on the stripe totally spreads on the stripe (channel morphology) if the contact angle is smaller than the approximate value of 38°.

Note that the value of 38° has a physical meaning: if we assume that the liquid in morphology II is a long cylinder of length L', and if we neglect the two ends, the interfacial free energy is approximately

$$E = \gamma S_{\mathrm{LG}} - \gamma_{SG} S_{\mathrm{base}} + \gamma_{\mathrm{SL}} S_{\mathrm{base}} \tag{3.46}$$

corresponding to the energy of the liquid–gas surface plus the energy of the base (where solid–gas contact has been replaced by solid–liquid contact). Using Young's law, we find

$$E = \gamma \left(S_{\mathrm{LG}} - \cos\theta_{lyo} S_{\mathrm{base}} \right). \tag{3.47}$$

Using the relations

$$\begin{aligned} S_{\mathrm{base}} &= LL' \\ S_{\mathrm{LG}} &= LL' \frac{\theta}{\sin\theta} \\ S_{\mathrm{cross-section}} &= \frac{L^2 \left(\theta - \sin\theta\cos\theta\right)}{4\sin^2\theta} \end{aligned} \tag{3.48}$$

where θ is the contact angle on the side of the cylindrical droplet. After substitution in (3.47), and introduction of the volume V, we deduce the expression of the surface energy as a function of the side contact angle θ:

$$E = \gamma \left[\frac{4V}{L} \sin\theta \frac{\theta - \sin\theta\cos\theta_{lyo}}{\theta - \sin\theta\cos\theta} \right]. \tag{3.49}$$

Stability of the interface is obtained by minimizing the surface energy relative to the angle θ, i.e., writing $\frac{\partial E}{\partial \theta} = 0$, and we find the implicit relation

$$\theta_{lyo} = \arccos\left(\frac{\theta}{2\sin\theta} + \frac{\cos\theta}{2} \right). \tag{3.50}$$

The Laplace law indicates that the energy is always smaller than that obtained for the value of $\theta = \pi/2$, corresponding to a minimum curvature radius; and the maximum value of θ_{lyo} for having channel-type morphology is

$$\theta_{lyo} = \arccos\left(\frac{\pi}{4}\right) \approx 38^\circ$$

which is the value found by the numerical simulations. Note that 38° is quite a small value for a contact angle, and channel-type morphology requires very hydrophilic surfaces for water droplets.

The case of two parallel stripes has also been investigated [14], showing four different possible morphologies (Fig. 3.36). These morphologies are in direct similarity with the preceding morphologies for the case of a single stripe.

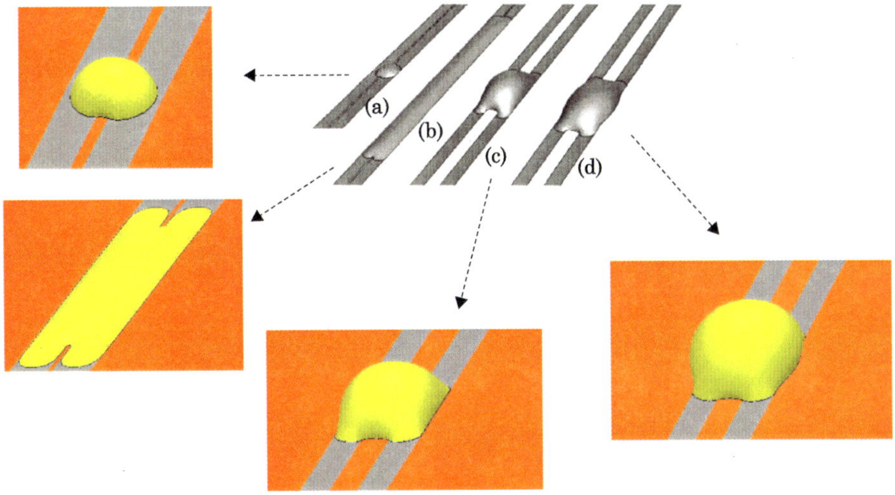

Figure 3.36 Different morphologies of a droplet on two lyophilic bands: results of a numerical simulation [14].

Morphology (a) corresponds to a small volume of liquid for which the drop is nearly spherical with the exception of the trace of the middle lyophobic line. If this lyophobic band is very thin, the drop is unstable and moves towards one of the two lyophilic bands. In morphology (b), the stripes are close and the contact angle is small so that the liquid spreads, almost like the channel case (II). When the volume of liquid is increased and the contact angle is larger, the drop bulges and can either be pinned on the two outer limits of the lyophilic bands (c) or spread over onto the lyophobic surfaces.

3.2.8 Droplet Pierced by a Micro-wire (catena)

A catena is sometimes used as a zero potential contact for open electrowetting on dielectrics (EWOD) systems (Fig. 3.37) [17]. The catena should be electrically conducting and is usually made of gold. The presence of a catena should not modify too much the functioning of the microsystem by its action on the droplets. In this section, we investigate the effect of the catena on the droplet.

It has been observed that the catena modifies the behavior of the drop. In the case of a hydrophobic contact a water drop adopts a position tangential to the catena. After adding surfactants, the contact becomes hydrophilic and the drop moves to a position centered on the wire (Fig. 3.38).

Why does the presence of a catena modify the position and shape of drops? A very simple approach is to consider the capillary forces on the contact lines between the catena and the drop.

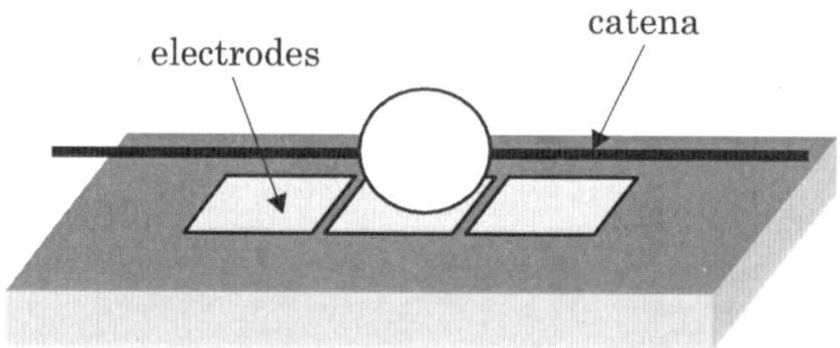

Figure 3.37 Schematic view of an open EWOD microsystem with catena.

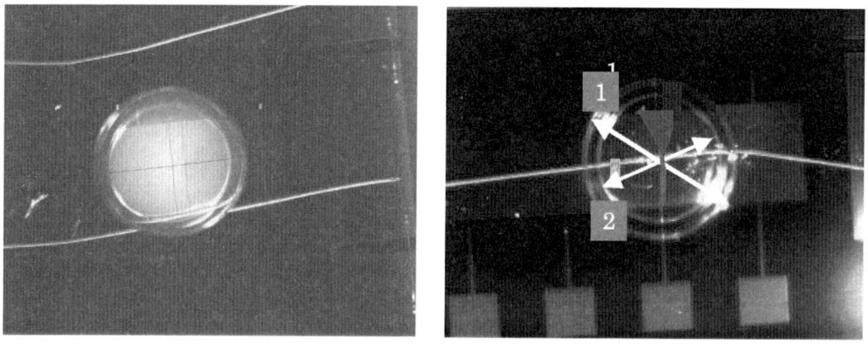

Figure 3.38 Left: the micro-drop is tangent to the catena due to a hydrophobic contact; right: the catena intersects the drop in its middle when the contact is hydrophilic due to the addition of surfactants: 1-diameter of the droplet; 2-diameter of contact circle. (Photo Ph Clementz)

In the case of a hydrophilic catena, these forces can be schematically represented as shown in Fig. 3.39; their resultant is a force that pushes the drop into a symmetry position on the catena. The same reasoning may be done for a hydrophobic contact and the resultant of the forces is directed in the opposite direction and the drop moves aside to stop at an equilibrium position when the catena is tangent to the drop surface.

As a conclusion, the position of the catena above the electrodes has to be carefully chosen depending on its wettability. A hydrophilic catena should be located above the median of the electrode row, whereas a hydrophobic catena is best located above the edge of the electrode row. A catena at a non-optimized position can block the drop motion.

This approach is confirmed by the results of the Evolver numerical simulation (Fig. 3.40).

3.3 Drops on Inhomogeneous Surfaces

The reasoning used to establish Young's law supposed a perfectly flat homogeneous surface. This is somewhat an abstraction, and surfaces—even

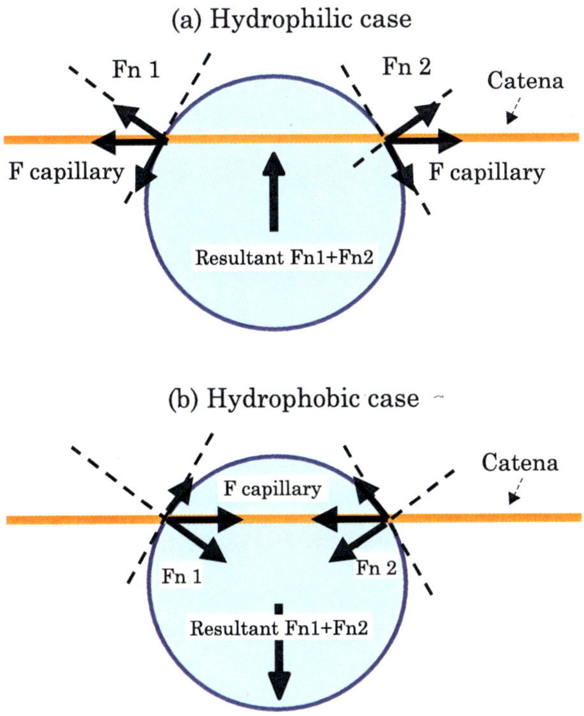

Figure 3.39 In the case of a hydrophilic contact (a), the resultant of the normal forces moves the drop until it becomes symmetrical for the catena. In the hydrophobic case (b), the droplet moves till it is tangent to the catena.

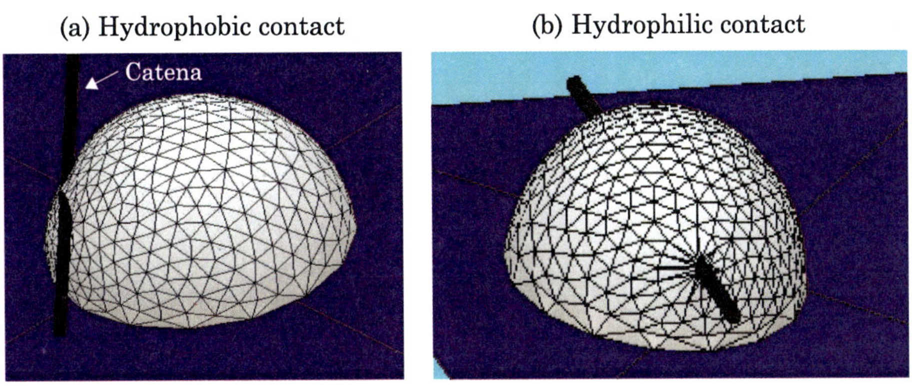

Figure 3.40 Numerical simulation of a droplet pierced by a catena using the Surface Evolver software, (a) the contact of the liquid and the catena is hydrophobic and the catena tangents the drop; (b) the contact is hydrophilic and the catena intersects the droplet symmetrically.

when carefully microfabricated have some roughness and may not be chemically homogeneous. The following sections deal with rough and chemically

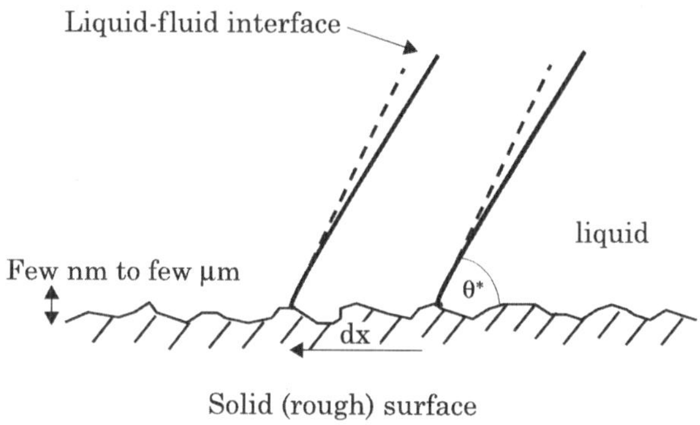

Figure 3.41 Contact on a rough surface.

inhomogeneous surfaces. It will be shown that Young's law should be corrected to take into account the imperfections of the surface.

3.3.1 Wenzel's Law

It has been observed that roughness of the solid wall modifies the contact between the liquid and the solid. But the effect of roughness on the contact angle is not intuitive. It is a surprising—and also a very useful observation as we will see later on—that roughness amplifies the hydrophilic or hydrophobic character of the contact.

Suppose that θ^* is the angle with the surface with roughness and θ the angle with the smooth surface (in both cases, the solid, liquid, and gas are the same). One very important point here is that we have made the implicit assumption that the size of the roughness is very small, so that the molecules of the liquid are macroscopically interacting with a plane surface but microscopically with a rough surface. This explains why we can use the unique angle of contact θ^*.

Suppose a very small displacement of the contact line (Fig. 3.41). Then the work of the different forces acting on the contact line is given by

$$dW = \sum \vec{F}.d\vec{l} = \sum F_x dx = (\gamma_{\mathrm{SL}} - \gamma_{SG})\, r dx + \gamma_{\mathrm{LG}} \cos\theta^* dx \quad (3.51)$$

where r is the roughness (rdx is the real distance on the solid surface when the contact line is displaced by dx). Note that by definition, $r > 1$. Thus the change in energy is

$$dE = dW = (\gamma_{\mathrm{SL}} - \gamma_{SG})\, r dx + \gamma_{\mathrm{LG}} \cos\theta^* dx. \quad (3.52)$$

In fact, if we imagine that the drop finds its equilibrium state after the small perturbation dx, it finally stops at a position where its energy is minimum, so

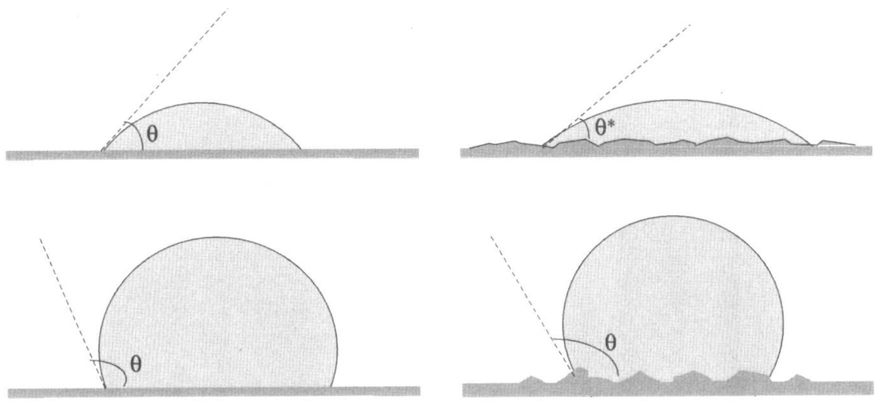

Figure 3.42 Contact of a liquid drop on a rough surface.

that

$$\frac{dE}{dx} = 0$$

and we obtain the relation

$$\gamma_{\mathrm{LG}} \cos \theta^{*} = (\gamma_{SG} - \gamma_{\mathrm{SL}})\, r. \tag{3.53}$$

If we recall that Young's law for a smooth surface

$$\gamma_{\mathrm{LG}} \cos \theta = \gamma_{SG} - \gamma_{\mathrm{SL}}$$

then we obtain Wenzel's law

$$\cos \theta^{*} = r \cos \theta. \tag{3.54}$$

Taking into account that $r > 1$, this relation implies that

$$|\cos \theta^{*}| > |\cos \theta| \,. \tag{3.55}$$

We can deduce that if θ is larger than 90° (hydrophobic contact), then $\theta^{*} > \theta$ and the contact is still more hydrophobic due to the roughness. If θ is smaller than 90° (hydrophilic contact), then $\theta^{*} < \theta$ and the contact is still more hydrophilic (Fig. 3.42). In conclusion, surface roughness increases the wetting character.

An important observation at this stage is that the scale of the roughness on the solid surface is *very small compared to that of the drop* [11]. Indeed, if not, it would not be possible to define a unique contact angle; the drop would no

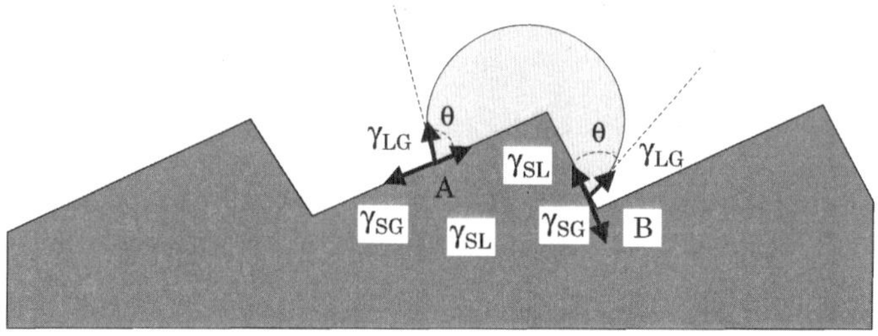

Figure 3.43 Large scale roughness: schematic view of a drop located on an angle of the solid surface. The position of the drop might not be stable.

longer be axisymmetrical, and the contact could be sketched as in Fig. 3.43 (the position of the drop might not be stable).

3.3.2 Cassie–Baxter Law

The same analysis of Wenzel was done by Cassie for chemically inhomogeneous solid surfaces. As for Wenzel's law, the same requirement of small size heterogeneities compared to interaction size between liquid and solid wall applies. For simplicity we analyze the case of a solid wall constituted by microscopic inclusions of two different materials. If θ_1 and θ_2 are the contact angles for each material at a macroscopic size, and f_1 and f_2 are the surface fractions of the two materials (Fig. 3.44), then the energy to move the interface by dx is

$$dE = dW = (\gamma_{\mathrm{SL}} - \gamma_{SG})_1 f_1 dx + (\gamma_{\mathrm{SL}} - \gamma_{SG})_2 f_2 dx + \gamma_{\mathrm{LG}} \cos\theta^* dx. \quad (3.56)$$

The equilibrium is obtained by taking the minimum of E

$$\gamma_{\mathrm{LG}} \cos\theta^* = (\gamma_{SG} - \gamma_{\mathrm{SL}})_1 f_1 + (\gamma_{SG} - \gamma_{\mathrm{SL}})_2 f_2$$

and by comparison with Young's law, we obtain the Cassie–Baxter relation

$$\cos\theta^* = f_1 \cos\theta_1 + f_2 \cos\theta_2. \quad (3.57)$$

This relation may be generalized to a more inhomogeneous material

$$\cos\theta^* = \sum_i f_i \cos\theta_i. \quad (3.58)$$

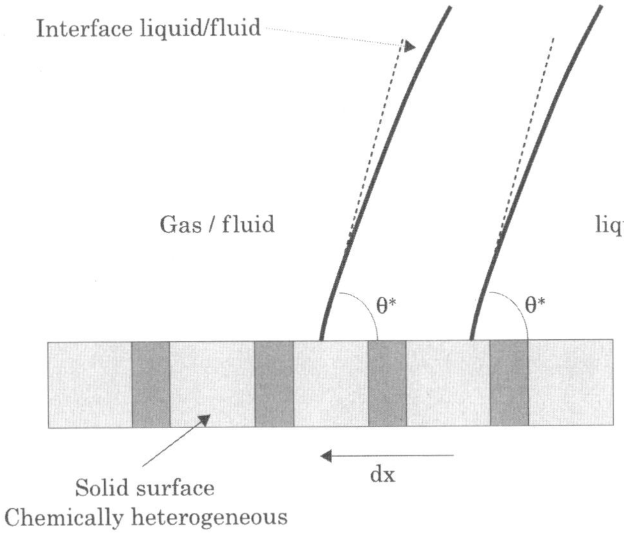

Figure 3.44 Displacement of the contact line of a drop on an inhomogeneous solid surface.

Note that

$$f_1 + f_2 = 1 \quad \text{or} \quad \sum_i f_i = 1.$$

The Cassie–Baxter relation shows that the cosine of the contact angle on a microscopically inhomogeneous solid surface is the barycenter of the cosine of the contact angles on the different chemical components of the surface.

The Cassie–Baxter law explains some unexpected experimental results. Sometimes—if not enough care was taken during micro-fabrication—a micro-fabricated surface may present chemical inhomogeneity and the wetting properties are not those that were intended. For example, if a uniform layer of Teflon is deposited on a rough substrate, the surface should become hydrophobic. However, if the layer is too thin, the Teflon layer may be porous and the coating inhomogeneous; the wetting properties are then modified according to the Cassie–Baxter law and the gain in hydrophobicity may not be as large as expected.

As for Wenzel's law, an important remark at this stage is that the scale of change of the different chemical materials of the solid surface is *very small compared to that of the drop* [11]. Indeed, if not, it would not be possible to define a unique contact angle any more. This latter type of inhomogeneity is related to drop pinning or to drop motion depending of the morphology of the inhomogeneity.

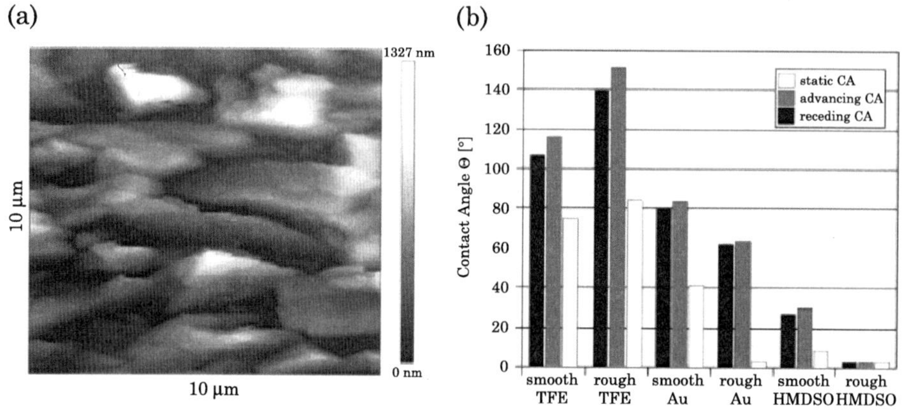

Figure 3.45 (a) Tin crystals on top of a silicon substrate [18]; (b) CVD deposition of TFE or gold renders the surface hydrophobic or hydrophilic. In the legend, CA stands for Contact Angle. From [18]. © 2003 Elsevier; reprinted with permission.

3.3.3 Contact on Fabricated Surfaces: The Transition Between the Wenzel and Cassie Laws

3.3.3.1 Introduction

We have seen in the preceding sections that the contact angle of a liquid on a solid surface depends on the roughness and the chemical homogeneity of the surface; this contact angle is given by the Wenzel and Cassie laws. These laws have been established at a mesoscopic scale. For example, Wenzel's law shows that rugosities of the surface increase the hydrophobic/hydrophilic characteristics of a surface. In biotechnology, the materials used are limited—PDMS, Teflon, SU8, glass, silicon, gold. Plastics (Teflon, SU8, PDMS) are generally hydrophobic whereas glass and metals are hydrophilic and silicon neutral. A way to force the surface to be hydrophobic/hydrophilic is to chemically deposit a hydrophobic/hydrophilic coating on top of the substrate. For example, Uelzen and Mueller [18] have developed a technique that consists of growing microscopic tin crystals on a silicon substrate—in order to increase the roughness (Fig. 3.45)—and then a hydrophobic coating of TFE or a hydrophilic coating of gold is deposited by CVD (Chemical Vapor Deposition). The result is an enhanced hydrophobic or hydrophilic contact.

Another example of obtaining a superhydrophobic substrate is to deposit a coating of CF_4–H_2–He (Fig. 3.46) by plasma discharge [19]. The hydrophobic character increases with the numbers of discharges. Note that in this case, the increase in roughness of the surface is moderate. The rugosities created by the plasma coating are only 10 nm (Fig. 3.47). This method can be applied to very different substrates, like silicon, gold, or even cotton.

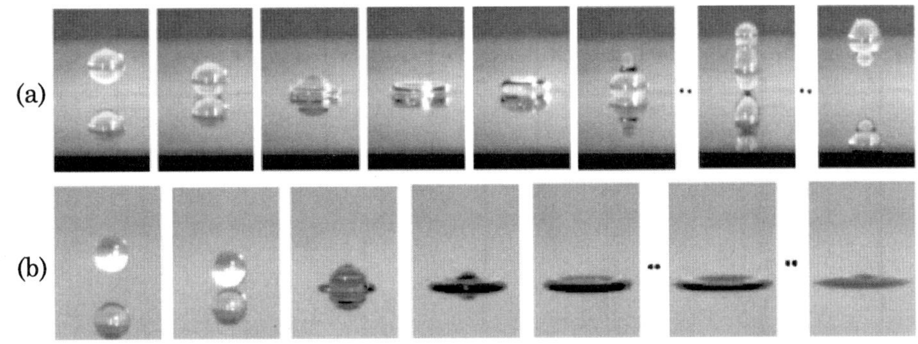

Figure 3.46 A flat gold plate treated with CF_4–H_2–He plasma deposition becomes very hydrophobic. (a) A water droplet bounces back from the surface treated with CF_4–H_2–He plasma deposition, (b) whereas it spreads on the original gold surface; reprinted with permission from [19]. © 2005 American Chemical Society.

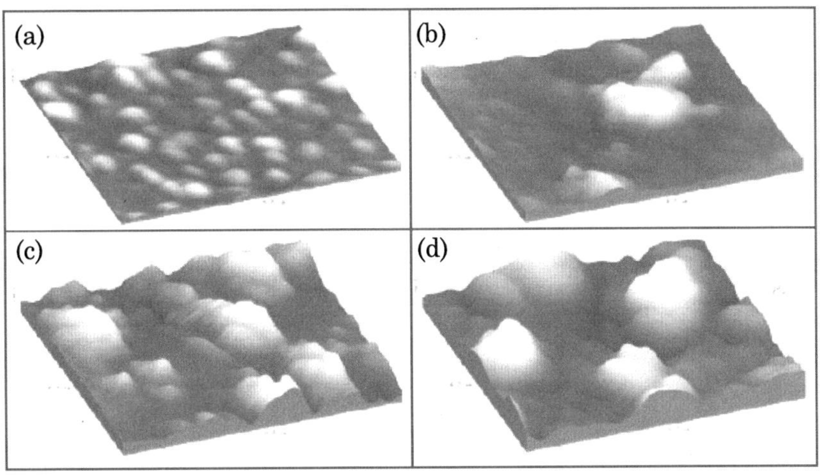

Figure 3.47 AFM image of a gold surface treated with CF_4–H_2–He plasma discharges. The number of discharges is (a) 0, (b) 3, (c) 7, and (d) 11. After 10 discharges, the surface is superhydrophobic and the rugosities are still very small, of the order of 10 nm; reprinted with permission from [19]. © 2005 American Chemical Society.

Another way to increase the hydrophilic/hydrophobic character of a surface is to artificially create patterned microstructures in order to increase the roughness of the surface—by creating micro-pillars or grooves. Fig. 3.48 shows an example of patterning a silicon surface with micro-pillars [20].

If Wenzel's law is applicable at such a roughness scale, it is expected that the hydrophilic/hydrophobic character will be increased. So the question is: can Wenzel's formula, taking into account a roughness based on the shape of the micro-structures, be used to derive the contact angle? The answer is not that straightforward. It has been observed that the droplet does not always contact

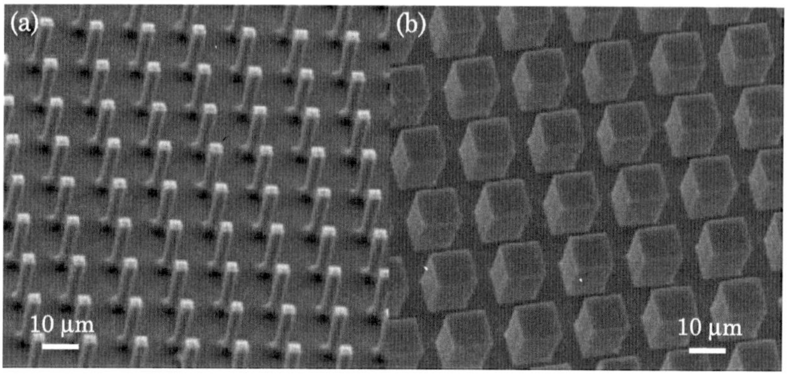

Figure 3.48 Microfabricated pillars of silicon. From [20]. © 2006 Elsevier; reprinted with permission.

Figure 3.49 Left: droplet penetrating between the pillars, right: droplet sitting on top of the pillars (fakir effect). θ_W is the Wenzel contact angle and θ_C is the Cassie contact angle.

the bottom plate and sometimes stays on top of the pillars—which is called the "fakir effect" (Fig. 3.49). In such a case, should not the Cassie law, based on a juxtaposition of solid surface and air, have been used? And what is the limit between a Wenzel droplet and a Cassie droplet? All these questions are discussed next.

3.3.3.2 Contact Angle on a Microfabricated Substrate. Case of Hydrophobic Contact

The contact angle of a sessile drop sitting on microfabricated pillars has been the subject of many investigations recently. As we have seen previously, Young's law defines the contact angle on the substrate material

$$\cos\theta = \frac{\gamma_{SG} - \gamma_{\mathrm{SL}}}{\gamma_{\mathrm{LG}}}. \tag{3.59}$$

If the drop penetrates between the pillars, one can write the Wenzel angle as

$$\cos\theta_W = r\cos\theta \tag{3.60}$$

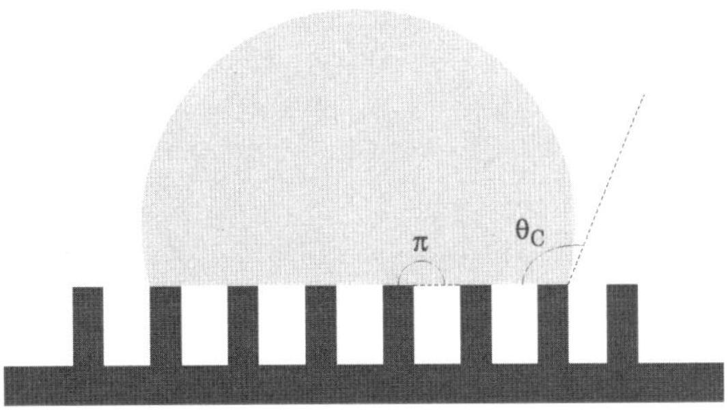

Figure 3.50 Sketch of a Cassie drop (fakir effect). The interface between the pillars is roughly horizontal.

where θ_W is the "Wenzel" contact angle and r the roughness of the surface. If the drop stays on top of the pillars, one can write the Cassie law under the form

$$\cos\theta_C = f\cos\theta + (1-f)\cos\theta_0 \tag{3.61}$$

where θ_C is the "Cassie" contact angle, θ_0 the contact angle with the layer of air, and f the ratio of the contact surface (top of the pillars) to the total horizontal surface. If the pillars are not too far from each other, the value of θ_0 is roughly $\theta_0 = \pi$ (see Fig. 3.50).

Equation (3.61) then simplifies to

$$\cos\theta_C = -1 + f(1+\cos\theta). \tag{3.62}$$

It is usual to plot the two relations (3.60) and (3.62) in the diagram of Fig. 3.50 [21–26]. In such a representation, the two equations correspond to two straight lines, the first one with a slope r, and the second one with a slope f. The two lines intersect, because

$$r = \frac{S_{\text{total}}}{S_{\text{horizontal}}} > f = \frac{S_{\text{top}}}{S_{\text{horizontal}}}.$$

The two lines intersect at a Young contact angle θ_i defined by $\theta_C = \theta_W$, so that

$$\cos\theta_i = \frac{f-1}{r-f}. \tag{3.63}$$

In the diagram of Fig. 3.51, for a given Young angle, there are two contact angles. The question is now: which one is the real one? From energy considerations—for example by using Laplace's law—it can be deduced that

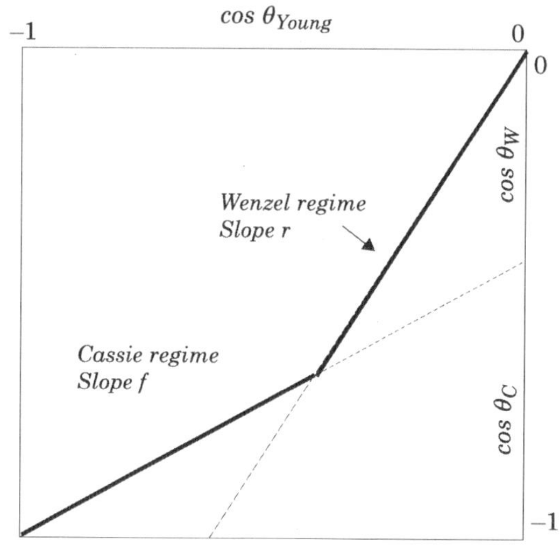

Figure 3.51 Plot of the Wenzel and Cassie laws for a sessile droplet sitting on a surface textured with micro-pillars. Using the Laplace law, it can be seen that the energy of the drop is lower when the contact angle is the smallest. The physical situation then corresponds to the continuous lines, with a Wenzel drop at small Young contact angle and a Cassie drop for large Young contact angle.

the real contact angle is the smaller one, so when the Young contact angle is not very hydrophobic ($\theta < \theta_i$), the contact corresponds to a Wenzel regime and the drop wets the whole surface. When the Young contact angle is more hydrophobic ($\theta > \theta_i$), the drop is in a Cassie regime and sits on top of the pillars.

Note that the situation we have just described does not always correspond to reality. It happens that a droplet is not always in its lowest energy level and that droplets are sometimes in metastable regimes. One example was given by Bico, et al. [21–23]. It is that of a drop deposited by a pipette on a pillared surface. Even if it should be in a Wenzel regime and the droplet should penetrate between the pillars, it stays on top of the pillars. It needs an impulse, mechanic, electric, or acoustic, to join the expected Wenzel regime.

Transition from a Cassie droplet to a Wenzel droplet can be obtained by electrowetting actuation. Krupenkin, et al. [27] have shown that a droplet sitting on pillars in a Cassie regime, i.e., sitting only on the top of the pillars, sinks down as soon as electrowetting actuation is turned on (Fig. 3.52).

3.3.3.3 Example of a Square Lattice

Take the case of a lattice of square pillars disposed along a square grid; this case has been studied by Patankar [25,26], Zhu et al. [28]. Suppose the rugosities are constituted of square pillars of size $a \times a$, height H and spacing b similar to that of Fig. 3.48 and symbolized in Fig. 3.53.

Figure 3.52 Sessile droplet sitting on micropillars: (a) when no electric actuation, the drop sits on top of the pillars; (b) electrode actuated with 22 V electric potential: the droplet wets the flat part of the substrate. The wire at the top is the zero potential electrode. Reprinted with permission from [27]. © 2005 American Chemical Society.

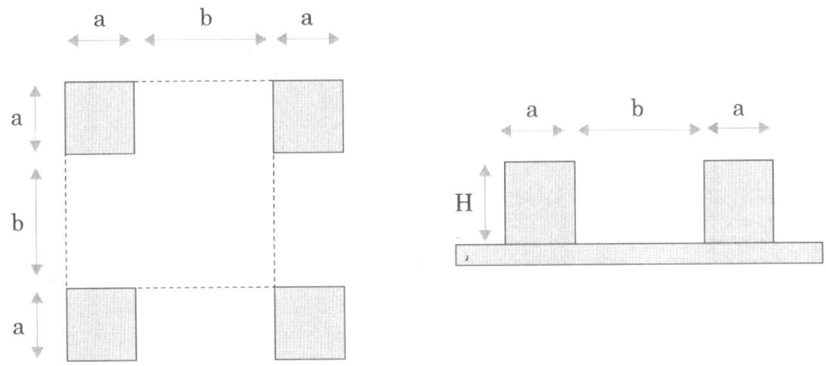

Figure 3.53 Top and cross-sectional schematic sketches of a square pillars patterning.

By considering one "motif" in Fig. 3.53, the roughness r is determined by

$$r = \frac{S_{\text{total}}}{S_{\text{horizontal}}} = \frac{(a+b)^2 + 4aH}{(a+b)^2} = 1 + \frac{4A}{(a/H)} \tag{3.64}$$

where

$$A = \left(\frac{a}{a+b}\right)^2.$$

On the other hand, the Cassie coefficient f is given by

$$f = \frac{S_{\text{top}}}{S_{\text{horizontal}}} = \frac{a^2}{(a+b)^2} = A. \tag{3.65}$$

Substitution of (3.64) in (3.60) and (3.65) in (3.62) leads to the Wenzel and Cassie formulas for square pillars in a square "motif"

$$\cos\theta_W = \left[1 + \frac{4\,A}{(a/H)}\right]\cos\theta \tag{3.66}$$

and

$$\cos\theta_C = -1 + A\,(1+\cos\theta)\,. \tag{3.67}$$

Using (3.66), (3.67), and (3.63), one finds that the shift from Wenzel to Cassie droplet occurs when

$$\cos\theta_i = \frac{-1}{1+4\frac{aH}{b^2}}. \tag{3.68}$$

Let us estimate the influence of the parameters a, H, and b in (3.68).

If the height of the pillars is decreased, $H \rightarrow 0$, and we find $r \rightarrow 1$ using (3.64), meaning that the roughness induced by the pillars has disappeared and, using (3.68), $\theta_i \rightarrow \pi$. Reducing the height of the pillars increases the transition angle between Cassie and Wenzel regimes and favors the Wenzel regime.

When the pillars are thin, $a \rightarrow 0$, and we find $A \rightarrow 0$ and, using (3.65), $f \rightarrow 0$. Then, using (3.63), we have the relation $\cos\theta_i \approx \frac{-1}{r}$. Because $r \rightarrow 1$ from (3.64), we conclude that the Wenzel regime is predominant.

Finally, when the pillars are very close to one another, $b \rightarrow 0$, using (3.68) we find that $\theta_i \rightarrow 0$ and we observe a Cassie regime. A droplet on a hydrophobic surface does not penetrate small holes in the wall. A membrane pierced with very small holes has to be hydrophilic for water to penetrate in the holes.

3.3.3.4 Super-Hydrophobicity

A surface is said to be super-hydrophobic when the contact angle of aqueous liquid is close to 180°. In nature, some tree leaves in wet regions of the globe have super-hydrophobic surfaces in order to force water droplets to roll off the leaves, preventing rotting of the leaves, as shown in Fig. 3.54 taken from Barthlot, et al. [29].

Super-hydrophobicity requires either a super-hydrophobic Young contact angle or a Cassie regime. Super-hydrophobicity corresponds to $\theta_{\text{real}} = \pi$ and $\cos\theta_{\text{real}} = -1$. It cannot be obtained in the Wenzel regime except if $r = 1$ and

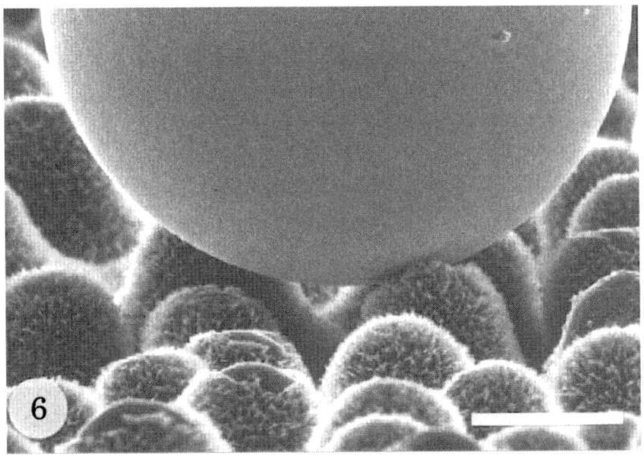

Figure 3.54 Mercury droplet on the papillose adaxial epidermal surface of Colocasia esculenta demonstrating the effect of roughness on wettability. Due to the decreased contact area between liquid and surface, air is enclosed between the droplet and the leaf, resulting in a particularly strong water-repellent surface. The bar is 20 µm. From [29]. © 1997 Springer; reprinted with permission.

$\theta_{\text{Young}} = \pi$. Otherwise, it can only be approached in a Cassie regime when the value of f is very small (Fig. 3.54). Using (3.65) f can be cast in the form

$$f = \frac{a^2}{(a+b)^2} = \frac{1}{\left(1+\frac{b}{a}\right)^2}. \tag{3.69}$$

From (3.69), f is small when $b/a \gg 1$. In such a case, the pillars are very thin and the contact area of the droplet with the solid substrate is very small. The best situation for super-hydrophobicity for a geometrically textured surface is f as small as possible and r as large as possible (Fig. 3.55). We have already seen the condition for f. Using the approximation $b/a \gg 1$ in the expression of the roughness r, we find $r \approx 1 + 4\frac{aH}{b^2}$. This latter relation shows that if H is sufficiently high, r will be large.

3.3.3.5 Irregularly Patterned and Fractal Surfaces

In the introduction of this section, we have seen that an irregular coating was obtained by growing crystals on a surface. In this case, the rugosities were not important and the Wenzel regime was the regime for any water droplet on this surface. In this section, we analyze the situation when the rugosities are irregular and sufficient to obtain the Cassie regime.

The first example is that of Fan et al. [30] who made pillars grow from a surface (Fig. 3.56) using a GLAD technique. GLAD (glancing angle deposition) is a physical vapor deposition technique along an angle.

The contact of the water with the pillars has roughly the shape described in Fig. 3.57.

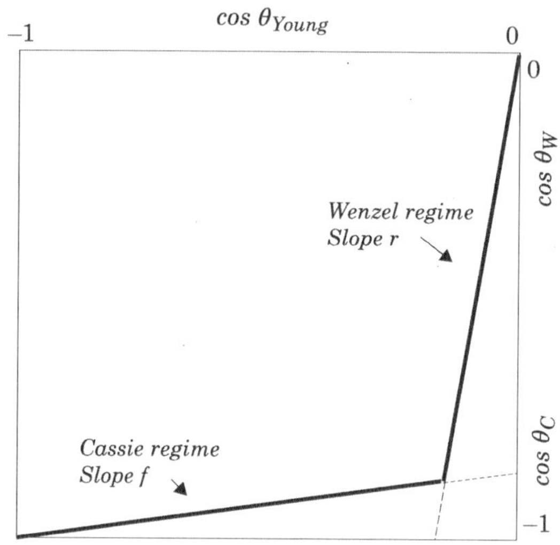

Figure 3.55 Super-hydrophobicity requires a Cassie/Wenzel diagram with a very small coefficient f and a large coefficient r.

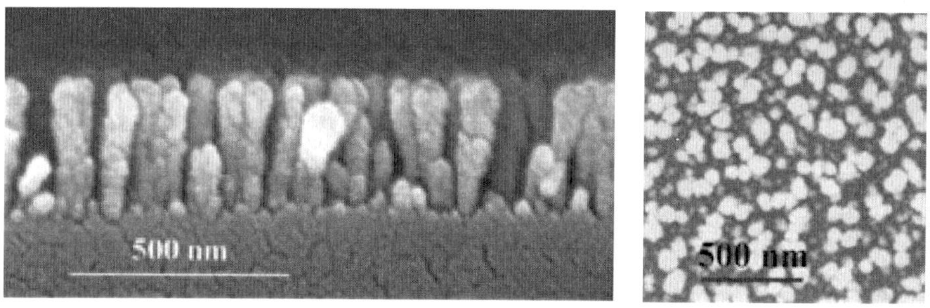

Figure 3.56 Irregular pillars obtained by Fan et al. From [30]. © 2004 IOP; reprinted with permission.

Figure 3.57 Sketch of droplet sitting on pillars similar to those of Fig. 3.55.

Typically this is a Cassie droplet. However, due to the roughness of the tops of the pillars, the Cassie contact angle can then be defined by an improvement of relation (3.61), taking into account the local roughness r^* in the area of contact.

$$\cos\theta_C = r^* f \cos\theta + (f - 1). \tag{3.70}$$

In the case where the tops of the pillars are spherical, Bico et al. [22] have shown that (3.70) becomes

$$\cos\theta_C = -1 + f(1 + \cos\theta)^2. \tag{3.71}$$

This situation combines the effects of Cassie and Wenzel regimes. Note that equation (3.72) shows that the Cassie regime is not linear in $\cos\theta$ but quadratic.

Similarly, Onda et al. and Shibuichi et al. [31,32] have fabricated fractal surfaces (Fig. 3.58) using AKD (alkylketene dimer). These surfaces may be called biomimetic because they resemble the structure of hydrophobic leaf surfaces which are also fractal. Using fractal dimensions they have deduced a relation similar to the Wenzel law

$$\cos\theta_W = \left(\frac{L}{l}\right)^{D-2} \cos\theta \tag{3.72}$$

where L and l are the largest and smallest fractal size limits of surface and D is the fractal dimension ($D = 2.3$ in their case). It is typically a case of super-hydrophobicity since the coefficient $r = (L/l)^{D-2}$ is large and the value of f is very small.

3.3.3.6 Contact Angle on a Microfabricated Substrate. Case of Hydrophilic Contact

In the preceding sections, we have dealt with non-wetting textured surfaces. Here we examine the case of a hydrophilic (wetting) textured surface. This case refers to the theory of impregnation [23]: a droplet on a rough wetting surface has a smaller contact angle than the Young contact angle, according to Wenzel's law. However, it has been observed that in some cases, spreading occurs, i.e., a part of the liquid forms a film on the substrate.

If we keep the same notations, r the roughness of the surface defined by (3.64) and f the Cassie ratio defined by (3.65), one can define a critical contact angle by

$$\cos\theta_{\text{crit}} = \frac{1 - f}{r - f}. \tag{3.73}$$

Figure 3.58 Left: AFM image of a highly fractal surface made of AKD; right: the contact angle is larger than 174°. Reprinted with permission from [31,32]. © 1996 American Chemical Society.

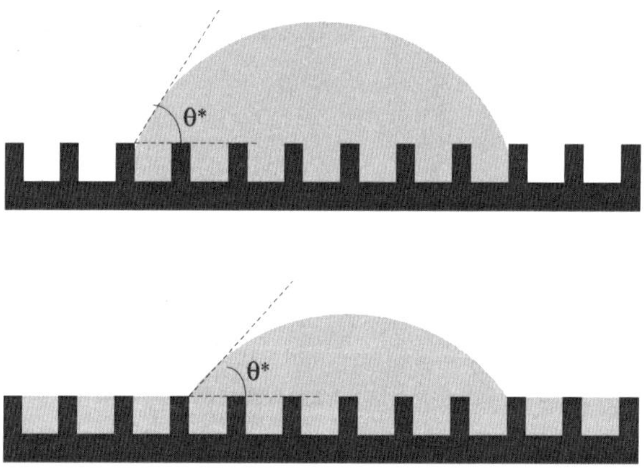

Figure 3.59 The two possible morphologies of a droplet on a wetting textured surface.

If the Young contact angle θ is such that $\theta < \theta_{\text{crit}}$, then the liquid wets the surface, i.e., a liquid film spreads on the surface. The two possible morphologies are shown in Fig. 3.59.

Relation (3.73) is very similar to the case of the hydrophobic critical angle defined by (3.63). Then, for a flat surface $r \to 1$ and the surface is wetted only if the Young contact angle is $\theta = 0$. For a porous medium (r infinite), equation (3.73) indicates that wetting occurs when $\theta < \pi/2$. More generally, equation (3.73) defines a critical angle comprised between 0 and $\pi/2$.

In the Wenzel regime ($\theta > \theta_{\text{crit}}$) the contact angle is

$$\cos\theta^* = r\cos\theta \tag{3.74}$$

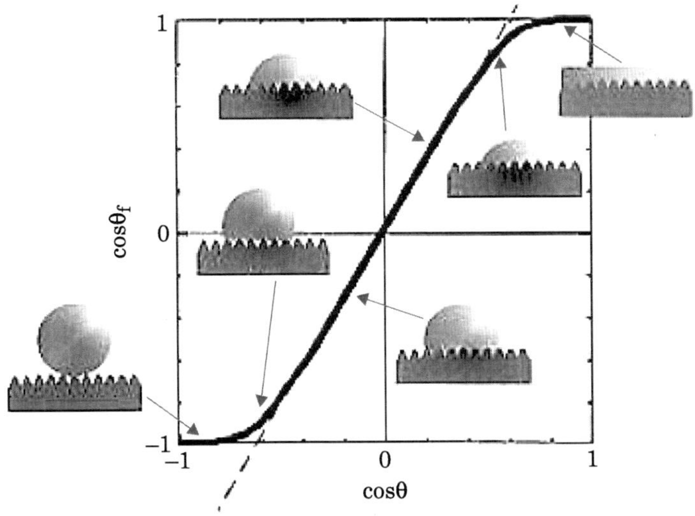

Figure 3.60 Plot of the relation between $\cos(\theta_{\text{real}})$ and $\cos(\theta_{\text{Young}})$ for patterned surfaces. Reprinted with permission from [32]. © 1996 American Chemical Society.

and, above the critical value θ_{crit}, a film forms, but the remaining droplet has a contact angle defined by

$$\cos\theta^* = 1 - f\,(1 - \cos\theta)\,. \tag{3.75}$$

This expression shows that the presence of a film improves the wetting ($\theta^* < \theta$), but it is not possible to induce a wetting transition (total wetting) by texturing a solid: equation (3.75) shows that complete wetting $\theta^* = 0$ requires $\theta = 0$.

3.3.3.7 Conclusion/Discussion

A complete diagram of wetting transitions is shown in Fig. 3.60 [31,32]. On the hydrophobic side, if the Young angle θ is such that $\theta > \theta_i$ where θ_i is defined by $\cos\theta_i = \frac{f-1}{r-f}$ (equation (3.63)), the droplet stays on the pillar tops (fakir effect) producing a super-hydrophobic situation. If $\pi/2 < \theta < \theta_i$ the droplet is in the Wenzel regime, completely in contact with the surface of the pillars, with a contact angle larger than the Young contact angle. On the hydrophilic side, if the Young contact angle θ is such that $\pi/2 > \theta > \theta_{\text{crit}}$ the droplet is in the Wenzel regime, with a real contact angle θ^* smaller than θ. For $\theta < \theta_{crit}$ where θ_{crit} is defined by equation (3.73), the liquid spreads between the pillars and leaves a droplet above the pillars with a very small contact angle.

3.4 Drops Moving by Capillarity

Macroscopic flows are usually moved by hydrodynamic forces like pressure, or mechanical forces like pistons or moving parts. The same types of forces can also be employed to actuate microfluidic flows. Peristaltic micro-pumps and syringe pumps are the microscopic versions of pumps and pistons. However, at the microscopic scale there are other forces to move fluids that are not efficient at the macroscopic scale. These forces are electro-osmosis and capillarity. In particular capillarity is widely used for actuating droplets.

In the following, we deal with capillarity as a force to displace droplets. This problem is extremely difficult. In order to simplify the approach, we do not focus on the dynamics of the droplets but on the equilibrium state. In doing so, we assume slow motion and confine the study to quasi-static displacements.

3.4.1 Drop Moving Over a Transition of Wettability

It is a well established fact that micro-drops can be put into motion by hydrophilic and hydrophobic forces. Let's suppose for instance that a water droplet is placed on a perfectly smooth horizontal plane (Fig. 3.61) at the boundary between two different chemical coatings: hydrophilic on one side and hydrophobic on the other. In such a case, we have the following scheme for the contact forces (as we don't know exactly what is the shape of the drop at the very instant it is placed on the surface, we have drawn an approximate shape, close to—but not exactly—a circle; however the reasoning will stand, whatever the shape):

From this scheme, it results that the droplet moves towards the hydrophilic surface. If L_1 and L_2 are the contact lines respectively in the hydrophilic and hydrophobic planes, and θ_1 and θ_2 the contact angles, the force acting on the drop is

$$\begin{aligned} F_x &= \int_{L_1} (\gamma_{SG} - \gamma_{\mathrm{SL}})_1 \, (\vec{i}.\vec{dl}) - \int_{L_1} (\gamma_{SG} - \gamma_{\mathrm{SL}})_2 \, (\vec{i}.\vec{dl}) \\ &= \int_{L_1} \gamma_{\mathrm{LG}} \cos\theta_1 (\vec{i}.\vec{dl}) - \int_{L_2} \gamma_{\mathrm{LG}} \cos\theta_2 (\vec{i}.\vec{dl}) < 0. \end{aligned} \tag{3.76}$$

So the resulting force is directed towards the left in the scheme of Fig. 3.26, and the drop moves to the left. The motion stops when the resultant of the contact forces is zero, i.e., when the drop is entirely on the hydrophilic region, as shown in Fig. 3.62.

Experimental evidence confirms the preceding analysis. In Fig. 3.63, a drop is deposited with a micropipette on a flat horizontal surface, at the boundary of two regions with different contact angles

The preceding analysis is numerically confirmed by a calculation with the Evolver code (Fig. 3.64). We can start with any volume of water spread over a

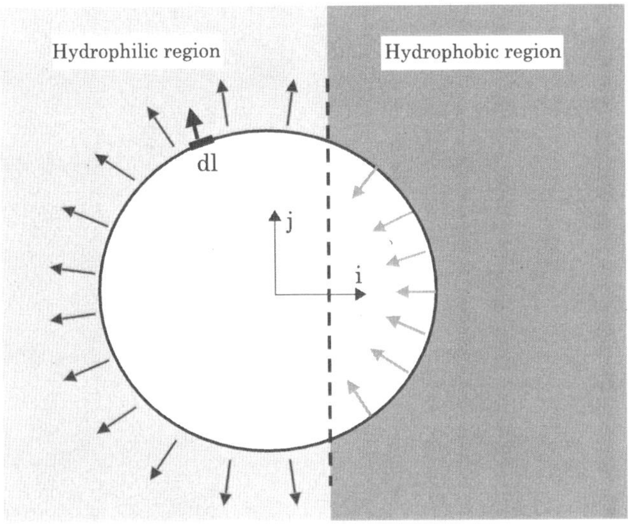

Figure 3.61 Schematic view of a water droplet standing above a hydrophilic/hydrophobic contact. There is a resulting force directed towards the hydrophilic region.

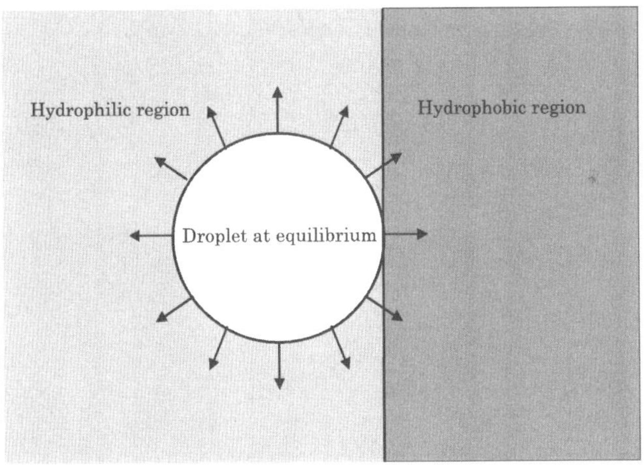

Figure 3.62 The water droplet is at equilibrium when it is entirely located on the hydrophilic region.

hydrophilic/hydrophobic boundary. After a few iterations, the drop is formed but it is not at equilibrium because of the global force directed towards the hydrophilic region. The drop evolves to find its equilibrium location which is located just at the boundary of the transition line, on the hydrophilic side.

Note that the direction of the motion of a water drop is towards the hydrophilic region, whereas the motion of an oil droplet would be towards the hydrophobic region in the same geometrical conditions. Note also that the drop behavior would be similar if the transition between the two regions were

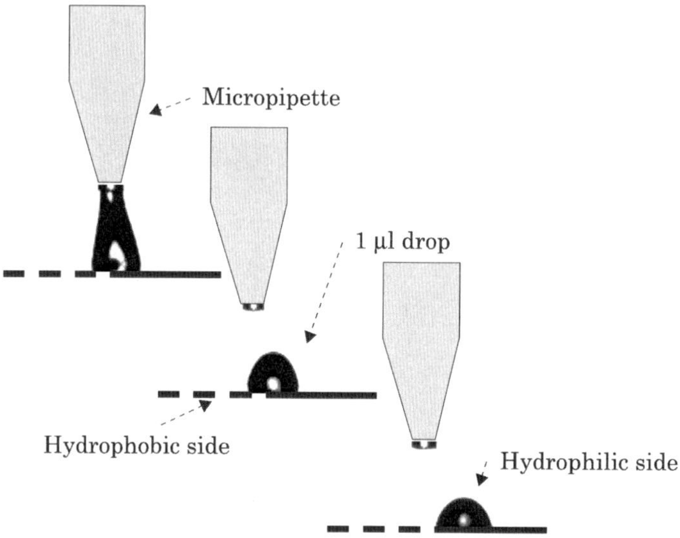

Figure 3.63 Experimental view of the relocation of a micro-drop (1 µl) deposited on a hydrophilic/hydrophobic boundary.

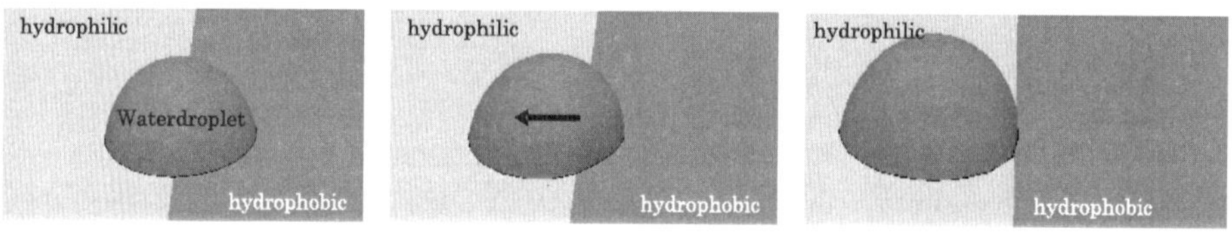

Figure 3.64 Motion of a drop towards the hydrophilic plane (simulation with Surface Evolver). The micro-drop is initially deposited over the hydrophilic/hydrophobic transition line and it is not an equilibrium state. Drop moves to find an equilibrium state on the hydrophilic plate, just at the boundary between the two regions.

not abrupt, but smooth, and if there were a wettability gradient between the two regions [33]. In the next section, we investigate the effect of a gradient of wettability.

3.4.2 Drop Moving Uphill

Capillary forces may be sufficient to make micro-drops move upwards on an inclined plane. Chaudhury et al. [34] have shown that a droplet can go up a slightly inclined plate presenting a wettability gradient (Fig. 3.65). They have shown that a surface having a spatial gradient in its surface free energy was capable of causing drops of water placed on it to move uphill. This motion was the result of an imbalance in the forces due to surface tension acting on the liquid–solid contact line on the two opposite sides ("uphill" or "downhill")

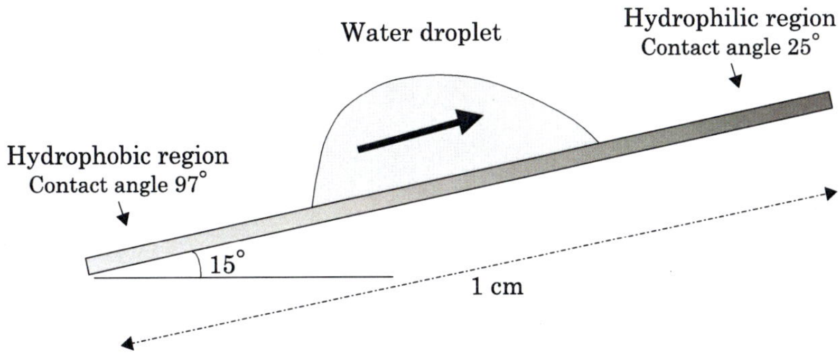

Figure 3.65 An oil drop may run uphill towards the more hydrophobic region.

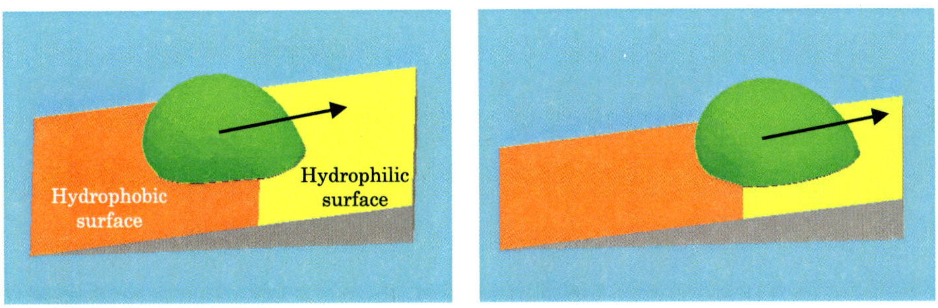

Figure 3.66 Droplet moving up a plane at a wettability transition (simulation made with Surface Evolver).

of the drop. The required gradient in surface free energy was generated on the surface of a polished silicon water by exposing it to the diffusing front of a vapor of decyltrichlorosilane, Cl3Si(CH2)9CH3. The resulting surface displayed a gradient of hydrophobicity (with the contact angle of water changing from 97 degrees to 25 degrees) over a distance of 1 centimeter. When the wafer was tilted from the horizontal plane by 15-degrees, with the hydrophobic end lower than the hydrophilic, and a drop of water (1 to 2 microliters) was placed at the hydrophobic end, the drop moved toward the hydrophilic end with an average velocity of approximately 1 to 2 millimeters per second.

A similar motion can be obtained at a wettability transition on a tilted plane. If a droplet is initially deposited over a wettability transition, it moves uphill from the lyophobic region (hydrophobic for a water droplet) towards the lyophilic region (hydrophilic for a water droplet). Fig. 3.66 shows that the droplet is not stable when located on the transition line and moves uphill. The slope angle a micro-drop can move uphill depends on the size of the droplet (in case gravity intervenes) and the contact angles on both sides of the transition line.

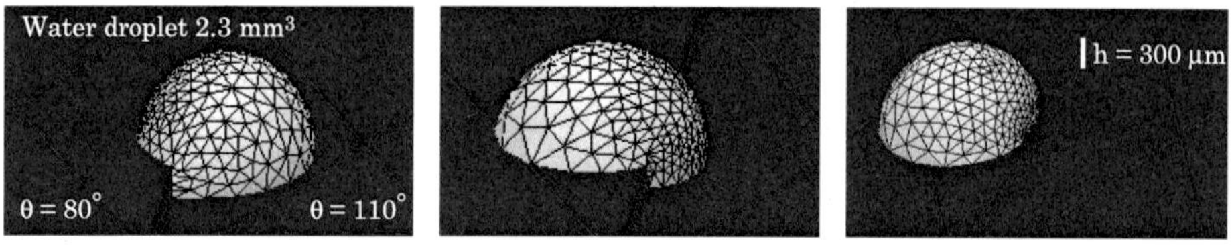

Figure 3.67 Motion of a drop up a step towards the hydrophilic plane (simulation made with Surface Evolver). Left: the droplet is located on the lower surface, with a part of it overlapping on the upper plane; middle: the droplet is not in an equilibrium state, and moves up the step pulled by hydrophobic forces of the upper plate; right: motion continues until an equilibrium state is reached when drop is entirely on the upper plate.

3.4.3 Drop Moving up a Step

Another interesting demonstration of the power of capillary forces may be seen by making a drop move up a step. In such a case, a micro-drop of water is initially located on a step at the boundary of a hydrophilic region (on top of the step) and a hydrophobic region (at the base of the step). The calculation with the Evolver code shows that the drop progressively moves towards the hydrophilic region, even if this region is located at a higher level (Fig. 3.67). Capillary forces dominate gravity in this example.

In biotechnology, wells, cusps and grooves are used to confine micro-drops. This confinement must be efficient. The preceding analysis has shown that attention should be given to the wettability of surfaces and that the combination of a non-wetting bottom of the well/groove and a wetting upper surface can result in poor confinement.

3.4.4 Drop Moving Over a Gradient of Surface Concentration of Surfactant

In the preceding sections, the capillary forces inducing droplet motion were linked to a local change of contact angle of the substrate. In this section, we show how chemical reactions between the liquid of the droplet and the substrate can create droplet motion. To this extent the experiment of Dos Santos et al. [35] is characteristic. Suppose that a droplet of n-alkanes containing silane molecules is placed on a hydrophilic surface. It is well known that silane molecules form dense grafted monolayers on silicon or glass, rendering the surface hydrophobic. If we deposit such a droplet on a glass surface and initiate a motion by pushing it with a pipette, then the droplet continues to move on the substrate. It moves in nearly linear segments and changes its direction each time it encounters a hydrophobic barrier (Fig. 3.68). The droplet cannot cross its own tracks.

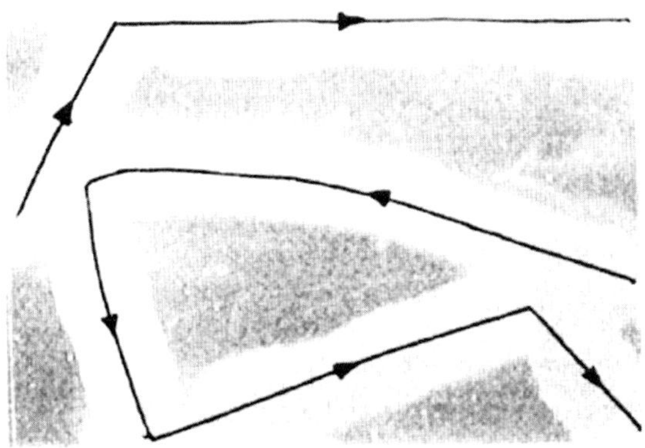

Figure 3.68 Free running n-alkane droplet on a horizontal hydrophilic substrate. From [35]. © 1995 APS; reprinted with permission.

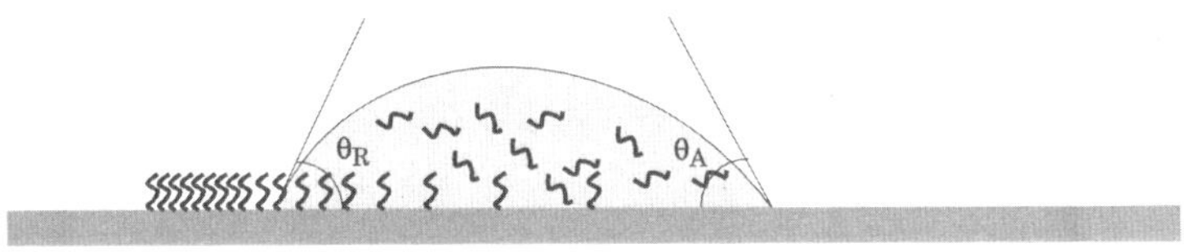

Figure 3.69 Advancing and receding Young contact angles are not identical. The hydrophobic monolayer changes the receding Young contact angle.

The motion is governed by an imbalance of capillarity: the advancing contact line has a hydrophilic Young contact angle. On the other end, molecules of silane concentrate at the vicinity of the receding contact line and form a hydrophobic layer. The Young angle at the receding line is then a hydrophobic contact angle (Fig. 3.69). As a result, the droplet moves in a straight line, except if it encounters surface defects or an existing hydrophobic trail.

3.4.5 Conclusion

We have seen that capillary forces are often powerful enough to induce motion of micro-drops. However, we have considered perfect, smooth surfaces, and have neglected the effect of hysteresis—local change of contact angle—and its extreme form which is droplet pinning—blockage of a droplet due to defects of the surface. In reality, a drop does not move as soon as there is a gradient of wettability. It moves as soon as the gradient of wettability is sufficient for the capillary forces to dominate opposite reaction forces like gravity, hysteresis, and pinning. In the following section, we concentrate on the forces preventing droplet motion linked to the state of the surface.

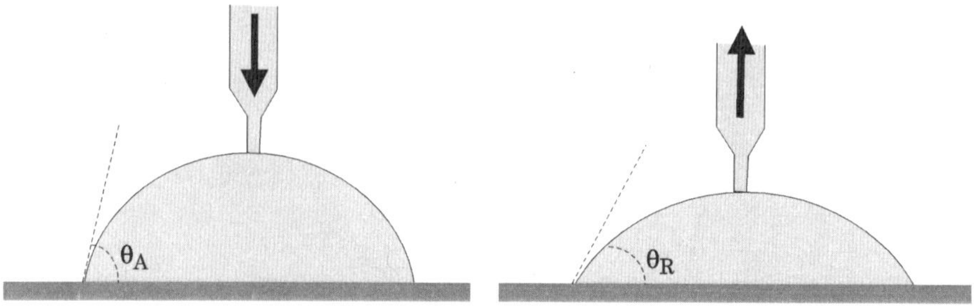

Figure 3.70 Contact angle is not the same when the droplet grows or recedes due to hysterisis of the triple line.

3.5 Contact Angle Hysteresis

The concept of hysteresis for triple contact lines is difficult to define with accuracy. Indeed, it is not clear how to set the boundaries for such a concept. Let us start with a droplet sitting on a perfectly horizontal, clean, smooth plane. In such a case, as we have seen in Chapter 2, the contact angle of the liquid is defined by Young's law

$$\cos\theta = \frac{\gamma_{SG} - \gamma_{\mathrm{SL}}}{\gamma}.$$

However, no surface is ideally perfect; microscopic defects—morphological as well as chemical—are usually present. We have also seen in the preceding section that the Young contact angle is modified if the surface is rough (Wenzel's law) or chemically heterogeneous (Cassie's law). Some authors have associated the change of contact angle according to Wenzel's or Cassie's law to hysteresis [36], however, the concept of hysteresis is preferentially associated to *local* defects [1]. These *local* defects can be isolated point defects, surface defects, or line defects. A point defect can be an isolated roughness due to adherence of a microparticle; a surface defect can be a stain of a chemical product; a line defect can be a long chain of polymer adhering to the surface.

Suppose we have a plane surface with dispersed local point defects. Let us deposit a droplet of water on the surface with a pipette in a first phase, and in a second phase, remove the water with the pipette (Fig. 3.70).

We observe that the contact angle is not the same during the two phases. If θ_A denotes the contact angle during the first—advancing—phase and θ_R the contact angle during the second—receding—phase, we have

$$\theta_A \geq \theta_{\mathrm{Young}} \geq \theta_R. \tag{3.77}$$

Fig. 3.71 shows typical evolution of the advancing and receding contact angles corresponding to the experiment of Fig. 3.70.

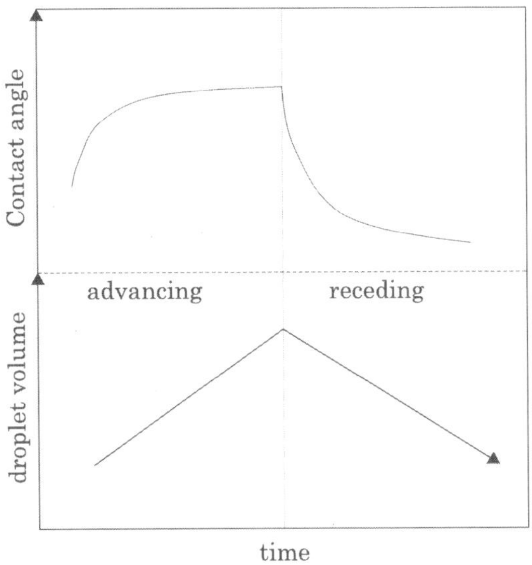

Figure 3.71 Typical time evolution of the contact angle during the growing and shrinking phase of a droplet on a flat substrate according to Wege et al. [37], Lam et al. [38] and Tavana et al. [39].

Intuitively, the advancing front has a larger contact angle because it is locally "slowed down" by inhomogeneities on the surface, whereas the receding front has a smaller contact angle because it is locally "pulled back" by these same inhomogeneities. Note that the same type of behavior is observed during the electrowetting process, with a similar method. More details will be given in Chapter 5.

Values of the change of contact angle are very difficult to predict, because they depend on many parameters. This situation is very different from the case of a plug flow where the Hoffmann–Tanner law links the advancing and receding contact angles to the plug velocity.

Cubaud and Fermigier [40] proposed a very instructive approach by showing that, in the presence of surface defects, the contact line can be separated between a three-phase—far from the defect or heterogeneity—and a four-phase contact line at the contact of the defect. An image of local hysteresis at extremely low speed of motion is shown in Fig. 3.72.

The contact line Λ is the union of the set of the triple contact lines Λ_3 and of the set of four-phase lines Λ_4. This can be analytically described by

$$\Lambda = g(t)\Lambda_3 + (1 - g(t))\Lambda_4 \tag{3.78}$$

where $g(t)$ is a function of time and surface geometry. For a quasi-static motion, with a small capillary number $Ca = \frac{\mu U}{\gamma} \ll 1$, the contact angle varies

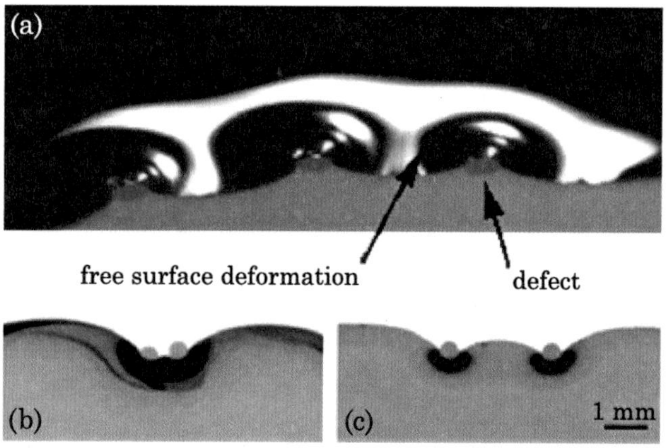

Figure 3.72 Image of an interface slowly moving over defects. Far from the defect, the contact angle is equal to the Young contact angle θ_3, close to the defect the interface is twisted, and the contact angle varies between θ_3 and that of the defect θ_4.
From [40]. © 2004 Elsevier; reprinted with permission.

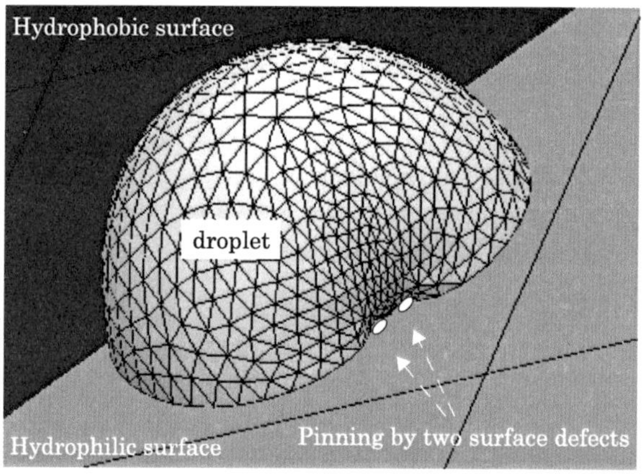

Figure 3.73 Evolver simulation of a water droplet encountering defects during its motion from a hydrophobic surface towards a hydrophilic surface. This case similar to that of Fig. 3.72.

between the Young contact angle θ_3 of the triple line and the contact angle θ_4 corresponding to the defect. Fig. 3.72 shows an enlarged view of the interface when it advances past defects. The deformation of the interface and the modified contact angle can be seen in black in the figure.

This situation can be simulated by the Surface Evolver numerical software. These defects can be geometrical as well as chemical. In the model, the defect is a local change of value of the wettability. As in the approach of Cubaud et al.,

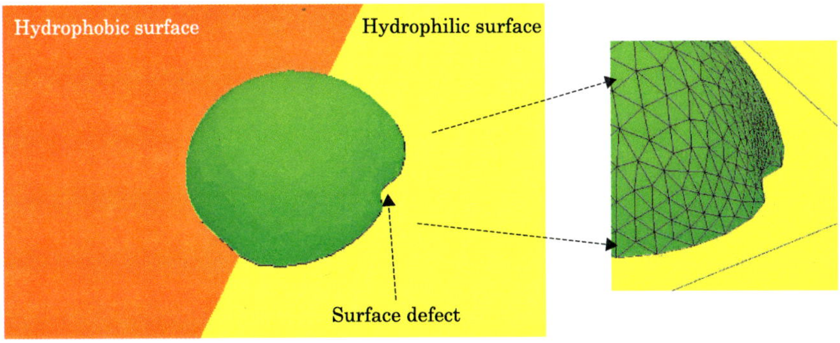

Figure 3.74 Pinning of a drop moving from a hydrophobic area towards a hydrophilic surface due to a defect of the surface.

the contact angle is locally modified by the defect (Fig. 3.73), resulting in a loss of capillary force.

If we imagine now that there are many defects, we can easily understand that the global contact angle is an average between θ_3 and θ_4, resulting in a larger advancing contact angle because θ_4 is larger than θ_3.

In the next section, we investigate droplet pinning, which is the extreme form of hysteresis where the droplet is totally blocked in its motion by defects on the solid surface.

3.6 Droplet Pinning

Young's law determines the contact angle at a triple line between a liquid, a fluid, and a solid. It is important to keep in mind that Young's law is valid only if there are no other forces applied on the triple line. In digital microfluidics, there are cases where Young's law is violated: this is the case in hysteresis, as we have seen in the previous section, and this is also the case in droplet "pinning". In the following we show that droplets can be pinned by "point like" defects or by wettability boundary lines.

3.6.1 Droplet Pinning on a Surface Defect

Local chemical and/or geometrical defects locally modify the contact angle. If the defect is sufficiently important, or if there are sufficient numbers of defects, the droplet cannot move even if capillary—or electro-capillary—forces are applied on it. This phenomenon is called pinning. We show in Fig. 3.74 a numerical simulation of pinning of a droplet during its motion from a hydrophobic to a hydrophilic substrate.

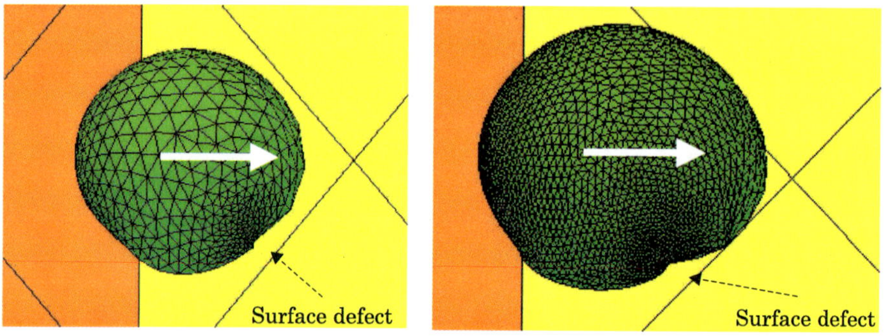

Figure 3.75 Depending on its location, a surface defect does not always prevent the water droplet from moving to the hydrophilic surface.

Note that an important defect does not always pin the droplet. Depending on the relative location of the defect, the droplet may be pinned or just deviated in its motion, as is shown in Fig. 3.75.

3.6.2 Droplet Pinning at a Wettability Separation Line

Pinning may also occur at a transition line between two surfaces with different chemical coatings, inducing a sharp transition of wettability [41]. We have already seen this phenomenon in section 3.2.9. This effect is also known as "canthotaxis", and is due to the fact that, when the contact line reaches the separation line, we have a four-phase contact line. An analysis of the boundary conditions shows that there is equilibrium as long as the contact angle is comprised between the Young angles on both side θ_1 and θ_2 as shown in Fig. 3.76 [41,42]

$$\theta_1 \leq \theta \leq \theta_2. \tag{3.79}$$

Equation (3.79) can be written in terms of surface energy

$$\arccos\left(\frac{\gamma_{S1G} - \gamma_{S1L}}{\gamma}\right) \leq \theta \leq \arccos\left(\frac{\gamma_{S2G} - \gamma_{S2L}}{\gamma}\right) \tag{3.80}$$

where the indices $S1$ and $S2$ denote the left and right solid surfaces. If the external constraint is such that θ continues to increase, the triple line is suddenly de-pinned and the liquid is released and invades the lyophobic surface. On the right surface, the contact angle is the Young contact angle θ_2.

An interesting investigation of the pinning along a sinusoidal line has been done by Ondarçuhu [41] using the theory of elasticity of the interface by de Gennes [1]. This approach is presented in Chapter 5 for the pinning of an interface on a jagged electrode boundary.

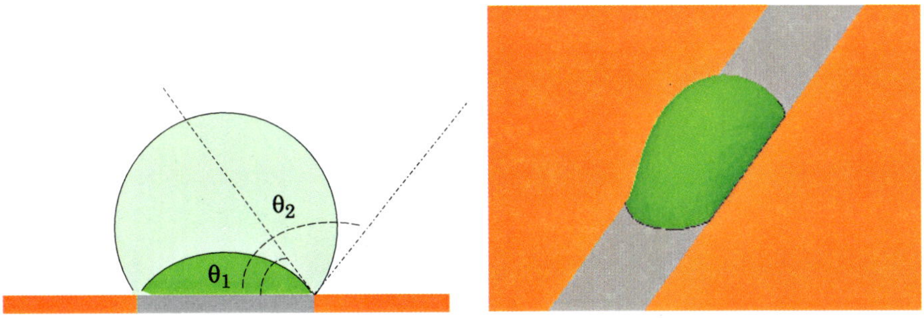

Figure 3.76 Quadruple contact line pinning on a wettability boundary. Left: vertical cross section of the droplet with the limiting contact angles θ_1 and θ_2. Right: an Evolver simulation of the droplet.

3.6.3 Droplet Pinning on an Edge

Pinning—or attachment—can also occur on an edge [43]. Suppose there is a straight edge, and the interface coming to contact this edge (Fig. 3.77). The contact line on the angle stays fixed as long as the contact angle is not forced over the limit $\alpha+\theta$, where α is the angle between the two planes. The condition for pinning is then

$$\theta \leq \phi \leq \alpha + \theta \tag{3.81}$$

where ϕ is the contact angle, and θ the Young contact angle with the solid surfaces. In the case where the two planes have a different chemical surface, the Young contact angles can be denoted θ_1 and θ_2, and the condition (3.81) becomes

$$\theta_1 \leq \phi \leq \alpha + \theta_2.$$

We have already discussed the possible morphologies of liquid in a microgroove. Let us remark here that the interface of the liquid can stay pinned to the rim of the groove (Fig. 3.78). Pinning occurs when, in a cross section, the interface does not bulge above the limit imposed by the Young contact angle of the top plane (limit 1 in Fig. 3.78) or decreases below the wedge limit (limit 2 in Fig. 3.78).

3.7 The Effect of Surfactants

3.7.1 Introduction

As we mentioned in Chapter 2, surfactants like Triton or Tween are often added to biological liquids to prevent the colloidal suspension contained in the

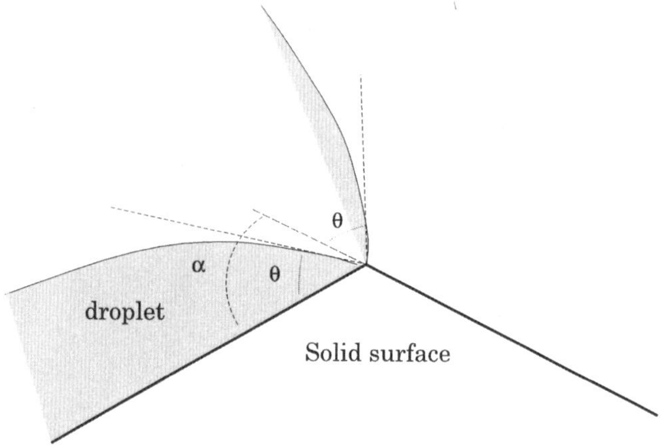

Figure 3.77 Droplet pinning on an edge: the droplet is pinned as long as the contact angle varies between the natural Young contact angle θ to the value $\theta + \alpha$. Above this value, the interface moves over the right plane and the droplet is released.

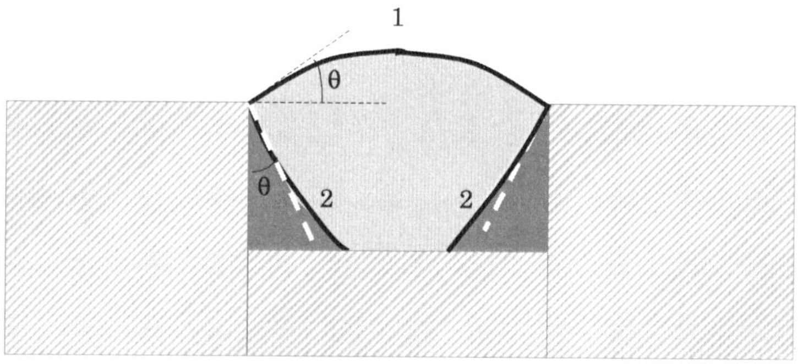

Figure 3.78 Pinning of liquid in a groove: the interface stays attached to the rim as long the volume of liquid is such that it is comprised between the two limits 1 and 2 defined by the continuous black lines.

droplet from sticking to the solid surfaces and aggregating. This adding of surfactants has the consequence of reducing the surface tension of the liquid with the air γ and also with the solid surface γ_{SL}. According to Young's law

$$\gamma \cos\theta = \gamma_{SG} - \gamma_{\mathrm{SL}}.$$

At the same time the surface tension γ and the contact angle θ are changed by the addition of surfactants. On one hand, the surface tension γ is reduced by the migration of the surfactants to the interface which results in a reduction of the interfacial energy. On the other hand, the contact angle θ is reduced indirectly because of the deposition of surfactants on the solid surface changing the surface tension between the solid and the liquid γ_{SL}. Fig. 3.79 results from

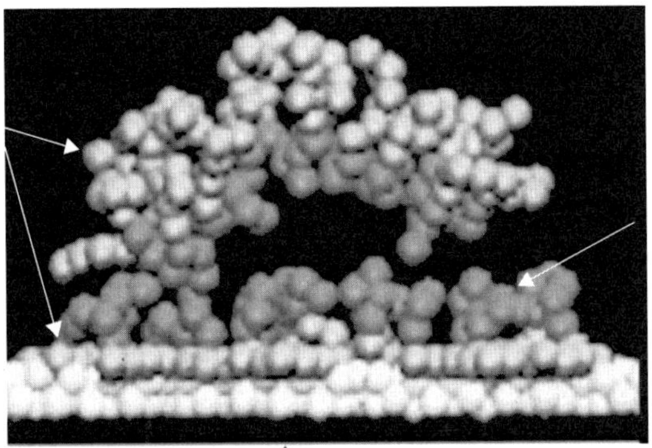

Figure 3.79 Molecular dynamics simulation of a water droplet showing the surfactant molecules at the interface and on the substrate. Reprinted with permission from [44]. © 2005 American Chemical Society.

a molecular dynamics simulation [44] and shows the two preferential locations of the surfactants.

Some surfactants reduce the contact angle by 10 or 20 degrees, but are not sufficient to obtain a complete wetting of the surface. Some others result in a total wetting of the liquid. The following section illustrates the effect of two types of surfactants: type 1 is linear chain molecules and type 2 is T-shape molecules.

3.7.2 From Partial Wetting to Total Wetting

All surfactants are not equally effective in changing droplet wettability. The morphology of the surfactant molecule at the nanoscopic scale is determining. Shen et al. [44] have simulated the effect of flexible linear chain surfactants—like alkyl chains, and rigid T-shape surfactants—like trisiloxane (Fig. 3.80).

T-shape surfactants work better than flexible linear surfactants because they facilitate the formation of a liquid film. As shown in Fig. 3.80, at the bottom left corner, T-shape molecules rearrange between the liquid interface and the solid interface to form a "sandwich" laminating a water film.

3.8 Marangoni Convection

3.8.1 Introduction

Depending on the nature of the liquid, sitting droplets can be at rest with no internal convective motion, or can be agitated by internal motion, even if

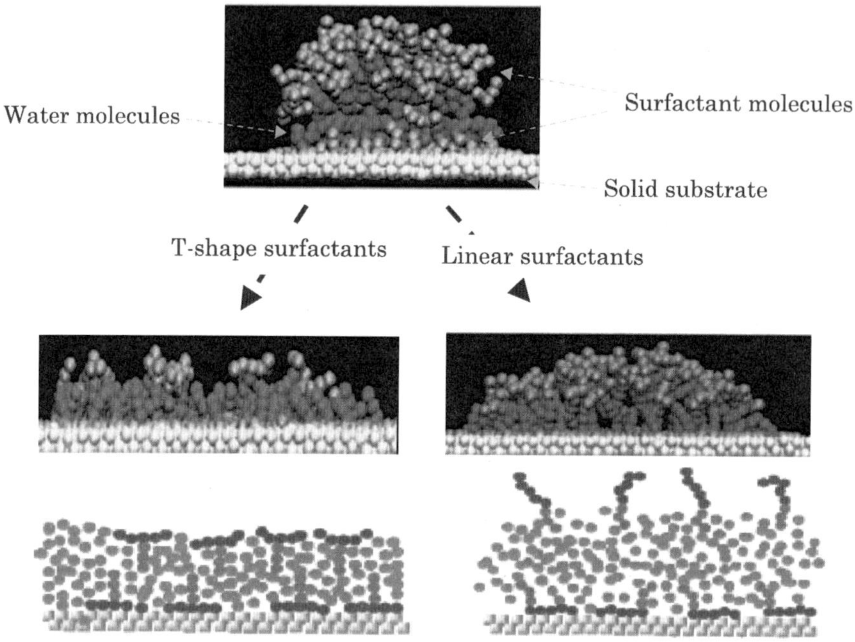

Figure 3.80 MD numerical simulations of a droplet spreading on a substrate under the action of T shape surfactants and flexible linear chain surfactants. Reprinted with permission from [44]. © 2005 American Chemical Society.

the external shape remains unchanged. High viscosity liquids and ionic liquids behave similarly to gels (see Chapter 7); they seldom show internal convection, at least when they are not moving on the substrate, whereas water droplets are often agitated by convective motions. There are two reasons for these motions: Marangoni convection and evaporation (the two phenomena might be linked, as we will see in Section 3.9.4).

In Chapter 2, we have seen that Marangoni convection is associated to a gradient of surface tension. This gradient of surface tension may be caused by temperature, concentration of surfactants, and even by an electric field. The general formulation proposed by Colin [45] uniting these three possible sources of surface tension gradient is

$$d\gamma = \frac{\partial \gamma}{\partial T} dT + \frac{\partial \gamma}{\partial c} dc + \frac{\partial \gamma}{\partial E} dE. \tag{3.82}$$

These three types of convective motions can be found in microfluidics systems.

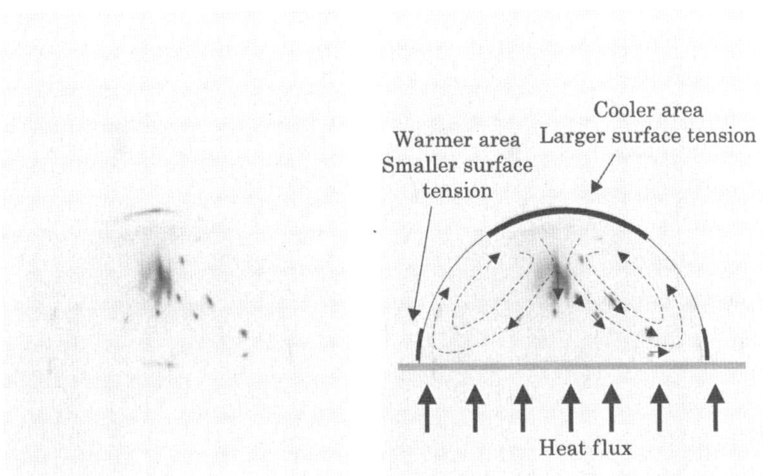

Figure 3.81 Observation of Marangoni convection in a sessile droplet obtained by heating an ionic liquid droplet from below (photo Ph. Dubois, CEA/LETI).

3.8.2 Marangoni Convection Due to Thermal Non-uniformity

In Chapter 2, we have seen that surface tension depends on temperature. To the first order, this dependency is

$$\gamma = \gamma_0 (1 + \alpha T) \tag{3.83}$$

where $\alpha = -\frac{1}{T_C}$ and T_C is the critical temperature in kelvin. The coefficient α being negative, the value of the surface tension decreases with temperature.

Very often the temperature of the solid substrate is not the same as that of the surrounding air. If the substrate is warmer than the air—which is the case, for example, during heating phases of PCR reactions—the liquid droplet sitting on the substrate will experience a warmer region near the substrate and a cooler region near the apex of the droplet. Thus the surface tension will be lower near the substrate and larger at the apex, resulting in a tangential force on the interface. If this tangential force is sufficient, it induces a circulation inside the droplet, as shown in Fig. 3.81. Thus, even if the external shape of the droplet does not change, there can be internal motions induced by a Marangoni effect. The internal motion is particularly strong near the interface.

Marangoni convection occurs if the variation of the surface tension force dominates the viscosity forces. A dimensionless number—the Marangoni number—determines the strength of the convective motion [46]

$$Ma = \frac{\Delta \gamma R}{\rho \nu \alpha} \tag{3.84}$$

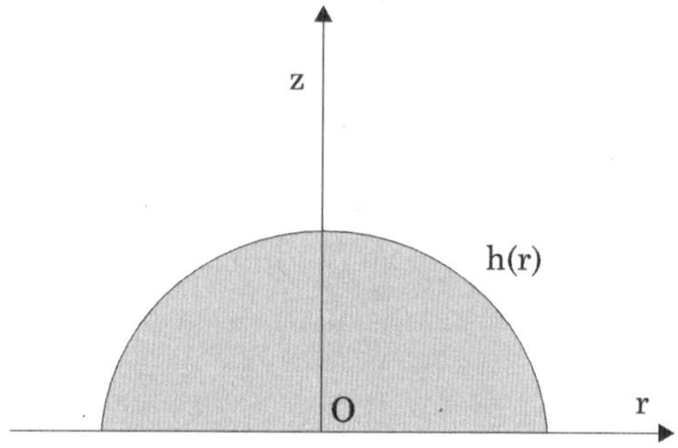

Figure 3.82 A sessile spherical cap droplet on the substrate in an axisymmetrical coordinate system.

where R is the radius of the spherical cap, r the density of the liquid, ν the kinematic viscosity, α the thermal diffusivity, and Δ_γ the variation of surface tension on the interface. The Marangoni number represents the ratio between the tangential stress and the viscosity.

Thermal Marangoni convection in a droplet is numerically modeled with standard fluid mechanics approaches. First, because of the small dimensions of the droplet, the internal velocities are moderate (1 cm/s is an upper limit) and the capillary number defined by

$$Ca = \frac{\mu \bar{u}}{\gamma} \tag{3.85}$$

is then small, much smaller than 1. In such a case, the shape of the droplet is not changed by the internal convective motions. Because the Bond number is also small, the droplet stays spherical all the time (Fig. 3.82). Then we can use an axisymmetric coordinate system to describe the geometry.

Suppose we have a steady-state regime. Neglecting the inertial terms, the equations governing this steady-state convective motion are the Stokes equations [47]

$$\begin{aligned}
\frac{1}{r}\frac{\partial (r u_r)}{\partial r} + \frac{\partial u_z}{\partial z} &= 0 \\
\mu \left[\frac{\partial}{\partial r} \left(\frac{1}{r} \frac{\partial (r u_r)}{\partial r} \right) + \frac{\partial^2 u_r}{\partial z^2} \right] &= \frac{\partial P}{\partial r} \\
\mu \left[\frac{\partial}{\partial r} \left(\frac{1}{r} \frac{\partial (r u_z)}{\partial r} \right) + \frac{\partial^2 u_z}{\partial z^2} \right] &= \frac{\partial P}{\partial z}
\end{aligned} \tag{3.86}$$

and the temperature field is determined by the energy equation

$$\rho C_p \vec{u}.\nabla T + k \Delta T = 0. \tag{3.87}$$

At the substrate boundary, the velocity is zero $u_r = u_z = 0$. On the interface we have the tangential stress condition

$$\tau|_{z=h} = \frac{d\gamma}{ds} = -\alpha\gamma_0 \frac{\partial T}{\partial s} \tag{3.88}$$

where s is the curvilinear coordinate of the interface.

3.8.3 Marangoni Convection Due to Concentration Gradient

Surface concentration of macro-molecules modifies locally the value of the surface tension. This is particularly true for surfactants. If the concentration reaches the CMC (critical micelle concentration), the surface is uniformly saturated and there are no possible surface gradients. However, below the CMC threshold, the surface is not saturated; if the surface concentration in surfactants is not homogeneous, the surface tension is not uniform. The dependency of the surface tension on the concentration is given to the first order by

$$\gamma = \gamma_0 (1 + \beta C) \tag{3.89}$$

inducing the tangential stress

$$\tau = \frac{d\gamma}{ds} = \beta\gamma_0 \frac{\partial C}{\partial s}.$$

A Marangoni effect may appear, assuming the Marangoni number defined by $Ma = \frac{\Delta\gamma R}{\rho\nu D}$ is large enough (D is the coefficient of diffusion). A schematic view of the Marangoni convection caused by surface concentration is shown in Fig. 3.83.

Unlike thermal convection, the concentration Marangoni convection is usually a transient phenomenon. As soon as the surface concentration becomes uniform, the motion stops.

An experimental observation of the concentration Marangoni convection has been made by Raccurt et al. [48,49]. The authors used an electrowetting microdevice. A droplet with surfactants is initially at rest on an electrode. If suddenly the interface is increased by the actuation of a neighboring electrode, the newly created surface has a very low concentration in surfactants. A gradient of surface tension then appears and a Marangoni motion progressively equalizes the surface tension (Fig. 3.84). The proof of the relaxing of the surface tension is the relaxing of the curvature of the interface.

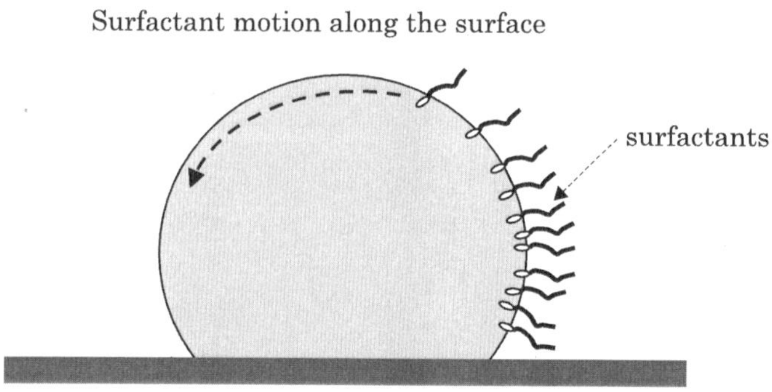

Figure 3.83 Schematic view of surfactant molecules moving along the surface from the regions of large surface concentration to the regions of smaller surface concentration.

Figure 3.84 Internal motion of the fluid during extrusion from a reservoir in an EWOD microsystem. First an electrowetting motion transports very quickly a part of the droplet on the neighboring electrode. Then a Marangoni motion starts and progressively uniformizes the surface tension. The motion is followed using fluorescent beads of 9 µm diameter. (Photo O. Raccurt).

3.8.4 Marangoni Convection Due to an Electric Field

The formula of Colin (3.82) introduces a variation of the surface tension as a function of the electric field, resulting in an electric Marangoni convection. This type of convection is currently the object of debate, and its effect on droplets is a subject of research. Strictly speaking, when an electric field is applied on an interface, it induces stresses on the interface. In a droplet, it has been observed that the tangential stress may be important and induces Marangoni type convection. Fig. 3.85 shows the motion of fluorophores at a droplet interface when an electric field is actuated by an electrode located in the solid substrate below the droplet [49].

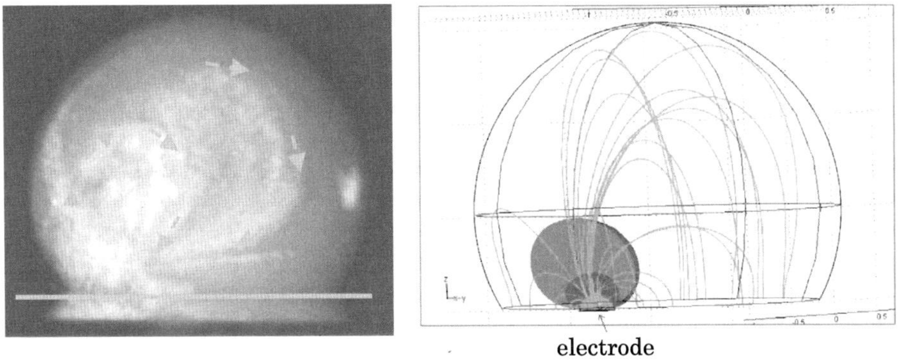

Figure 3.85 Left: convective motion induced by an electric field applied in the droplet; right: simulation of the electric field lines and iso-potentials in the droplet (numerical program COMSOL).

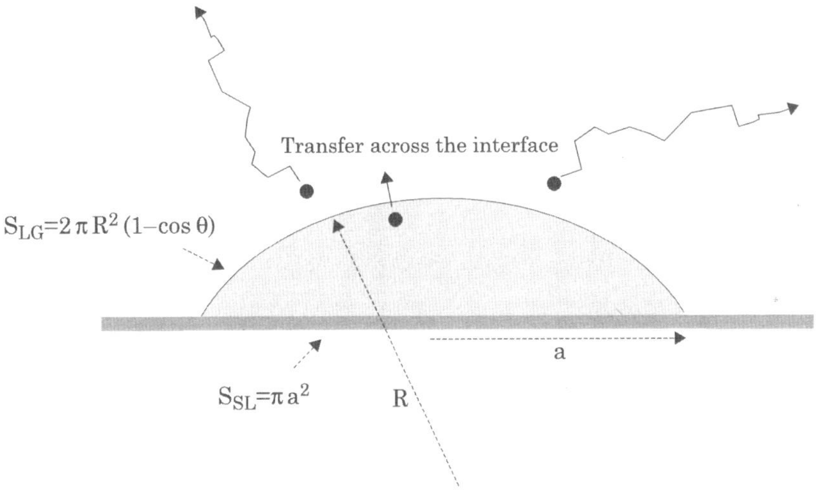

Figure 3.86 The two mechanisms during evaporation (assuming that there are no convective motions) evaporation is limited by diffusion in the gas phase rather than by the transfer rate across the interface.

3.9 Evaporation

A drawback when working with aqueous droplets—or other volatile liquids—in a non-regulated atmosphere is evaporation. Since the volume of liquid in a droplet is very small, evaporation time is also small, and it is usual to see droplets rapidly vanishing. Of course, evaporation can be suppressed by controlling the ambient atmosphere; if the vapor content is set to maintain the atmosphere at dew (saturation) point, evaporation is suppressed. Note that the problem does not appear when using some organic liquids (silicone oil, etc.) or ionic liquids.

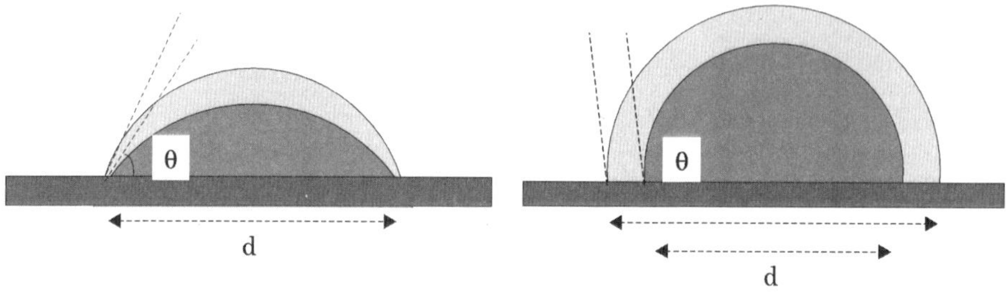

Figure 3.87 The two schemes for sessile droplet evaporation [50] depending on the wettability of the surface.

One would expect that the evaporation process of liquid droplets was described many years ago. Indeed, Wittaker, Morse, and Langmuir have set the basis for a theory in the 1910's; however, it is just recently that a more complete understanding of the evaporation process of micro-drops has been announced.

3.9.1 Evaporation of Sessile Droplets

In this section we investigate evaporation of sessile droplets. We assume that there is no Marangoni type convection inside the droplet and that the surrounding gas is at rest. Liquid evaporation is strongly associated with the diffusion of the vapor away from the liquid/gas interface (Fig. 3.86). If the gas is saturated with vapor (dew point for water), evaporation is stopped.

3.9.1.1 Experimental Observations

It has been observed experimentally [50] that wetting and non-wetting droplets do not evaporate in the same way: in the case of a wetting droplet the contact radius remains constant during the evaporation process—except at the very end, when the volume of liquid is vanishing.;uring the process, the contact angle decreases gradually. It is as if the contact line was pinned on the initial contact line. In the case of a non-wetting droplet, it is the contact angle that remains constant—except at the very end of the evaporation process (Figs. 3.87–3.89); during the process the contact radius decreases gradually.

In both cases, the mass rate of liquid loss is proportional to the height of the drop [50–53]; in the case of a wetting substrate the contact angle decreases linearly with time, whereas in the case of a non-wetting substrate the droplet radius decreases as the square root of time [54]. In the following section, we derive expressions for the contact angle change with time in the wetting case, and for the contact radius change with time in the non-wetting case.

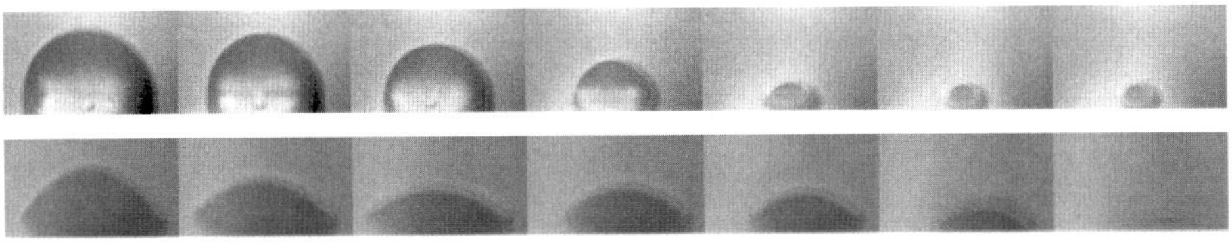

Figure 3.88 Evaporation of a water droplet on silanized silicon (top) and on silicon oxide (bottom): recall that the silanization process makes the surface hydrophobic, whereas the silicon oxide is hydrophilic. In the first case (top) the contact angle remains constant; in the second case (bottom) the contact radius remains constant.

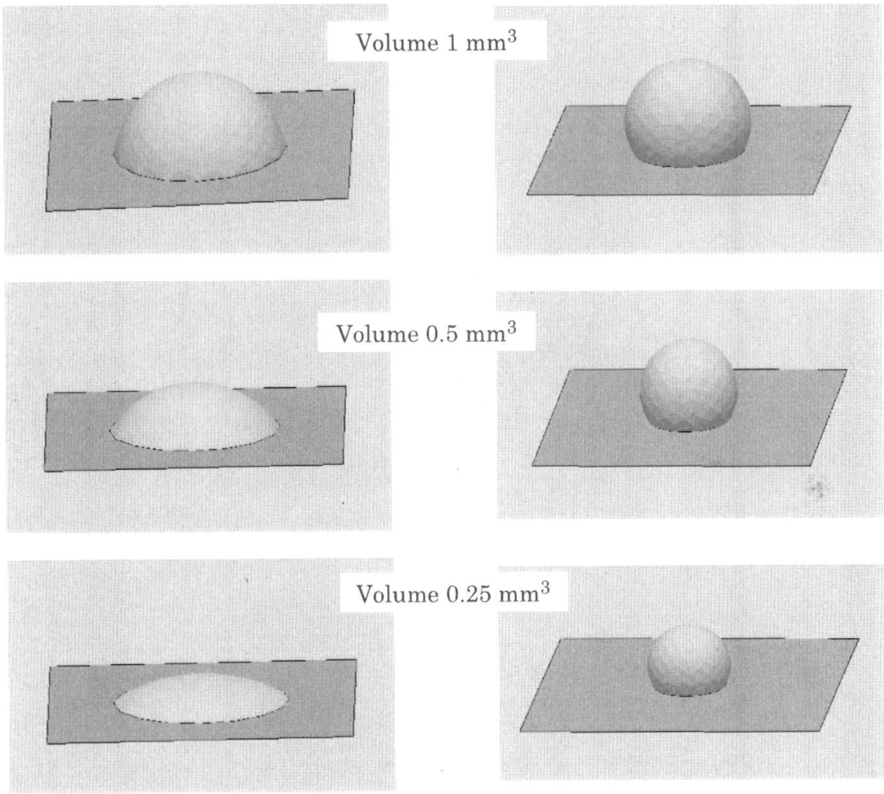

Figure 3.89 Evaporation of a water droplet on hydrophilic and hydrophobic surfaces using the Evolver numerical program: in the case of a hydrophilic contact, the triple contact line has been pinned and the base radius is constant; in the case of a hydrophobic contact, the contact angle is constant.

3.9.1.2 Theoretical Model

In the case of a complete sphere, i.e., a spherical droplet in space, the approach was done by Birdi et al. [55]; using Fick's law, they obtained the evaporation

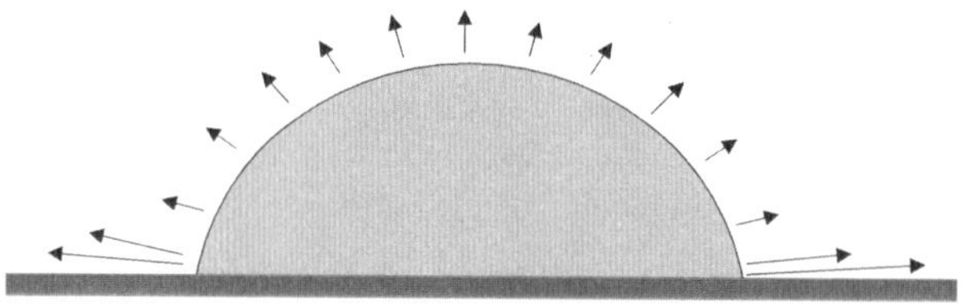

Figure 3.90 Evaporation flux along the surface droplet according to Hu [62].

rate as

$$\frac{dm}{dt} = -D \int \nabla c . d\vec{S} \tag{3.90}$$

where D is the diffusion coefficient of the vapor and c its concentration. Equation (3.90) can be developed under the form

$$\frac{dm}{dt} = -D \int \frac{\partial c}{\partial n} dS. \tag{3.91}$$

Note that, due to the spherical isotropy, the gradient of vapor concentration is uniform and using the boundary conditions: $c = c_\infty$ as $r \to \infty$, and $c = c_0$ at $r = r_d$ where r_d is the radius of the liquid drop, and taking into account that the diffusion profile around a sphere varies as $1/r$, the concentration gradient can be approximated by [55]

$$\frac{\partial c}{\partial n} = -\frac{c_0 - c_\infty}{r_d}. \tag{3.92}$$

The evaporation rate then becomes

$$\frac{dm}{dt} = 4\pi r_d D \left(c_0 - c_\infty\right) \tag{3.93}$$

showing that the rate of evaporation is proportional to the radius of the spherical drop. In the case of spherical cap geometry, equation (3.90) still applies

$$\frac{dm}{dt} = \rho \frac{dV}{dt} = -D \int \nabla c . dS = -D \int \frac{\partial c}{\partial n} dS \tag{3.94}$$

where ρ is the volumic mass of the vapor and the integral is taken on the surface of the spherical cap. It can be shown that the assumption of a radially constant gradient is not strictly verified, but only approximately. The evaporation flux is radially uniform except close to the triple line, as shown in Fig. 3.90.

3.9.1.2.1 Wetting (lyophilic) case: $\theta < 90°$

Using an alternative expression for the spherical cap surface

$$S_{LG} = 2\pi R^2 (1 - \cos\theta) = 2\pi R h$$

the rate of reduction of volume due to evaporation can be approximated by

$$\begin{aligned}\frac{dV}{dt} &= -\frac{D(c_0 - c_\infty)}{\rho}\int \frac{dS}{R}\\ &= -\frac{D(c_0 - c_\infty)}{\rho}\int_0^h \frac{2\pi R}{R} dh = -\frac{2\pi D(c_0 - c_\infty)}{\rho} h.\end{aligned} \tag{3.95}$$

Equation (3.95) can then be written under the form

$$\frac{dV}{dt} = -\lambda h \tag{3.96}$$

where

$$\lambda = -\frac{2\pi D(c_0 - c_\infty)}{\rho}.$$

Relation (3.96) shows that the rate of mass loss is proportional to the height of the droplet. Using the expression of the droplet volume as a function of the contact angle

$$V = \frac{\pi R^3}{3}\left(2 - 3\cos\theta + \cos^3\theta\right) = \frac{\pi r_b^3}{3}\frac{\left(2 - 3\cos\theta + \cos^3\theta\right)}{\sin^3\theta}$$

we derive

$$\begin{aligned}\frac{dV}{dt} &= \frac{\partial V}{\partial\theta}\frac{d\theta}{dt} = \frac{\pi r_b^3}{3}\frac{d}{d\theta}\left[\frac{\left(2 - 3\cos\theta + \cos^3\theta\right)}{\sin^3\theta}\right]\frac{d\theta}{dt}\\ &= \pi r_b^3 \frac{(1-\cos\theta)^2}{\sin^4\theta}\frac{d\theta}{dt}.\end{aligned} \tag{3.97}$$

After substitution in (3.95), using the relation $h = r_b \tan(\theta/2)$ and the trigonometric relation $\tan(\theta/2) = (1 - \cos\theta)/\sin\theta$, we obtain the following differential equation

$$\frac{d\theta}{dt} = -\frac{\lambda h}{\pi r_b^3 \frac{(1-\cos\theta)^2}{\sin^4\theta}} = -\frac{\lambda \sin^3\theta}{\pi r_b^2 (1 - \cos\theta)}. \tag{3.98}$$

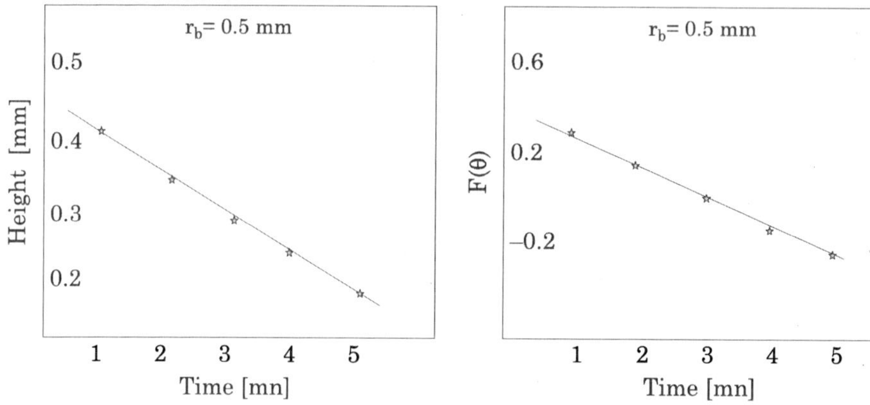

Figure 3.91 Left: time variation of the droplet height; right: time variation of the function $F(\theta)$. Droplet radius of 0.5 mm. Data from [52].

We can separate the variables in equation (3.98) to obtain

$$\frac{(1-\cos\theta)}{\sin^3\theta}d\theta = -\frac{\lambda}{\pi r_b^2}dt.$$

Integration of this differential equation gives

$$F(\theta) = \ln\left(\tan\left(\frac{\theta}{2}\right)\right) + \frac{(1-\cos\theta)}{\sin^2\theta} = -\frac{\lambda}{\pi r_b^2}(t-t_0) \tag{3.99}$$

where t_0 is a constant of integration. Equation (3.99) is implicit; however it has been shown that for a wetting angle in the interval $30° < \theta < 90°$, the function $F(\theta)$ is linear

$$F(\theta) = -1.592 + 1.632\theta. \tag{3.100}$$

Upon substitution of (3.100) in (3.99) we find a linear decrease of the contact angle with time during evaporation (λ is a positive number)

$$\theta = \theta_0 - \frac{\lambda}{1.632\pi r_b^2}(t-t_0).$$

This law fits the experimental data extremely well in the range 30°–90°, as shown in Fig. 3.91.

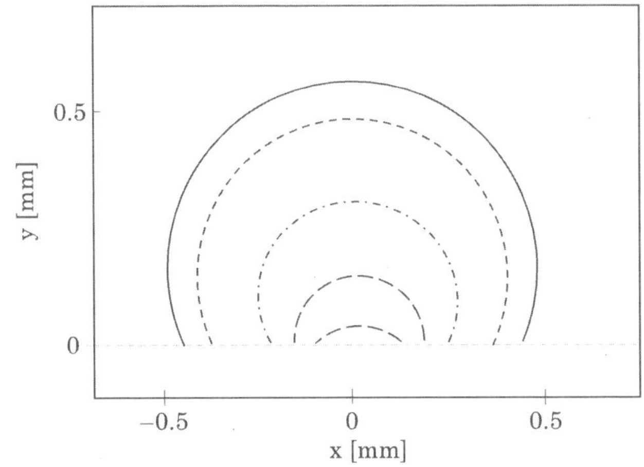

Figure 3.92 Vertical profile of evaporating droplet; case of a lyophobic surface and an initial contact radius of 0.48 mm: the contact angle remains approximately constant except at the end of the evaporation process. Reprinted with permission from [54]. © 1998 American Chemical Society.

3.9.1.2.2 Non-wetting (lyophobic) case: $\theta > 90°$

This case has been studied by McHale et al. [54]. It can be shown that equation

$$\frac{dV}{dt} = -\lambda h$$

still holds. But from this point on, the treatment of the equation changes: this time the contact radius is not kept constant, but the contact angle remains approximately constant (Fig. 3.92)

Then

$$\frac{dV}{dt} = \frac{\partial V}{\partial r}\frac{dr}{dt} = -\lambda h.$$

After some algebra, similar to that of the preceding case, we obtain the differential equation

$$r\frac{dr}{dt} = -\frac{\lambda \sin^2\theta}{\pi\,(1-\cos\theta)\,(2+\cos\theta)} \tag{3.101}$$

which can be solved to give

$$r^2 = r_0^2 - \frac{2\lambda \sin^2\theta}{\pi\,(1-\cos\theta)\,(2+\cos\theta)}t. \tag{3.102}$$

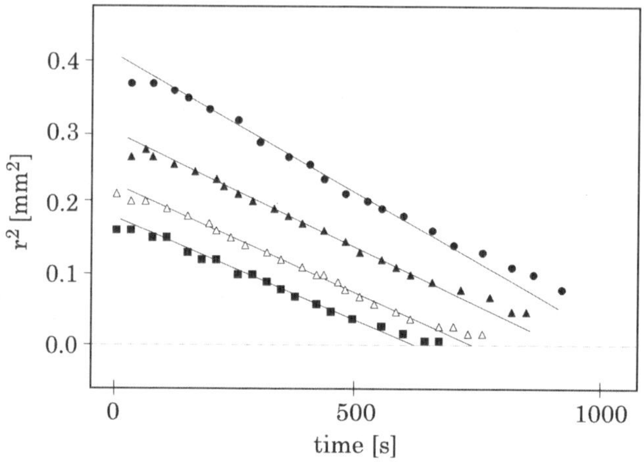

Figure 3.93 Time variation of the contact radius r for water droplets on Teflon (hydrophobic case), showing the proportionality of r^2 with time except at the very end of the evaporation process. Reprinted with permission from [54]. © 1998 American Chemical Society.

Equation (3.102) shows that the square of the contact radius is proportional to the time. This is confirmed by McHale's experimental results (Fig. 3.93).

3.9.1.3 Discussion

In equations (3.99) or (3.102) the variation rate is related to the diffusion coefficient of vapor in the surrounding gas, through the coefficient λ. This can be used as a method for estimating the diffusion coefficient D. From the curves of Fig. 3.92, one can estimate the diffusion coefficient as 2.32×10^{-5} m^2/s, whereas the *CRC Handbook of Chemistry and Physics* indicates a value of 2.56×10^{-5} m^2/s for the same conditions of temperature and pressure.

We have seen in the preceding paragraphs that the evaporation behavior of wetting and non-wetting droplets is quite different. There is no clear explanation for this difference at this time. According to McHale et al., the difference could be attributed to the fact that, in the non-wetting case, there is a ring of saturated vapor trapped near the contact line, maintaining constant conditions and a constant Young angle during the evaporation process. It is likely also that in the wetting case, impurities could gather in the sharp corner near the triple line, leading to the pinning of this contact line.

In the following section, we show that it is indeed the case for colloidal suspensions.

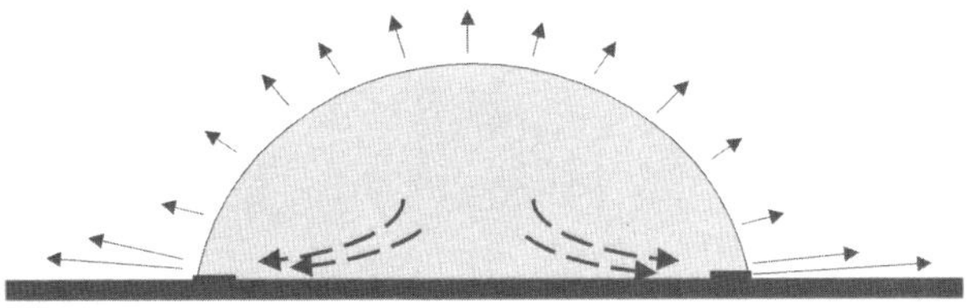

Figure 3.94 Sketch of an evaporating droplet: the two segments near the contact line represent the ring where colloidal particles are deposited. The evaporation rate is stronger near the contact line provoking a radial convective motion from the centre towards the contact line. Here it is supposed that there is no additional motion due to Marangoni convection [60].

3.9.2 Evaporation Rings

Most of the time, the micro-drops we are interested in are laden with particles, macromolecules, or polymers. A typical feature of evaporation of these droplets is evaporation rings. Ring formation is well documented in the literature; it has been studied by numerous authors like Deegan [56], Deegan et al. [57], Zheng et al. [58], Popov [59].

It seems that during the evaporation of a liquid droplet, the liquid at the edge of the droplet is "pinned" to the underlying surface by the deposited particles. This pinning prevents the droplet from shrinking. It implies that the footprint of the droplet remains constant. The evaporated liquid at the edge is replenished by liquid from the bulk of the droplet. This means that there is a flow of liquid moving towards the edge of the droplet. If the liquid droplet contains particles, like in a droplet of coffee, these particles will be transported outwards by the flow and deposited near the contact line (Fig. 3.94). Finally, when all the liquid has evaporated, the colloidal particles will form a ring of stain (Fig. 3.95).

In digital microfluidics these rings are a drawback. They modify locally the surface energy, and it becomes difficult to displace new droplets over these evaporation rings. Careful—and sometimes difficult—washing of the surface is needed to restore the quality of the surface.

3.9.3 Evaporation Stains

Evaporation rings are not a ubiquitous feature of droplet evaporation. Rings form only if the particles or macromolecules are sufficiently large to sediment or adhere to the solid surface in the corner near the triple line. If the molecules are very small, they are agitated by the convective motion and do not deposit in the corners. As a result, the center part of the droplet—the last region to dry—has more solute deposition than that of the border (Fig. 3.96 left). Chen et al. [61]

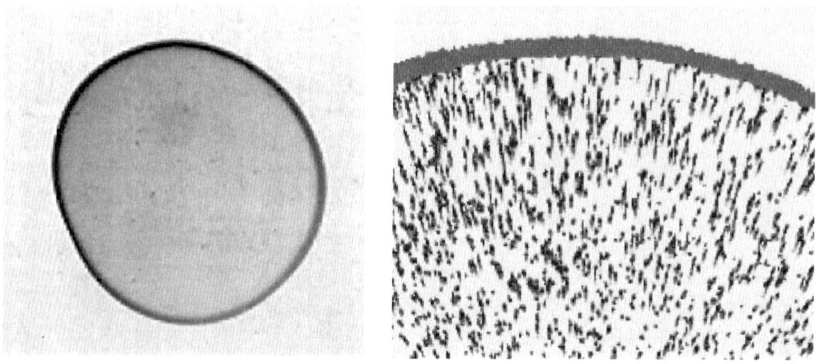

Figure 3.95 Evaporation rings left by an evaporating droplet containing microparticles. From [56]. © 2000 APS; reprinted with permission.

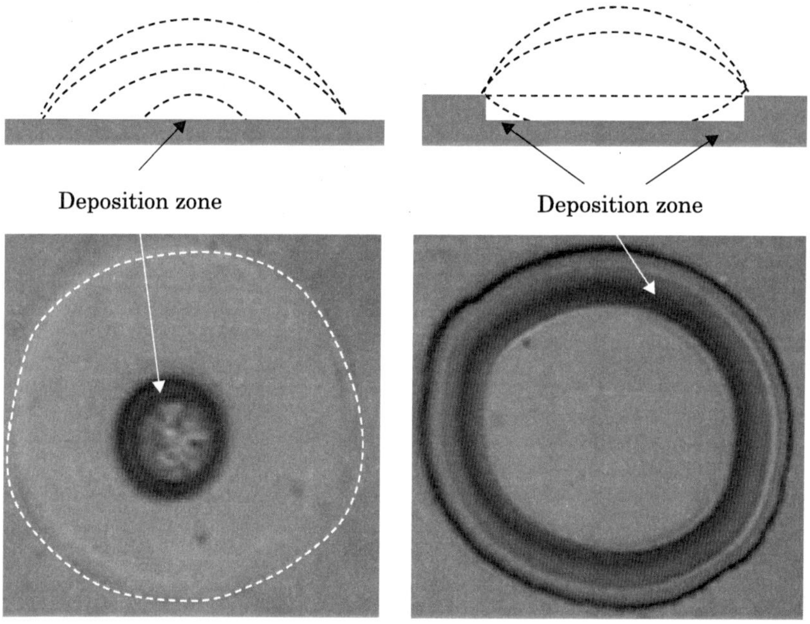

Figure 3.96 Top: schematic of a drying droplet; the droplet sitting on a smooth horizontal plane reduces progressively to a small central region, whereas the droplet in a micro-well reduces to a ring adjacent to the rim of the well. Bottom: experimental view of the deposit. From [61]. © 2006 Elsevier; reprinted with permission.

have experimentally shown the difference between a droplet evaporating on a smooth horizontal plane and in a micro-well. Due to pinning on the edges of the micro-well, the last volume of liquid to evaporate is a ring adjacent to the rim of the well (Fig. 3.96 right).

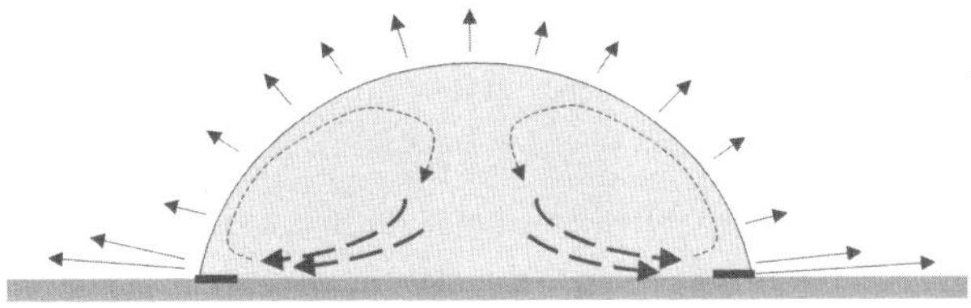

Figure 3.97 Schematic of the velocity field in an evaporating droplet when Marangoni convection superposes with the evaporating flow field [47].

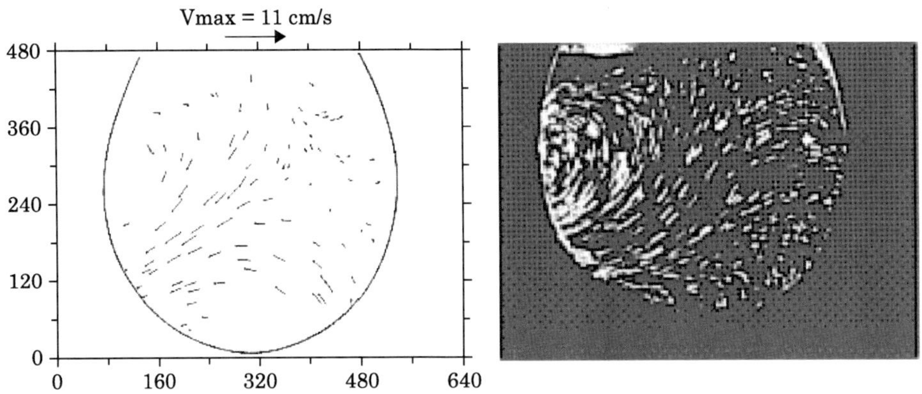

Figure 3.98 Marangoni convection due to thermal instabilities at the surface of an evaporating droplet. From [46]. © 1996 APS; reprinted with permission.

3.9.4 Marangoni Convection During Evaporation

Besides the formation of rings or stains at the contact of the solid substrate during evaporation, another typical feature of droplet evaporation is Marangoni convection. One can distinguish two cases: that of a sessile droplet and that of a pendant droplet. In the first case, the evaporation rate is larger at the contact of the surface, resulting in a lower temperature of the interface and a larger surface tension. A stable Marangoni pattern forms in the droplet, as shown in Fig. 3.97.

The case of a pendant drop is different [46]. There are no surface regions permanently at a lower temperature; it is a very instable situation: when a cold spot forms on the surface, it is a focal point for convective motion; as soon as the temperature loss is compensated by the convective transport, the intensity of the Marangoni convection vanishes, and the same process appears again at a new cold spot. Fig. 3.98 shows the Marangoni motion converging towards a hot spot.

3.9.5 The Use of Droplet Evaporation in Biotechnology, Biology, and Nanoassembly

Evaporation is a somewhat slow process compared to molecular motion, and the receding contact line is geometrically linear at the scale of a group of molecules. Thus, particles and molecules have time to rearrange during the receding of the triple line. This feature has been used for specific applications in modern biotechnology, biology, and nanoassembly.

In this section we present some interesting applications of droplet evaporation.

3.9.5.1 Drop Evaporation as a Tool for DNA Stretching

In genomics—the study of the sequences, functions, and interactions of genes—DNA stretching is of great importance. Usually, DNA strands bunch up and stretching them is a necessary step before any observation of the DNA chain, and when very precise immobilization is needed [62,63]. This process is called molecular combing. One of the first methods—and still a very commonly used one—to comb DNA was proposed by Bensimon et al. [63]. This method takes advantage of liquid droplet evaporation (Fig. 3.99). A DNA strand attached by one end to a glass surface stretches progressively as the gas/liquid interface retreats during evaporation. When all the liquid has vanished, the DNA strand forms a linear segment on the glass plate.

3.9.5.2 Drop Evaporation as a Tool for Nanoassembly

Evaporation of a liquid in a microchannel is a process that is, at the same time, not too fast or too slow. It can be taken advantage of for coating the internal surfaces of microfluidic systems. Such coatings may be used to change the chemical composition of the walls, or to mask a boundary of wettability between two zones of the microsystem (composite microsystems).

If the volatile liquid is laden with the appropriate particles, the evaporating liquid deposits the particles in an ordered pattern along the solid walls of the microsystem. Vengallatore et al. [64] have shown how to coat the interior silicon walls of a microsystem with a layer of spherical polystyrene or silica beads by evaporating the carrier fluid (Fig. 3.100). An annealing treatment—heating of the coating for a brief moment—may be performed to stabilize the coating.

The result is shown in Fig. 3.101. Particles predominantly assemble in hexagonal close-packed configurations, and, if sufficient precautions are taken, they can coat uniformly the internal walls of the microsystem.

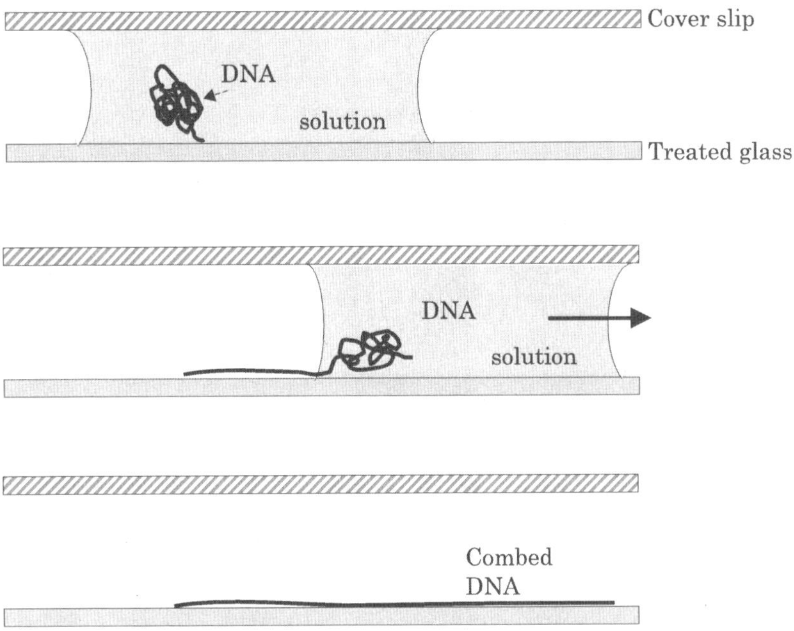

Figure 3.99 Combing DNA using droplet evaporation.

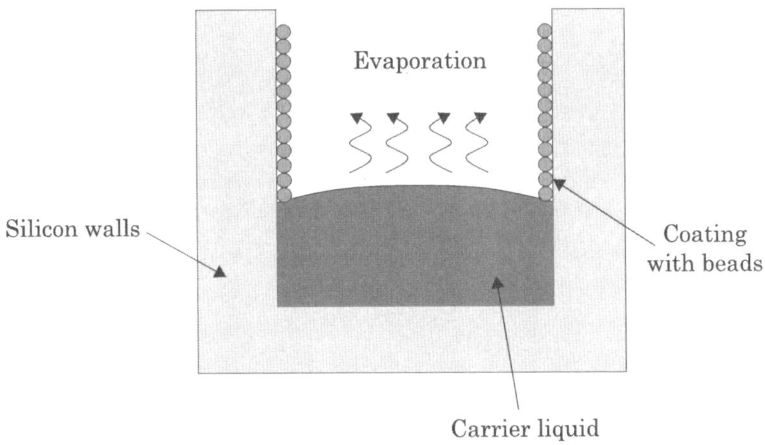

Figure 3.100 Schematic illustration of convective colloidal self-assembly: evaporation of the colloidal suspension leads to particle assembly on the internal surfaces.

3.9.5.3 Conclusion

Deposition of macromolecules or particles using droplet evaporation is of current interest. In the literature, there have been recent publications on this topic. Let us cite here the work of Bormashenko et al. [65] who used different polymer concentrations in an evaporating droplet of solvent to obtain patterned

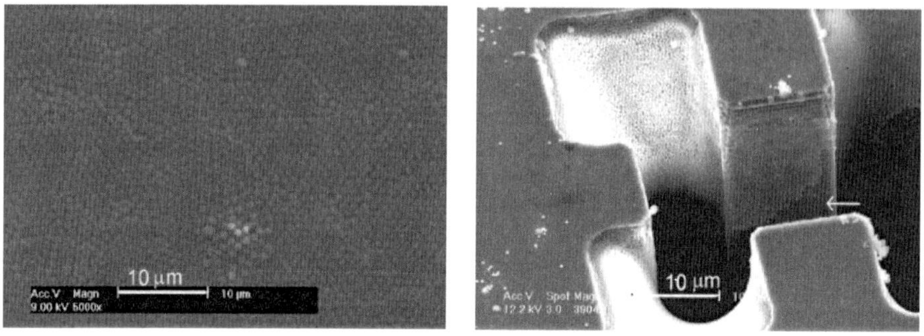

Figure 3.101 Left: coating of a surface with monodisperse polystyrene beads. Right: nearly uniform coating of a microfabricated channel with monodisperse beads.
From [64]. © 2004 Elsevier; reprinted with permission.

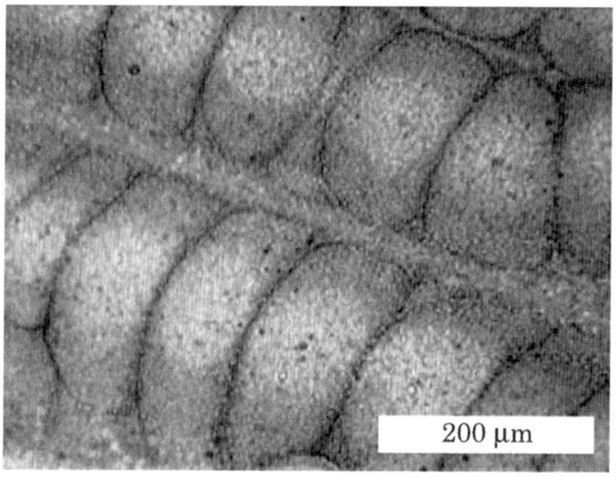

Figure 3.102 Polymer coating of a quartz surface obtained by evaporation of a dichloromethane droplet containing PC polymers. From [65]. © 2006 Elsevier; reprinted with permission.

coating (Fig. 3.102) and the publication of Schnall-Levin et al. [66] who showed the patterns made by the rearrangement of particles during evaporation.

3.10 Summary

This chapter has shown that droplet behavior shows very particular features, compared to usual fluid mechanics. At this scale, the interfacial forces are dominant and govern the shape and position of a droplet in a microsystem. These interfacial forces are dependant on the nature of the liquid, on the colloids that are present in this liquid, on the surface properties of the substrate, and on the geometry of the device.

Chapters 2 and 3 constitute the basis for the understanding of capillarity on micro-drops. We have seen that capillarity by itself can be sufficient to displace droplets or to pin them. However, these capillary forces are fixed, and their intensity or location cannot easily be modified. Two types of forces can be relatively easily superposed on capillary effects in order to have a more flexible system: electric and acoustic forces. In the case of electric forces, we will speak of electrocapillarity or electrowetting. In the next chapters, we investigate the numerous applications where these forces are superposed to capillarity to handle droplets.

References

1. P.-G. de Gennes, F. Brochard-Wyart and D. Quéré, *Drops, Bubbles, Pearls, Waves*, Springer, New-York, 2004.
2. K. Brakke, "The Surface Evolver," *Exp. Math.*, Vol. 1, p. 141, 1992.
3. J.W. Harris and H. Stocker, *Handbook of Mathematics and Computational Science*, Springer-Verlag, New York, 1998. p. 107.
4. J.G. Hocking and G.S. Young, *Topology*, Dover, New York, 1988.
5. P. Concus and R. Finn, "On the behavior of a capillary surface in a wedge," *PNAS*, Vol. 63, no. 2, pp. 292–299, 1969.
6. K. Brakke, "Minimal surfaces, corners, and wires," *J. Geom. Anal.*, Vol. 2, pp. 11–36, 1992.
7. J.-C. Baret, M. Decré, S. Herminghaus and R. Seemann, "Electroactuation of fluid using topographical wetting transitions," *Langmuir*, Vol. 21, pp. 12218–12221, 2005.
8. R. Seemann, M. Brinkmann, E.J. Kramer, F. Lange and R. Lipowsky, "Wetting morphologies at microstructured surfaces," *PNAS*, Vol. 102, pp. 1848–1852, 2005.
9. R. Lipowsky, M. Brinkmann, R. Dimova, C. Haluska, J. Kierfeld and J. Shillcock, "Wetting, budding, and fusion—morphological transitions of soft surfaces," *J. Phys.: Condens. Matter*, Vol. 17, pp. S2885–S2902, 2005.
10. C.C. Meek and P. Pantano, "Spatial confinement of avidin domains in microwell arrays," *Lab on a Chip*, Vol. 1, pp. 158–163, 2001.
11. J. Berthier and P. Silberzan, *Microfluidics for Biotechnology*, Artech House Publishers, 2005.
12. H. Gau, S. Herminghaus, P. Lenz and R. Lipowsky, "Liquid morphologies on structured surfaces: From microchannels to microchips," *Science*, Vol. 383, pp. 46–49, 1999.
13. P. Lenz and R. Lipowsky, "Morphological transitions of wetting layers on structured surfaces," *Phys. Rev. Letters*, Vol. 80, no. 9, pp. 1920–1923, 1998.
14. M. Brinkmann and R. Lipowsky, "Wetting morphologies on substrates with striped surface domains," *Journal of Applied Physics*, Vol. 92, no. 8, pp. 4296–4306, 2002.
15. A. Klingner and F. Mugele, "Electrowetting-induced morphological transitions of fluid microstructures," *Journal of Applied Physics*, Vol. 95, no. 5, pp. 2918–2920, 2004.
16. H. Gau, S. Herminghaus, P. Lenz and R. Lipowski, *Science*, Vol. 283, pp. 46–49, 1999.
17. Y. Fouillet, H. Jeanson, D. Jary and C. Vauchier, Moving droplets with microcatenaries, *Proceedings of MTAS 2003, Seventh International Conference on Micro Total Analysis Systems*, Squaw Valley, California, USA, 5–9 October, 2003.
18. Th. Uelzen and J. Müller, "Wettability enhancement by rough surfaces generated by thin film technology," *Thin Solid Films*, Vol. 434, pp. 311–315, 2003.
19. Seong H. Kim, Jeong-Hoon Kim, Bang-Kwon Kang and Han S. Uhm, "Superhydrophobic CFx coating via in-line atmospheric RF plasma of He–CF_4–H_2," *Langmuir*, Vol. 21, pp. 12213–12217, 2005.

20. Liang Zhu, Yanying Feng, Xiongying Ye and Zhaoying Zhou, "Tuning wettability and getting superhydrophobic surface by controlling surface roughness with well-designed microstructures," *Sensors and Actuators, A Physical*, Vol. 130–131, pp. 595–600, 2006.
21. J. Bico, U. Thiele and D. Quéré, "Wetting of textured surfaces," *Colloids and Surfaces A: Physicochemical and Engineering Aspects*, Vol. 206, pp. 41–46, 2002.
22. J. Bico, C. Marzolin and D. Quéré, "Pearl drops," *Europhys. Lett.*, Vol. 47, no. 2, pp. 220–226, 1999.
23. J. Bico, C. Tordeux and D. Quéré, "Rough wetting," *Europhys. Lett.*, Vol. 55, no. 2, pp. 214–220, 2001.
24. N.A. Patankar, "Transition between superhydrophobic states on rough surfaces," *Langmuir*, Vol. 20, pp. 7097–7102, 2004.
25. N.A. Patankar and Y. Chen, Numerical simulation of droplet shapes on rough surfaces, *Proceedings of the 2002 Nanotech Conference*, Puerto-Rico, pp. 116–119, April 21–25, 2002.
26. N.A. Patankar, "On the modelling of hydrophobic contact angles on rough surfaces," *Langmuir*, Vol. 19, pp. 1249–1253, 2003.
27. T.N. Krupenkin, J. Ashley Taylor, Tobias M. Schneider and Shu Yang, "From Rolling Ball to Complete Wetting: The Dynamic Tuning of Liquids on Nanostructured Surfaces," *Langmuir*, Vol. 20, pp. 3824–3827, 2004.
28. L. Zhu, Y.Y. Feng, X.Y. Ye and Z.Y. Zhou, Tuning wettability and getting superhydrophobic surfaces by controlling surface roughness with well-designed microstructures, *Transducers'05, 13rd International Conference on Solid-State Sensors, Actuators and microsystems*, Seoul, Korea, 5–9 June, 2005.
29. W. Barthlott and C. Neinhuis, "Purity of the sacred lotus, or escape from contamination in biological surfaces," *Planta*, Vol. 202, pp. 1–8, 1997.
30. J.-G. Fan, X.-J. Tang and Y.-P. Zhao, "Water contact angles of vertically aligned Si nanorod arrays," *Nanotechnology*, Vol. 15, pp. 501–504, 2004.
31. T. Onda, S. Shibuichi, N. Satoh and K. Tsujii, "Super water-repellent surfaces," *Langmuir*, Vol. 12, no. 9, pp. 2125–2127, 1996.
32. S. Shibuichi, T. Onda, N. Satoh and K. Tsujii, "Super water-repellent surfaces resulting from fractal structures," *J. Phys. Chem.*, Vol. 100, pp. 19512–19517, 1996.
33. N. Moumen, R.S. Subramanian and J. McLMaughlin, "Experiment on the motion of drops on a horizontal solid surface due to a wettability gradient," *Langmuir*, Vol. 22, pp. 2682–2690, 2006.
34. M.K. Chaudhury and G.M. Whitesides, "How to make water run uphill," *Science*, Vol. 256, pp. 1539–1541, 1992.
35. Domingues Ondarçuhu, "Free-running droplets," *Phys. Rev. Lett.*, Vol. 75, no. 16, pp. 2972–2975, 1995.
36. S.M.M. Ramos, E. Charlaix and A. Benyagoub, "Contact angle hysterisis on nano-structured surfaces," *Surface Science*, Vol. 540, pp. 355–362, 2003.
37. H.A. Wege, J.A. Holgado-Terriza, J.I. Rosales-Leal, R. Osorio, M. Toledano and M.A. Cabrerizo-Vilchez, "Contact angle hysterisis on dentin surfaces measured with ADSA on drops and bubbles," *Colloids and Surfaces, A: Physicochemical and Engineering Aspects*, Vol. 206, pp. 469–483, 2002.
38. C.N.C. Lam, R. Wu, D. Li, M.L. Hair and A.W. Neumann, "Study of the advancing and receding contact angles: liquid sorption as a cause of contact angle hysterisis," *Advances in Colloid and Interface Science*, Vol. 96, pp. 169–191, 2002.
39. H. Tavana and A.W. Neumann, "On the question of rate-dependence of contact angles," *Colloids and Surfaces, A: Physicochemical Engineering Aspects*, Vol. 282–283, pp. 256–262, 2006.
40. T. Cubaud and M. Fermigier, "Advancing contact lines on chemically patterned surfaces," *Journal of Colloid and Interface Science*, Vol. 269, pp. 171–177, 2004.

41. T. Ondarçuhu, "Total or partial pinning of a droplet on a surface with chemical discontinuity," *J. Phys. II France*, Vol. 5, pp. 227–241, 1995.
42. P. Lenz and R. Lipowsky, "Morphological transitions of wetting layers on structured surfaces," *Phys. Rev. Lett.*, Vol. 80, no. 9, pp. 1920–1923, 1998.
43. A. Buguin, L. Talini and P. Silberzan, "Ratchet-like topological structures for the control of microdops," *Appl. Phys. A.*, Vol. 75, pp. 207–212, 2002.
44. Yangyang Shen, A. Couzis, J. Koplik, C. Maldarelli and M.S. Tomassone, "Molecular dynamics study of the influence of surfactant structure on surfactant-facilitated spreading of droplets on solid surfaces," *Langmuir*, Vol. 21, pp. 12160–12170, 2005.
45. S. Colin, *Microfluidique*, Hermes Science, Paris, 2004.
46. J.J. Hegseth, N. Rashidnia and A. Chai, "Natural convection in droplet evaporation," *Phys. Rev. E*, Vol. 54, no. 2, pp. 1640–1644, 1996.
47. Hua Hu and R.G. Larson, "Analysis of the effects of Marangoni stresses on the microflow in an evaporating sessile droplet," *Langmuir*, Vol. 21, pp. 3972–3980, 2005.
48. O. Raccurt, J. Berthier and P. Clementz, Time dependant surface tension modification due to surfactant during electrowetting basic droplet operation, *Euromech Colloquium 472 on Microfluidics and Transfer*, Grenoble, France, 6–8 September, 2005.
49. O. Raccurt, J. Berthier, Ph. Clementz, M. Borella and M. Plissonnier, "On the influence of surfactants in electrowetting," *J. Micromec. Microeng.*, Vol. 17, pp. 1–7, 2007.
50. Jun Fukai, Hirotaka Ishizuka, Yosuke Sakai, Masayuki Kaneda, Masamichi Morita and Atsushi Takahara, "Effects of droplet size and solute concentration on drying process of polymer solution droplets deposited on homogeneous surfaces," *International Journal of Heat and Mass Transfer*, Vol. 49, pp. 3561–3567, 2006.
51. R.A. Meric and H.Y. Erbil, "Evaporation of sessile drops on solid surfaces: pseudospherical cap geometry," *Langmuir*, Vol. 14, pp. 1915–1920, 1998.
52. S.M. Rowan, M.I. Newton and G. McHale, "Evaporation of micro-droplets and the wetting of solid surfaces," *J. Phys. Chem.*, Vol. 99, pp. 13268–13271, 1995.
53. Hua Hu and R.G. Larson, "Evaporation of a sessile droplet on a substrate," *J. Phys. Chem. B*, Vol. 106, pp. 1334–1344, 2002.
54. G. McHale, S.M. Rowan, M.I. Newton and M.K. Banerjee, "Evaporation and the wetting of a low-energy solid surface," *J. Phys. Chem. B.*, Vol. 102, pp. 1964–1967, 1998.
55. K.S. Birdi, D.T. Vu and A. Winter, "A study of the evaporation rate of small water drops placed on a solid surface," *J. Phys. Chem.*, Vol. 93, pp. 3702–3703, 1989.
56. R.D. Deegan, "Pattern formation in drying drops," *Phys. Rev. E*, Vol. 61, no. 1, pp. 475–485, 2000.
57. R.D. Deegan, O. Bakajin, T.F. Dupont, G. Huber, S.R. Nagel and T.A. Witten, "Capillary flow as the cause of ring stains from dried liquid drops," *Nature*, Vol. 389, pp. 827–829, 1997.
58. Rui Zheng, Y.O. Popov and T.A. Witten, "Deposit growth in the wetting of an angular region with uniform evaporation," *Phys. Rev. E*, Vol. 72, p. 046303, 2005.
59. Y.O. Popov, "Evaporative deposition patterns: Spatial dimensions of the deposit," *Phys. Rev. E*, Vol. 71, p. 036313, 2005.
60. Hua Hu and R.G. Larson, "Analysis of the microfluidic flow in an evaporating sessile droplet," *Langmuir*, Vol. 21, pp. 3963–3971, 2005.
61. Chin-Tai Chen, Fan-Gang Tseng and Ching-Chang Chieng, "Evaporation evolution of volatile liquid droplets in nanoliter wells," *Sensors and Actuators A*, Vol. 130–131, pp. 12–19, 2006.
62. J.F. Allemand, D. Bensimon, L. Jullien, A. Bensimon and V. Croquette, "pH-dependent specific binding and combing of DNA," *Biophysical Journal*, Vol. 73, pp. 2064–2070, 1997.
63. D. Bensimon, A.J. Simon, V. Croquette and A. Bensimon, "Stretching DNA with a receding meniscus: experiments and models," *Phys. Rev. Lett.*, Vol. 74, pp. 4754–4757, 1995.

64. S. Vengallatore, Y. Peles, L.R. Arana and S.M. Spearing, "Self-assembly of micro- and nanoparticles on internal micromachined silicon surfaces," *Sensors and Actuators A*, Vol. 113, pp. 124–131, 2004.
65. E. Bormashenko, R. Pogreb, A. Musin, O. Stanevsky, Y. Bormashenko, G. Whyman, O. Gendelman and Z. Barkay, "Self-assembly in evaporated polymer solutions: Influence of the solution concentration," *Journal of Colloid and Interface Science*, Vol. 297, pp. 534–540, 2006.
66. M. Schnall-Levin, E. Lauga and M. Brenner, "Self-assembly of spherical particles on an evaporating sessile droplet," *Langmuir*, Vol. 22, pp. 4547–4551, 2006.

4 Electrowetting Theory☆

4.1 Introduction

In the presence of a direct or alternating-current electric field, electric charges gather at the interface between a conductive and a non-conductive (dielectric) material. The same electric field that induces these charges then couples with the interfacial electric charges to exert an interfacial force, and if the interface is deformable—like that of a conductive liquid and a non-conductive fluid or gas—this force can distort the interface. These induced interfacial electric forces are especially strong on a liquid–gas interface at the vicinity of the contact line, as the electric field often becomes singularly large at the sharp wedges near the contact line. The physics describing the electric forces on interfaces of conducting liquids and on triple contact lines is called "electrowetting".

As early as 1857, Gabriel Lippmann [1] observed a change in the capillary rise of mercury in the presence of electric charges (see Chapter 2 and Jurin's law), and advanced the principle of electro-capillarity as described by an equation which is now called the Lippmann law. This equation contains the principle of electrowetting, and more specifically electrowetting on dielectrics (EWOD), a technique aimed at activating and transporting liquid droplets on a solid substrate. This technique was pioneered by Berge in the 1990s [2,3] who combined the Lippmann law with the Young law to obtain the now-called Lippmann–Young law

$$\cos\theta = \cos\theta_0 + \frac{C}{2\gamma_{LG}}V^2 \tag{4.1}$$

where C is the capacitance of the dielectric layer separating the bottom electrode from the liquid. Relation (4.1) predicts the change of contact angle of a sessile droplet with the intensity of the electric field, as shown in Fig. 4.1.

A similar phenomenon occurs when considering the capillary rise of a liquid between two vertical plates (see Jurin's law in Chapter 2). We have seen that the capillary rise is given by

$$h_{cap} = \frac{2\gamma_{LG}\cos\theta_0}{\rho g D} \tag{4.2}$$

where D is the distance between the plates, and ρ the density of the liquid. If electrodes are embedded inside the two vertical plates (Fig. 4.2), the liquid

☆This Chapter was written in collaboration with Hsueh-Chia Chang (Bayer Professor of Engineering, Department of Chemical and Biomolecular Engineering, University of Notre-Dame, Indiana, USA)

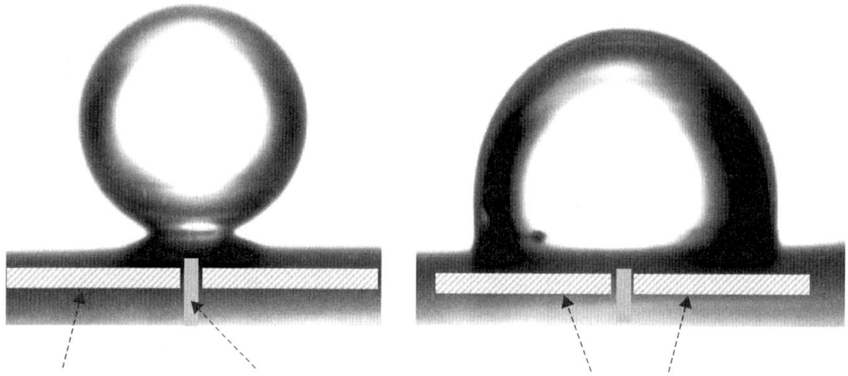

Figure 4.1 Left: in absence of electric charges, a droplet of water shows a very hydrophobic contact with the solid substrate. Right: the contact angle of the water with the substrate decreases when the electrode is actuated.

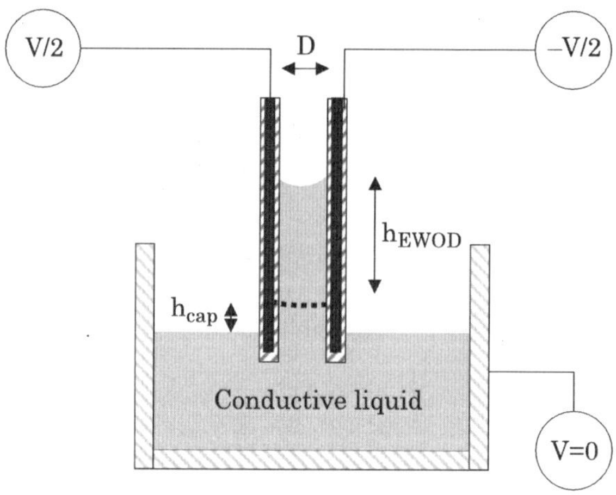

Figure 4.2 Electrocapillary rise of a conductive liquid between two plates.

rise depends on the applied voltage. It can be shown that the "electrocapillary" rise, i.e., the additional vertical rise, is given by a similar expression

$$h_{\mathrm{EWOD}} = \frac{2\gamma_{\mathrm{LG}}\left(\cos\theta - \cos\theta_0\right)}{\rho g D} = \frac{C}{\rho g D}\left(\frac{V}{2}\right)^2 \tag{4.3}$$

where C is the capacitance of the dielectric layers.

As depicted in Figs. 4.1 and 4.2, the electrowetting effect may seem like a simple physical phenomenon. We shall see in this chapter that the detailed physics of electrowetting is in fact quite complex. The interpretation of Lippmann's law is delicate; and the causes of some phenomena—like saturation—are still debated. In this chapter, we present the theoretical background of electrowetting leading to the key relation also called the Lippmann–Young law for static electrowetting.

Although it is not the only way to transport droplets—we shall see in Chapter 8 the principle of acoustic actuation—the electrowetting effect is quite effective when used to move, merge, mix, and separate droplets, and has been a major promoter of the recent developments of microsystems using digital microfluidics. The engineering aspects and applications of the electrowetting effect will be presented in Chapter 5.

4.2 Theoretical Background

The derivation of the Lippmann law, i.e., the dependency of the contact angle according to the applied voltage, can be done by the following different approaches: the classical thermodynamic approach, the energy minimization approach, and the electromechanical approach. All these approaches are detailed by Mügele et al. in their review paper [4].

4.2.1 Thermodynamic Approach

Although the interfacial tension in Lippmann's law is the gas–liquid interfacial tension, it is the solid–liquid interface that is relevant here—the electric field must penetrate this interface. We suppose a perfectly smooth solid surface at the contact of a conductive liquid or an electrolyte. We first follow the original Lippmann's approach: the solid is a metal directly in contact with the liquid and the potential difference is small enough so that no electric current is flowing through the liquid (no hydrolysis or any other Faradaic reaction at the metal surface). Upon applying an elementary electric field, an elementary potential difference builds up at the interface and an electric double layer forms in the liquid at the contact of the surface. Gibbs' interfacial thermodynamics which relates the normal force density to a gradient in the interfacial tension then yields

$$d\gamma_{\mathrm{SL}}^{\mathrm{eff}} = -\rho_{\mathrm{SL}} dV \tag{4.4}$$

where γ^{eff} denotes the effective surface tension at the liquid–solid interface, ρ_{SL} the field-induced surface charge density in counter-ions, and V the electric potential. If we make the Helmholtz simplifying assumption that the counter-ions are all located at a fixed distance d_H from the surface (d_H is of the order of a few nanometers), the double layer has a fixed specific capacitance (capacitance per unit area) in response to the charging by the normal electric field

$$C_H = \frac{\varepsilon_0 \varepsilon_l}{d_H} \tag{4.5}$$

where ε_l is the relative permittivity of the liquid and ε_0 is the permittivity of vacuum: $\varepsilon_0 = 8.8541878176 \times 10^{-12}$ F/m. Direct integration of (4.4) yields

$$\begin{aligned}\gamma_{\mathrm{SL}}^{\mathrm{eff}}(V) &= \gamma_{\mathrm{SL}} - \int_{V_{\mathrm{pzc}}}^{V} \rho_{\mathrm{SL}} dV \\ &= \gamma_{\mathrm{SL}} - \int_{V_{\mathrm{pzc}}}^{V} C_H V dV = \gamma_{\mathrm{SL}} - \frac{C_H}{2}\left(V - V_{\mathrm{pzc}}\right)^2 \end{aligned} \tag{4.6}$$

where V_{pzc} is the potential at no charge: spontaneous charges appear at the surface of a solid when immersed into electrolyte solutions at zero voltage, and the potential at no charge is the voltage required to compensate this spontaneous charging. In assigning V in (4.6) to the voltage of the electrode relative to a distant reference electrode, we are assuming that most of the voltage drop occurs in the double layer and there is no Ohmic loss in the conducting liquid. This occurs when the double layer is fully charged and in the AC case, it stipulates that the AC frequency is lower than the inverse charging time of the double layer, $D/d_H^2 = \sigma/\varepsilon_0\varepsilon_l$ where D is the ion diffusivity and σ the conductivity of the liquid. Equation (4.6) is the original Lippmann's law for electrolytes in contact with a metal electrode. However, it should be emphasized that this derivation is truly valid only at an isolated solid–liquid interface and not at a contact line: nowhere in the reasoning does the contact line appear. It is also valid only for a metallic electrode surface. The analysis of the situation where a dielectric layer exists between the liquid and the electrode will be done later in this section. We have, in fact, estimated only the change in the electrode–liquid interfacial tension in the presence of an electric field. Nevertheless, using Young's law, we can translate the change in $\gamma_{\mathrm{SL}}^{\mathrm{eff}}$ to a change in the contact angle at a contact line. Young's law applied successively at zero potential and at potential V can be written

$$\begin{aligned}\gamma_{\mathrm{SG}} - \gamma_{\mathrm{SL}} &= \gamma_{\mathrm{LG}} \cos\theta_0 \\ \gamma_{\mathrm{SG}} - \gamma_{\mathrm{SL}}^{\mathrm{eff}}(V) &= \gamma_{\mathrm{LG}} \cos\theta \end{aligned} \tag{4.7}$$

where θ, θ_0 are respectively the actuated and non-actuated contact angles. Upon subtraction of these two equations and substitution in (4.6), by assuming that (4.6) derived for a bulk interface can be applied at the contact line, we obtain the Lippmann–Young law [2]

$$\cos\theta = \cos\theta_0 + \frac{C_H}{2\gamma_{\mathrm{LG}}}\left(V - V_{\mathrm{pzc}}\right)^2. \tag{4.8}$$

Equation (4.8) shows that the contact angle decreases with an increase of the applied voltage. However, direct applications of Lippmann's law to a conducting liquid contacting a metallic surface are of little use because of the limitation of the voltage due to hydrolysis phenomena. For water, $d_H \sim 2$ nm,

$\varepsilon_l \sim 80$, $\gamma_{SL} \sim 0.040$ N/m, and the maximum voltage difference is of the order of 0.1 V, so that the relative change of the value of the surface tension is $\Delta\gamma_{SL}/\gamma_{SL} \sim 2\%$. In terms of contact angle, using $\gamma_{LG} = 0.072$ N/m, we find $(\cos\theta - \cos\theta_0) < 0.01$.

It is obvious that larger potentials are required to produce noticeable changes of the contact angle. In order to avoid hydrolysis, a thin dielectric layer can be used to insulate the liquid from the electrode [3]. In such a case, the former reasoning no longer applies, and we have to consider that the double layer is in series with the dielectric film. The specific capacitance is decreased by the presence of the dielectric (usually the electrolyte has a double layer capacitance much larger than that of the dielectric film). Because the capacitors are in series, it is the smaller dielectric layer capacitance that should now be used in Lippmann's law. However, this smaller capacitance of the dielectric film is more than compensated by the much larger working voltages and a much larger contact angle distortion can be produced. This technique is called electrowetting on dielectric (EWOD). In this new configuration, the electric double layer builds up at the surface of the insulator. The system now comprises two capacitors in series, namely the double layer at the solid surface—specific capacitance C_H—and the dielectric layer—specific capacitance C_D given by

$$C_D = \frac{\varepsilon_0 \varepsilon_D}{d}. \tag{4.9}$$

Comparing C_H—equation (4.5)—and C_D—equation (4.9), we find

$$\frac{C_D}{C_H} = \frac{\varepsilon_D}{\varepsilon_l}\frac{d_H}{d}.$$

This relation shows that $C_D \ll C_H$ because $d_H \ll d$ and $\varepsilon_D < \varepsilon_l$. Thus, the total capacitance C, given by the relation

$$\frac{1}{C} = \frac{1}{C_D} + \frac{1}{C_H}$$

can be approximated by $C \approx C_D$ by neglecting the double-layer capacitance. This relation shows that the voltage drop occurs within the dielectric layer (in comparison the voltage drop in the double layer can be neglected), and equation (4.6) is replaced by

$$\gamma_{SL}^{\text{eff}}(V) = \gamma_{SL} - \frac{C}{2}V^2 = \gamma_{SL} - \frac{\varepsilon_0 \varepsilon_D}{2d}V^2 \tag{4.10}$$

where the potential at no charge V_{pzc} has been neglected assuming that the insulating layer does not give rise to spontaneous adsorption of charge. Using

Young's law, we find the Lippmann–Young law for electrowetting on dielectric

$$\cos\theta = \cos\theta_0 + \frac{C}{2\gamma_{LG}}V^2. \tag{4.11}$$

The last term on the right hand side of equation (4.11) is dimensionless, and is sometimes called the electrowetting numberη

$$\eta = \frac{C}{2\gamma_{LG}}V^2 = \frac{\varepsilon_0\varepsilon_D}{2\gamma_{LG}}V^2. \tag{4.12}$$

For a dielectric like Teflon or parylene 1 μm thick, and a droplet of water, the coefficient $\varepsilon_0\varepsilon_D/2\gamma_{LG}$ is of the order of 10^{-4} V^{-2}. In order to obtain a substantial change in contact angle, applied electric potentials should be of the order of 30 to 80 V. We will see in Section 4.4 that the contact angle cannot go to zero by increasing the potential V, as predicted by (4.11), and that there is a limit (called saturation limit) above which an increase of potential does not result in a decrease of the value of the contact angle.

Equation (4.11) has been verified—for small to medium voltages—by using the experimental device of Fig. 4.3. Once the actuation is switched on, the contact angle decreases and the drop changes to a more hydrophilic shape. The relation between the actuated contact angle and the applied voltage follows indeed the Lippmann–Young equation, if the voltage is sufficiently far from to the saturation value. Fig. 4.3 corresponds to a droplet of water sitting on a 1 μm dielectric coated with Teflon. The plot of Fig. 4.4 (right) shows the quadratic dependency of $\cos\theta - \cos\theta_0$ versus the voltage V: if we plot $\cos\theta - \cos\theta_0$ versus V^2, the curve is well fit by a straight line, according to the Lippmann–Young law (Fig. 4.5).

In this section, we have shown that the electrowetting effect can be viewed as a change of the value of the effective solid–liquid surface tension due to the apparition of electric charges either in the double layer, or at the dielectric surface. This change in solid–liquid interfacial tension results in a change of the contact angle. The variation of the effective solid–liquid surface tension is proportional to the square of the potential drop between the electrode and the liquid, as stated by the Lippmann law. Finally, using Young's law we deduce the Lippmann–Young law representing the change in contact angle associated with an applied voltage.

In the following section, we present another approach based on energy minimization, leading directly to the Lippmann–Young equation. We show that the change of droplet contact angle results from a competition between dielectric energies which want to expand the liquid–solid area so as to decrease the charge concentration versus the surface tension energies which want to minimize the free surface of the droplet (see Chapter 2, Section 2.10).

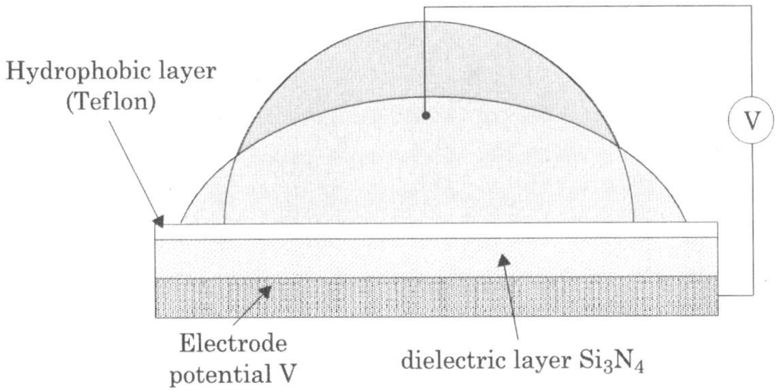

Figure 4.3 Scheme of the electrowetting set up used to verify Lippmann–Young equation. The specific capacitance C of the system is the sum of the specific capacitances of the different layers between the electrode and the liquid. The zero potential electrode may be placed anywhere in the conducting drop. Upon actuation of a voltage V, the droplet spreads on the substrate. The value of the contact angle depends on the value of the actuation potential V.

4.2.2 Energy Minimization Approach

Following the work of Shapiro and coworkers [5,6], this section presents an energy minimization approach to derive Young's law as well as the Lippmann–Young law. The derivation is rigorous and exact and ignores the true mechanism that distorts the contact angle: change in bulk solid–liquid interfacial tension or a focused Maxwell force at the solid–liquid interface. The only approximations reside within an incomplete knowledge of the energies or within the assumption that the drop is a truncated perfect sphere of radius R and contact angle θ (see Chapter 3, Section 3.2). We are reminded that the drop has a spherical shape if the characteristic Bond number is much smaller than 1 corresponding to a situation where gravitational forces are unimportant (very small drop).

Mathematically, the energy of the droplet is a function $E = E(R, \theta; p)$ where R is the droplet radius, θ the contact angle, and p is a set of system parameters (such as the applied voltage, dielectric constants, ion concentrations, temperature, etc). Thus the energy depends on the geometry (R, θ) and system parameters. First, note that the system assumes its equilibrium shape when the energy $E(R, \theta; p)$ is at a minimum. The derivative of the energy with respect to R and θ is then zero

$$dE = \frac{\partial E}{\partial R}(R, \theta; p)\, dR + \frac{\partial E}{\partial \theta}(R, \theta; p)\, d\theta = 0. \tag{4.13}$$

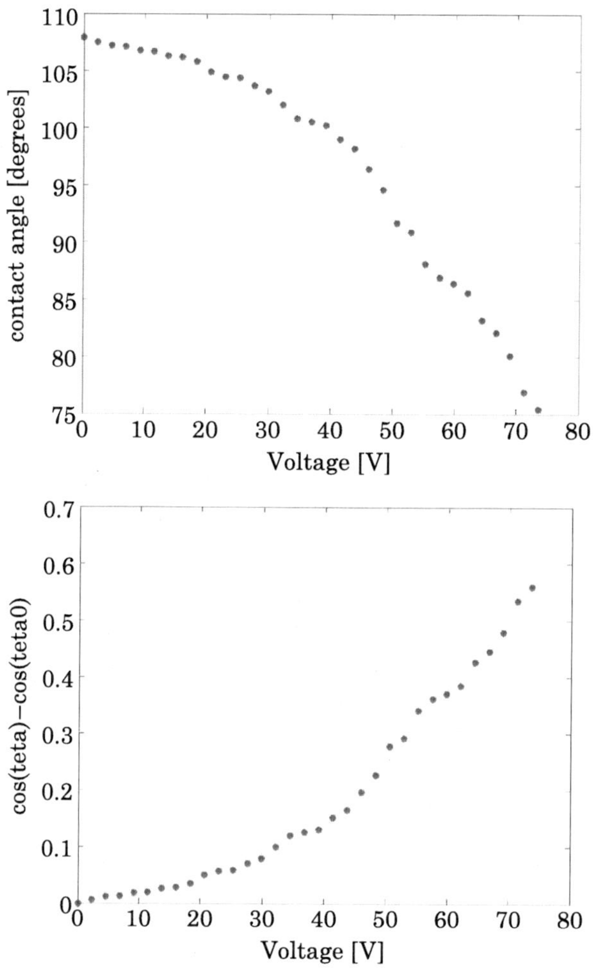

Figure 4.4 Lippmann–Young Relation between contact angle and applied voltage (courtesy Dr. O. Raccurt, CEA/LETI); top: contact angle θ versus V; bottom: $\cos\theta - \cos\theta_0$ versus V.

However, R and θ are not independent since the droplet volume is constant. They are linked by

$$V(R,\theta) = \frac{\pi}{3}R^3\left(2 - 3\cos\theta + \cos^3\theta\right) = \pi R^3\left(\frac{2}{3} - \frac{3\cos\theta}{4} + \frac{\cos 3\theta}{12}\right). \tag{4.14}$$

By putting $dV = 0$, we find

$$dR = RG(\theta)\,d\theta = R\left[-\frac{2\cos^2(\theta/2)\cot(\theta/2)}{2+\cos\theta}\right]d\theta. \tag{4.15}$$

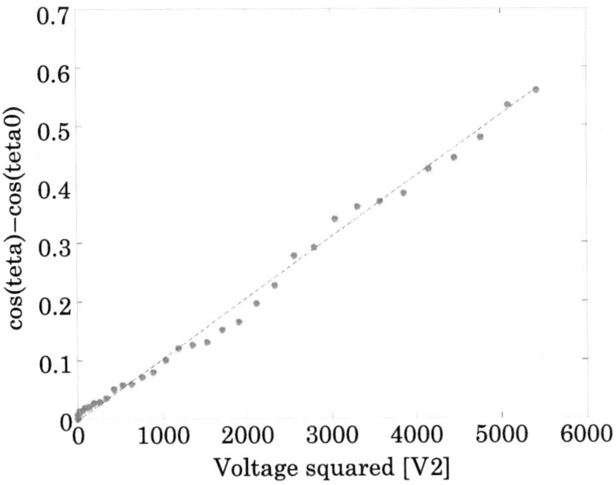

Figure 4.5 Linear dependency between $\cos\theta - \cos\theta_0$ and V^2.

Substituting (4.15) in (4.13) yields

$$\frac{dE}{d\theta} = -R\left[\frac{2\cos^2(\theta/2)\cot(\theta/2)}{2+\cos\theta}\right]\frac{\partial E}{\partial R} + \frac{\partial E}{\partial \theta} = 0. \tag{4.16}$$

Equation (4.16) constitutes the base for the derivation of Young's and Lippmann–Young laws.

4.2.2.1 Derivation of Young's Law

In this case, the only energy we consider is the interfacial energy. We have seen in Chapter 3 that the droplet interfacial energy is given by

$$E_{\text{int}} = \gamma_{\text{LG}} A_{\text{LG}} + (\gamma_{\text{SL}} - \gamma_{\text{SG}})\, A_{\text{SL}} \tag{4.17}$$

where the subscripts S, L, and G denote solid, liquid, and gas phases, respectively. A denotes the interfacial area and γ the interfacial tension. The interfacial areas for a spherical cap have been derived in Chapter 3. We have

$$\begin{aligned} A_{\text{LG}} &= 2\pi R^2(1-\cos\theta) \\ A_{\text{SL}} &= \pi R^2 \sin^2\theta. \end{aligned} \tag{4.18}$$

Finally, the interfacial energy can be cast under the form

$$E_{\text{int}} = \pi R^2\left[2\gamma_{\text{LG}}(1-\cos\theta) + (\gamma_{\text{SL}} - \gamma_{\text{SG}})\sin^2\theta\right]. \tag{4.19}$$

Now, if we plug the expression of E_{int} given by (4.19) into (4.16)

$$\frac{dE_{\text{int}}}{d\theta} = -R\left[\frac{2\cos^2(\theta/2)\cot(\theta/2)}{2+\cos\theta}\right]\frac{\partial E_{\text{int}}}{\partial R} + \frac{\partial E_{\text{int}}}{\partial \theta} = 0$$

we recover, after some algebra, the well-known Young's law

$$\gamma_{\text{LG}}\cos\theta = \gamma_{\text{SG}} - \gamma_{\text{SL}}. \tag{4.20}$$

4.2.2.2 Derivation of Lippmann–Young Law

Consider now the droplet of Fig. 4.3. In Section 4.2.1, we found that almost all of the electric energy is located in the dielectric. Then the electric energy for an elementary volume of the dielectric dv is given by

$$dE_{\text{diel}} = \frac{1}{2}\left(\vec{E}.\vec{D}\right)dv \tag{4.21}$$

where $\vec{D}$ is the polarizability vector field, i.e., electric displacement vector. $\vec{D}$ represents the induced dipole moment in the dielectric per unit volume. For an ideal dielectric, this moment is linearly related to the electric field

$$\vec{D} = \varepsilon_0\varepsilon_D\vec{E} \tag{4.22}$$

and the electric energy in the elementary volume is

$$dE_{\text{diel}} = \frac{1}{2}\varepsilon_0\varepsilon_D\left|\vec{E}\right|^2 dv. \tag{4.23}$$

If we neglect the edge effect, i.e., the distortion of the electric field in the dielectric at the vicinity of the triple contact line (Fig. 4.6), we have a uniform electrical field

$$\vec{E} = \{0, 0, -V/d\}. \tag{4.24}$$

This assumption is correct if the droplet is sufficiently large and might break down if the trend to miniaturizing the EWOD devices continues (at the present time, a typical dimension of a droplet in EWOD devices is 500 μm).

Using (4.24), we can integrate (4.23) to find the electric energy stored in the dielectric

$$E_{\text{diel}}(R,\theta) = \frac{1}{2}\varepsilon_0\varepsilon_D\left(\frac{V}{d}\right)^2 A_{\text{SL}}d = \frac{\varepsilon_0\varepsilon_D V^2}{2d}\pi R^2\sin^2\theta. \tag{4.25}$$

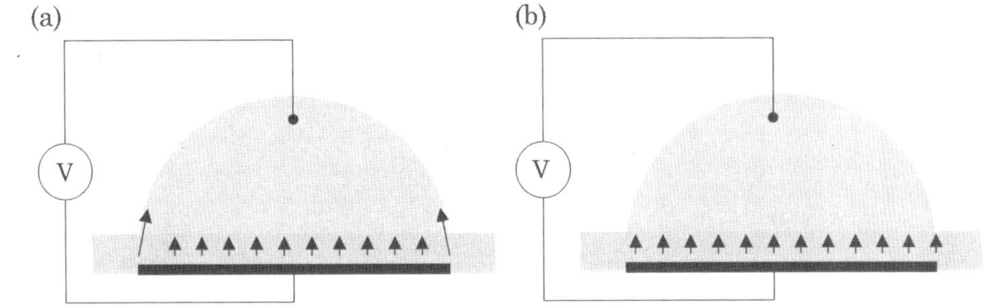

Figure 4.6 (a) The electric field is larger near the triple line due to edge effects; (b) if the contact surface of the liquid is sufficiently large, we can neglect the edge effect.

The total electric energy of the system is the difference between the electric energy stored in the dielectric and the energy in the external charging source [7]. It can be shown that the energy stored in the external charging source is minus twice the energy stored in the dielectric. The basic reason is that every time the drop shape changes, the charged volume immediately under the solid–liquid contact area changes, and a packet of charge ΔQ must be received from or pushed back against the fixed voltage source. This requires a work $W = V\Delta Q = -E$. The energy stored in the charge source is then twice the energy stored in the dielectric but with an opposite sign. Hence, the total energy (surface plus electric energy) of the system is

$$E = R^2 \left[\left(\gamma_{\mathrm{SL}} - \gamma_{\mathrm{SG}} - \frac{\varepsilon_0 \varepsilon_D V^2}{2d} \right) \pi \sin^2\theta + \gamma_{\mathrm{LG}} 2\pi \left(1 - \cos\theta\right) \right]. \quad (4.26)$$

Upon substitution of (4.26) into (4.16), and observing that the derivation is similar to that of equation (4.19), replacing the term $(\gamma_{\mathrm{SL}} - \gamma_{\mathrm{SG}})$ by $\left(\gamma_{\mathrm{SL}} - \gamma_{\mathrm{SG}} - \varepsilon_D V^2/2d\right)$, we recover the Lippmann–Young equation

$$\cos\theta - \left(\frac{\gamma_{\mathrm{SG}} - \gamma_{\mathrm{SL}}}{\gamma_{\mathrm{LG}}} + \frac{\varepsilon_0 \varepsilon_D V^2}{2\gamma_{\mathrm{LG}} d} \right) = 0 \quad (4.27)$$

which can be equivalently expressed as

$$\cos\theta = \cos\theta_0 + \frac{\varepsilon_0 \varepsilon_D V^2}{2\gamma_{\mathrm{LG}} d}. \quad (4.28)$$

This energy minimization approach explicitly considers the dielectric layer and the dielectric energy within it. However, like the thermodynamic approach of Section 4.2.1, it does not provide a physical picture of how the contact angle is reduced in mechanical terms. Whether such a distortion is solely due to a change in interfacial tension or some unique force at the contact line remains

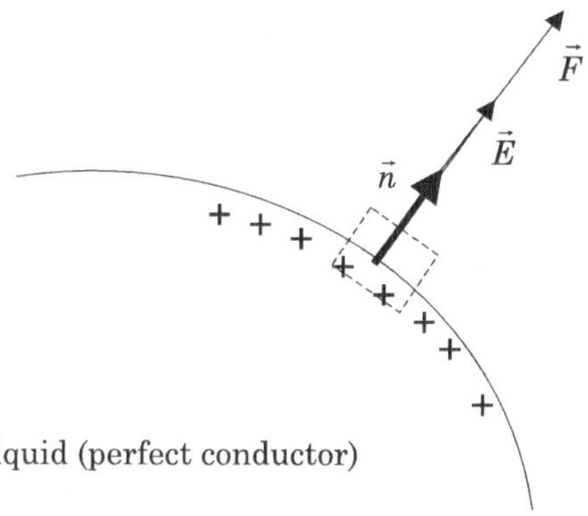

Figure 4.7 Electric force acting at the interface of a conducting liquid.

unresolved. While such a distinction may not be necessary for the contact angle, it becomes paramount when electrowetting dynamics needs to be scrutinized. With viscous effects, energy is not conserved and one cannot use the energy minimization approach. The actual force at the contact line must be captured. In the following section, we consider the electric forces exerted on the liquid explicitly, although we shall limit ourselves to static wetting problems.

4.2.3 Electromechanical Approach

The two preceding approaches do not give any insight into the electric forces acting on the droplet, particularly why a Maxwell force at the solid–liquid interface has the same effect as reducing the gas–liquid interfacial tension. These two approaches only address the static problem. If one wants to address the dynamic problem, which would require the inclusion of viscous effects, the forces exerted by the electric field on the interface have to be explicitly determined. The electromechanical approach was introduced by Jones et al. [8,9], Kang [10], and recently reviewed by Zeng and Korsmeyer [11].

In the case of liquids, the body force density due to an electric field is given by the Korteweg–Helmholtz relation

$$\vec{f_k} = \sigma_f \vec{E} - \frac{\varepsilon_0}{2} E^2 \nabla \varepsilon_f + \nabla \left[\frac{\varepsilon_0}{2} E^2 \frac{\partial \varepsilon_f}{\partial \rho} \rho \right] \tag{4.29}$$

where ρ is the mass density of the liquid, ε_f the permittivity of the liquid, and σ_f the electric charges. The last term in equation (4.29) describes electrostriction and can be neglected here as the mass density of the liquid remains constant. The force acting on a volume element dV is obtained by integrating equation (4.29). It can be shown that it is equivalent to integrating the momentum flux density, i.e., the Maxwell stress tensor

$$T_{ik} = \varepsilon_0 \varepsilon \left(E_i E_k - \frac{1}{2}\delta_{ik} E^2 \right) \tag{4.30}$$

along the surface of dV. In (4.30), the notation E^2 corresponds to $\left|\vec{E}\right|^2$ and δ_{ik} is the Kronecker delta function: $\delta_{ik} = 0$ if $i \neq k$ and $\delta_{ii} = 1$; and $i, k = x, y, z$. Surface integration of (4.30) is much easier than volume integration of (4.29). In general, the net force acting on the liquid volume element is

$$F_i = \oint_\Omega T_{ik} n_k dA \tag{4.31}$$

where we use the Einstein summation convention (summation on the repeated indices). At the surface of a perfectly conducting liquid, on the gas side, the electric field is perpendicular to the surface (Fig. 4.7), and related to the surface density of electric charges by Gauss' law

$$\sigma_s = \varepsilon_0 \vec{E}.\vec{n} \tag{4.32}$$

where $\vec{n}$ is the outward unit normal vector. Moreover, the electric field vanishes in the conducting liquid. If we consider the x, y, z axis system such as the x-axis is aligned with n, the electric field is $E = (E_x, 0, 0)$ in the gas domain, and $E = (0, 0, 0)$ in the liquid domain. In the gas domain, the Maxwell tensor is

$$[T] = \varepsilon \begin{pmatrix} \frac{1}{2}\varepsilon_0 E_x^2 & 0 & 0 \\ 0 & -\frac{1}{2}\varepsilon_0 E_x^2 & 0 \\ 0 & 0 & -\frac{1}{2}\varepsilon_0 E_x^2 \end{pmatrix} \tag{4.33}$$

and vanishes in the liquid domain

$$[T] = [0]. \tag{4.34}$$

We can now integrate (4.31); the cross terms xy, yz, and zx are all zero, the forces in the y (respectively z) direction cancel out, and we find that the only non-vanishing contribution is a force directed along the outward normal $\vec{n}$

$$\frac{\vec{F}}{\delta A} = P_e \vec{n} = \frac{\varepsilon_0}{2} E^2 \vec{n} = \frac{\sigma_s}{2} \vec{E} \tag{4.35}$$

where δA is an elementary surface area of the interface. In (4.35), P_e is the electrostatic pressure defined by $P_e = \varepsilon_0 / 2E^2$. The electrostatic pressure P_e acts

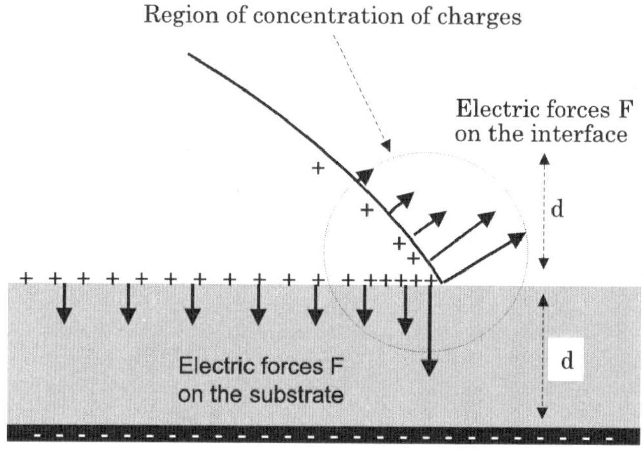

Figure 4.8 Schematic of the electric charge distribution in the vicinity of the triple contact line. Electric charges are located at the solid–liquid interface and at the liquid–gas interface, approximately within a distance d from the solid surface. In reality, this is a very simplified view since the liquid interface is distorted by the electric forces very close to the triple contact line.

on the liquid surface, and brings a negative contribution to the total pressure within the liquid. The liquid interface is distorted by the electric forces acting on it, according to equation (4.35). They depend on the distribution of charges σ_f at the surface. Electric charges at the liquid–gas interface are located close to the triple contact line, as sketched in Fig. 4.8, within a distance equal to the dielectric thickness d.

The problem of the surface distortion has been solved by Kang [10] and Vallet et al. [12] using the Schwarz–Christoffel conformal mapping [13]. First, note that the electric potential ϕ satisfies the Laplace equation $\nabla^2\phi = 0$ and consequently is a harmonic function. The theory of analytic and harmonic functions show that there exists a conformal mapping (the Schwarz–Christoffel mapping) that transforms the functions E and ϕ for a half plane into the same functions for a wedge (Fig. 4.9). This mapping shows that the field and charges concentrate at the tip of the wedge. In fact, as is the case of any harmonic function near a geometric singularity, the electric field is singular at a wedge (see for example Thamida and Chang [14]) but the Maxwell pressure is integrable and produces a finite Maxwell force at the contact line after (4.35) is integrated.

$$
\begin{aligned}
F_{\text{horizontal}} &= \frac{\varepsilon_0 \varepsilon_D V^2}{2d} \\
F_{\text{vertical}} &= \frac{\varepsilon_0 \varepsilon_D V^2}{2d} \frac{1}{\tan\theta}.
\end{aligned}
\tag{4.36}
$$

We notice first that the electric forces distribution on the liquid–gas interface caused by the Maxwell-stress is limited to a very small region close to the triple

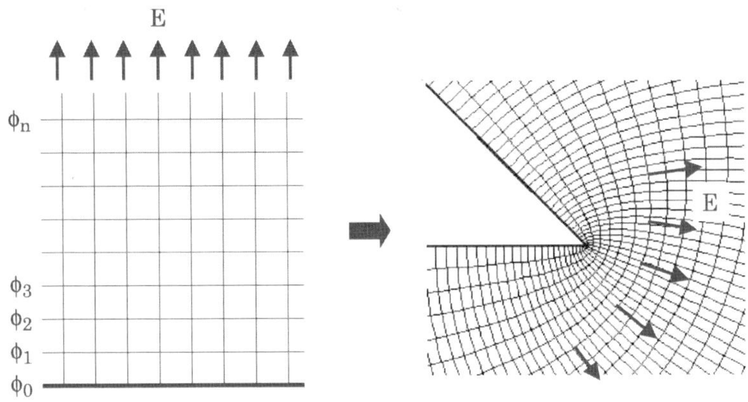

Figure 4.9 Principle of the Schwarz–Christoffel conformal mapping. The two orthogonal fields E and ϕ are transformed from a simple half plane geometry (with an evident solution) to the geometry of a wedge. Remark the concentration of the electric field aroung the tip of the wedge.

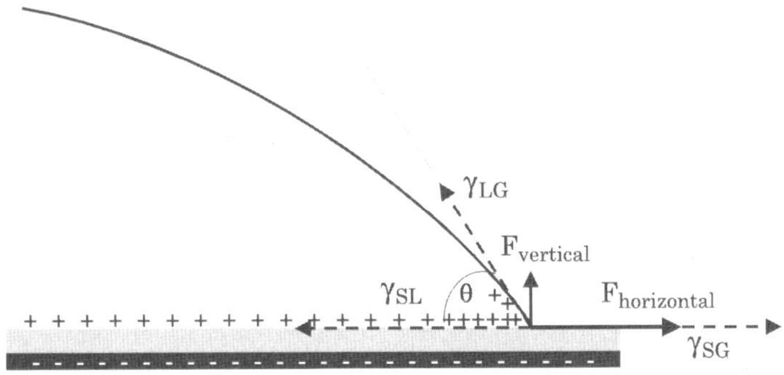

Figure 4.10 Resultant of the electric forces on the liquid–gas interface.

line. At a macroscopic scale, we can picture the situation as in Fig. 4.10, and assume that the forces are point forces acting on the triple line.

Second, using relation (4.36), we can make the balance of the forces—electric plus capillary—in the horizontal direction at the triple line, as has been done for deriving Young's equation, and we recover once again Lippmann's equation

$$\cos\theta = \cos\theta_0 + \frac{\varepsilon_D V^2}{2\gamma_{LG} d}. \tag{4.37}$$

The electromechanical approach shows that the Lippmann term (electrowetting number η) comes directly from the Maxwell stress on the liquid–gas interface near the contact line, but its origin can be traced back to the thin solid dielectric layer and the wedge geometry of the liquid. In fact, since the liquid has a much higher permittivity (and conductivity) than the gas, it is the wedge

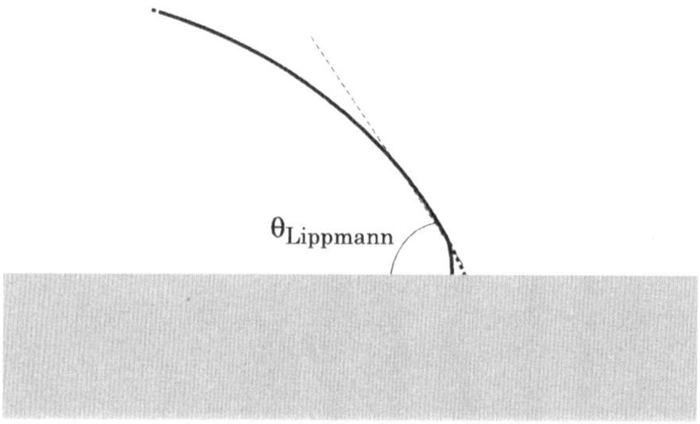

Figure 4.11 Deformation of the interface at high voltage [15].

shape of the gas phase that is responsible for distortion of the contact angle, as is clearly shown by the Schwarz–Christoffel transformation. The Lippmann term can also be called electrostatic pressure. Another consequence of equation (4.36) is that there is a vertical force on the liquid interface near the triple line. This vertical force increases quickly when the contact angle decreases. The theory even predicts that this force goes to infinity when the contact angle goes to zero. This could be an explanation for the phenomenon of the contact angle saturation (the contact angle cannot go to zero), usually accompanied by ejection of nanodrops (see Section 4.4). This could also be the reason why, extremely close to the solid surface, the shape of the interface is distorted [14], especially when the applied voltage is important (Fig. 4.11).

4.3 Lippmann–Young Law and the Electrocapillary Equivalence

As seen from the previous sections, Lippmann's law has many physical interpretations. Although they all lead to the same mathematical formula, they often have different physical origins and have different physical implications. In the thermodynamic approach, Lippmann's law is related to a change in the solid–liquid apparent interfacial tension

$$\gamma_{SL}(V) = \gamma_{SL,0} - \frac{C}{2}V^2 \tag{4.38}$$

and substitute Young's law

$$\begin{aligned} \gamma_{SG} - \gamma_{SL,0} &= \gamma_{LG}\cos\theta_0 \\ \gamma_{SG} - \gamma_{SL}(V) &= \gamma_{LG}\cos\theta \end{aligned} \tag{4.39}$$

to obtain the Lippmann–Young law

$$\cos\theta = \cos\theta_0 + \frac{1}{2}\frac{C}{\gamma_{LG}}V^2. \tag{4.40}$$

Another slightly different way of writing the Lippmann–Young law, making the electric and capillary forces appear, is

$$\gamma_{LG}\cos\theta - \gamma_{LG}\cos\theta_0 = \frac{1}{2}CV^2. \tag{4.41}$$

We see here that the voltage can be continuous or alternating. In the case of alternating voltage, the value of the voltage V in equations (4.40) and (4.41) should be the rms (root mean square) value

$$V_{\text{rms}} = \frac{\sqrt{2}}{2}V$$

and

$$V_{\text{rms}}^2 = \frac{1}{2}V^2.$$

The first interpretation of the Lippmann–Young law stems from equation (4.40): when actuated, the contact angle of the droplet changes resulting in an "*apparent*" change of capillarity. It is very important here to note the word "apparent": in reality, the contact angle at the wall is not the Lippmann contact angle (Fig. 4.11) but still the Young contact angle; however, at some—not well defined—distance to the surface, the macroscopic contact angle has the value given by the Lippmann–Young law [15]. Besides, one should be very careful with dynamic motion as viscous effects with other length scales and dynamic contact angle conditions would invalidate the current static theories. Indeed, the situation is quite different during rapid dynamic motion of droplets under EWOD actuation: the contact angles are not the equilibrium contact angles. For example, in Fig. 4.12 we show the observed dynamic contact angles during the motion of a droplet under EWOD actuation [16]. So this interpretation is not formally rigorous, but it is very convenient in an engineering approach to explain many *static or quasi-static* phenomena occurring in EWOD microsystems [17–20]. For the first analysis of dynamic (or spontaneous) electrowetting, see Yeo and Chang [21], the review [22], and a related treatment of electrokinetic displacement of bubbles by Takhistov et al. [23].

The second, more rigorous interpretation of the Lippmann–Young law stems from equation (4.38). In Chapter 2 we have seen that the capillary line force on a triple line with contact angle θ_0 is given by

$$f_{\text{cap}} = \gamma_{LG}\cos\theta_0. \tag{4.42}$$

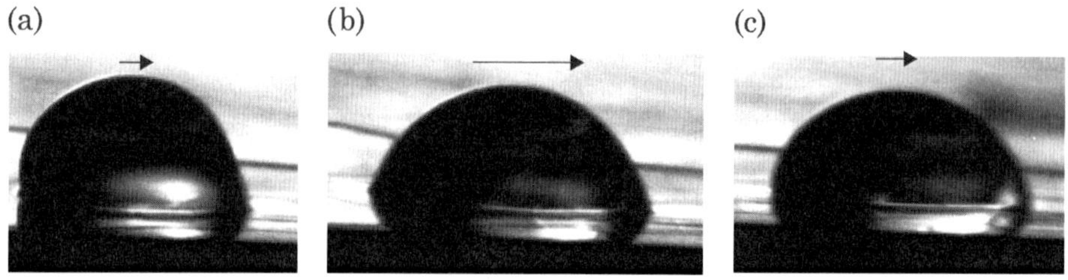

Figure 4.12 Photograph of a droplet during its motion: (a) phase 1, the droplet starts its motion; (b) droplet in full motion, (c) end of motion (courtesy R. Bavière, CEA/LETI).

This line force is directed in the solid plane perpendicularly to the triple contact line and oriented outward. Following Jones et al. [8,9], Kang [10], and Zeng and Korsmeyer [11], we have seen that the line tension electrostatic force is

$$f_{\mathrm{EWOD}} = \frac{C}{2} V^2. \tag{4.43}$$

The Lippmann–Young law—under the form (4.41)—can then be written as

$$f_{\mathrm{tot}} = f_{\mathrm{cap}} + f_{\mathrm{EWOD}}. \tag{4.44}$$

Under such a form, the electrowetting effect is to add an electric force to the capillary force on the triple line. A characteristic example of the interpretation of the Lippmann–Young law by electric charge distribution and the Maxwell-stress tensor has been pointed out by Yeo and Chang [21]: they showed that a droplet can be stretched using two parallel line electrodes, instead of planar electrodes (Fig. 4.13). Clearly, in such a case, it is impossible to use the "apparent" contact angle interpretation.

4.4 Saturation

4.4.1 Saturation of the Electrowetting Effect

It has been observed that the domain of validity of the Lippmann–Young equation is limited to small and medium electric potentials. We have set up an experimental micro-device to explore the Lippmann–Young relations for different fluids (Fig. 4.3). In this experimental setup, the zero potential electrode is located directly in the substrate below the droplet. Contact angles have been determined by image analysis of side views of the droplet using a Digidrop®analysis apparatus.

Relation (4.40) has been experimentally checked for different biological buffers (Fig. 4.14) on an EWOD substrate consisting of a 300 nm Si_3N_4

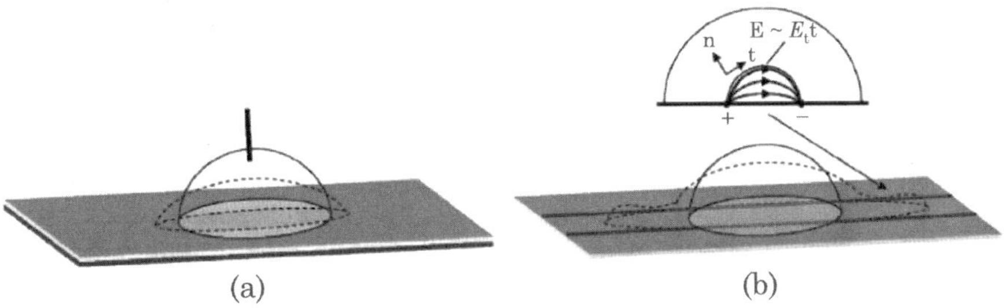

Figure 4.13 Two different configurations for electrowetting experiments: (a) Static change in the macroscopic contact angle obtained using dielectric film coated planar electrodes; (b) electrowetting effect produced using parallel line electrode configuration. The inset shows a schematic representation of the cross section of the bulk drop and electrowetting film showing that the electric field lines are predominantly tangential to the interface. From [21]. © 2006 APS; reprinted with permission.

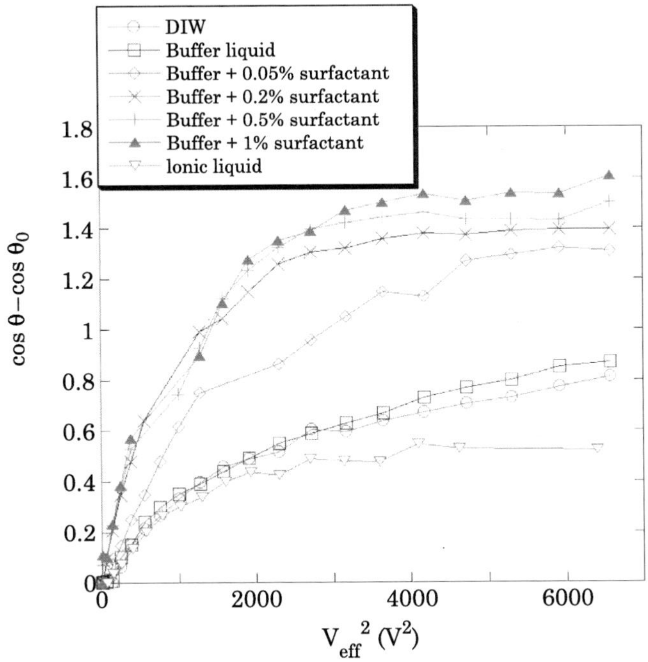

Figure 4.14 Plots of $(\cos\theta - \cos\theta_0)$ as a function of V^2 for different fluids (ionic liquid, de-ionized water and buffers with 0, 0.05, 0.2, 0.5 and 1% surfactant). The drops are immerged in silicone oil.

dielectric layer and a 1 μm Teflon coating. The experimental results show that the Lippmann–Young relation is valid for electric potentials below 40 to 50 volts; above this value there is a saturation effect and the contact angle does not decrease any more with the increasing voltage. Hence, there is an upper limit to the electrowetting force.

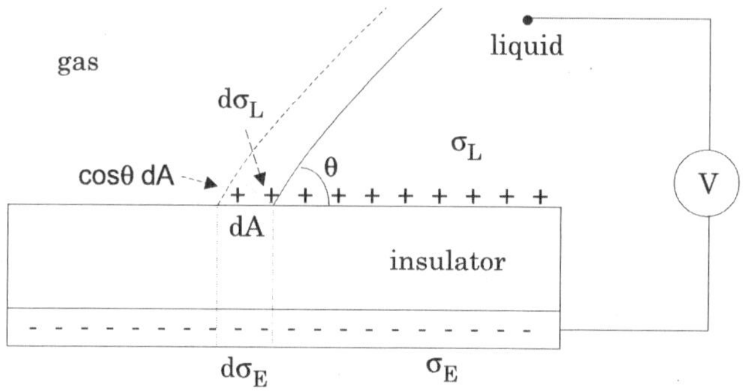

Figure 4.15 Schematic of the electric charges during electrowetting. No trapped charges.

4.4.2 Possible Explanations for the Saturation Effect

At the present time, different explanations have been proposed for the saturation limit. We have already seen a possible explanation with Kang's formula for the electric force—equation (4.36). However, there is no consensus on the real cause of saturation, and the debate is still open. The saturation effect is important because it limits the electrowetting force acting on a droplet and constitutes a bottleneck for the development of scale down electrowetting devices. In the following sections we present different explanations proposed for the saturation effect.

4.4.2.1 Trapping of Charges

Verheijen and Prins [24] proposed a model of saturation based on the trapping of charges in the insulator when the voltage becomes sufficiently large. This trapping of charges reduces the density of charges at the solid–liquid surface and weakens the electrowetting effect. Saturation is obtained if, above the saturation voltage, the solid–liquid density of charges does not increase any more; in such a case the electric charges accumulate only in the dielectric.

In order to investigate the effect of trapping of charges, we follow the approach of Verheijen and Prins which consists in using the virtual displacement principle to derive the expression of the free energy of the system. In the following, the development is done first in the case of no trapped charges, then in the case of charge trapping.

4.4.2.1.1 Virtual displacement: no charge trapping and Lippmann–Young equation

When an electric voltage is set between the electrode and the liquid, a charge density σ_L builds up in the liquid phase close to the liquid–solid interface. This charge density induces an image charge density σ_E on the metal electrode (Fig. 4.15). An increase of surface charges provokes a displacement of the interface. Suppose there is an infinitesimal displacement dA. The free energy of the system can be written

$$dE = \gamma_{SL} dA - \gamma_{SG} dA + \gamma_{LG} dA \cos\theta + dU - dW \tag{4.45}$$

where U is the capacitive energy stored in the dielectric and dW the work performed by the voltage source. Note that equation (4.45) collapses to the Young equation in the absence of an externally applied voltage: at equilibrium, $dE/dA = 0$, and we have $0 = \gamma_{SL} - \gamma_{SG} + \gamma_{LG} \cos\theta$, which is Young's equation.

The energy stored in the dielectric is derived in Section 4.2.2.2, equation (4.21)

$$\begin{aligned} \frac{dU}{dA} &= \frac{1}{2} dED = \frac{1}{2} d\left(\frac{V}{d}\right)(\varepsilon_0 \varepsilon_D E) \\ &= \frac{1}{2} d\left(\frac{V}{d}\right)\left(\varepsilon_0 \varepsilon_D \frac{V}{d}\right) = \frac{1}{2} d\left(\frac{V}{d}\right)\sigma_L = \frac{1}{2} V \sigma_L. \end{aligned} \tag{4.46}$$

In (4.46), D is the electric displacement $D = \varepsilon_0 \varepsilon_D E$, and we have used Gauss's law to obtain $\varepsilon_0 \varepsilon_D E = \sigma_L$. Using the same argument as that of Section 4.2.2.2 [7], the energy of the voltage source is twice that of the dielectric

$$\frac{dW}{dA} = V \sigma_L. \tag{4.47}$$

Upon substitution of (4.46) and (4.47) into (4.45), and setting $dE/dA = 0$, we recover once again the Lippmann–Young equation

$$\gamma_{LG} \cos\theta = \gamma_{SG} - \gamma_{SL} + \frac{\varepsilon_0 \varepsilon_D}{2d} V^2. \tag{4.48}$$

4.4.2.1.2 Trapping of charges and saturation effect

Suppose now that, at sufficiently high voltage, charges are pulled into the dielectric by electric forces. These charges may become trapped if the interactions of the ions with the solid are stronger than that with the liquid. Suppose for simplicity of the reasoning that these charges are trapped at a depth d_2 from the solid surface (Fig. 4.16).

The work performed by the voltage source is still given by (4.47). However, the energy stored in the dielectric is different than that produced by (4.46) because the charges are not located at the same place, and also because of the presence of trapped charges in the dielectric under the gas. If V_T denotes the potential at the level of the trapped charges on the gas side and V_T^L in the liquid side, we can write

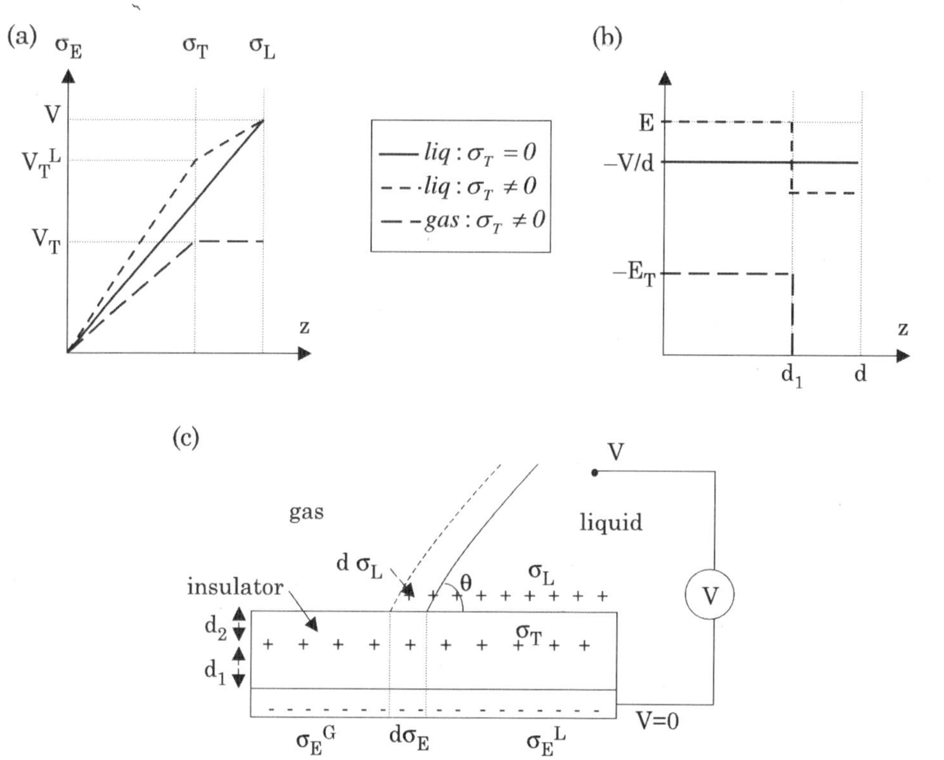

Figure 4.16 Schematic of the electric charges during electrowetting with and without trapped charges. (a) Potential profiles, (b) electric field, (c) schematic of the trapped charges. Remark that the bottom electrode is at a zero potential and the liquid at the potential V.

$$\frac{dU_L}{dA} = \frac{1}{2}d_1E_1D_1 + \frac{1}{2}d_2E_2D_2 = \frac{1}{2}d_1\frac{V_T^L}{d_1}(\sigma_T + \sigma_L) + \frac{1}{2}d_2\frac{V - V_T^L}{d_2}\sigma_L$$
$$= \frac{1}{2}V_T^L\sigma_T + \frac{1}{2}V\sigma_L. \tag{4.49}$$

On the gas side

$$\frac{dU_G}{dA} = \frac{1}{2}d_1\frac{V_T}{d_1}\sigma_T = \frac{1}{2}V_T\sigma_T. \tag{4.50}$$

Hence

$$\frac{dU}{dA} = \frac{dU_L}{dA} - \frac{dU_G}{dA} = \frac{1}{2}V_T^L\sigma_T + \frac{1}{2}V\sigma_L - \frac{1}{2}V_T\sigma_T. \tag{4.51}$$

Setting $dE/dA = 0$, equation (4.45) becomes

$$\frac{dE}{dA} = 0$$
$$= \gamma_{\mathrm{SL}} - \gamma_{\mathrm{SG}} + \gamma_{\mathrm{LG}} \cos\theta + \frac{1}{2} V_T^L \sigma_T + \frac{1}{2} V \sigma_L - \frac{1}{2} V_T \sigma_T - V \sigma_L. \quad (4.52)$$

Let us express V_T^L, σ_T and σ_L using Gauss' law

$$V_T^L = V_T + \frac{d_1}{d}(V - V_T)$$
$$\sigma_T = \varepsilon_0 \varepsilon_D \frac{V_T}{d_1}$$
$$\sigma_L = \varepsilon_0 \varepsilon_D \frac{V - V_T^L}{d_2} = \varepsilon_0 \varepsilon_D \frac{V - V_T}{d}.$$

After substitution in (4.52) and, we obtain

$$\gamma_{\mathrm{LG}} \cos\theta = \gamma_{\mathrm{SG}} - \gamma_{\mathrm{SL}} + \frac{\varepsilon_0 \varepsilon_D}{2d}(V - V_T)^2 \quad (4.53)$$

which is similar to the Lippmann–Young law, except that the potential is $V - V_T$ instead of V. Let us remark here that V_T is a function of V. At saturation, i.e., for $V > V_{\mathrm{sat}}$ (or $V < -V_{\mathrm{sat}}$), the contact angle remains constant; from (4.53) we deduce that $V - V_T$ must be constant at saturation (hence $V_T = V + cst$); consequently the density of surface charges σ_L must also be constant at saturation. We then have the diagram of Fig. 4.17: the trapped potential V_T is proportional to V outside the interval $[-V_{\mathrm{sat}}, V_{\mathrm{sat}}]$ and vanishes inside this interval; whereas σ_L is constant outside the interval and proportional to V inside the interval.

Verheijen and Prins have experimentally observed that trapped charges exist, but they found that it was very difficult to find a quantitative estimate for V_T.

4.4.2.2 Zero Surface-liquid Energy Limit

This model was pioneered by Peykov, Quinn, Ralston, and Sedev [25,26], and we will refer to it in the following by the abbreviation "PQRS model." The PQRS model considers that the saturation phenomenon is a thermodynamic limit of stability. In Section 4.2.1, we presented the thermodynamic approach to the electrowetting effect; in this approach, the effective solid–liquid surface tension decreases with the voltage according to equation

$$\gamma_{\mathrm{SL}}^{\mathrm{eff}}(V) = \gamma_{\mathrm{SL}} - \frac{1}{2} C V^2 \quad (4.54)$$

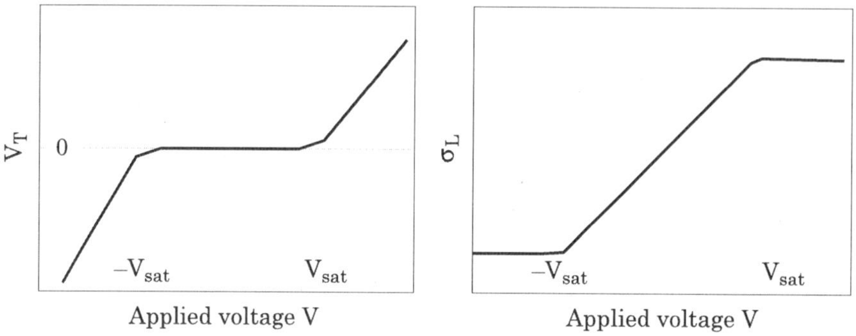

Figure 4.17 Left: trapped charge potential V_T as a function of applied potential V; right: surface charges σ_T as a function of the applied potential.

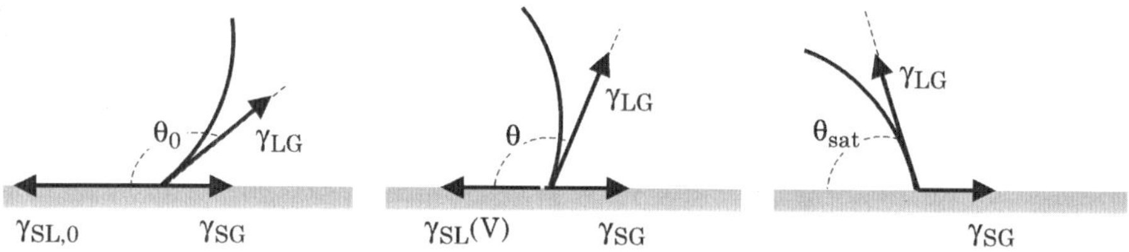

Figure 4.18 Sketch of the different contact angles depending on the applied voltage. Left: at zero potential, the contact angle is determined by the classical Young law; middle: the contact angle decreases when the applied voltage increases; right: the lower limit of the contact angle is obtained when the solid–liquid surface tension vanishes.

and the voltage dependent Young's law can be cast under the form

$$\cos\theta\,(V) = \frac{\gamma_{SG} - \gamma_{SL}^{\text{eff}}(V)}{\gamma_{LG}}. \tag{4.55}$$

Relation (4.55) is plotted in Fig. 4.18 for different values of the applied voltage. For a zero voltage, the force balance is that defined by the classical Young's law. As the voltage increases, the effective solid–liquid surface tension decreases, and the contact angle decreases according to equation (4.55). The lower limit for the effective solid–liquid surface tension is zero; at this point, the projection of γ_{LG} is equal and opposite to γ_{SG}. When this value is reached, the minimum contact angle is obtained. This minimum value is the saturation contact angle θ_{sat}.

According to equation (4.55), we have, at saturation

$$\cos\theta_{sat} = \cos\theta\,(V_{sat}) = \frac{\gamma_{SG}}{\gamma_{LG}} \tag{4.56}$$

which is equivalent to

$$\theta_{\text{sat}} = \arccos\left(\frac{\gamma_{SG}}{\gamma_{LG}}\right). \tag{4.57}$$

Using equation (4.54), we derive

$$V_{\text{sat}} = \left(\frac{2\gamma_{SL,0}}{C}\right)^{\frac{1}{2}} = \left(\frac{2\left(\gamma_{SG} - \gamma_{LG}\cos\theta_0\right)}{C}\right)^{\frac{1}{2}}. \tag{4.58}$$

Formulas (4.57) and (4.58) require the knowledge of the solid surface tension γ_{SG}. We have seen in Chapter 2 that this surface tension is rather difficult to determine, except in the case of apolar surfaces for which Zisman's criterion produces a good approximation [27]. In the particular case of EWOD, the coating of the substrate is realized with apolar materials like Teflon, parylene, PET, PTFE, etc., and we can reasonably use the value of the wetting surface tension (Zisman's criterion) for γ_{SG} in equations (4.57) and (4.58). The PQRS model predicts values of the saturation contact angle in reasonably good agreement with the experimental observations, although it is sometimes somewhat overestimated.

4.4.2.3 Electric Resistance of the Liquid

This approach has been proposed by Shapiro and coworkers [5,6]. In their model, the droplet liquid is not perfectly conducting; it presents a small electric resistance that depends on the shape of the droplet. When the droplet spreads under the action of the applied electric potential, the resistance of the liquid increases: saturation is then seen as a geometrical effect.

The electrical diagram for the droplet-substrate system is shown in Fig. 4.19. In each domain (respectively liquid and dielectric) the electric potential ϕ is given by the equation

$$\nabla.\vec{j} = \nabla\left(\sigma\vec{E}\right) = \nabla\left(\sigma\nabla\phi\right) = \nabla\left(\frac{1}{\rho}\nabla\phi\right) = 0 \tag{4.59}$$

where j is the electric current density, σ the electric conductivity (unit S/m), and ρ the electric resistivity (unit Ω/m). The boundary conditions are (Fig. 4.20)

- $\phi = V$ at the top electrode
- $\phi = 0$ at the bottom electrode (4.60)
- $\nabla\phi.\hat{n} = 0$ at the droplet surface

where $\hat{n}$ is the normal unit vector.

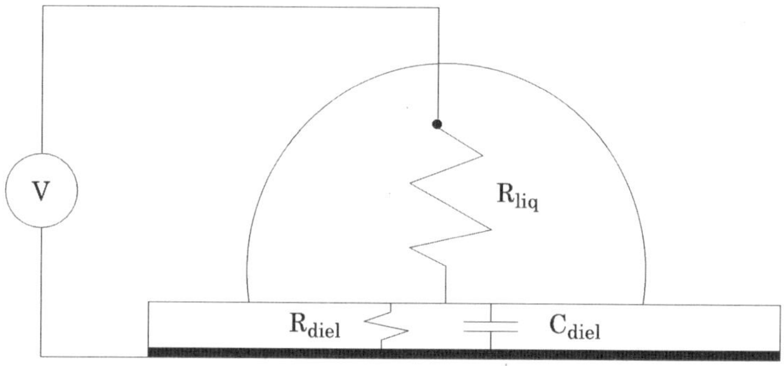

Figure 4.19 Electric diagram of the system showing a small amount of liquid resistance and a large dielectric solid resistance.

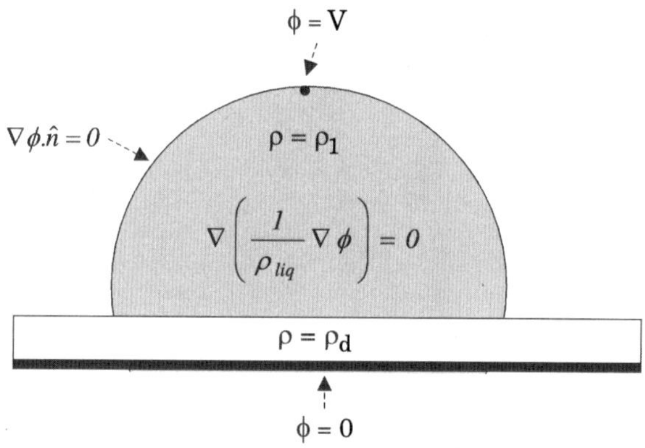

Figure 4.20 Boundary conditions and domain equations for the calculation of the electric potential in the slightly resistive droplet.

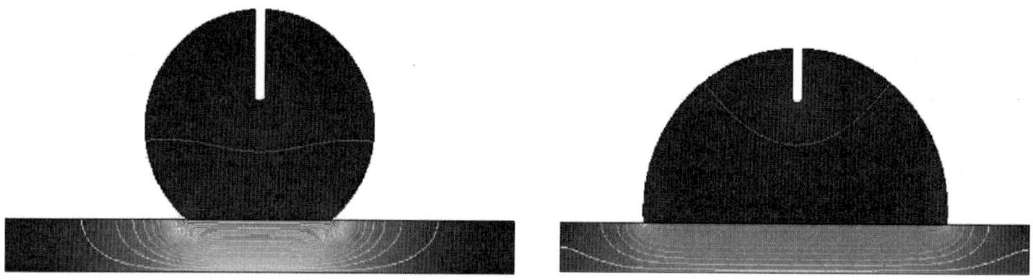

Figure 4.21 The amount of stored electrical energy decreases with the contact angle. Left: large contact angle; right: smaller contact angle. The contour lines correspond to the electric potential (calculation performed with the COMSOL numerical software).

Equation (4.59) with boundary conditions (4.62) can easily be solved using a finite element technique (Fig. 4.21). The results show that the strength of the electric field inside the solid dielectric, and thus the amount of stored electrical energy, decreases with the contact angle. As the droplet spreads,

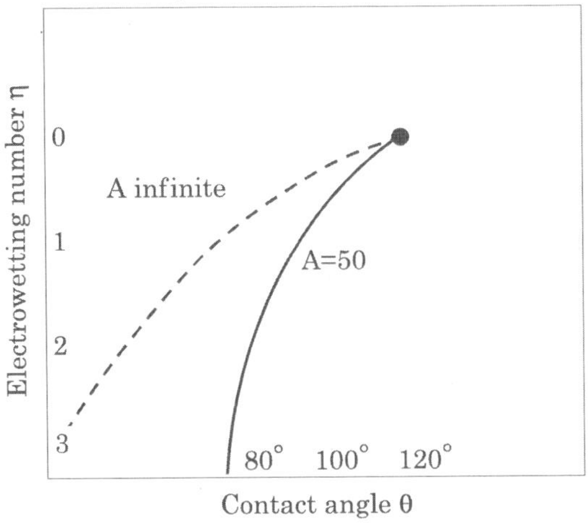

Figure 4.22 When the electrowetting number increases, the contact angle cannot go to zero if there is a sufficient resistivity of the liquid (A is the ratio $\rho_d d/\rho_l R_0$, A infinite corresponds to a perfectly conducting liquid).

there is progressively a longer liquid path from the top electrode to the bottom electrode near the triple line. If there were no electric resistance in the liquid, the norm of the electric field would remain the same. Globally, the resistance of the droplet increases when the droplet spreads. Using Shapiro's energy approach, and by analogy with equation (4.26), it can be shown that the energy of the system can be cast under the form

$$\frac{E(\theta)}{\gamma_{LG} R_0^2} = r(\theta)^2 \left[\left(\frac{\gamma_{SL} - \gamma_{SG}}{\gamma_{LG}} \right) \pi \sin^2\theta + 2\pi (1 - \cos\theta) \right] - \frac{1}{2} \left(\frac{\varepsilon_D V^2}{\gamma_{LG} d} \right) r(\theta)^2 \frac{d}{R_0} F\left(\theta, \frac{\rho_d}{\rho_l}, \frac{d}{R_0} \right) \tag{4.61}$$

where R_0 is the initial radius, $r(\theta) = R(\theta)/R_0$, ρ_l and ρ_d the liquid and dielectric resistivity, and F a function that can be determined numerically. The problem depends now on three nondimensional parameters:

- the surface tension coefficient $\Gamma = \frac{\gamma_{SL} - \gamma_{SG}}{\gamma_{LG}}$
- the electrowetting number $\eta = \frac{\varepsilon_D V^2}{\gamma_{LG} d}$
- the ratio of the solid to liquid resistivity $A = \frac{\rho_d}{\rho_l} \frac{d}{R_0}$

The difference with Section 4.2.3 is the supplementary term due to the resistivity of the liquid.

The function F contains the information on the decrease in energy when the contact angle decreases. It can be shown numerically, using formula (4.61), that the contact angle does not go to zero if the liquid has a sufficient resistivity (Fig. 4.22). Above the saturation limit, the capacitive energy does not increase,

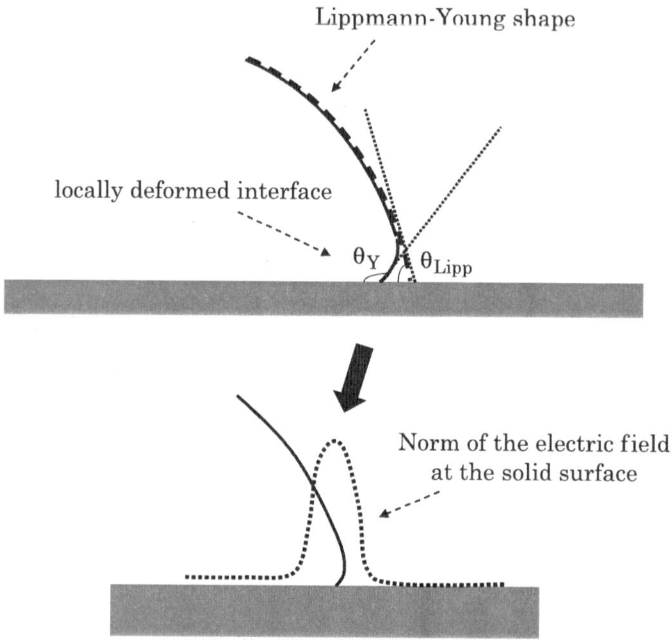

Figure 4.23 Top: at high voltage, the liquid interface is distorted under the action of the electric field. Bottom: the magnitude of the electric field may be 10 times larger at the contact line at high voltage.

and the increase in electric energy is balanced by the increase of the resistive loss. The saturation is then caused by the global resistance of the liquid when it spreads on the solid surface.

Shapiro's approach is very sensitive to the ratio of the resistivities. The latter has been used as a fitting parameter and is difficult to measure precisely.

4.4.2.4 Other Models and Discussion

All the preceding models neglect the "fringe effect", i.e., the increase of the electric field near the triple line due to the discontinuity of material properties at the triple line. Papathanasiou and Boudouvis [15] have shown that the magnitude of the electric field increases up to 10 times in the dielectric at the level of the triple line for an external applied potential of the order of the saturation voltage (in their calculation, the liquid has been assumed to be perfectly conducting). Besides, the liquid interface is distorted by the action of the electric field in the vicinity of the contact line (Fig. 4.23). Such an increase of the electric field results in the dielectric breakdown and charge leakage (see Section 4.6.2). In their view, saturation is closely associated to dielectric breakdown.

Vallet and co-workers [12] have observed that, close to saturation, the luminescence of a region neighboring the triple line suddenly increases (Fig. 4.24).

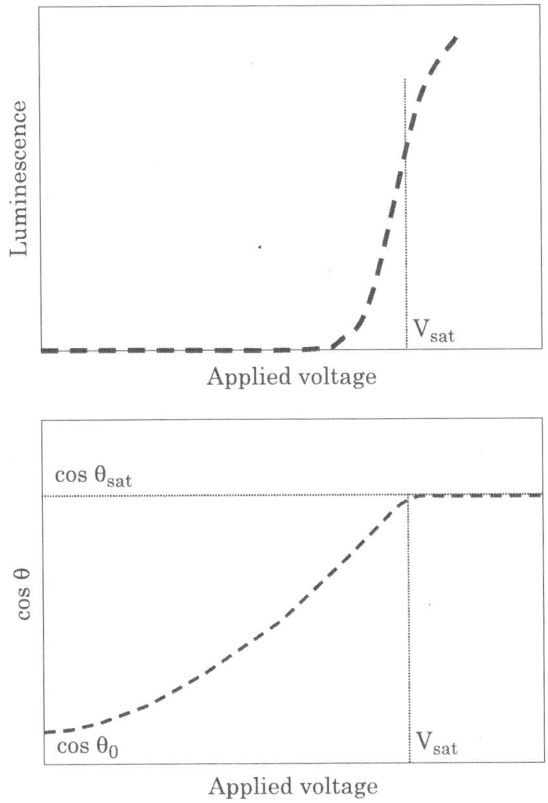

Figure 4.24 Top: luminescence versus applied voltage; bottom: saturation curve. The comparison of the two curves shows that luminescence is correlated to saturation.

They remark that if the air is ionized, electrowetting stops working: the principle of electrowetting is that charges are adsorbed in the conducting liquid at the solid–liquid interface. These charges exert a lateral pressure outwards on the wetting line, inducing the spreading of the drop. If air gets ionized, there is a leakage of the charges and the pressure on the triple line cannot increase with the applied voltage. The theory proposed by Vallet et al. assumes that the surrounding fluid is a gas. But what if the surrounding fluid is silicone oil, for instance?

We have seen that many different explanations for the contact angle saturation for electrowetting have been proposed. These explanations are based on very different physical phenomena, like charge trapping, zero effective solid–liquid surface tension, increase of the liquid electric resistance, leakage due to dielectric breakdown, or air ionization near the triple line. Obviously, the saturation phenomenon is not yet clearly explained. However, independently of the physical explanation, we show in the next section that we can give a general analytical expression for the Lippmann–Young law taking into account the saturation phenomenon. This expression will be useful in the modeling of electrowetting.

4.4.3 Modified Lippmann–Young Law

We have seen that, for low potentials, the relation between the contact angle θ and the applied voltage V is given by the Lippmann–Young law. This law shows a linear dependence between $\cos\theta$ and V^2. At large electric potentials, the contact angle has an asymptote. This situation is similar to that of paramagnetism where the induced magnetic moment is a linear function of the magnetic field $M = \chi H$, when the magnitude of the field is small, and has a saturation limit when the magnitude of the field increases [19,20]. The work pioneered by Langevin in the year 1905 has led to the definition of the so-called Langevin, or sometimes Langevin-Debye, function defined by

$$L(x) = \coth(3x) - \frac{1}{3x}. \tag{4.62}$$

Why is this function especially adapted to the situation? First let us look at Langevin's function at the origin. Due to the term $1/3x$, we have to work out a Taylor's expansion at the origin. Remember that Taylor's expansions at the origin for hyperbolic sine and cosine are

$$ch(3x) = 1 + \frac{(3x)^2}{2!} + \frac{(3x)^4}{4!} + O(x^6)$$
$$sh(3x) = x + \frac{(3x)^3}{3!} + \frac{(3x)^5}{5!} + O(x^7).$$

Then, for $x \to 0$, the Langevin function can be approximated by

$$L(x) \approx \coth(3x) - \frac{1}{3x} \approx \frac{1 + \frac{(3x)^2}{2!} + \frac{(3x)^4}{4!} + O\left(x^6\right)}{3x + \frac{(3x)^3}{3!} + \frac{(3x)^5}{5!} + O(x^7)} - \frac{1}{3x}.$$

After some algebra we find

$$L(x) \approx \coth(3x) - \frac{1}{3x} \approx \frac{1}{3x}\left[\frac{1 + \frac{(3x)^2}{2!} + \frac{(3x)^4}{4!}}{1 + \frac{(3x)^2}{3!} + \frac{(3x)^4}{5!}} - 1\right]$$
$$\approx \frac{1}{3x}\left[\left(1 + \frac{(3x)^2}{2!}\right)\left(1 - \frac{(3x)^2}{3!}\right) - 1\right] \approx \frac{1}{3x}\left[(3x)^2\left(\frac{1}{2} - \frac{1}{6}\right)\right] \approx x.$$

Thus, when $x \to 0$, $L(x) \approx x$.

Note that there is no term in x^2 in the expansion of L; the next term is $9x^3/4$. This is very important,because it means that the Langevin's function is very

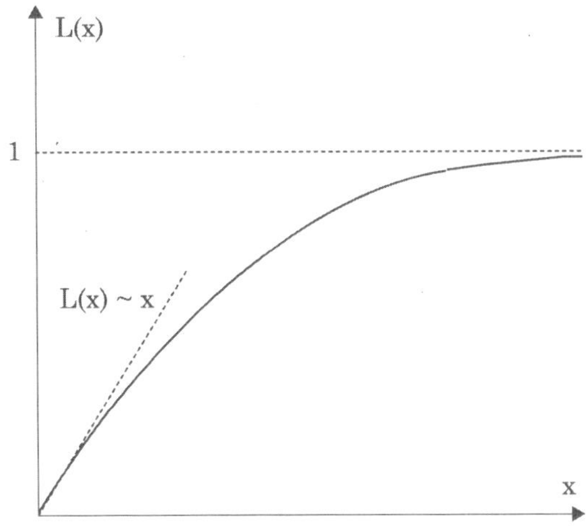

Figure 4.25 Graph of the Langevin function.

close to the linear function $L(x) = x$ at the origin. This is one of the reasons for the good fit of Langevin's law at the origin.

Next, let us look at the behavior of Langevin's function at infinity. The calculation is straightforward:

$$\lim_{x \to \infty} L(x) = \lim_{x \to \infty} \coth(3x) = \lim_{x \to \infty} \frac{e^{3x} + e^{-3x}}{e^{3x} - e^{-3x}} = \lim_{x \to \infty} \frac{e^{3x}}{e^{3x}} = 1.$$

In conclusion, Langevin's function is closely approximated by the linear relation $L(x) = x$ at the origin and has the asymptote $L(x) = 1$ when $x \to \infty$ (Fig. 4.25).

In the case of electrowetting, the Lippmann–Young law can be written under the form

$$\frac{\cos\theta - \cos\theta_0}{\cos\theta_S - \cos\theta_0} = \frac{CV^2}{2\gamma(\cos\theta_S - \cos\theta_0)}. \tag{4.63}$$

Hence, the law that takes into account the saturation can be written

$$\frac{\cos\theta - \cos\theta_0}{\cos\theta_S - \cos\theta_0} = L\left(\frac{CV^2}{2\gamma(\cos\theta_S - \cos\theta_0)}\right) \tag{4.64}$$

where L is the Langevin function $L(X) = \coth(3X) - 1/3X$, and θ_s is the saturation angle. Equation (4.64) satisfies equation (4.63) for small and moderate values of the potential V. At large potentials, it also satisfies the saturation asymptote. Relation (4.64) may be called "modified" or "extended" Lippmann–Young law. It has been verified that this function fits

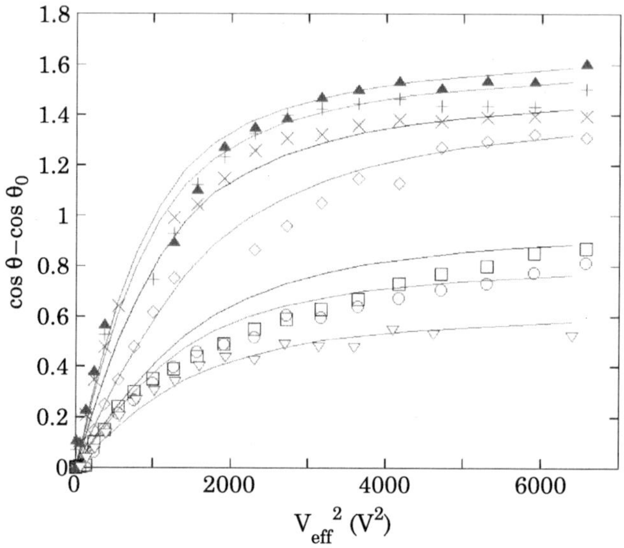

Figure 4.26 Fit of the experimental results for $\cos\theta - \cos\theta_0$ versus V^2 obtained by Langevin's functions. Note that in this particular case, we have considered the rms value of the voltage, because the electric alimentation was AC.

the experimental results [19]. Fig. 4.26 shows the fit between the experimental points and the "modified" Lippmann law.

A plot of the capillary force $\gamma_{LG}(\cos\theta - \cos\theta_0)$ versus the electric potential squared is shown in Fig. 4.27. At low voltage, all the theoretical curves and experimental points collapse on the line $CV^2/2$ showing the validity of the approach.

Fig. 4.27 shows that the different liquids—placed on the same EWOD microsystem—respond very differently to an electrowetting actuation. Indeed, equation (4.43) states that the electrowetting force is

$$f_{\mathrm{EWOD}} = \frac{C}{2}V^2 = \gamma_{LG}(\cos\theta - \cos\theta_0). \tag{4.65}$$

At saturation,

$$f_{\mathrm{EWOD,max}} = \frac{C}{2}V_{\mathrm{sat}}^2 = \gamma_{LG}(\cos\theta_{\mathrm{sat}} - \cos\theta_0). \tag{4.66}$$

According to Fig. 4.27, it is the saturation phenomenon that separates the different liquids regarding the electrowetting force, and there is a very important difference in the maximum electrowetting force between the different liquids. However, it seems that this observation has not yet been taken into account to give a supplementary insight into the saturation phenomenon.

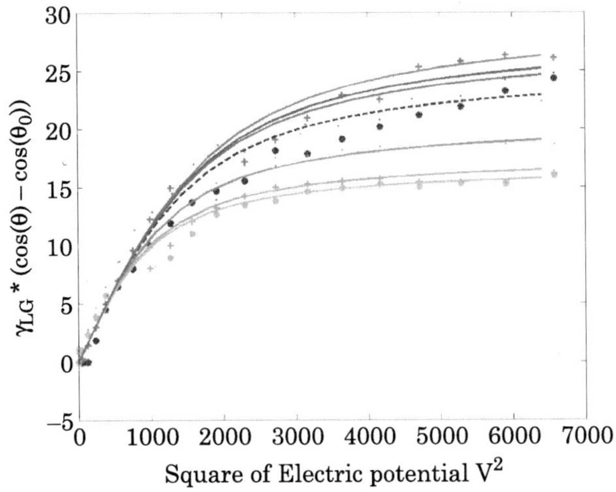

Figure 4.27 At low voltages, all curves collapse on the same straight line defined by its slope $C/2$. At high voltages, the curves separate to merge with their asymptote $\gamma_{LG}(\cos\theta_{sat} - \cos\theta_0)$. Continuous lines correspond to the modified Lippmann–Young law and dots to experimental observations.

4.5 Hysteresis

4.5.1 Introduction

Capillary hysteresis has been presented in Chapters 2 and 3. Hysteresis is defined as the deviation of the contact angle from its theoretical (and mean) value due to physical phenomena like microscopic surface defects and roughness. During dynamic motion of an interface, dynamic hysteresis refers to the advancing and receding contact angles.

Hysteresis is also observed in electrowetting. When we established the Lippmann–Young law in Section 4.2, the value of the macroscopic contact angle was in reality the average between an advancing and a receding value. If we start with a non-actuated droplet and we increase the value of the voltage, the droplet spreads. The contact angle is then an advancing contact angle. Conversely, when the voltage decreases, the droplet regains its initial shape and the observed contact angles are the receding contact angles. The advancing and receding contact angles differ (Fig. 4.28).

Another manifestation of electrowetting hysteresis occurs in EWOD microsystems, during the motion of a droplet on a line paved with electrodes (we present droplet motion in detail in the next chapter). Below a minimum actuation voltage V_{min}, the droplet does not move. In the following section we produce the relation between hysteresis and minimum actuation voltage.

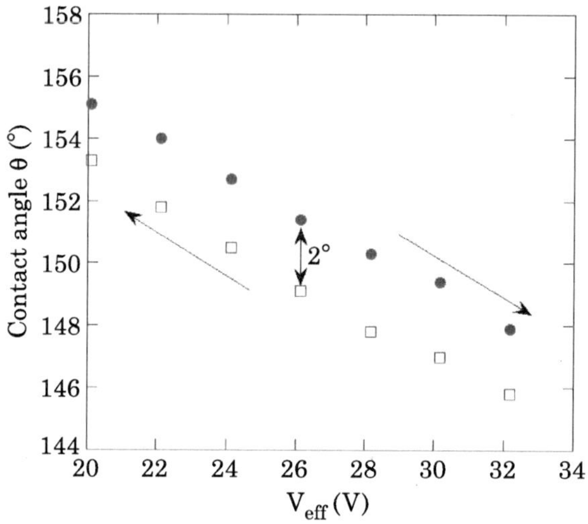

Figure 4.28 Experimental evidence of electrowetting hysterisis. Case of a sessile droplet of de-ionized water immersed in silicone oil (Brookfield) and contacting a SIOC substrate. (Courtesy Dr. O. Raccurt.). The arrows show the advancing and receding phases.

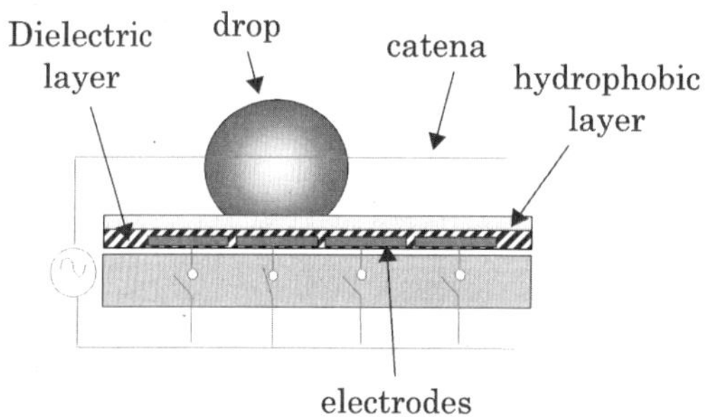

Figure 4.29 Left: principle of an open EWOD design. In the picture, the catena (zero potential) is placed above the substrate, but it can be also embedded into the substrate. Right: prototype of open EWOD microsystem (from [28]).

4.5.2 Hysteresis and Minimum Actuation Potential

4.5.2.1 Introduction

Let us consider the example of electrowetting on dielectric microsystems (EWOD) schematized in Fig. 4.29. It is observed that a droplet of conductive liquid does not move from one electrode to the next as soon as an electric actuation is applied. A minimum voltage threshold is required in order to obtain the motion of the drop [24]. This minimum electric potential—noted V_{min}—depends on the nature of the liquid/fluid/solid triplet.

In this section, we relate the value of the minimum potential to the hysteresis contact angle. We show that the force balance on the droplet produces an

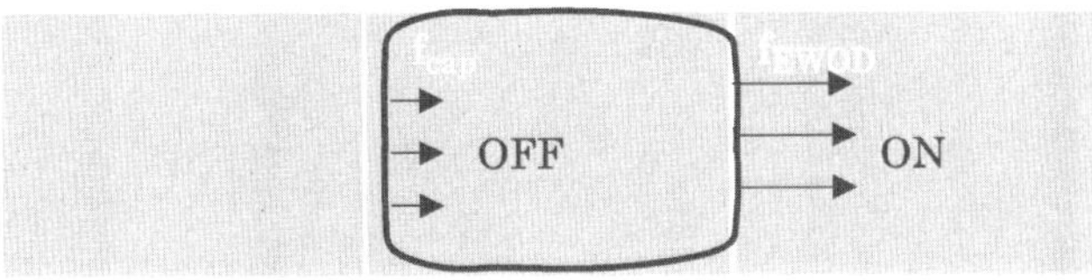

Figure 4.30 Sketch of a droplet starting to move towards the actuated electrode.

implicit relation linking the minimum potential $V_{\min}$ to the value of the hysteresis contact angle α.

Experiments have been conducted using different substrates (Teflon and SiOC) and different conductive liquids and surrounding gas/fluids (de-ionized water in oil or air, biological buffers with surfactants in oil or air, ionic liquids in air). It will be shown that the results of the models are in meaningful agreement with the experimental results.

4.5.2.2 Capillary and Electrowetting Forces

Our starting point is the Lippmann–Young law

$$\cos\theta - \cos\theta_0 = \frac{C}{2\gamma}V^2. \tag{4.67}$$

At the onset of the motion, there is no dynamic effect; hence we will interpret the Lippmann–Young law as a pseudo-capillary effect, according to Section 4.3. A droplet starts to move under an "*apparent wettability gradient*" between an actuated and a non-actuated electrode. We remind here that the electrowetting line force density on a triple line is given by

$$f_{\text{EWOD}} = \frac{C}{2}V^2 = \gamma\left(\cos\theta - \cos\theta_0\right). \tag{4.68}$$

This line force acts on that part of the triple line located above the actuated electrode (Fig. 4.30). On the part of the triple line located on the initial non-actuated electrode, the forces are just the capillary line forces. Usually, the substrate is hydrophobic, so the forces are exerted in the same direction as the electrowetting forces.

We recall that the capillary line force acts on the triple line in the plane of the substrate, perpendicularly to the triple line. In the case of a hydrophilic contact, $\cos\theta > 0$ and the line force points away from the liquid, whereas, in the case of hydrophobic contact, the line force is negative and points inside the droplet. The EWOD line force, as stated by Kang (Section 4.2.3), acts on the triple line—at least extremely close to the triple line—and has a component located in the plane of the substrate, perpendicular to the triple line, and a vertical component.

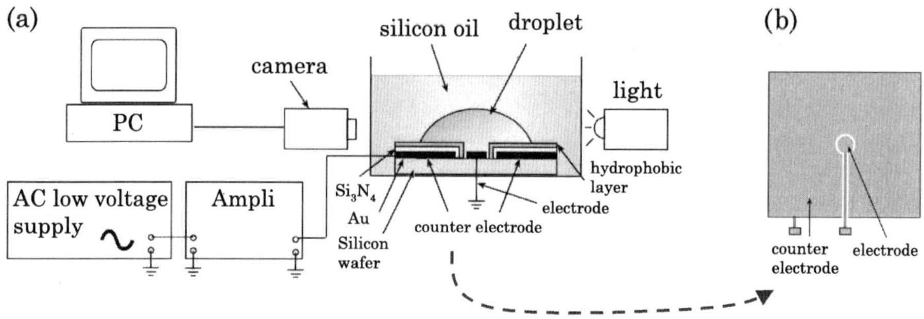

Figure 4.31 Schematic view of the experimental set up. (a) View of the experimental set up with electronics and monitoring; (b) view of the electrowetting pad from top.

If the surface were perfect, and using the pseudo capillary equivalence for the electrowetting force, the drop would immediately move as soon as the neighboring electrode is actuated [29,30]. However, experiments have shown that there is an electric potential threshold below which the droplet does not move, i.e., there is a pseudo-gradient of wettability below which the droplet does not move.

4.5.2.3 Minimum Potential of Actuation

4.5.2.3.1 Measure of hysteresis contact angle

As we have seen in Chapter 3, a special experimental device is used to measure contact angle hysteresis [31]. Liquid is pumped in and out of the drop by a motorized syringe. The contact angle during each of the phases is observed by photographing the droplet from its side. The contact angle hysteresis is the difference of contact angle between the advancing and receding motion.

The particular case of electrowetting is still more difficult to comprehend. We have set up an experimental device similar to that of Tavana and Neumann; however in the case of electrowetting, liquid is not pumped in or out of the drop, but the contact line spreads or shrinks following an increase or a decrease of the actuation voltage. Fig. 4.31 shows the experimental setup. The substrate is similar to a usual EWOD substrate, with electrodes, Si_3N_4 dielectric layer, and hydrophobic coating of SiOC or Teflon. The second counter-electrode is located in the middle of the chip.

Experiments have been performed using different liquids, substrates, and surrounding fluids. Fig. 4.28 shows the experimental results for a microdrop of de-ionized water immersed in silicone oil and placed on a SiOC substrate. The contact angle is not the same when the droplet is spreading on the substrate (advancing) or receding from the substrate. After having performed different plots using different substrates and liquids, we have noticed that the plot of Fig. 4.28 is typical. The vertical shift between the two curves defines the

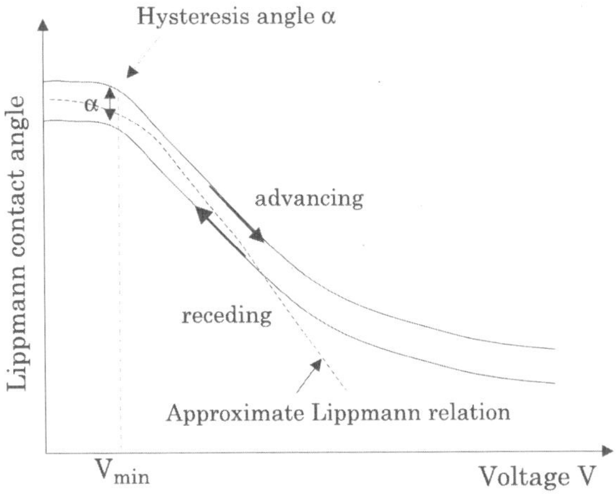

Figure 4.32 Sketch of the hysterisis curves. Actuated contact angles are not identical if the voltages are increasing or decreasing. Advancing and receding curves are approximately shifted by the hysterisis angle α.

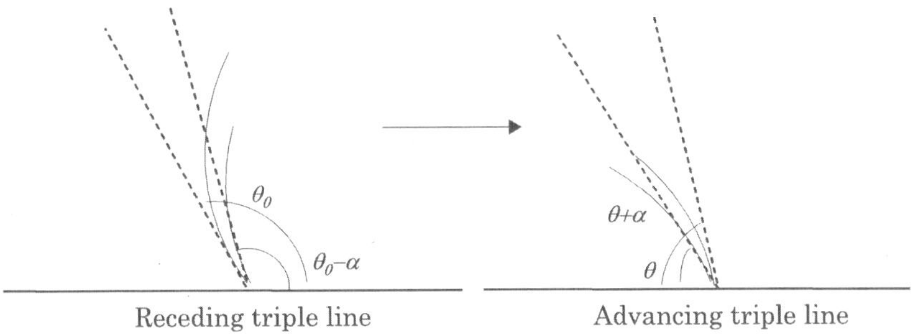

Figure 4.33 Sketch of the advancing and receding contact angles with and without the hysterisis angle.

electrowetting hysteresis angle. A schematized view of the phenomenon is given in Fig. 4.32.

The notion of contact angle hysteresis is very approximate. In Fig. 4.32, we have simplified somewhat the situation: in reality the shift between the two curves is not strictly constant. It shows some variations, not only when the electric potential changes, but also with the experimental conditions, and between two similar chips. Observed values of the hysteresis contact angle are given in Table 4.1.

4.5.2.3.2 Model

The analysis of the preceding section suggests that the contact angle may vary between its advancing and receding values (Fig. 4.33). We assume here that the two contact angles are the actuated and nonactuated Young contact angles plus or minus the hysteresis angle.

Table 4.1 Experimental Observations of the Hysterisis Angle

Liquids	Surrounding fluid	Substrate	Hysteresis angle
De-ionized water	air	Teflon	7–9°
Biologic buffer 0.01%Tween	air	Teflon	4–5°
Ionic liquids [tmba][NTf_2]	air	Teflon	3–4°
Ionic liquids [$C_{10}H_{23}NO_3$][NTf_2]	air	Teflon	3–4°
Ionic liquids [bmin][PF_4]	air	Teflon	6–7°
Ionic liquids [bmin][PF_6]	air	Teflon	7–8°
De-ionized water	Silicone oil	SIOC	1.5–2°
Biologic buffer 0.01% Tween	Silicone oil	SIOC	1–1.5°
Biologic buffer 0.1% Tween	Silicone oil	SIOC	0.5–1°

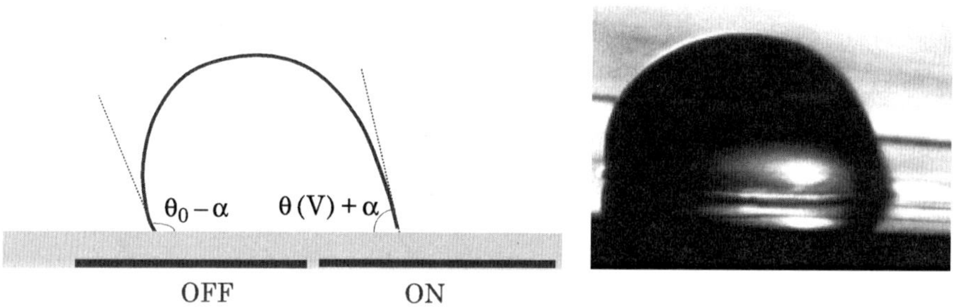

Figure 4.34 Left: sketch of the droplet at onset of actuation. If $V > V_{\min}, \theta(V) + \alpha < \theta_0 - \alpha$ and the drop moves to the right under the action of capillary and electrocapillary forces. Right: photograph of a droplet starting its motion (courtesy Dr. R. Bavière).

The static advancing and static receding limit contact angles are then $\theta + \alpha$ and $\theta_0 - \alpha$ where θ is the actuated contact angle and θ_0 the nonactuated contact angle. This notation stems from the Hoffman-Tanner law [32] indicating that the advancing and receding contact angles are respectively larger and smaller than their Young values. The minimum actuation potential is then the potential required for obtaining a net positive electrocapillary force, i.e.,

$$\theta\left(V_{\min}\right) + \alpha \leq \theta_0 - \alpha. \tag{4.69}$$

This relation is illustrated by the sketch of Fig. 4.34.

Using equation (4.68), we calculate the forces on a droplet placed on a row of electrodes. We demonstrate first that the real shape of the droplet on the electrodes has no importance. Fig. 4.35 shows different forms of contact lines. Suppose any form—even non-physical—of the contact line of the droplet as sketched in Fig. 4.36.

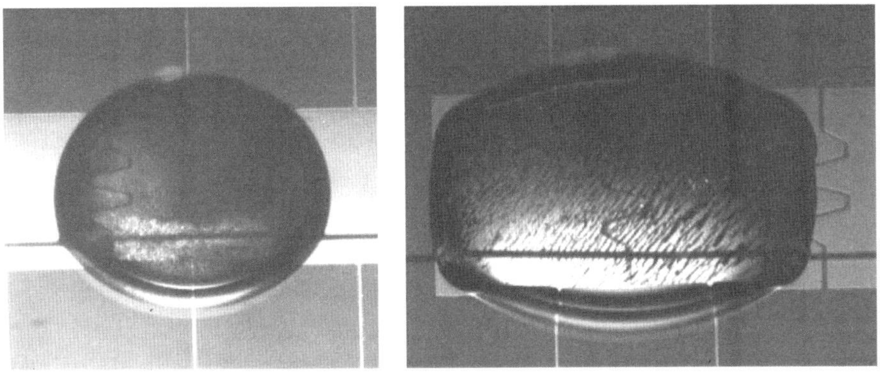

Figure 4.35 Different shapes of droplets on EWOD electrodes.

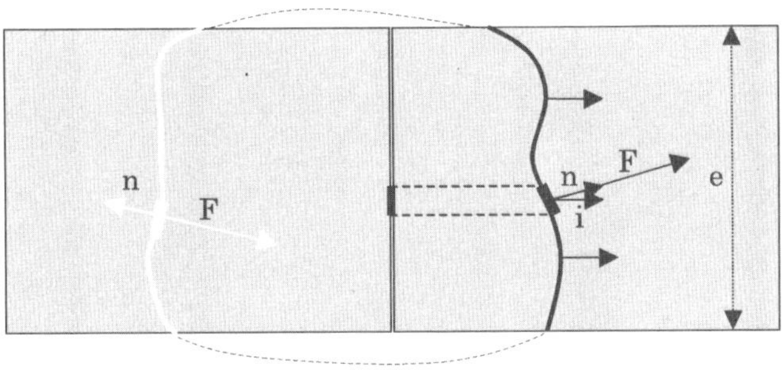

Figure 4.36 Sketch of the contact of a drop with the substrate.

The electro-capillary force in the direction x (unit vector i) on the hydrophilic electrode is given by

$$F_x = \int_L \gamma \cos \theta dl \vec{n}.\vec{i} \tag{4.70}$$

where dl is a unit element of the contour line, and n the normal unit. Equation (4.70) can be integrated

$$F_x = \gamma \cos \theta \int_L dl \vec{n}.\vec{i} = e\gamma \cos \theta. \tag{4.71}$$

Equation (4.71) shows that the shape of the triple line above an electrode has no effect on the capillary force. If we observe that the x-direction force on the triple line outside the electrodes vanishes, we conclude that the x-direction capillary force on the droplet, whatever its shape, is

$$F_x = e\gamma \left(\cos \theta - \cos \theta_0\right). \tag{4.72}$$

This force remains constant during the motion of the droplet between two electrodes. Now, if we take into account the contact angle hysteresis, we obtain the advancing and receding capillary forces

$$\begin{aligned} F_{a,x} &= e\gamma\cos(\theta+\alpha) \\ F_{r,x} &= -e\gamma\cos(\theta_0-\alpha). \end{aligned} \tag{4.73}$$

It is assumed that the contact angle hysteresis is small compared to the actuated and nonactuated contact angles. Because α is small compared to the Young's values of the contact angles, equation (4.73) can be rearranged

$$\begin{aligned} F_{a,x} &= e\gamma\left[\cos\theta - \alpha\left(\cos\theta + \sin\theta_a\right)\right] \\ F_{r,x} &= -e\gamma\left[\cos\theta_0 - \alpha\left(\cos\theta_0 - \sin\theta_0\right)\right]. \end{aligned} \tag{4.74}$$

The total capillary force is then

$$F_x = e\gamma\left[\cos\theta - \cos\theta_0\right] - e\gamma\alpha\left[\cos\theta + \sin\theta - \cos\theta_0 + \sin\theta_0\right]. \tag{4.75}$$

The first term on the right hand side of (4.75) is the usual "Lippmann" force—equation (4.67). The second term is a resistance force depending on the value of the hysteresis contact angle. It can be shown that this second term is always negative, because $\cos\theta - \cos\theta_0 > 0$. A consequence is that hysteresis reduces the capillary force, in accord with the analysis of the first section.

Because the minimum potential corresponds to the linear part of the Lippmann–Young relation, the "Lippmann" force can be expressed using equation (4.67)

$$F_{\mathrm{EWOD}} = \frac{eC}{2}V^2. \tag{4.76}$$

A criterion for drop displacement is then

$$\frac{eC}{2}V^2 - e\gamma\alpha\left[\cos\theta + \sin\theta - \cos\theta_0 + \sin\theta_0\right] > 0. \tag{4.77}$$

Without hysteresis ($\alpha = 0$) the drop would move even with an infinitely small electric actuation. Taking into account the hysteresis ($\alpha \neq 0$), condition (4.77) shows that the minimum electric potential is given by

$$V_{\min}^2 = \frac{2\gamma}{C}\alpha\left[\cos\theta(V_{\min}) + \sin\theta(V_{\min}) - \cos\theta_0 + \sin\theta_0\right]. \tag{4.78}$$

Using the Lippmann–Young law, equation (4.78) can be cast under the form

$$\frac{C}{2\gamma}V_{\min}^2 = \frac{\alpha}{1-\alpha}\left[\sin\theta(V_{\min}) + \sin\theta_0\right]. \tag{4.79}$$

Equation (4.79) is somewhat cumbersome, because it is an implicit equation due to the fact that θ depends on V. Equation (4.79) can be solved numerically, taking into account that the term $\sin\theta(V_{\min})$ is obtained via the Lippmann–Young equation (4.68). In the case where $V_{\min}$ and α are sufficiently small, equation (4.79) can be simplified

$$V_{\min} \approx 2\sqrt{\frac{\gamma\alpha\sin\theta_0}{C}}. \tag{4.80}$$

A large capacitance, a low liquid surface tension, a large (very hydrophobic) Young angle, and a small value of the hysteresis angle minimize the value of the voltage required to move droplets.

4.5.2.4 Experiments—Comparison With the Numerical Results

Experiments have been performed on open EWOD systems using two types of substrates: a 1.2 μm hydrophobic layer of SiOC on top of a 400 nm dielectric layer of Si_3N_4, and a 400 nm Teflon layer on top of the same Si_3N_4 layer.

The values of $V_{\min}$ have been obtained by increasing the potential by 0.5 V until the droplet starts to move. Measurements have been repeated on many electrodes of the same chip with back and forth motions. De-ionized water, biologic buffers in air or silicone oil, and ionic liquids in air have been successively tested. However carefully one operates, there remains a discrepancy of about 2.5 to 5 V. Note that there is necessarily a small bias in the approach because the observed values of $V_{\min}$ have been obtained on a different microsystem than that for the hysteresis angle, although the fabrication process is the same.

The capacitance C is defined by the relation

$$C = \frac{\varepsilon_0}{\sum_i \frac{d_i}{\varepsilon_i}}$$

where d_i is the thickness of the underlying layers (the layer of water may be neglected), and ε_i their relative electric permittivity. Using the values of Table 4.2, the capacitances of both systems have been calculated ($C_{\text{SiOC}} = 0.026\ \text{mF/m}^2$ and $C_{\text{Teflon}} = 0.045\ \text{mF/m}^2$).

In Fig. 4.37, we compare measured minimum actuation potentials with calculated ones. Calculated values have been obtained by solving equation (4.79) with the hysteresis values of Tablc 4.1. Because there are two limit values for the hysteresis angle α and an experimental range for $V_{\min}$, the corresponding

Table 4.2 Values of Thickness and Relative Permittivity for the Different Usual Materials

Material	Thickness	Relative permittivity
SiOC	1.2 μm	3.36
Teflon	400 nm	1.9
Si_3N_4	400 nm	7.8
Teflon	1000 nm	2.2
Water (electric double layer)	<30 nm	80

values on the figures are contained in a rectangular box. Calculated values of the minimum potential agree rather well with the experimentally measured potentials.

It can be observed that high surface tension droplets like DIW (de-ionized water) surrounded by air require an important V_{min}, of the order of 30 to 40 V, whereas lower surface tension liquids like ionic liquids require lower values of V_{min} (15 to 25 V). Water and biologic buffers immersed in silicone oil have at the same time a low surface tension and a small hysteresis due to the oil film beneath the droplet, thus the corresponding minimum potential is quite small, of the order of 5 V. The results correspond to former observations described in the literature. The reduction of hysteresis due to a liquid film at the solid surface has already been reported [32,33]. The surface of the film is smooth and homogeneous, and the situation is close to that observed by Zisman who found no hysteresis for polymer coated surfaces [27].

4.6 Working Range of EWOD Devices

4.6.1 Maximum Potential

Equation (4.79) or (4.80) gives an expression of the minimum actuation that is required to move droplets [34]. On the other hand, there exists also a maximum actuation voltage—noted V_{max}—above which the electrocapillary force on a drop does not increase further due to saturation phenomenon (see Section 4.4). For the moment, we do not take into account dielectric breakdown voltage—which is closely related to saturation, according to Papathanasiou et al.; this point will be treated later in Section 4.8. Hence $V_{\text{max}} = V_{\text{sat}}$. Thus, for a given type of EWOD micro-device [35–37] characterized by its capacitance C and its surface properties, and for a given electrically conductive liquid immersed in a surrounding non-conductive gas or fluid, the electrowetting principle will only be effective between the two limits V_{min} and V_{max}.

As we have seen in Section 4.4, different models are proposed for the saturation phenomenon. It is very convenient here to use the PQRS model

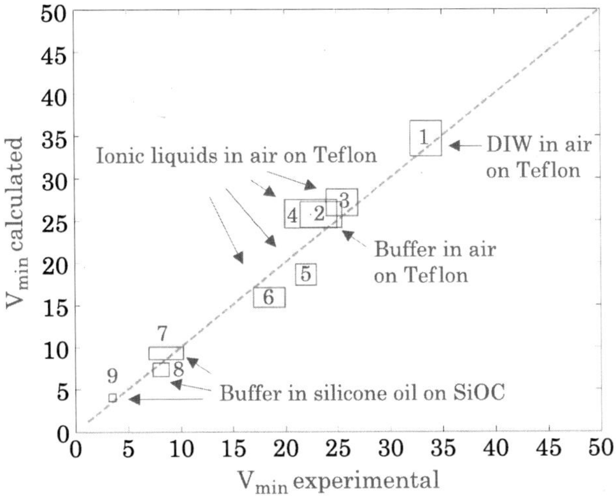

Figure 4.37 Comparison between calculated and measured values of the minimum potential: biological buffers immerged in silicone oil have the smallest $V_{\min}$. 1: De-ionized water, 2: biologic buffer $NH_4HCO_3 + 0.05\%$ Tween, 3:[bmim][PF_6], 4: [bmim][BF_4], 5: [$C_{10}H_{23}NO_3$][NTf_2], 6: [tmba][NTf_2], 7: deionized water, 8: biologic buffer $NH_4HCO_3 + 0.05\%$ Tween, 9: biologic buffer $NH_4HCO_3 + 0.1\%$Tween. For 1 to 6, the surrounding fluid is air and the substrate is Teflon. For 7–9, the surrounding fluid is silicone oil and the substrate is SiOC. The values of γ_c are 47 mN/m for Teflon/air and 20 mN/m for SIOC/silicone. From [34]. © 2007 Elsevier; reprinted with permission.

to express $V_{\max}$ as indicated in (4.58)

$$V_{\max} = \left(\frac{2\gamma_{SL,0}}{C}\right)^{\frac{1}{2}} = \left(\frac{2\left(\gamma_{SG} - \gamma_{LG}\cos\theta_0\right)}{C}\right)^{\frac{1}{2}}. \tag{4.81}$$

The advantage of this model is to produce an analytical relation for $V_{\max}$ which, combined with the $V_{\min}$ model, indicates the maneuverability interval for any droplet in open EWOD systems. The maximum actuation potential depends on the microfabrication of the chip through the capacitance C and the surface tension of the substrate γ_c, and on the interfacial properties between the liquid and the surrounding fluid, through the term $\gamma\cos\theta_0$.

Using the same data as above, we have compared the calculated and experimental values of $V_{\max}$ in Fig. 4.38.

4.6.2 Working Range

It is interesting to bring together equations (4.79) and (4.81) defining the values of $V_{\min}$ and $V_{\max}$. The minimum potential corresponds to the potential required to overcome the contact angle hysteresis and displace drops by EWOD; the maximum potential is linked to the saturation limit. The domain for EWOD

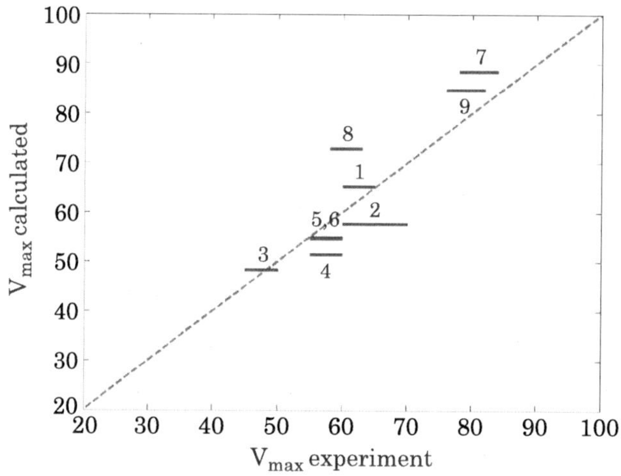

Figure 4.38 Comparison for $V_{\max}$ between experimental results (continuous line) and model. The numbers refer to the same data as in the preceding figure. From [34]. © 2007 Elsevier; reprinted with permission.

workability is then given by the following relation

$$\frac{\alpha}{1-\alpha}\gamma\left[\sin\theta(V_{\min})+\sin\theta_0\right]<\frac{CV^2}{2}<\gamma_C-\gamma\cos\theta_0. \tag{4.82}$$

At this stage, we notice that the interval $[V_{\min}, V_{\max}]$ depends on the capacitance C as shown in Fig. 4.39. Microdevices with thinner layers of dielectric have a larger capacitance and require lower level of actuation. There is a clear advantage in trying to increase the specific conductance of the dielectric layer: the actuation requires lower potentials as well for $V_{\min}$ than for $V_{\max}$.

Equation (4.82) can be equivalently written

$$\begin{aligned} f_{\min} &= \frac{\alpha}{1-\alpha}\gamma\left[\sin\theta(V_{\min})+\sin\theta_0\right]<f_{\text{EWOD}}<\gamma_C-\gamma\cos\theta_0 \\ &= f_{\max} \end{aligned} \tag{4.83}$$

showing the range of electrowetting force. The optimal solution would be the lowest possible f_{min} and a largest possible $f_{\max}$. First, let us analyze the conditions for obtaining the smallest possible $f_{\min}$. The hysteresis angle α is usually small, and the left side of (4.83) can be approximated by $2\alpha\gamma\sin\theta_0$. For aqueous liquids, a very hydrophobic contact angle θ_0 is optimal for having a very low minimum force for droplet motion, because $\sin\theta_0 \sim (\pi-\theta_0)\sim 0$. On the other hand, the maximum electrowetting force is $f_{\max}=\gamma_{\text{SL0}}$. The hydrophobic substrate that maximizes γ_{SL0} is the most suitable. If we choose the most hydrophobic situation possible in order to satisfy the $f_{\min}$ constraint, we have approximately $\cos\theta_0\sim -1$, and the maximum force is $f_{\max}\sim\gamma_c+\gamma=\gamma_{\text{SL0}}$.

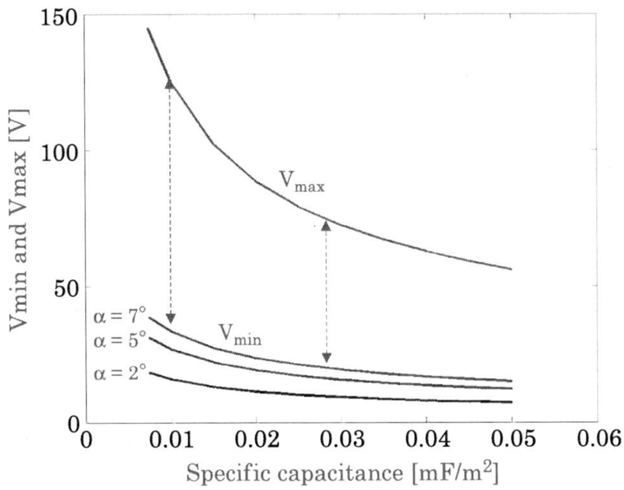

Figure 4.39 Interval V_{min}–V_{max} as a function of the capacitance C.

From an experimental standpoint, water droplets in silicone oil show a large working interval: on one hand, they present a small minimum force to move droplets because the value of θ_0 is large and the contact angle hysteresis α is small, due to an oil liquid film between the droplet and the substrate. On the other hand, the value of γ_{SL} is increased due to the presence of the oil film separating the hydrophobic substrate and the liquid. The result can be generalized; aqueous liquids surrounded with silicone oil instead of air perform better in EWOD microsystems.

Using a typical value for the specific capacitance $C \sim 0.03\ \mathrm{mF/m^2}$, relation (4.83) shows that the minimum electro-capillary forces lie in the range 1–20 μN, and the maximum forces in the range 15–40 μN.

4.6.3 Conclusion

We have presented a model for the minimum actuation potential in EWOD microsystems that incorporates contact angle hysteresis and produces the value of the threshold electric potential to displace drops. On the other hand, the maximum potential is approximately given by the zero interfacial tension criterion of the Peykov–Quinn–Ralston–Sedev model. These two values define the workability range of an EWOD-based microsystem for any combination of droplet liquid, surrounding fluid, and solid substrate.

Consequently, and in accordance with the experimental observations, it is shown why some triplets liquid/substrate/surrounding fluid are better adapted to EWOD technique than others. In the next chapters, we give examples of different liquids that can be displaced by EWOD (aqueous liquids, ionic liquids, blood, milk, etc.). As there is a growing interest in downsizing

EWOD devices, particular attention should be given to reducing the minimum actuation potential. The model shows that the largest possible non-actuated (hydrophobic for aqueous liquids) contact angle is recommended for minimizing the working voltage. To this extent, superhydrophobic substrates are currently investigated [38–41]; we shall present these substrates later in Section 4.7.4.

So far, we have not considered the phenomenon of dielectric breakdown. In this section, the maximum voltage of actuation has been calculated on the base of the saturation effect. We will see in the next section that dielectric breakdown is a major concern in electrowetting, and reinforce the importance of the choice of the substrate in EWOD microsystems.

4.7 Materials and Substrates

In this section, we investigate the characteristics of the solid substrates regarding their specific capacitance and their dielectric breakdown limit. It will be seen that dielectric breakdown is a considerable limit to the electrowetting force in EWOD microsystems. We then examine some new concepts aimed at increasing the hydrophobic character of the substrate in order to reduce the actuation voltages.

4.7.1 Capacitance

The specific capacitance of a substrate can be deduced by two different approaches. First, we use the Lippmann–Young plots; second, we perform a direct calculation.

4.7.1.1 Determination of the Specific Capacitance via the Lippmann–Young Relation

One important remark here is that the values of the coefficient $C/2\gamma$ are the slope of the linear part of the Lippmann–Young plots at low voltage. Fig. 4.40 shows the linear relations between $\cos(\theta) - \cos(\theta_0)$ and V^2 for different liquids on the same substrate. If we know at least one of the values of the surface tension γ, we can deduce the value of the capacitance C. Conversely, we then can deduce the values of the surface tensions of the other liquids. If we know the values of all the surface tensions, we obtain a set of consistent values for C, and the average value can be considered as the real specific capacitance.

4.7.1.2 Direct Determination of the Specific Capacitance

In a vertical cross section, the substrate comprises successively an electrode, a layer of dielectric, and a hydrophobic layer (Fig. 4.41). In the horizontal

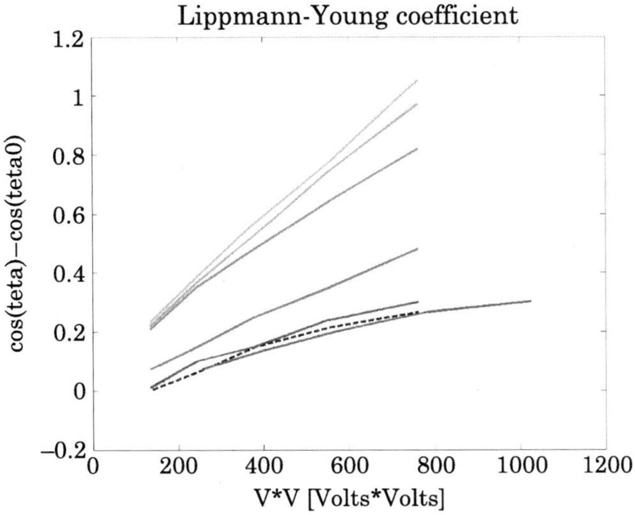

Figure 4.40 Linear part of the Lippmann–Young plots for different liquids. The dotted line corresponds to de-ionized water $\gamma_{LG} = 30$ mN/m immersed in silicone oil. The other surface tensions for biologic buffers are then 28.4, 19.9, 13.4, 10.7 and 9.9 mN/m. The surface tension for ionic liquid is 42.6 mN/m. The value of the capacitance is $C = 20$ μFarads/m^2.

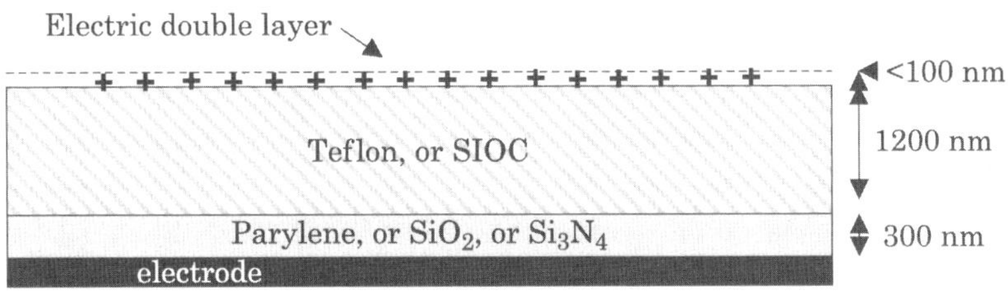

Figure 4.41 Total capacitance includes the contribution of the dielectric layer (Parylene, SiO_2, or Si_3N_4), the hydrophobic layer (Teflon, SIOC, etc.); the contribution of the electric double layer can be neglected.

plane, the electrodes are separated by an electrically insulating spacer. The dielectric layer guarantees the electric insulation; materials commonly used for dielectric are parylene, SiO_2, Teflon, and Si_3N_4. The hydrophobic layer contributes to increase of the amplitude of the change of contact angle between non-actuated and actuated states. Hydrophobic layers are often made of Teflon and spread by spin coating; sometimes they are made of SiOC spread by plasma deposition.

From a general standpoint, the effective capacitance of n parallel layers is given by

$$\frac{1}{C} = \sum_{i=1,n} \frac{1}{C_i}. \tag{4.84}$$

If ε_i and d_i respectively denote the dielectric constant and thickness of the layer i, the capacitance of layer i is given by

$$C_i = \frac{\varepsilon_0 \varepsilon_i}{d_i} \tag{4.85}$$

where ε_0 is the permittivity of vacuum, equal to 8.854×10^{-12} F/m (C^2/Jm). Substitution of (4.85) into (4.84) yields

$$C = \frac{\varepsilon_0}{\sum\limits_{i=1,n} \frac{d_i}{\varepsilon_i}}. \tag{4.86}$$

In Table 4.2, we have indicated approximate values of the dielectric constants for different materials frequently used in the composition of the substrate. The values of Table 4.2 show that the contribution of the electric double layer to the total capacitance is negligible. It is less than 3/1000. Hence, the double layer can be taken out of equation (4.86) and we obtain

$$\frac{1}{C} = \frac{1}{C_{\text{dielectric}}} + \frac{1}{C_{\text{hydrophobic}}}. \tag{4.87}$$

An order of magnitude of the capacitance of current EWOD microsystems is $C \sim 2.2 \times 10^{-5}$ F/m^2. According to equation (4.87), the capacitance is increased by reducing the thicknesses of the layers of the substrate. However, there is a limit to this reduction given by the dielectric breakdown limit. In the following section we develop the notion of breakdown of a dielectric.

4.7.2 Dielectric Breakdown

Breakdown of dielectric occurs when the electric field in the dielectric exceeds a limit value called the critical electric field, denoted here E_{BD}. Above this value, the material is disrupted. This threshold is also called the theoretical dielectric strength of a material. It is an intrinsic property of the bulk material. At breakdown, the electric field frees bound electrons. If the applied electric field is sufficiently high, free electrons may become accelerated to velocities that can liberate additional electrons during collisions with neutral atoms or molecules in a process called avalanche breakdown. Breakdown occurs quite abruptly (typically in nanoseconds), resulting in the formation of an electrically conductive path and a disruptive discharge through the material. For solid materials, a breakdown event severely degrades, or even destroys, its insulating capability. The detailed physical explanation of dielectric breakdown is not the subject of this book and is well documented in the literature [42]. For a

Table 4.3 Values of the Breakdown Voltage for some Usual Materials. Note that the values indicated here are given for perfect materials. Real values are usually less than that

Material	Paraffin	Oil	Teflon	Glass	Mica
Critical electric filed E_{BD}[MV/m or V/μm]	10	15	59	100	197

dielectric of thickness d, the critical electric field E_{BD} is related to the dielectric breakdown voltageV_{BD} by

$$V_{\mathrm{BD}} = dE_{\mathrm{BD}}. \tag{4.88}$$

Indications of the value of the critical electric field are given in Table 4.3 for some of the usual materials. In a preceding section, we developed the notion of saturation potential. Typically, saturation potentials are of the order of 80 V for a chip of capacitance $C \sim 2.2 \times 10^{-5}$ F/m^2, obtained with a total dielectric/insulating layer of Teflon of 1.5 μm thickness, approximately. Thus the electric field at saturation is of the order of 55 V/μm. This value is just below the breakdown value of Teflon. In other words, these commonly used chips have been designed to function unto the saturation potential.

However, dielectric breakdown is sometimes observed at lower values of the potential, as shown in Fig. 4.42. What could be the reason for this breakdown? This is presently a subject of studies, but it seems that breakdown frequently occurs when there are defaults in the substrate surface or when objects like cells or proteins adhere to the substrate. A possible explanation could be the anomalous value of the electric field at the vicinity of geometrical inhomogeneities.

The contact of an object with the substrate is sketched in Fig. 4.42.

Assuming that the liquid is perfectly conductive and that the object is insulating—which is a coarse assumption in the case of a cell or a protein, and if ε_1 and ε_d denote the relative permittivity of the object and the solid dielectric, respectively, the electric potential in the domain defined in Fig. 4.43 is given by the Laplace equation [10,15]

$$\varepsilon_1 \nabla^2 \phi_1 = \varepsilon_d \nabla^2 \phi_d. \tag{4.89}$$

From a theoretical point of view, the electric potential can be obtained through the Schwarz–Christoffel conformal mapping, in a similar mathematical approach as that of Section 4.2.3. If the angle of the wedge is "mathematically" sharp, there is a pole of the transform at the tip of the wedge; using Cauchy's theorem, the value of the potential field is theoretically infinite at the pole. However, sharp angles are not physical, and an electrostatic numerical calculation immediately shows that there is local peak in the magnitude of the

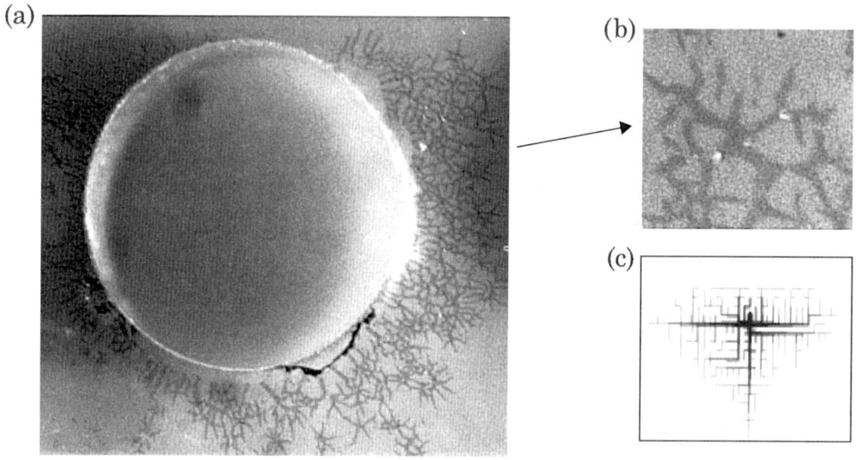

Figure 4.42 (a) Dielectric breakdown at the vicinity of a spherical object; the breakdown is materialized by the formation of cracks having the shape of tree branches ; (b) close up on the "tree effect"; (c) electrostatic breakdown model showing the growth of "failure tree" due to electron avalanche. From [43]. © 2003 IOP; reprinted with permission.

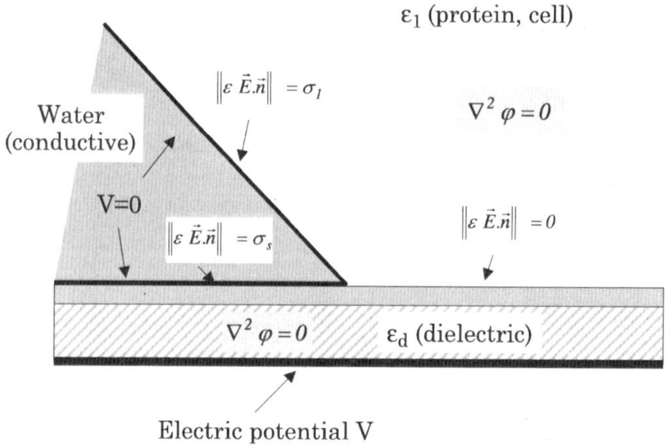

Figure 4.43 Electric scheme of the contact of an object with the substrate.

electric field at the tip of the wedge formed by the substrate, the conducting liquid, and the object, as shown in Fig. 4.44 (left). In our case, we have supposed the object perfectly insulating, which is an approximation. In reality, the peak will be less than that predicted by the preceding approach; nevertheless there will be an anomalous value of the magnitude of the electric field in the vicinity of the contact line.

This anomalous, localized value of the electric field may increase the voltage above the dielectric breakdown voltage [42]. This could be the explanation why dielectric breakdown is sometimes observed when cells or proteins adhere to the substrate (Fig. 4.44), or when the surface get crackled after a long period of use.

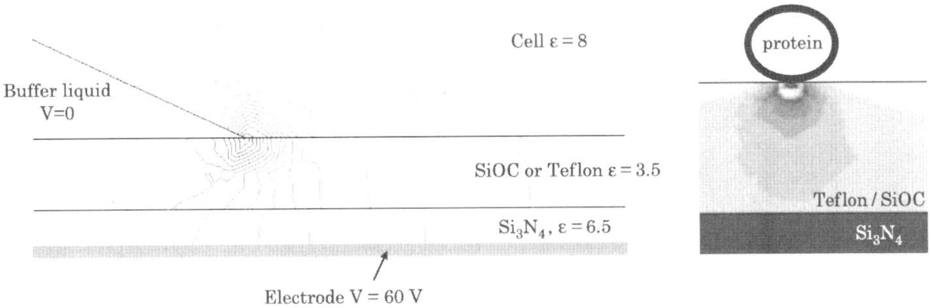

Figure 4.44 Left: contour plot of the magnitude of the electric field in the vicinity of the contact of a cell with the substrate (Comsol software): the magnitude of the electric field shows a peak at the tip of the wedge. Right: contour plot of the magnitude of the electric field below a protein sticking to the surface (Comsol software).

Reinforcement of the level of the dielectric breakdown voltage, without raising the value of the capacitance, by using a more resistant material seems possible, or introducing a buffer layer in the substrate [43]; however the introduction of a buffer layer in the dielectric is criticized by Kuiper and Hendriks [44]: their advice is that, if the dielectric is covered with a low dielectric constant hydrophobic topcoat, the field amplification in this topcoat may cause topcoat breakdown followed by dielectric breakdown. This question is presently debated and not yet solved.

4.7.3 Materials

In Section 4.6 we have seen that the maximum electrowetting line force that we can obtain in an EWOD system is limited by the saturation effect. However, there is another limit: the dielectric breakdown limit. Once the dielectric has broken down, electric charges leak through the dielectric, and the electrowetting effect vanishes.

It is quite easy to see that the condition brought by the DBV requires that the dielectric thickness be larger than a minimum value. Suppose we want to obtain an electrowetting force corresponding to a given $\Delta \cos\theta$ (smaller than $\Delta \cos\theta_{\mathrm{sat}}$) for a droplet of water. Let us start from the Lippmann–Young law written under the form

$$\cos\theta_a - \cos\theta_0 = \frac{1}{2\gamma}\frac{\varepsilon_0\varepsilon_D}{d}V^2$$

then the potential V can be expressed as a function of d

$$V = \sqrt{\frac{2\gamma d}{\varepsilon_0\varepsilon_D}(\cos\theta_a - \cos\theta_0)}. \tag{4.90}$$

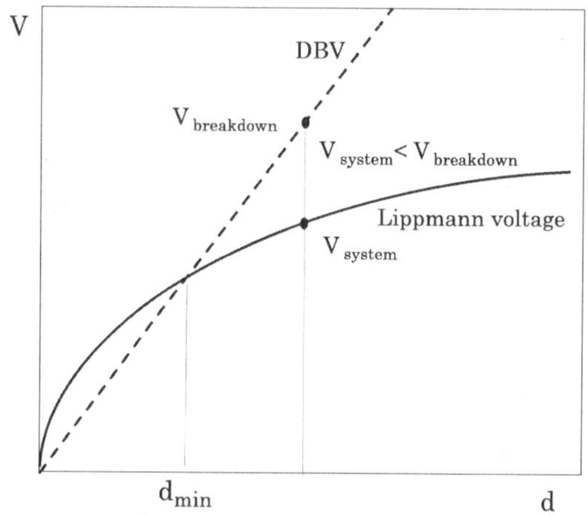

Figure 4.45 Minimum dielectric thickness: the working voltage must be smaller than the DBV; for a targeted $\Delta \cos\theta$ the minimum dielectric thickness must be larger than $d_{\min}$.

For a given Lippmann's angle θ_a, the voltage V varies like the square root of the dielectric thickness d; the corresponding graph has been plotted in Fig. 4.45 (continuous line). This relation has been experimentally verified by Saeki et al. [45]. On the other hand, relation (4.88) has also been plotted in Fig. 4.45 (dotted line). The two curves intersect at $d = d_{\min}$; the coordinates of the intersection point are [44]

$$d_{\min} = \frac{2\gamma\left(\cos\theta_a - \cos\theta_0\right)}{\varepsilon_0 \varepsilon_D E_{\mathrm{BD}}^2}$$
$$V_{\min} = \frac{2\gamma\left(\cos\theta_a - \cos\theta_0\right)}{\varepsilon_0 \varepsilon_D E_{\mathrm{BD}}}. \tag{4.91}$$

The dielectric thickness must then be larger than $d_{\min}$ for the system to function with the specific requirements $(\cos\theta_a - \cos\theta)$. Below this lower limit for the dielectric thickness, the voltage required to obtain the contact angle difference $(\cos\theta_a - \cos\theta)$ will result in the breakdown of the substrate.

If we find a better dielectric withstanding a larger critical electric field E_{BD}, the slope of the DBV is then increased, and the minimum thickness is reduced (Fig. 4.46). The minimum dielectric thickness depends also on the desired amplitude of $\Delta \cos\theta$. If the system can function with a smaller $\Delta \cos\theta$, the dielectric thickness can be further reduced.

We have already noticed that the thickness d of the dielectric is usually chosen so that the saturation voltage V_{sat} corresponds to the dielectric breakdown voltage V_{BD}. The explanation is straightforward: using the PQRS saturation

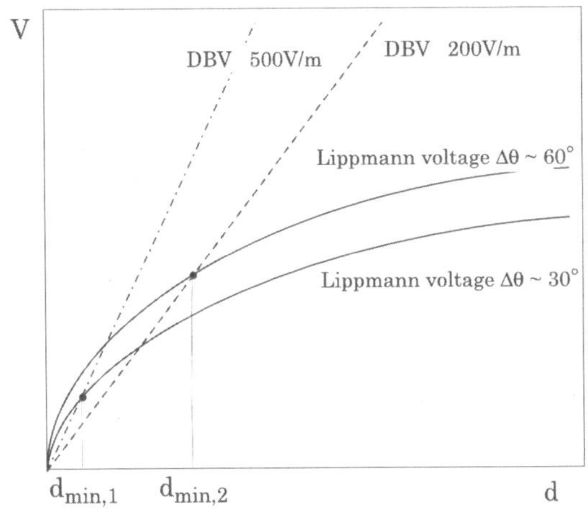

Figure 4.46 The minimum substrate thickness depends on the amplitude of the contact angle variation $\Delta\cos\theta$ (limited by $\Delta\cos\theta_{sat}$) and on the DBV of the substrate.

model—equation (4.58), we find

$$V_{sat} = \sqrt{\frac{2\gamma_{SL0}d}{\varepsilon_0\varepsilon_D}} = \sqrt{\frac{2d\gamma_{LG}\left(\cos\theta_{sat} - \cos\theta_0\right)}{\varepsilon_0\varepsilon_D}}. \tag{4.92}$$

The right term of equation (4.92) is a direct consequence of the PQRS saturation model, in which the following relation yields

$$\gamma_{SL0} = \gamma_{LG}\left(\cos\theta_{sat} - \cos\theta_0\right).$$

We obtain the plot of Fig. 4.47: the whole range of contact angles can be reached, unto the saturation limit. Note that we have not considered here the fringe effect described by Papathanasiou and Boudouvis [15] which results in a lower limit for the breakdown voltage (see Section 4.4.2.4).

4.7.4 Existing Substrates and Their Characteristics

The different characteristics of substrates found in the literature are shown in Table 4.4. At the present time, the usual EWOD microsystems use low voltages (no more than 80 V) in order to reduce their electric power. In order to avoid fragility problems and the consequences of surface defects, and to achieve robustness, dielectric thickness is usually of the order of 1 μm.

Recently there has been a trend toward using thinner dielectric layers in order to increase the specific capacitance. However, this trend has been limited by the dielectric breakdown. When the dielectric layer is very thin, any

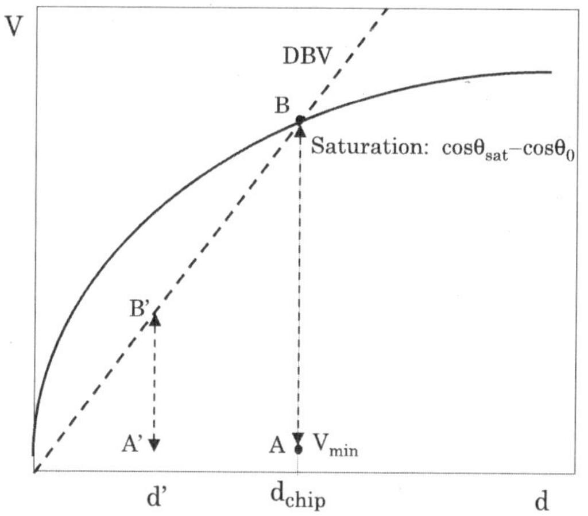

Figure 4.47 The working interval for a EWOD microchip with a substrate thickness d_{chip} is the segment comprised between the points A and B. A corresponds to the voltage V_{min}. If the thickness of the substrate was $d' < d_{chip}$, then the whole range $(\cos\theta_{sat} - \cos\theta)$ could not be covered because of early breakdown of the dielectric. The optimal theoretical thickness is then d_{chip}.

Table 4.4 Different Characteristics of Substrate Found in the Literature

Dielectric material	Dielectric thickness	Dielectric relative permittivity	Top coat (hydrophobic)	Top coat thickness	Top coat relative permittivity	Maximum voltage [V]	Maximum $\Delta\theta$	References
Parylene	20 μm	2.65	Teflon AF 1600	20 nm	2.2	250	60°	[36,46,47]
Parylene	10 μm	2.65	Teflon AF 1600	30 nm	2.2	250	60°	[24,46,47]
Teflon AF 1600	5.8 μm	2.2	–	–	–	200	60°	[45,47]
Si_3N_4	300 nm	7.8	Teflon AF 1600	1 μm	2.2	70	45°	[20,28]
Si_3N_4	300 nm	7.8	SiOC	1.2 μm	3.4	80	45°	[20,28]
Parylene	700 nm	2.65	Teflon AF 1600	200 nm	2.2	55	?	[47,48]
SiO_2	100 nm	4.0	Teflon AF 1600	20 nm	2.2	15	40°	[33,47]
BST	70 nm	?	Teflon	70 nm	2.2	10	35°	[36,47]

microfabrication defect leads to dielectric breakdown; besides a charge trapping mechanism could lead to electric charge leakage; finally the sensitivity to objects adhering to the substrate (proteins, cells in biology) is very high.

4.8 Discussion: Special Substrates and New Concepts

At the end of this chapter, the question is: from the expression of the electrowetting law, what can we expect from EWOD microsystems? In other terms, what would be an optimal EWOD microsystem, keeping in mind that the trend is to scale down such microsystems to be able to handle smaller liquid volumes?

There are two possible answers depending on the specific applications: either we want to perform a list of operations and manipulations with the smallest possible electric supply—this is, for example, the case of portable systems in space or biotechnological applications, or we want to build the most effective system regardless of the cost of electric energy.

4.8.1 Low Voltages

We investigate here the conditions for working with low voltages. The approximate value of the minimal voltage has been derived in Section 4.5

$$V_{\mathrm{min}} \approx 2\sqrt{\frac{\alpha\gamma\sin\theta_0}{C}}. \tag{4.93}$$

Hence, the smallest possible hysteresis contact angle α is a first requirement, a very hydrophobic surface is a second requirement, and a large capacitance is a third requirement, i.e., a very thin dielectric. However, there is a limit to the reduction of the thickness of the dielectric. If we want to be able to operate unto saturation angle, the maximum voltage is given by

$$V_{\mathrm{max}} = \min\left\{V_{\mathrm{DB}}, V_{\mathrm{sat}}\right\}. \tag{4.94}$$

Using the results obtained in this chapter, relation (4.94) can be written

$$V_{\mathrm{max}} = \min\left\{dE_{\mathrm{DB}}, \sqrt{\frac{2\gamma_{\mathrm{SLO}}}{C}}\right\}. \tag{4.95}$$

The minimum dielectric thickness is then

$$d = \frac{2\gamma_{\mathrm{SLO}}}{\varepsilon_0\varepsilon_D E_{\mathrm{DB}}^2}. \tag{4.96}$$

Hence a material with a large critical electric field E_{DB} is necessary to reduce the dielectric thickness.

4.8.2 Electrowetting Forces

We consider now the electrowetting force; we will see in the following chapter that a large electrowetting force is needed to perform operations such as cutting droplets in two, dispensing of droplets from a reservoir, etc. Thus, we may want a system having the lowest possible minimum electrowetting force to displace droplets and the largest possible maximum electrowetting force.

The minimal force is

$$f_{\text{min}} \approx 2\alpha\gamma \sin\theta_0. \tag{4.97}$$

Thus a substrate with a low hysteresis and a super-hydrophobic behavior is optimal. On the other hand, the maximum force is the minimum of two forces: the force corresponding to the DBV and the saturation force

$$f_{\text{max}} = \frac{1}{2} C \min\left\{V_{\text{DB}}^2, V_{\text{sat}}^2\right\}. \tag{4.98}$$

Hence the maximum electric line-force is

$$f_{\text{max}} = \min\left\{\frac{1}{2}\varepsilon_0 \varepsilon_D d E_{\text{DB}}^2, \gamma_{\text{SLO}}\right\}. \tag{4.99}$$

Again, a material with a large critical electric field is recommended.

4.8.3 Superhydrophobic Microfabricated Substrates

It has been shown in the preceding chapter that superhydrophobic substrates are advantageous, because for one thing, their hysteresis level is low: droplets can be moved and maneuvered at very low actuation potential. This is a very important point since nominal actuation levels should be far from the breakdown voltage; hence the durability of the chip is increased. Electrowetting on textured superhydrophobic substrates is currently the object of many investigations, as well theoretical as experimental.

4.8.3.1 Theoretical Approach

We first focus on the theoretical approach, following Torkelli's approach [38]. Suppose a droplet deposed on a super-hydrophobic substrate, in a Cassie regime (see Chapter 3). The fraction of the liquid in contact with the substrate is denoted f and the fraction in contact with the air trapped below is $1 - f$. When the electrode is actuated, an electrowetting force appears. According to the thermodynamic approach, the diagram of the forces exerted on the droplet is sketched in Fig. 4.48.

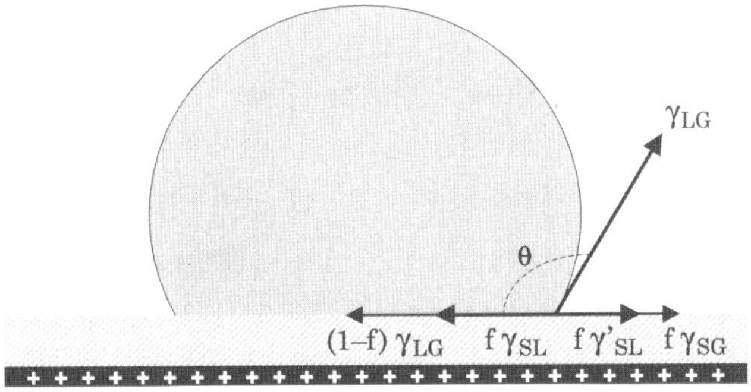

Figure 4.48 Schematic of the forces exerted on the droplet.

In the thermodynamical approach, the electrowetting effect results in a decrease of the apparent solid–liquid surface tension $\gamma_{SL} - \gamma'_{SL}(V)$. In the special case of superhydrophobic contact (Cassie regime), it affects only the wetted fraction f of the surface. We have seen before in this Chapter

$$\gamma'_{SL}(V) = \frac{CV^2}{2}.$$

Let us write the force balance

$$\gamma_{LG}\cos\theta + (1-f)\gamma_{LG} + f\gamma_{SL} = f\gamma'_{SL}(V) + f\gamma_{SG} \tag{4.100}$$

which can be cast under the form

$$\cos\theta = (f-1) + \frac{f\gamma'_{SL}(V)}{\gamma_{LG}} + f\frac{\gamma_{SG} - \gamma_{SL}}{\gamma_{LG}}.$$

Substitution of the Young equation yields

$$\cos\theta = (f-1) + \frac{f\gamma'_{SL}(V)}{\gamma_{LG}} + f\cos\theta_0.$$

And substitution of the Cassie–Baxter equation yields

$$\cos\theta = \cos\theta_C + f\frac{CV^2}{2\gamma_{LG}} \tag{4.101}$$

where θ_C is the Cassie angle. In conclusion, the Lippmann–Young equation can be transposed to super-hydrophobic substrate under the condition to correct the electrowetting force by the contact factor f. This factor f is small for super-hydrophobic substrate, so that the electrowetting force is reduced. But, on the

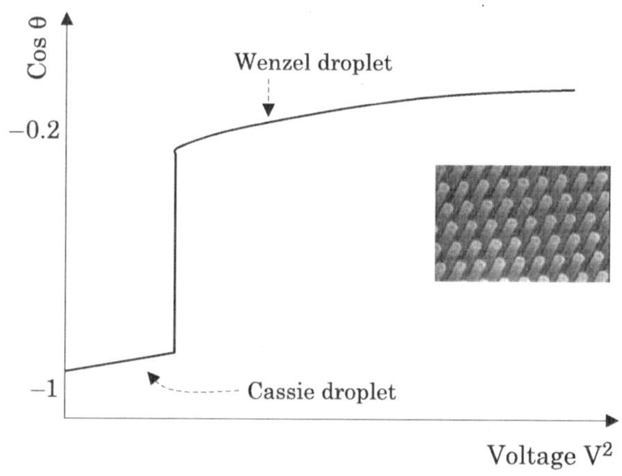

Figure 4.49 Electrowetting on super-hydrophobic nanostructured substrate: the droplet has initially a nearly 180° contact angle and sits on the top of the pillars (Cassie regime). For some value of the actuation voltage, the droplet sinks between the pillars (Wenzel regime). The mechanism is not reversible.

other hand, hysterisis has been considerably reduced. The droplet moves more easily on the electrodes of the EWOD-microsystem.

4.8.3.2 Experimental Approach

We have seen in Chapter 3 that nanostructured surfaces, if they are correctly designed, have a super-hydrophobic behavior. Krupenkin and coworkers [39] first experimented with such substrates for electrowetting.

The electrowetting effect is observed on such substrates (Fig. 4.49). However, the drawback is that there is no reversibility. In the initial state, or at a very low voltage, the droplet sits on top of the pillars in a Cassie regime. When the voltage increases above some level, the droplet sinks between the pillars; the droplet is now in a Wenzel regime. Decreasing the voltage does not bring the droplet back to its initial Cassie shape.

More recently, Dhindsa et al. [40] have improved Krupenkin's design. They use carbon nanofibers (CNF) coated with Al_2O_3 or parylene (Fig. 4.50) to form the pillars.

A saline droplet surrounded by air has the behavior described by Krupenkin. However, reversibility is achieved if the droplet is immersed in dodecane. Fig. 4.51 shows the dodecane or the air film between the droplet and the substrate. In both cases, when the voltage is actuated the droplet sinks between the pillars. But when the voltage is switched off, the film of dodecane reconstitutes and the droplet regains its initial shape on the top of the pillars. Such a design seems interesting for applications in optics; however, for biotechnology and microchemistry, the drawback might be the contamination of the film by macromolecules initially contained in the saline solution. To avoid

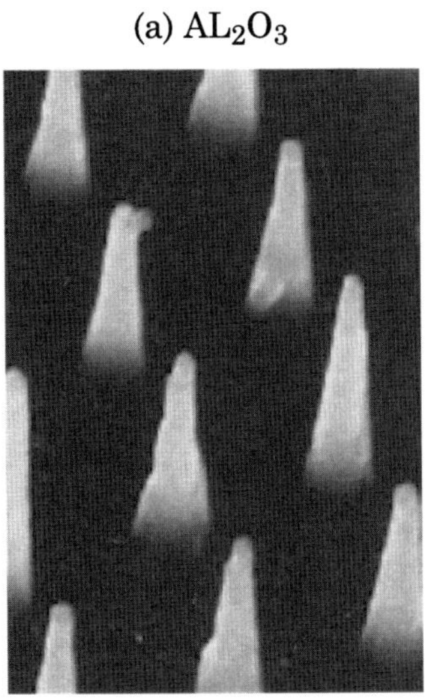

Figure 4.50 Vertically aligned CNFs covered with a layer of AL_2O_3 (a) and Parylene (b). Reprinted with permission from [40]. © 2006 American Chemical Society.

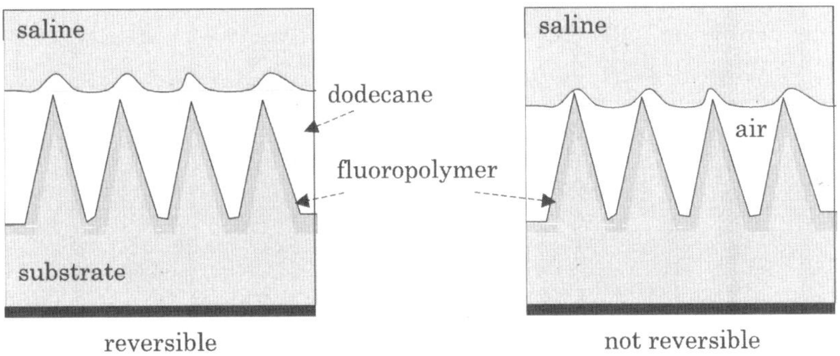

Figure 4.51 Sketch of the film of dodecane or air below the droplet (at zero voltage).

contamination, a recent proposal was to encapsulate the droplet [41] instead of nanostructuring the substrate. This is the topic of the next section.

4.8.4 Superhydrophobic Droplets

We have just seen that Krupenkin et al. [39] and Dhindsa et al. [40] have tested electrowetting on a substrate constituted of microfabricated pillars; however reversibility is not always achieved by such systems. The originality of the work of McHale et al. [41] is the transposition of the idea of superhydrophobicity to the droplet itself (Fig. 4.52). The starting point is that an aqueous droplet can be coated with hydrophobic particles. This can be

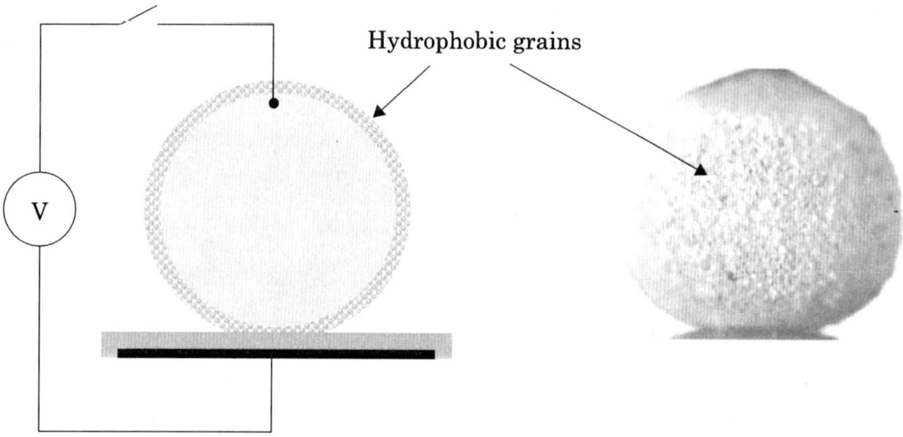

Figure 4.52 Left: sketch of the electrowetting device, the aqueous droplet is coated with hydrophobic grains (lycopodium or silica); right: a photograph of the droplet showing the granular coating of the interface. Reprinted with permission from [40]. © 2007 American Chemical Society.

shown theoretically by considering the surface energy of the droplet. Suppose we have a spherical grain at the surface of the liquid, as shown in Fig. 4.53 (left). The surface area A_{LP} is given by

$$A_{LP} = \pi R_S^2 \sin^2 \theta_e \tag{4.102}$$

where R_s is the radius of the spherical grain and θ_e its Young contact angle. On the other hand, the surface A_{PG} is

$$A_{PG} = 2\pi R_S^2 \left(1 + \cos \theta_e\right). \tag{4.103}$$

From an energy point of view, when a spherical grain attaches to the liquid surface, it replaces the solid–air interface area of the particle A_{PG} by a same area solid–liquid interface, and the liquid–air interface loses a surface area A_{LP}. Hence, the net change in surface free energy is

$$\Delta E = 2\pi R_S^2 \left(1 + \cos \theta_e\right) \left(\gamma_{\mathrm{SL}} - \gamma_{\mathrm{SG}}\right) - \pi R_S^2 \sin^2 \theta_e \gamma_{\mathrm{LG}}. \tag{4.104}$$

Using Young's law,

$$\cos \theta_e = \frac{\gamma_{\mathrm{SG}} - \gamma_{\mathrm{SL}}}{\gamma_{\mathrm{LG}}}$$

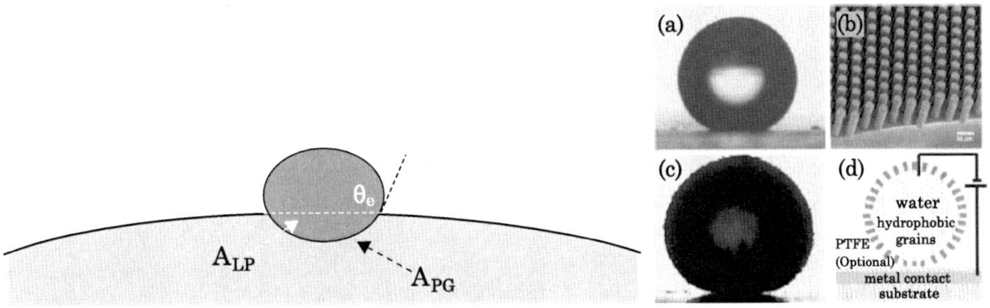

Figure 4.53 Left: schematic of the solid sphere on a liquid interface; right: a droplet coated with hydrophobic particles can have a hydrophobic behavior similar to that of a droplet on nanostructured pillars. Reprinted with permission from [41]. © 2007 American Chemical Society.

we obtain

$$\Delta E = -2\pi R_S^2 \gamma_{\mathrm{LG}} \left(1 + \cos\theta_e\right)^2 . \tag{4.105}$$

Relation (4.105) shows that ΔE is always negative or zero (zero only if $\theta_e = 180°$). As a result, it is always energetically favorable that grains spontaneously attach to the liquid–vapor interface, even if they are hydrophobic. If R_g denotes the radius of the grain, the length of a grain protruding into air is

$$d_g = R_g \left(1 - \cos\theta_e\right) . \tag{4.106}$$

Relation (4.106) shows that highly hydrophobic grains protrude substantially, enough to obtain a hydrophobic configuration similar to that of the nanostructure surface described in the preceding section (Fig. 4.53, right). It is shown that such an aqueous droplet coated with silica beads reacts to an electric actuation in the same manner as a simple water droplet on a super-hydrophobic substrate, and that the motion is reversible. In biotechnology, an advantage of such a system is the reduction of the possible contamination to the substrate.

4.9 Summary

In this chapter we have dealt with the theory of static electrowetting. We have derived the fundamental law of electrowetting, i.e., the Lippmann–Young law for low voltages, and analyzed the phenomena of saturation and dielectric breakdown for high voltages. We have seen that, in appearance, the Lippmann–Young law may seem easy to understand. However, the reality is much more complicated. In particular, the extension to dynamic wetting where the contact angle is a function of the wetting speed cannot be explained by this static law, unless the motion is very slow and a quasi-steady approximation can be made. There is still little theory that can predict the dynamic behavior of droplets in EWOD-actuated systems. Nevertheless, electrowetting devices have been successfully developed, principally based on the static and quasi-static assumption when possible [20], and empirically, when dynamic effects are important.

In the following chapter, we show how the electrowetting effect can be used to move, merge, separate, and mix droplets. All these steps are essential components of the digital droplet platform for lab-on-a-chip.

References

1. G. Lippmann, "Relations entre les phénomènes électriques et capillaires," *Ann. Chim. Phys.*, Vol. 5, p. 494, 1875.
2. B. Berge, "Electrocapillarity and wetting of insulator films by water," *Comptes rendus de l'Académie des Sciences, Séries II*, Vol. 317, pp. 157–163, 1993.
3. C. Quilliet and B. Berge, "Electrowetting: A recent outbreak," *Curr. Opin. Colloid Interface Sci.*, Vol. 6, pp. 34–39, 2001.
4. F. Mugele and J.-C. Baret, "Electrowetting: from basics to applications," *J. Phys.: Condens. Matter*, Vol. 17, pp. R705–R774, 2005.
5. B. Shapiro, H. Moon, R. Garell and C.J. Kim, "Equilibrium behavior of sessile drops under surface tension, applied external fields, and material variations," *J. Appl. Phys.*, Vol. 93, pp. 5794–5811, 2003.
6. N. Fortner and B. Shapiro, Equilibrium and dynamic behavior of micro flows under electrically induced surface tension actuation forces, Proceedings of the International Conference on MEMS, NANO and Smart Systems (ICMENS'03), 2003.
7. R.P. Feynman, R.B. Leighton and M. Sands, *The Feynman Lectures on Physics II*, Addison-Wesley, 1977. Section 8–2.
8. T.B. Jones, J.D. Fowler, Young Soo Chang and Chang-Jin Kim, "Frequency-based relationship of electrowetting and dielectrophoretic liquid microactuation," *Langmuir*, Vol. 19, pp. 7646–7651, 2003.
9. T.B. Jones, "An electromechanical interpretation of electrowetting," *J. Micromech. Microeng.*, Vol. 15, pp. 1184–1187, 2005.
10. Kwan Hyoung Kang, "How electrostatic fields change contact angle in electrowetting," *Langmuir*, Vol. 18, pp. 10318–10322, 2002.
11. Jun Zeng, T. Korsmeyer, "Principles of droplet electrohydrodynamics for lab-on-a-chip," *Lab. Chip.*, Vol. 4, pp. 265–277, 2004.
12. M. Vallet and B. Berge, "Limiting phenomena for the spreading of water on polymer films by electrowetting," *Eur. Phys. J. B*, Vol. 11, p. 583, 1999.
13. T.A. Driscoll and L.N. Trefethen, *Schwarz–Christoffel Mapping*, Cambridge University Press, 2002.
14. S. Thamida and H.-C. Chang, "Nonlinear electrokinetic ejection and entrainment due to polarization at nearly insulated wedges," *Phys. of Fluids*, Vol. 14, p. 4315, 2002.
15. A.G. Papathanasiou and A.G. Boudouvis, "A manifestation of the connection between dielectric breakdown strength and contact angle saturation in electrowetting," *Appl. Phys. Lett.*, Vol. 86, p. 164102, 2005.
16. Roland Bavière, Jérôme Boutet and Yves Fouillet, Dynamics of droplet transport induced by electrowetting actuation, Microfluidics and nanofluidics, article in print, 2007.

17. K. Adamiak, Numerical model for the contact angle in electrowetting, 5th International Meeting on Electrowetting, Rochester, NY, 31 May–2 June, 2006.
18. Ilju Moon and Joonwon Kim, "Using EWOD (electrowetting-on-dielectric) actuation in a micro conveyor system," *Sensors and Actuators A: Physical*, Vol. 130–131, pp. 537–544, 2006.
19. J. Berthier, Ph. Clementz, D. Jary, O. Raccurt, P. Claustre, C. Peponnet and Y. Fouillet, Mechanical behavior of micro-drops in EWOD systems: drop extraction, division, motion and constraining, Proceedings of the Nanotech 2005 Conference, Anaheim, 8–12 May, 2005.
20. J. Berthier, Ph. Clementz, O. Raccurt, D. Jary, P. Claustre, C. Peponnet and Y. Fouillet, "Computer aided design of an EWOD microdevice," *Sensors and Actuators A: Physical*, Vol. 127, pp. 283–294, 2006.
21. L. Yeo and H.-C. Chang, "Electrowetting films on parallel line electrodes," *Physical Review E*, Vol. 73, pp. 011605-1–011605-16, 2006.
22. L. Yeo and H.-C. Chang, "Static and spontaneous electrowetting," *Modern Phys. Lett. B.*, Vol. 19, p. 549, 2005.
23. P. Takhistov, A. Indeikina and H.-C. Chang, "Electrokinetic displacement of air bubbles in micro-channels," *Phys. of Fluids*, Vol. 14, p. 1, 2002.
24. H.J.J. Verheijen and M.W.J. Prins, "Reversible electrowetting and trapping of charge: model and experiments," *Langmuir*, Vol. 15, p. 6616, 1999.
25. V. Peykov, A. Quinn and J. Ralston, "Electrowetting: a model for contact angle saturation," *J. Colloid Polym. Sci.*, Vol. 278, pp. 789–793, 2000.
26. A. Quinn, R. Sedev and J. Ralston, "Influence of the electrical double layer in electrowetting," *J. Phys. Chem. B.*, Vol. 107, pp. 1163–1168, 2003.
27. W.A. Zisman, "Contact angle, wettability and adhesion," Advances in Chemistry Series, Vol. 43, Am. Chem. Soc., Washington, DC, p. 1, 1964.
28. Y. Fouillet, H. Jeanson, I. Chartier, A. Buguin and P. Silberzan, Déplacement de gouttes par un microcaténaire, Conférence SHF Microfluidique, Toulouse, France, décembre 2002.
29. M.K. Chaudhury and G.M. Whitesides, "How to make water run uphill," *Science*, 256, pp. 1539–1541, 1992.
30. N. Moumen, R.S. Subramanian and J. McLMaughlin, The Motion of a Drop on a Solid Surface due to a Wettability Gradient, Proceedings of the AIChe Annual Meeting, 2003.
31. H. Tavana and A.W. Neumann, "On the question of rate-dependence of contact angles," *Colloids Surf. A: Physiochem. Eng. Aspects*, Vol. 282–283, pp. 256–262, 2006.
32. M. Fermigier and P. Jenffer, "An experimental investigation of the dynamic contact angle in liquid–liquid systems," *J. Colloid. Interface Sci.*, Vol. 146, pp. 226–242, 1990.
33. E.B. Dussan, "On the spreading of liquids on solid surfaces: static and dynamic contact lines," *Annual Rev. Fluid Mech.*, Vol. 11, pp. 371–400, 1979.
34. J. Berthier, Ph. Dubois, Ph. Clementz, P. Claustre, C. Peponnet and Y. Fouillet, "Actuation potentials and capillary forces in electrowetting based Microsystems," *Sensors and Actuators*, Vol. 134, no. 2, pp. 471–479, 2007.
35. M.G. Pollack, R.B. Fair and D. Shenderov, "Electrowetting-based actuation of liquid droplets for microfluidic applications," *Applied Physics Letters*, Vol. 77, no. 11, pp. 1725–1726, September 2000.
36. H. Moon, S.K. Cho, R.L. Garrell and C.-J. Kim, "Low voltage electrowetting-on-dielectric," *Journal of Applied Physics*, Vol. 92, no. 7, pp. 4080–4087, October 2002.
37. S.K. Cho, H. Moon and C.-J. Kim, "Creating, transporting, and merging liquid droplets by electrowetting-based actuation for digital microfluidics circuits," *Journal of Microelectromechanical systems*, Vol. 12, no. 1, pp. 70–80, 2003.
38. Altti Torkkeli, Droplet microfluidics on a planar surface, Dissertation for the degree of Doctor of Science in Technology, 2003.

39. T.N. Krupenkin, J.A. Taylor, T.M. Schneider and Shu Yang, "From rolling ball to complete wetting: the dynamic tuning of liquids on nanostructured surfaces," *Langmuir*, Vol. 20, pp. 3824–2827, 2004.
40. M.S. Dhindsa, N.R. Smith, J. Heikenfeld, P.D. Rack, J.D. Fowlkes, M.J. Doktycz, A.V. Melechko and M.L. Simpson, "Reversible electrowetting of vertically aligned superhydrophobic carbon nanofibers," *Langmuir*, Vol. 22, pp. 9030–9034, 2006.
41. G. McHale, D.L. Hebertson, S.J. Elliott, N.J. Shirtcliffe and M.I. Newton, "Electrowetting of nonwetting liquids and liquid marbles," *Langmuir*, Vol. 23, pp. 918–924, 2007.
42. S.H. Kim, J.-H. Kim, B.-K. Kang and H.S. Uhm, "Superhydrophobic CFx coating via in-line atmospheric RF plasma of He-CF4-H2," *Langmuir*, Vol. 21, pp. 12213–12217, 2005.
43. S.J. Dodd, "A deterministic model for the growth of non-conducting electrical tree structures," *J. Phys. D: Appl. Phys.*, Vol. 36, pp. 129–141, 2003.
44. S.M. Lebedev, O.S. Gefle, Y.P. Pokholkov, E. Gockenbach, H. Borsi, V. Wasserberg, N. Abedi and J. Szczechowski, "Influence of high-permittivity barriers on PD activity in three-layer dielectrics," *J. Phys. D: Appl. Phys.*, Vol. 37, pp. 3155–3159, 2004.
45. S. Kuiper and B. Hendriks, Voltage reduction in electrowetting-on-dielectric, 5th International Meeting on Electrowetting, Rochester, NY, 31 May–2 June, 2006.
46. F. Saeki, J. Baum, Hyejin Moon, Jeong-Yeol Yoon, C.J. Kim and R. Garrell, "Electrowetting on dielectrics (EWOD): reducing voltage requirements for microfluidics," *Polym. Mater. Sci. Eng.*, Vol. 85, p. 12, 2001.
47. Y.-W. Chang and D.Y. Kwok, Electrowetting on dielectric: a low voltage study on self-assembled monolayers and its wetting kinetics, Proceedings of the 2004 International Conference on MEMS, NANO and Smart Systems (ICMENS'04), 2004.
48. E. Seyrat and R.A. Hayes, "Amorphous fluoropolymers as insulators for reversible low-voltage electrowetting," *Journal of Applied Physics*, Vol. 90, no. 3, pp. 1383–1386, 2001.

5 EWOD Microsystems☆

5.1 Introduction

Electrowetting micro-devices have been conceived using the principle of electrowetting discussed in the preceding chapter. Architecture of such micro-devices usually involves micro-fabricated electrodes embedded in a solid substrate, each electrode having roughly the characteristic size of a droplet (Fig. 5.1). The electrode is coated by a dielectric layer covered by a hydrophobic layer in order to increase the electrowetting effect on the droplet.

Successive actuation of electrodes is used to digitally activate targeted droplets. Initially, droplets are dispensed from a reservoir, then moved from one electrode to the next, merged with other droplets in order to introduce specific biochemical species for the desired reaction, then mixed for better reaction yield; after that, they are generally divided in order to regain the same initial volume and transported to the exit. All these operations are performed on a digital grid formed by electrodes that can be individually addressed, as sketched on Fig. 5.2.

In this chapter, we study how the basic operations like drop motion, merging, mixing, and division are performed. But first, we present the two types of electrowetting on dielectric (EWOD) microsystems, the "open" and "covered" EWOD microsystems.

5.2 Open and Covered EWOD Microsystems

Digital microfluidics is a promising way to manipulate biological targets like DNA, proteins, or cells in very small liquid volumes. The advantages of such devices are the use of lesser quantities of costly reagents, better biochemical reaction efficiency, and shorter operating times. Digital microfluidics based on EWOD is a fast developing technology [1–3]. The basic manipulations of drops can be achieved in such microsystems [4,5].

Two different types of EWOD-based microsystems have been developed: open systems where the sessile droplet is sitting freely on a horizontal solid substrate and covered systems where the droplet is confined between two plates (Figs. 5.1 and 5.2). Each of these systems has its own advantages. Drop dispense, motion, and splitting are easier in covered EWOD systems, whereas mixing and evaporation (for species concentration) are preferably performed in the

☆This Chapter was written in collaboration with Hsueh-Chia Chang (Bayer Professor of Engineering, Department of Chemical and Biomolecular Engineering, University of Notre-Dame, Indiana, USA).

Figure 5.1 Schematic view of digital microfluidic architecture (Courtesy Y. Fouillet CEA-LETI).

open configuration. With some hybrid designs, these two types of EWOD are interfaced such that droplet motion between a covered and an open region is possible under electrowetting actuation [6].

5.2.1 Open EWOD Microsystems

Historically, open EWOD systems have been first built with catenae (Figs. 5.3 and 5.4); the potential difference for droplet actuation is thus established between the electrode embedded in the substrate and the catena, which is used as the reference electrode. Catenae are difficult to mount in the microsystem and need to be carefully aligned. New concepts of actuation have been recently developed, like "buried" catenae or even no catenae at all. In this section, we discuss both open EWOD systems, with catenae and without.

5.2.1.1 Systems with catenae

The catena has exactly the same role as the reference electrode contacting the liquid in the electrowetting test system (Fig. 5.4). Its horizontal position, parallel to the surface of the substrate, allows for the droplet to move from one electrode to the next alongside the catena.

We have already introduced the capillary effect of a catena on a droplet in Chapter 3. We stress here that catenae should be carefully located, parallel to the substrate, aligned and centered with the electrode row; otherwise the motion could be blocked. Indeed, if there is an angle between the catena and the direction of the motion (direction of the electrode row) the droplet may

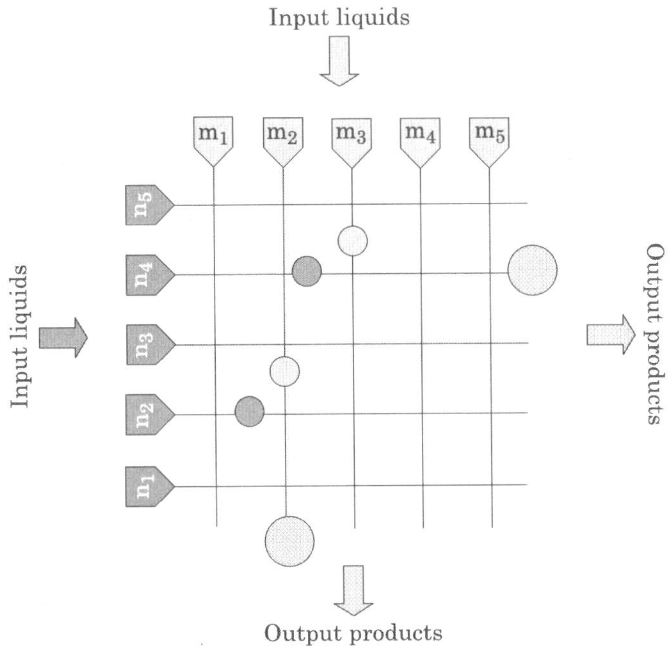

Figure 5.2 Schematic view of a digital microfluidics system designed to perform programmed operations on droplets.

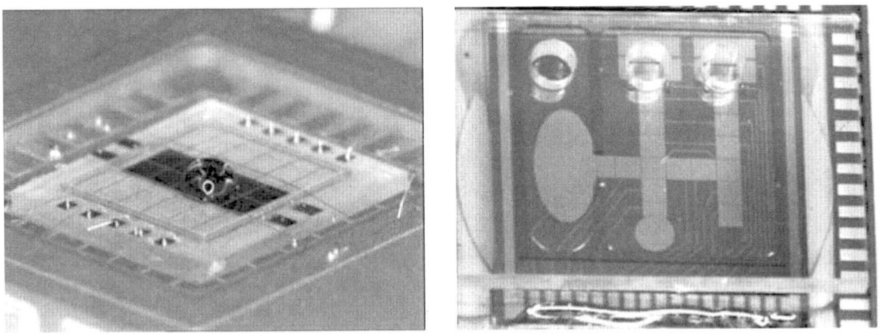

Figure 5.3 Left: open EWOD microsystem [7]; the drop sits on an electrode paved substrate; right: covered EWOD microsystem; the droplets are squished between two horizontal solid plates; the holes at the top of the picture are used for the injection of the liquid. (Both photographs courtesy C. Peponnet and Y. Fouillet, CEA-LETI).

be pinned. This is, for instance, the case of Fig. 5.5.b. The spherical droplet cannot move past the change of direction of the catena.

However, if catenae are correctly placed, they do not substantially perturb the behavior of droplets (an exception may be very hydrophilic liquids like ionic liquids). Usual electrode sizes are in the range 500 μm – 1 mm. Studies are being conducted to reduce this size. In consequence the droplets in

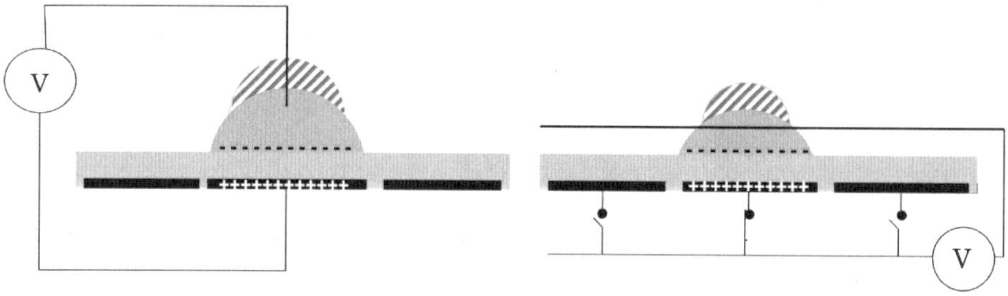

Figure 5.4 Left: electrowetting test system described in Chapter 3; right: open EWOD system with catena. The figure shows the two positions of the droplet depending on the actuation of the electrodes.

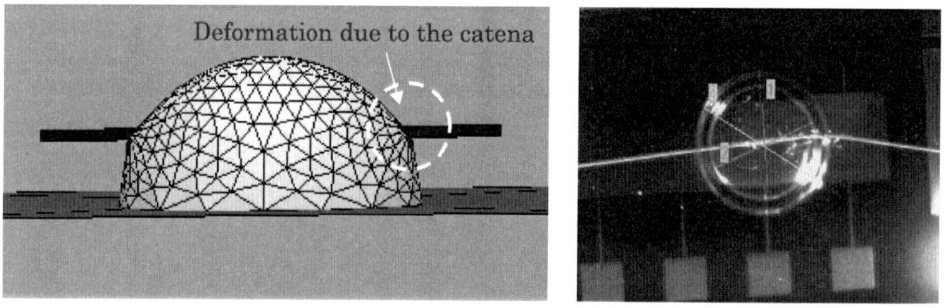

Figure 5.5 (a) Left: surface Evolver calculation of the shape of a droplet pierced by a hydrophilic catena [8]; deformation stays localized to a small area close to the wire.
(b) Right: experimental view confirming the negligible deformation caused by the catena (photo CEA-LETI).

EWOD microsystems have a Bond number $Bo = \Delta \rho g R^2/\gamma$ smaller than 0.05. Therefore, surface tension prevails over gravity, and sessile drops in open EWOD microsystems can be assumed spherical. They remain nearly spherical even if a catena is used as an electrode. Fig. 5.5 shows the effect of a hydrophilic catena piercing the droplet (usually catenae are metallic wires—very often made of gold, and they have a hydrophilic contact with aqueous solutions).

5.2.1.2 Systems without catenae

Catenae bring complexity to the system: they must be connected to the electric source, and they must be carefully aligned with the electrodes. Two solutions are currently being developed: (1) "buried" catena, i.e., the catena is buried in the substrate, (2) no catena at all, the next electrode being used as the reference electrode. The first case uses the same principle as that described in the preceding section (Fig. 5.6); the droplet is at the same potential than the catena.

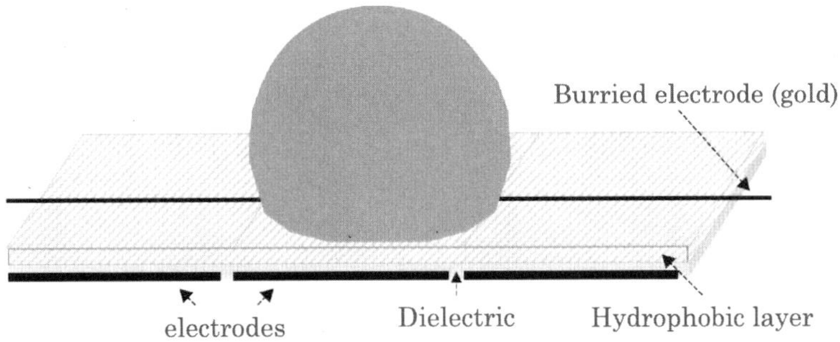

Figure 5.6 Principle of a buried reference electrode.

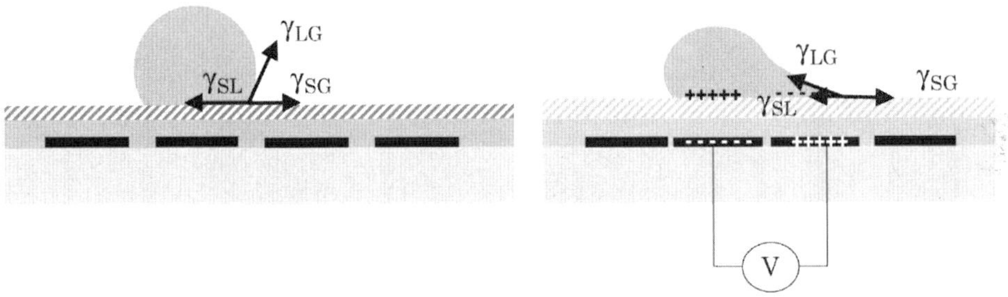

Figure 5.7 Electric charge distribution in the substrate and droplet when a potential difference is established between the droplet electrode and the next: the droplet spreads onto the two electrodes. Remark that we have adopted the notation where the effect of the electrowetting effect is translated in a decreased solid–liquid surface tension.

The second case uses a slightly different principle (Fig. 5.7): the electric field is established between one electrode and the next. One electrode is at the potential +V, the other at potential 0. However, the potential of the droplet is a floating potential. More details concerning the way these designs act on droplets will be given in Section 5.3.

5.2.2 Covered EWOD Microsystems

In a covered EWOD system, the catena is replaced by a top plate covering the droplet (Fig. 5.8). The free surface, i.e., the surface area between liquid and surrounding fluid (usually air, sometimes silicone oil), is reduced in this design. The substrate at the bottom is similar to that of open EWOD systems. The top plate is constituted by a plate electrode usually made of ITO coated with a thin layer of Teflon. Thus, without electric actuation, the contact is hydrophobic with both plates, as shown in Fig. 5.8.

In a covered EWOD system, the droplet follows closely the shape of the electrodes (Fig. 5.9). In such a configuration, the vertical gap is very small compared to the horizontal dimension of the droplet, and the surface area of

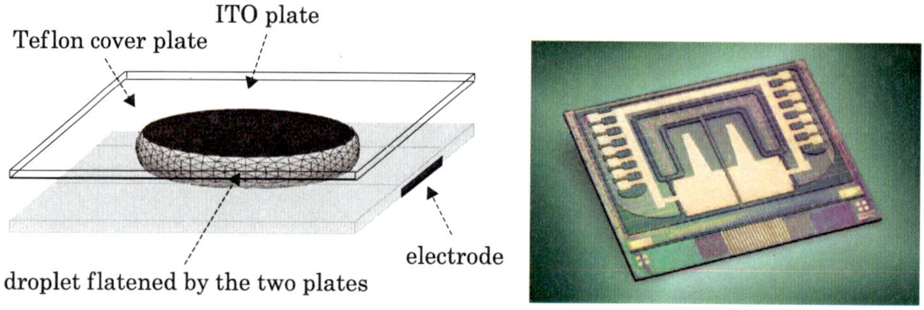

Figure 5.8 Schematic view of a covered EWOD microsystem; (a) droplet squished between two plates; (b) covered EWOD microsystem fabricated by the LETI (courtesy C. Peponnet and Y. Fouillet, CEA-LETI).

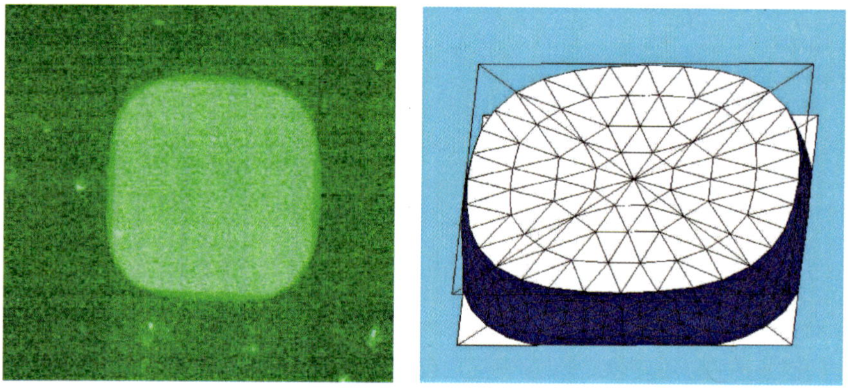

Figure 5.9 Left: photograph of a "square" droplet in a covered EWOD microchip; right: the droplet simulated with Evolver software.

the free interface (liquid–air) is very small compared to that of the solid–liquid interface. Consequently, the energy of the liquid–air interface is much smaller than that of the solid–liquid interface and the droplet adopts the shape of the underlying electrode. This is why a droplet can adopt a nearly "square" shape.

5.3 Droplet Motion

In Chapter 4, we analyzed how EWOD produces forces on droplets. In this section, we show that these forces can be used to move droplets. First we show how electrowetting actuation can force a droplet out of its equilibrium state, second we present some dynamic aspects of droplet motion, and finally we analyze the effect of different technological solutions on the motion of droplets.

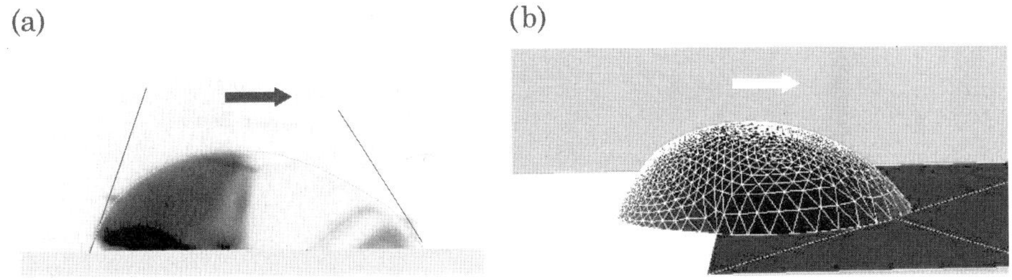

Figure 5.10 (a) View of a droplet of ionic liquid, moving to the right; (b) numerical simulation with Surface Evolver assuming a slow motion.

5.3.1 Principle of Droplet Motion—Quasi-static Aspects and Departure From Equilibrium

The principle of droplet motion due to a difference—or a gradient—of wettability is well known [9,10]. We showed in Chapter 3 that, when a droplet is located on the boundary between a lyophobic region and a lyophilic region, it moves to the lyophilic region, at least if hysteresis of contact angle is not too large. The situation is similar in the case of electrowetting (Fig. 5.10). If a conductive droplet is located on the boundary between an actuated and a non-actuated electrode, an electrowetting force is applied on the contact line located above the actuated electrode, and a capillary force is exerted on the contact line located on the lyophobic surface. The resultant of the forces parallel to the surface is directed towards the electrically actuated region. The droplet is out of equilibrium, and if the resulting force is sufficient to overcome hysteresis, the droplet moves.

In the preceding section, we presented the usual EWOD designs: open EWOD with catenae, open EWOD without catenae, and covered EWOD systems. Fig. 5.11 shows how the electric charges are induced at the liquid interface resulting in an electrowetting force acting on the droplet. Jones [11] has shown that the electrowetting force acting on a droplet positioned above different substrate with specific capacitanceC_i and a voltage difference between the electrode and the droplet V_i is

$$F_x = \sum_i \frac{\partial}{\partial x}\left(\frac{1}{2}C_i V_i^2\right). \tag{5.1}$$

We can deduce the electrowetting force in the different configurations of Fig. 5.11. The electrowetting force divided by the width of the electrode F_x/e is

$$f_x = \frac{F_x}{e} = \frac{1}{2}CV^2 \tag{5.2}$$

for cases (a) and (c). In case (d), the force is twice that of case (c)

$$f_x = \frac{F_x}{e} = 2\left[\frac{1}{2}CV^2\right] = CV^2. \tag{5.3}$$

Note that case (d) requires twice the electric energy of case (a) or (c). Configuration (b) is more complex because the potential in the droplet is not fixed [12,13]. We have approximately the relation

$$f_x = \frac{F_x}{e} \leq \frac{1}{4}CV^2 \tag{5.4}$$

showing that the electrowetting force is much smaller in this configuration.

5.3.2 Dynamic Aspect—Velocity of Motion

5.3.2.1 Scaling Numbers

In Chapter 3, we see that the balance between the gravity and surface tension forces was characterized by the Bond number [14]

$$Bo = \frac{g\Delta\rho l^2}{\gamma}. \tag{5.5}$$

When a droplet is in motion, other forces, like electrowetting, viscous and inertia forces are acting on the droplet. We have seen that the ratio of the electrowetting forces to the surface tension forces is given by the electrowetting number

$$\eta = \frac{CV^2}{2\gamma}. \tag{5.6}$$

Let us now consider the ratio between inertia and surface tension. It can be expressed by the non-dimensional Weber number

$$We = \frac{\rho v^2 l}{\gamma} \tag{5.7}$$

where v is the velocity of motion of the droplet, ρ its density, and l a characteristic length (in our case, the contact radius). If the Weber number is larger than the critical Weber number ($We_{\text{crit}} = 1.1$), the droplet is very distorted by inertia forces and break-up may occur [15]. At a very low Weber number, inertia is negligible; for a moderate Weber number, inertia deforms the

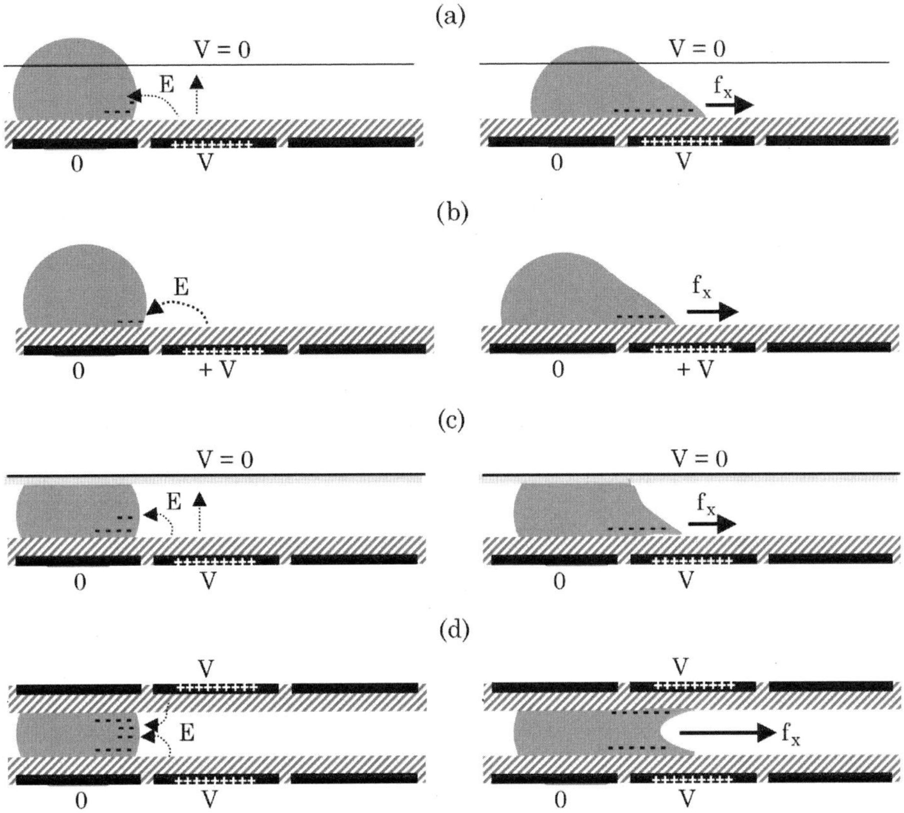

Figure 5.11 Electric scheme of the four different EWOD designs: (a) open EWOD with catena: an electrowetting force is exerted on the right boundary of the droplet; (b) open EWOD without catena: the neighboring electrode is the zero potential electrode; (c) covered EWOD: configuration similar to (a), the electrowetting force exerted on the right boundary of the droplet pulls the droplet onto the neighboring electrode; (d) the electrowetting force can be doubled if the cover plate is fabricated and actuated in the same way as the bottom plate.

droplet without disrupting it. Next, we consider the non-dimensional Ohnesorge number defined by

$$Oh = \frac{\mu}{\sqrt{\rho l \gamma}} \tag{5.8}$$

where μ is the dynamic viscosity of the liquid. This number represents the ratio of viscous forces to the inertia forces times the surface tension forces. Above the critical value $Oh_{\text{crit}} = 0.03$ [16], viscosity plays an important role during the manipulation of the droplet, for example, for droplet merging. The well known capillary number represents the ratio of the viscous forces to the surface

Table 5.1 Maximum Values of the Dynamic Non-dimensional Numbers

	Surface tension [mN/m]	Dynamic viscosity [kg/m/s]	Density [kg/m^3]	Maximum velocity [mm/s]	Typical dimension [µm]	We	On	Ca	Re
Water	72	0.001	1000	100	1000	0.13	4.10^{-3}	0.0013	100
Water + surfactant	40	0.001	1000	100	1000	0.20	5.10^{-3}	0.0020	100
Ionic liquid	40	0.1	1200	20	1000	0.012	0.5	0.05	0.2

tension forces

$$Ca = \frac{\mu v}{\gamma}. \tag{5.9}$$

The capillary number is linked to the Weber and Ohnesorge numbers by the relation

$$Ca = Oh\sqrt{We}. \tag{5.10}$$

The Reynolds number of the droplet represents the ratio of the internal inertia forces to viscosity forces; it is expressed by

$$Re = \frac{\rho v l}{\mu} \tag{5.11}$$

assuming that the velocity of the droplet motion v is a satisfactory scale of the internal velocity of the droplet. The Reynolds number is also a function of Weber and Ohnesorge numbers

$$Re = \frac{\sqrt{We}}{Oh}. \tag{5.12}$$

For micro-drops, the internal flow field is always laminar, i.e., the Reynolds number is always smaller than the critical Reynolds number. In conclusion, four independent non-dimensional numbers characterize the droplet behavior: Bo, η, We and Oh. Maximum values of the different dynamic non-dimensional numbers are shown in Table 5.1; these values correspond approximately to the maximum measured velocities (largest possible values of η).

Analysis of the values of Table 5.1 shows two different behaviors in common EWOD systems (approximate electrode size 500–1000 µm). On one hand, water-based liquids have, in comparison with surface tension forces, very small viscous effects during their motion, and small to moderate inertia effects. Conversely, ionic liquids show moderate to strong viscous effects (the Ohnesorge number

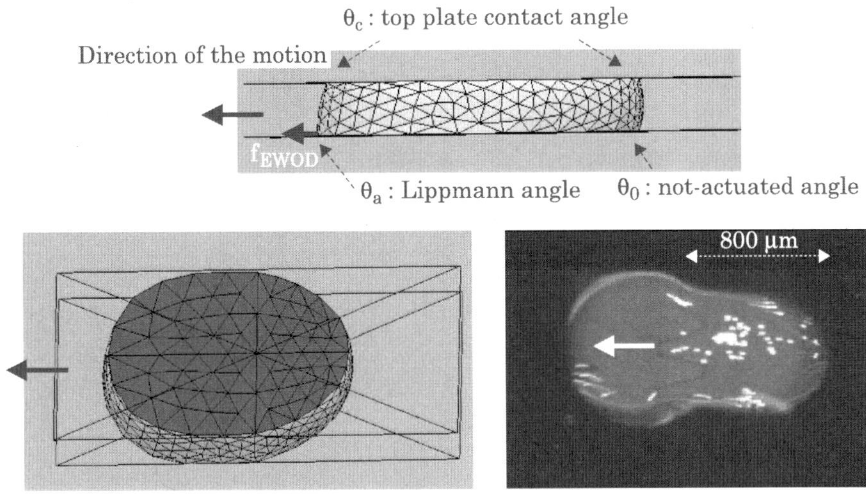

Surface Evolver quasi-static motion

dynamic image

Figure 5.12 Top: motion is triggered by dissymmetry of the electrocapillary forces on the triple lines. Bottom: the quasi-static model of Surface Evolver is compared to the real dynamic image of the motion. Discrepancy is mostly due to inertial effects.

may be larger than its critical value) and negligible inertia effects. In summary, most of the time, droplet motion in EWOD microsystems (of dimensions less than 1 mm) is essentially governed by surface tension forces and Young's constraints at the contact of the solid surface. Note that this will be even more the case when the size of the electrodes is reduced. *Thus the morphology of the electrodes is the key factor for the control of droplet motion and manipulation.*

Fig. 5.12 shows the effect of the Weber number on the motion of a water droplet. If inertia was totally negligible, the motion could be described by the electrocapillary equivalence (see Chapter 4) and modeled using a quasi-steady state approach (here, the Surface Evolver software). The reality shows that the droplet is slightly deformed by inertia forces. However, the discrepancy is not that important, and a model purely based on surface tension forces predicts correctly the possibility/impossibility of droplet motion, if not the accurate shape of the droplet during motion.

5.3.2.2 Droplet Velocity

In digital microfluidic systems, a droplet moves from one location on the chip to the next incrementally. Because inertia forces are small, the droplet accelerates very quickly, reaches a nearly steady velocity, and decelerates abruptly. In this section we focus on the "nearly steady state" velocity of the droplet.

Velocity depends on the strength of the electro-capillary forces. It is then expected that velocity of the droplet is directly related to this force. In reality,

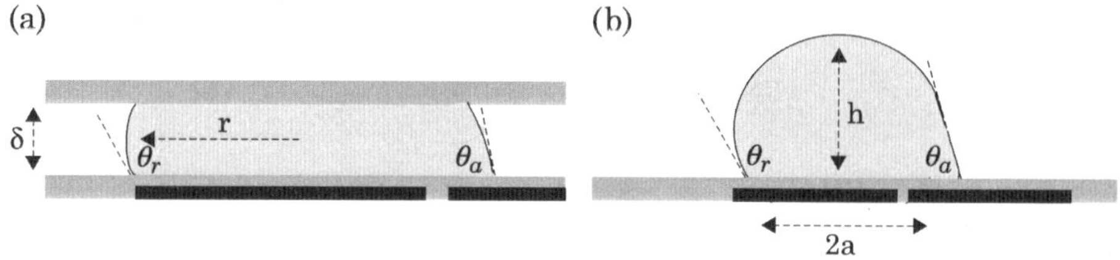

Figure 5.13 (a) Droplet moving in a covered EWOD configuration, (b) droplet moving in an open EWOD configuration.

the motion of micro-drops is a complex physical phenomenon involving surface tension, viscosity, dynamic contact angles, internal motion, etc., and is still a subject of research. In the following section, we present a very simple model proposed by Chen et al. [17] to make an estimate of the velocity of micro-drops during their motion in EWOD microsystems.

5.3.2.2.1 Model

We consider a droplet of volume *Vol* either sandwiched between two plates (covered EWOD) or sitting on a flat surface (open EWOD), as sketched in Fig. 5.13.

We can do the following simplified analysis: if the velocity of motion is constant, the electrowetting force balances the friction force due to viscous dissipation at portions of the wall in contact with the drop, assuming that the other resisting forces, such as contact-line resistance, are negligible compared to the viscous dissipation. Then, in the case of the covered system, the contact surface is assumed circular (which is not true in reality)

$$F_{\text{viscous}} \approx 2\pi r^2 \tau_w \approx \frac{12\mu \text{Vol}}{\delta^2} v_{\text{cov}} \tag{5.13}$$

where v_{cov} is the global velocity and τ_w is the wall friction, approximated with the Poiseuille profile

$$\tau_w \approx \frac{6\mu v_{\text{cov}}}{\delta}. \tag{5.14}$$

Using the capillary equivalence, we can express the electrowetting force by

$$F_{\text{elec}} \approx 2r\gamma_{\text{LG}} \left(\cos\theta_a - \cos\theta_r\right) \tag{5.15}$$

and by equating equations (5.13) and (5.15) we deduce the droplet velocity of motion

$$v_{\text{cov}} \approx \frac{\delta\gamma_{\text{LG}}}{6\pi\mu r} \left(\cos\theta_a - \cos\theta_r\right). \tag{5.16}$$

If the voltage is smaller than the saturation voltage, we can use the Lippmann–Young relation, and the velocity can be estimated by

$$v_{\text{cov}} \approx \frac{\delta}{12\pi\mu r} C V^2. \tag{5.17}$$

Equation (5.17) shows that the velocity of motion is proportional to the square of the electric potential.

In the case of the open EWOD system,

$$F_{\text{viscous}} \approx \pi a^2 \tau_w.$$

The shear stress τ_w can be approximately expressed by

$$\tau_w \approx \frac{5\mu v_{\text{open}}}{2h} \tag{5.18}$$

based on a peak velocity located at 2/5 of the drop height, according to numerical simulations. Then

$$F_{\text{viscous}} \approx \frac{5\pi\mu a^2}{2h} v_{\text{open}}. \tag{5.19}$$

The electrowetting force is

$$F_{\text{elec}} \approx 2a\gamma_{\text{LG}} \left(\cos\theta_a - \cos\theta_r\right) \tag{5.20}$$

and the droplet velocity is

$$v_{\text{open}} \approx \frac{4h}{5\pi\mu a} \gamma_{\text{LG}} \left(\cos\theta_a - \cos\theta_r\right). \tag{5.21}$$

In terms of voltage (far from the saturation limit), we have

$$v_{\text{open}} \approx \frac{2h}{5\pi\mu a} C V^2. \tag{5.22}$$

It is easy to see that the velocity of a droplet is larger in an open EWOD system. The velocity ratio $v_{\text{cov}}/v_{\text{open}}$ is given by

$$\frac{v_{\text{cov}}}{v_{\text{open}}} \approx \frac{5}{24} \frac{\delta}{r} \frac{a}{h}. \tag{5.23}$$

In (5.23), the term a/h is of the order of 1, and $\delta \ll r$, showing that the velocity of motion is much larger in an open configuration. We note

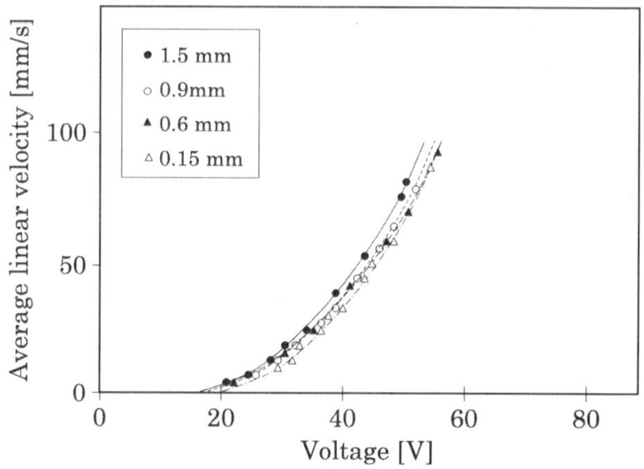

Figure 5.14 At low voltages, the velocity of motion varies as the square of the applied voltage. The size of the electrodes is indicated in the legend. From [19]; reproduced by permission of The Royal Society of Chemistry.

that this model assumes that the two dominant forces are the electrowetting and the friction forces, and supposes a Poiseuille flow inside the droplet. It neglects internal recirculation, the effect of the triple line, the deformation of the droplet during motion, etc. Much is still to be done to understand the motion of droplets under electrowetting actuation. Lets us mention that the first numerical simulations of Dolatabati et al. [18] agree on some points with Chen's model, but disagree about the effect of the vertical gap δ. However, it is a general observation, according to equations (5.16) and (5.21), that the velocity of motion is proportional to the electrowetting force and inversely proportional to the viscosity.

5.3.2.2.2 Experimental results

The variation of the velocity with the square of the voltage predicted by the preceding model has been checked by Pollack et al. [19] for low to moderate values of the applied voltage (Fig. 5.14). Besides, there is a very interesting point in Pollack's publication: the velocity is the same when the dimensions are reduced homothetically, i.e., when the ratio δ/r is kept constant. This is exactly what is predicted by equations (5.16) and (5.17).

At large voltages, the electrowetting force saturates and in consequence the velocity of motion is limited by an asymptote (Fig. 5.15). Using the modified Lippmann law (Chapter 4, Section 4.4.3) based on the Langevin function, equation (5.16) can be recast under the form

$$v_{\text{cov}} \approx \frac{\delta}{6\pi\mu r}\gamma_{\text{LG}}\left(\cos\theta_S - \cos\theta_0\right) L\left(\frac{CV^2}{2\gamma(\cos\theta_S - \cos\theta_0)}\right) \qquad (5.24)$$

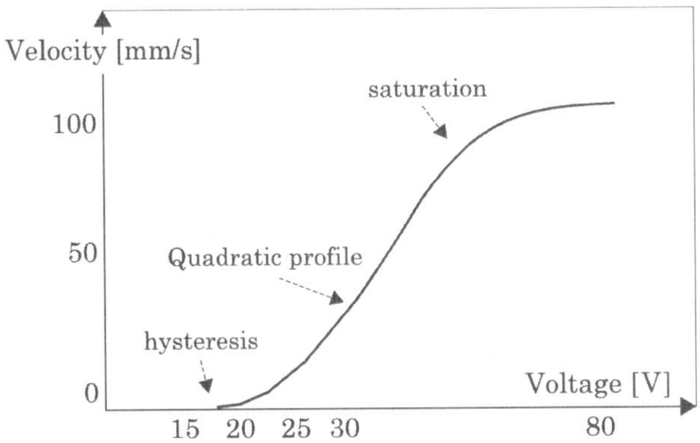

Figure 5.15 Relation between the velocity of motion and the voltage, showing saturation for large actuation potentials.

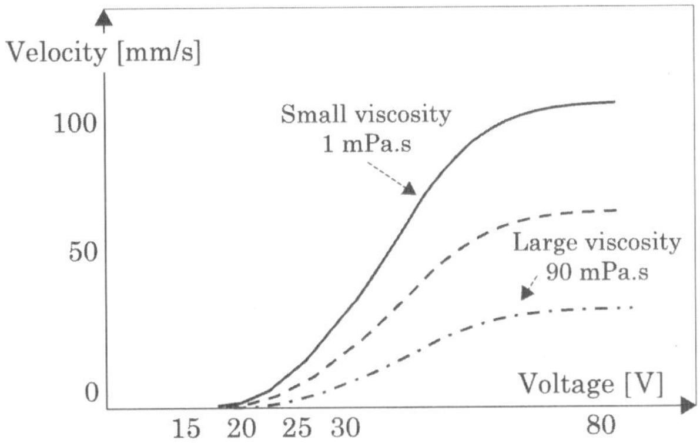

Figure 5.16 Droplet velocity is reduced by increasing the viscosity of the liquid.

and equation (5.21) under the form

$$v_{\text{open}} \approx \frac{4h}{5\pi\mu a}\gamma_{LG}\left(\cos\theta_S - \cos\theta_0\right) L\left(\frac{CV^2}{2\gamma(\cos\theta_S - \cos\theta_0)}\right) \tag{5.25}$$

where L is the Langevin function, and θ_s the saturation angle. These relations between the droplet velocity and the voltage then have the shape shown in Fig. 5.15. There is a constant asymptote at high voltages that represents the upper bound of the velocity. In conclusion, velocity of motion in electrowetting devices is limited by the saturation of the electrowetting force and very high velocities cannot be expected. This upper limit on drop velocity is typically 15 cm/s.

Experimental observations confirm that the velocity is reduced when the viscosity increases (Fig. 5.16). One expects that there is a viscous limit above which liquids cannot be moved in EWOD microsystems. In biology, viscous

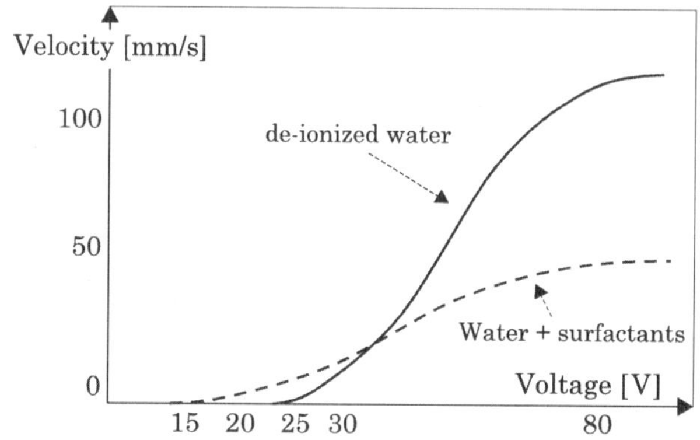

Figure 5.17 Observed velocities of a droplet as a function of applied voltage: velocities are higher when the surface tension is large, except at very low voltage.

(even non-Newtonian) liquids are sometimes used, like polysaccharides, blood, or alginates. Their viscosity increases with the concentration in monomers, dimmers, trimmers, etc. [20]. In Chapters 6 and 7, we will discuss the maneuverability of biological or chemical fluids. For instance, in chemical applications, it has been observed that ionic liquids can be displaced on EWOD chips, but at a small velocity (bottom curve of Fig. 5.16).

Let us finally analyze the effect of surfactants. Remember that in biology and biotechnology, surfactants are currently used for many reasons, the most important being to disperse aggregates and to reduce adhesion on the solid walls. A comparison of the droplet velocity between de-ionized water and biologic buffer is shown in Fig. 5.17. We observed that hysteretic effects are reduced when surfactants are present in the solution, but the droplet motion is much slower. The reduction of the hysteresis threshold in the case of biologic buffers can be attributed to the reduction of the quantity $\gamma_{LG} \sin \theta_0$ (Section 4.5), and the reduction of the velocity is certainly due to a smaller value of the solid–liquid surface tension γ_{SL}.

5.3.3 Shape of the Electrodes: Crenellated Electrodes

We have seen in the preceding section that, in electrowetting microsystems, micro-drops are displaced "digitally" on rows of electrodes. However, the motion of a droplet from one electrode to the next is not straightforward. Microfabrication imposes a gap separating the electrodes. This gap is usually of the order of 10 to 30 μm, depending on the precision of the lithography process, compared to an electrode size of the order of 800 μm. This gap creates a permanent hydrophobic region between two neighboring electrodes. If the droplet has a volume such that it is "contained" or "limited" by the boundaries of the electrode as shown in Fig. 5.18, it cannot move to the next electrode when the latter is actuated.

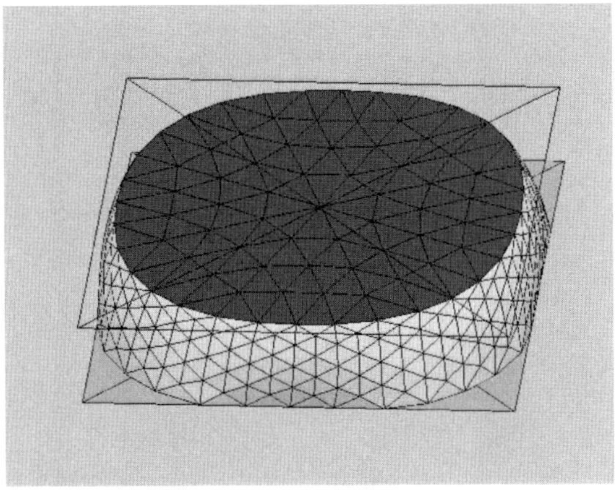

Figure 5.18 Microdrop confined by the electrode boundaries in a covered EWOD microsystem. Surface Evolver simulation [8].

This is frequently the case in covered EWOD microsystems where droplet volumes are carefully controlled, and the size of the electrodes determines the volume of liquid in each droplet within a margin of a few percent. In order to remedy the problem, jagged or crenellated electrodes have been designed as shown in Fig. 5.19. The idea behind such a design is that the droplet contact line with the electrode plane extends over onto the dents of the next electrode. As soon as the next electrode is actuated, electro-capillary forces act to produce the motion of the droplet. Such jagged electrodes require more complicated micro-fabrication, but are very efficient for droplet motion provided the geometry and dimension of the dents are correctly designed.

First, we show that, at the very beginning of the motion, the electrowetting force on a droplet is proportional to the length of the contact line located on the neighboring actuated electrode. This result proves that the electrode's jagged boundary plays a key role in electrowetting actuation, and that its design must be carefully considered for drop motion to commence. We then investigate in a second section the position of the contact line on a jagged electrode. From a theoretical standpoint, this position is related to the theory of partial/total pinning on wettability boundaries, pioneered by de Gennes [21] and Ondarçuhu [22]. We first review the theory and then show how a criterion for determining the size of the dents can be derived.

5.3.3.1 Electrowetting Force

In this section, we analyze the electrowetting force on the droplet at the very beginning of the motion. The electrowetting force is due to the Maxwell stress tensor [23–25] and, because the system is static at the onset of motion, can be translated as a capillary effect using the Lippmann–Young law. It has

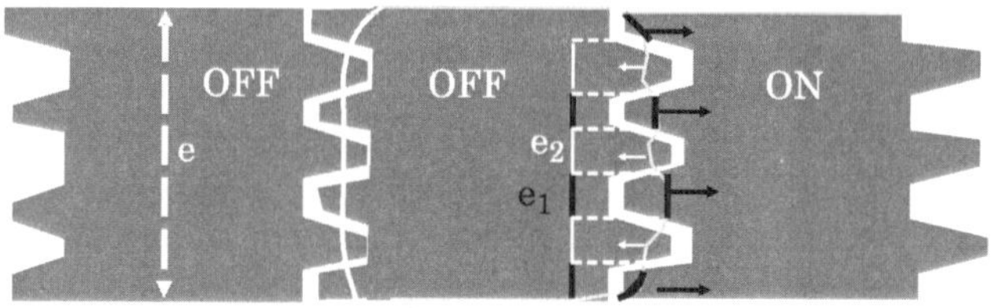

Figure 5.19 Sketch of the crenellated electrodes and position of the droplet. The force exerted by the neighboring electrode at the beginning of the displacement on the advancing contact line is symbolized by arrows.

been established [26] that the electrowetting force during full motion from one electrode to the next is given by

$$F_{e,\text{motion}} = e\gamma\left(\cos\theta_a - \cos\theta_{na}\right) \tag{5.26}$$

where e is the width of the electrode, and θ_a and θ_{na} the actuated and non-actuated contact angles. In Fig. 5.19, we show the position of the contact line at the very beginning of the motion.

In this case, the electrowetting force is

$$F_{e,\text{start}} = e_1\gamma\cos\theta_a + e_2\gamma\cos\theta_{na} - e\gamma\cos\theta_{na} \tag{5.27}$$

where e_1 and e_2 are the total cross dimensions corresponding to the part of the contact line on the actuated and non-actuated electrodes. The first term on the R.H.S. of (5.27) is the pulling force exerted on the liquid by the actuated neighbor electrode, and the second term is the force on that part of the contact line located on the non-actuated electrode (usually $\cos\theta_{na} < 0$, so that the force is directed to the right). The last term of (5.27) corresponds to the receding contact line. Note that

$$e_1 + e_2 = e.$$

After substitution in (5.26), we find

$$F_{e,\text{start}} = e_1\gamma\left(\cos\theta_a - \cos\theta_{na}\right). \tag{5.28}$$

Comparison of (5.28) and (5.26) shows that at the beginning of the motion, the force on the droplet is only

$$F_{e,\text{start}} = \frac{e_1}{e}F_{e,\text{motion}}. \tag{5.29}$$

We verify that if $e_1 = 0$, i.e., there is no part of the contact line overlapping on the neighboring electrode, no electrowetting force acts on the droplet. Relation (5.29) shows that it is important to have at rest a significant part of the contact line overlapping the next electrode.

5.3.3.2 Theory of Triple Contact Line Elasticity

The position and shape of an interface depends on two constraints: surface tension that tends to minimize the surface area, and the Young contact angles acting directly on the contact line. For a wavy contact line, these effects act oppositely. This is the basis of the elasticity theory for the contact line proposed by de Gennes [21] and applied by Ondarçuhu [22] to a sinusoidal contact line. Note that the elasticity theory differs from the line tension theory [27]: the elasticity theory corresponds to a "macroscopic" effect of the interface, whereas the line tension acts at a nanoscopic scale. In the case of this study, line tension is negligible compared to interfacial elasticity.

We just reproduce here the results of Ondarçuhu. The sinusoidal contact line is given by the expression

$$y = \eta(x) = \eta \cos(qx) \tag{5.30}$$

where η is the amplitude and q the wave vector. The results are obtained under the assumption that $\eta q \ll 1$. If the pinning is total, i.e., the contact line follows the sinusoidal boundary, the shape of the interface is obtained by solving the Laplace equation with the condition that far from the surface, the contact angle is an average contact angle θ_0. Then, the maximum and minimum contact angles θ_+ and θ_- can be derived. For the pinning to be total, these two values must be comprised in the interval $[\theta_1, \theta_2]$ corresponding to the contact angle with each surface (canthotaxis). It can be shown that four different morphologies of the contact line can be obtained. These morphologies are shown in Fig. 5.20.

The two lines in Fig. 5.20 are given by the relations

$$\begin{aligned} \eta q &< 1 - \frac{\theta_1}{\theta_0} \\ \eta q &< \frac{\theta_2}{\theta_0} - 1 \end{aligned} \tag{5.31}$$

where θ_0 is the uniform average angle of the interface at some distance of the contact line, θ_1 and θ_2 the Young contact angles with the two substrates. We note that in the case of total pinning, i.e., the contact line follows strictly the sinusoidal line, ηq is quite small.

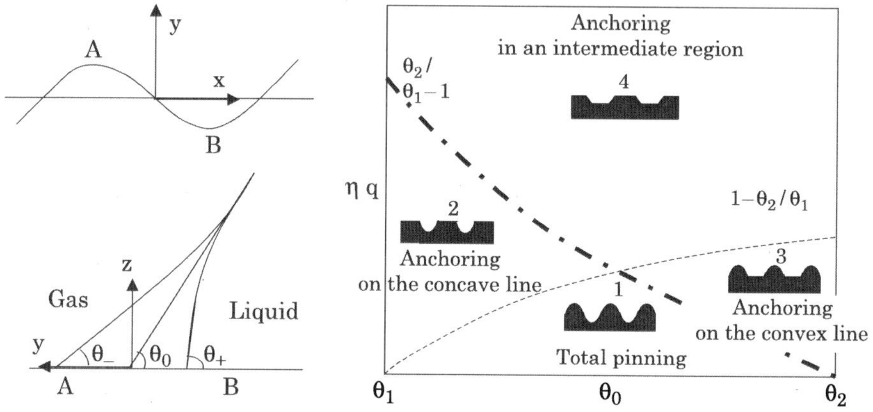

Figure 5.20 Left: schematic of the contact line and its deformation; top: the contact line in the (x, y) plane; bottom: profile of the interface in the (y, z) plane. Right: the four different types of pinning on a sinusoidal boundary line. θ_0 is the contact angle far enough from the contact line, where the interface ignores the details of the contact line. From [22]. © 1995 EDP; reprinted with permission.

5.3.3.3 A Criterion for the Dimensions of the Dents

If we approach the crenellated boundary line by a sinusoidal function, as shown in Fig. 5.21, the analysis of the preceding section applies. We note that the boundary line depends on the gap between the electrodes δ, so that the amplitude η is a function of the size of the "dents" λ and of the size of the gap δ. Approximately, we have

$$\eta \approx \frac{\lambda - \delta}{2}. \tag{5.32}$$

Relation (5.32) shows that if δ increases, the amplitude of the sinusoidal boundary line decreases. The sinusoidal function is then described by

$$\eta(y) = \eta \cos(qy) = \frac{\lambda - \delta}{2} \cos(qx) \tag{5.33}$$

where x is the horizontal coordinate and y the vertical coordinate according to Fig. 5.21.

In the case of electrowetting, we have to design the dents so that the contact line overlaps on the concave part of the sinusoidal curve. This allows the droplet to experience a capillary pull when the neighboring electrode is actuated. Referring to the preceding section, only domains 3 and 4 defined in Fig. 5.20 are acceptable and the following relation must be satisfied

$$\eta q > \frac{\theta_2}{\theta_0} - 1. \tag{5.34}$$

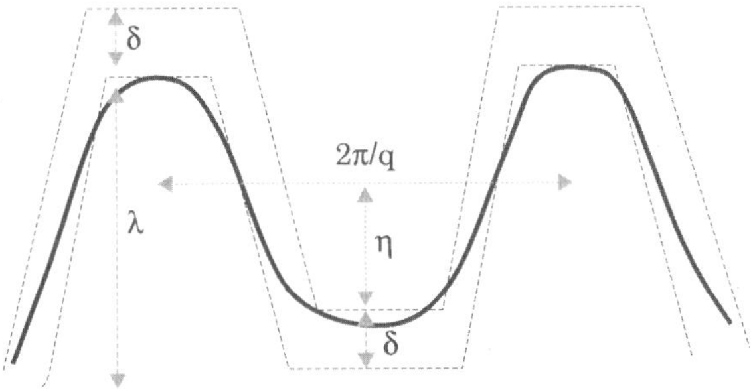

Figure 5.21 Crenellated electrodes can be approached by a sinusoidal boundary.

In equation (5.34) the exact value of θ_0 is not known. However if we choose the non-dimensional number ηq to be larger than

$$\eta q > \frac{\theta_2}{\theta_1} - 1 \tag{5.35}$$

we are assured that the contact line behavior lies within domain 4 describing the case of intermediate pinning. If we note e the electrode width and n the number of dents, we find the following relation for the dimensions of the crenellated electrodes [28]

$$\frac{\lambda - \delta}{2} = \eta > \left[\frac{\theta_2}{\theta_1} - 1\right]\frac{e}{2\pi n}. \tag{5.36}$$

Equation (5.36) can be written under the form

$$\lambda > \delta + \left[\frac{\theta_2}{\theta_1} - 1\right]\frac{e}{\pi n}. \tag{5.37}$$

In dimensionless form

$$\frac{\lambda}{e} > \frac{\delta}{e} + \left[\frac{\theta_2}{\theta_1} - 1\right]\frac{1}{\pi n}. \tag{5.38}$$

For aqueous liquids, typical values of the quantities appearing in equation (5.38) are: $\theta_1 = 80°$, $\theta_2 = 110°$, $e = 800$ µm, $\delta = 20\ \mu m$. The ratio λ/e as a function of the number of dents/peaks is given in Table 5.2.

We note that the values of Table 5.2 are somewhat overestimated since we have considered the most unfavorable value $\theta_0 = \theta_1$. The reality is probably closer to $\theta_0 = \frac{\theta_1+\theta_2}{2}$, so that (5.38) can be written

$$\frac{\lambda}{e} > \frac{\delta}{e} + \left[\frac{\theta_2 - \theta_1}{\theta_2 + \theta_1}\right]\frac{1}{\pi n}. \tag{5.39}$$

Table 5.2 Dimension of the Dents as a Function of the Number of Dents ($\theta_1 = 80°$, $\theta_2 = 110°$, $e = 800$ µm, $\delta = 20$ µm)

Number of dents	1	2	3	4	5	6
λ[µm]	115	68	52	44	39	36
λ/e	0.14	0.08	0.06	0.05	0.045	0.04

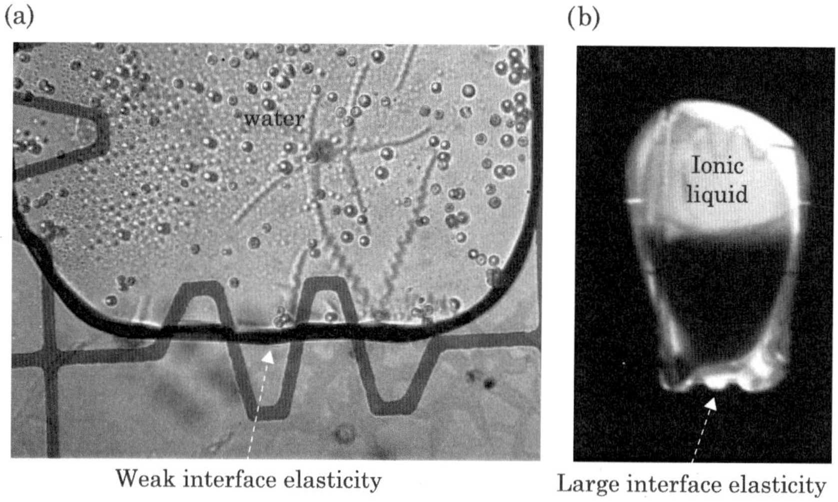

Figure 5.22 (a) View of a water droplet interface on a jagged boundary showing a very rigid interface. The dimensions of the dents are approximately 130 µm for an electrode size of 800 µm, which is largely above the threshold of 68 µm predicted by the elasticity theory; (b) view of a ionic liquid droplet: the interface at the bottom of the photograph is distorted by the presence of the dents, showing a large elasticity of the interface (photo courtesy Ph. Dubois, CEA-LETI).

Fig. 5.22 shows the positions of the interface of a water droplet (left) and an ionic liquid droplet (right) on a jagged electrode. Ionic liquids are most of the time used in chemical applications, and the physical behavior of such liquids will be detailed in Chapter 7, dedicated to chemical applications of EWOD. In both cases, the interface location corresponds to domain 4 in Fig. 5.20 (intermediate anchoring); in the case of water, the interface is slightly distorted by the crenellated shape of the electrodes; this is not the case for ionic liquids, for which the interface is close to total pinning. The results of the theory can be extended to larger values of ηq because the only morphology of interest in our case is morphology 4 (Fig. 5.20). For a given wave vector q, if the amplitude η is such that ηq satisfies (5.35) and $\eta q \ll 1$, a larger value of η will also produce a partial pinning situation.

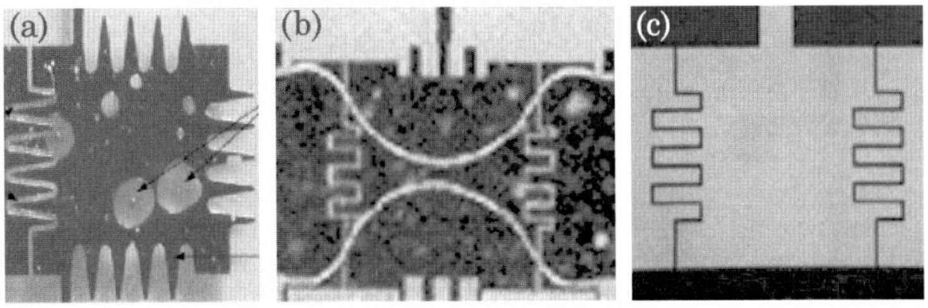

Figure 5.23 Jagged electrode designs from Duke University (a-left), UCLA (b-middle) and Pohang University (c-right). Reused with permission from [28]. © 2005 American Institute of Physics.

5.3.3.4 Examples of Jagged Electrode Designs

In the literature, many examples of crenellated electrodes can be found. We examine the value of the function

$$G(\lambda, e, \delta, n) = \frac{\frac{\lambda}{e}}{\frac{\delta}{e} + \left[\frac{\theta_2}{\theta_1} - 1\right] \frac{1}{\pi n}} \tag{5.40}$$

which must be larger than 1 according to (5.38). We set the values of the contact angles $\theta_1 = 80°$ and $\theta_2 = 110°$ corresponding approximately to water on Teflon.

Our first example is that of the electrode design by Duke University [29] shown in Fig. 5.23.a. In this geometry, the peaks/dents are very pronounced and we have $\lambda/e \sim 0.25$, $\delta/e \sim 1/30$, and $n = 4$. Substituting these values in (5.40) we find $G \sim 3.2$.

The second example is that of the UCLA EWOD microsystem [4,30] shown in Fig. 5.23.b. In such a case, the value of the function G is approximately: $G \sim 1.5$. The third example is that of the Pohang University microsystem (Fig. 5.23.c) [13]; for this system we find $G \sim 3.1$.

Another example is that of the EWOD microsystem of Fig. 5.24 designed by IMTEK [31]. In this case, the value of the function G is approximately equal to $G \sim 1.4$. The design of IEMN [32] in Fig. 5.24 corresponds to a value $G \sim 1.3$. We finally take the example of EWOD microsystems designed by LETI to displace biologic liquids and blood droplets (Fig. 5.25) [33]. The values of the function G are approximately 3 and 1.4.

The values of G for the different systems are compared in Table 5.3. We verify that in all cases the function G is larger than 1, and condition (5.38) is satisfied.

Some of the values of G in Table 5.3 are close to 1. However, there is probably a safety margin due to the fact that the values in the table would have been larger if we had used equation (5.39) instead of equation (5.38).

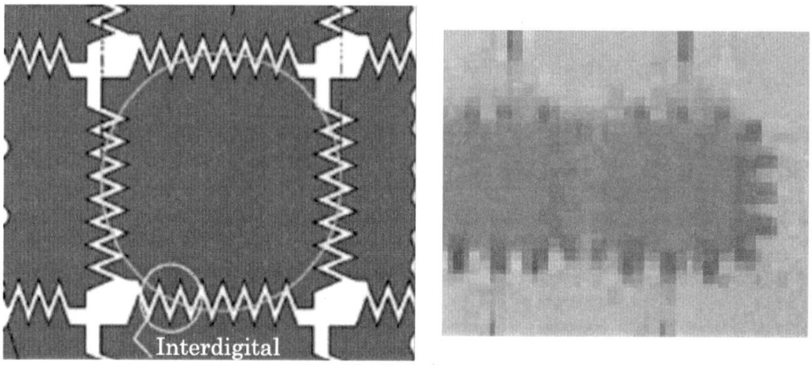

Figure 5.24 Jagged electrode design from IMTEK (left) and IEMN (right). Reused with permission from [28]. © 2005 American Institute of Physics.

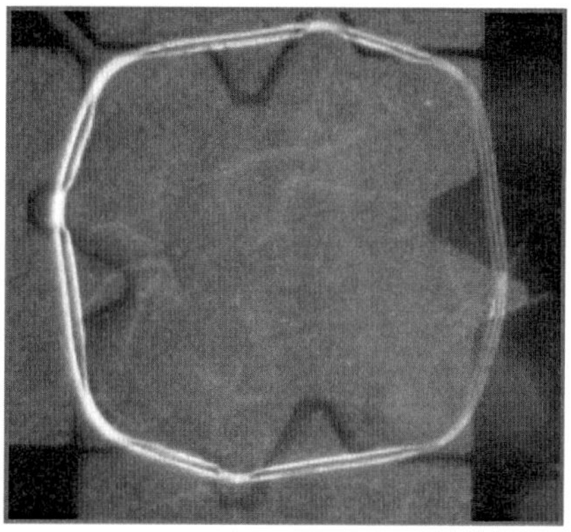

Figure 5.25 Jagged electrode design by the CEA/LETI.

5.3.3.5 Discussion

The theory of elasticity of the triple line has been used to derive a geometrical condition for the dimensioning of the dents of jagged electrodes in EWOD microsystems. This condition states that the non-dimensional ratio of the dent length to the dent width should be sufficiently large, larger than a threshold defined by the actuated and not-actuated contact angles. We have confirmed that this criterion is obeyed by the different electrode designs found in the literature. As a matter of fact, it was found that smaller dents would still have performed satisfactorily for water on Teflon. However, a safety margin is required for the use of combinations of liquids other than water/air systems, particularly for the case in which there is a larger difference between the actuated and equilibrium contact angles, e.g., water/oil or biological buffer/air systems.

An interesting aspect of the geometrical condition derived here concerns the scale down of EWOD micro-devices. The current approach is to homothetically

Table 5.3 Approximate Values of the Function F for Different EWOD Microsystems

Institute	Value of G
Duke University	3.2
UCLA	1.5
Pohang University	3.1
IMTEK	1.4
IEMN	1.3
LETI 1	3
LETI 2 (not shown here)	1.4

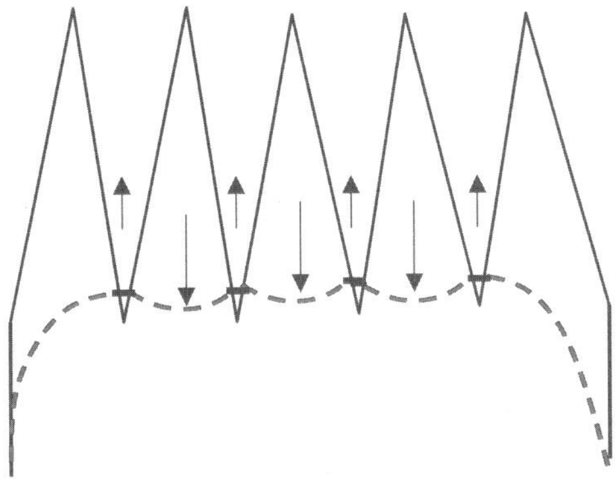

Figure 5.26 Small overlapping of the contact line in the case of a large wave vector.

scale down the size of the electrode and that of the dents; however, we anticipate that reducing the gap size by the same ratio as the reduction in electrode dimension will prove to be difficult. The condition derived here indicates that this effect could be partly compensated by increasing the numbers of dents. However, there is a limit to that increase: a large wave vector, i.e., a small wave length, is likely to result in a smaller length of the contact line overlapping the neighboring electrode (Fig. 5.26); in such a case, according to equation (5.28), the electrowetting force at the onset of actuation may be small and not sufficient to overcome hysteresis and/or pinning.

As a consequence, the scale down of EWOD microsystems will require high performances of the microfabrication process in order to eliminate, as much as possible, the surface defects and to decrease the gap dimension.

5.3.3.6 Numerical Simulation—Analysis of the Forces

As shown in the preceding section, the effect of jagged electrodes is to maintain a portion of the triple line overlapping the neighboring electrodes.

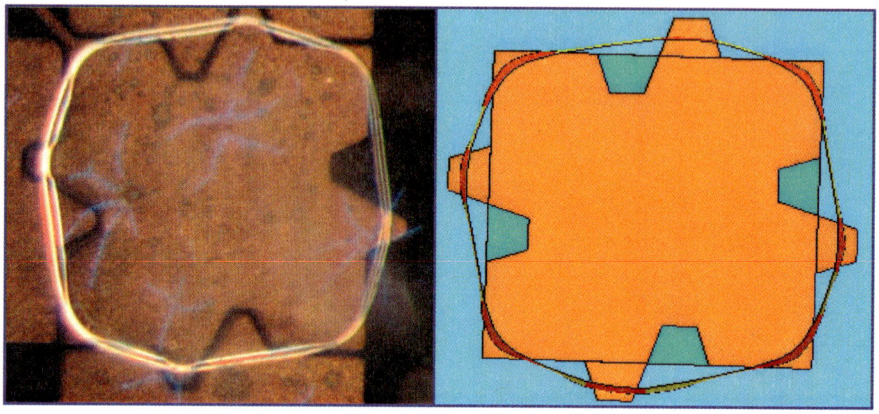

Figure 5.27 Comparison of the contour of the droplet between an experimental view (left) and a numerical simulation with Evolver software (right).

Fig. 5.27 shows a comparison between an experimental image of a droplet located on a actuated jagged electrode and the numerical modeling of the same droplet obtained with the numerical software Surface Evolver.

The action of the dents on the triple line appears clearly: the tips of the dents push the triple line outwards, whereas the holes pull the triple line inwards, resulting in an overlapping of the triple line on the neighboring electrodes. The interface adopts a position that minimizes the surface energy, so that the contact line seems to have a sort of tension that stretches it. This tension does not result from the line tension but on the constraints on the whole liquid–air interface. The numerical program Evolver, for instance, calculates the shape of the surface according to the energy minimization principle, and incorporating the Young's constraints on the contact line.

Fig. 5.27 corresponds to a completely static problem, and Fig. 5.28 to a quasi-static problem (the motion of the droplet is just starting). In Section 5.3.3.1, we have shown that the electrowetting force at the onset of motion is proportional to the transversal length of the triple line located on the actuated electrode. The portion of the contact in the actuated region pulls the contact line in the forward direction, whereas the portion on the non-actuated region pulls in the other direction. The numerical simulation shows in detail how a droplet overlapping the neighboring jagged electrode starts its motion (Fig. 5.28).

5.3.4 Motion from a Covered to an Open EWOD System

We have seen at the beginning of this chapter that two different types of EWOD microsystems have been developed: covered systems where the droplets are confined between two plates and open systems where the sessile droplet is sitting freely on a horizontal solid substrate. Each of these systems has its

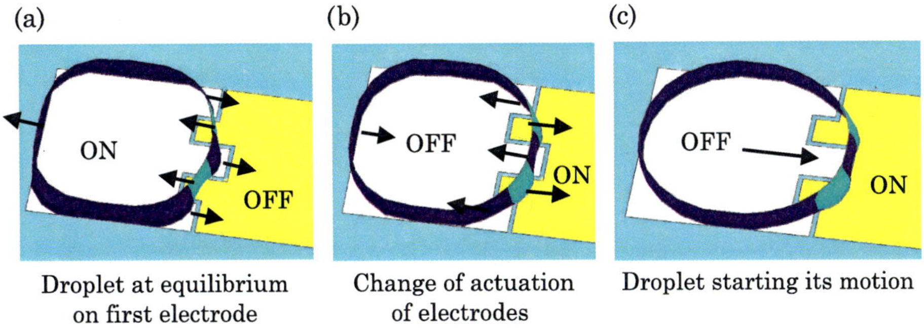

Figure 5.28 Numerical modeling of the motion of a droplet in a cover EWOD system with dented electrodes. The top cover has been dematerialized for clarity. (a) The droplet is at equilibrium, maintained on the first electrode. (b) the actuation has been shifted, and the resulting force on the droplet pulls the right triple line towards the second electrode. (c) the droplet accelerates its motion as soon a large part of the triple line has reached the actuated electrode.

own advantages. We will show later that drop dispensing and splitting are easier in covered EWOD systems whereas mixing and evaporation (for species concentration) are preferably performed in the open configuration [5,34]. Thus, the concept of a dual open/covered EWOD microsystem has been developed [6]. This concept relies on the fact that motion between a covered and open region—and backwards—is possible under electrowetting actuation. In this section, we analyze the possibilities of such a motion. The approach is performed in three steps. First, we use the Surface Evolver software [8] to model the displacement of the droplet from one region to the other. This simplified model assumes that the capillary and electrowetting forces are dominant over inertial and viscous forces (i.e., the Weber number and the Ohnesorge numbers are small). Second, from the Evolver results, and using the Laplace law, we deduce a very simple condition for droplet motion. Finally, we show a comparison between experimental results and the results of the model.

First, we notice that the Weber number and the Ohnesorge numbers are small. Hence, the electrowetting effect is considered equivalent to an electro-capillary effect according to the Lippmann–Young law

$$\cos\theta - \cos\theta_0 = \frac{C}{2\gamma_{\mathrm{LG}}} V^2 \tag{5.41}$$

where C is the capacitance (per unit area) of the dielectric layers, V the electric potential, and θ and θ_0 the actuated and non-actuated contact angles.

It has been shown that the minimization theory correctly predicts micro-drop behavior under electrowetting conditions [5]. As mentioned in Chapter 1, the Surface Evolver numerical program can be used to perform the minimization

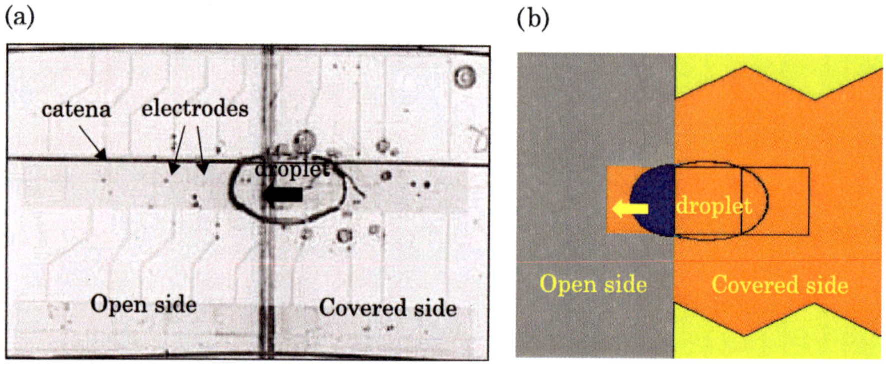

Figure 5.29 Left: experimental view of a droplet crossing the frontier between the covered region and the open region. The drop moves from right to left. Right: same motion predicted by a quasi-static approach using Surface Evolver software. From [6].

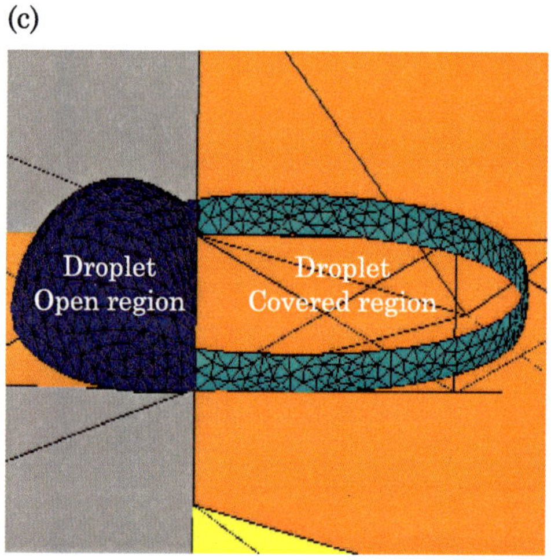

Figure 5.30 Close up view of the droplet crossing the covered/open boundary (the upper plate of the covered region has been dematerialized for clarity). From [6].

of the droplet surface energy under the Lippmann–Young conditions [35]

$$E = \gamma_{\mathrm{LG}} S_{\mathrm{LG}} - \gamma_{\mathrm{LG}} \int_S \int_{\mathrm{SL}} \cos\theta dA. \tag{5.42}$$

Typical features of water droplet motion calculated by the minimization approach are shown in Fig. 5.29.b and compared to experimental results (Fig. 5.29.a). A close-up view of the crossing of the covered/open boundary is shown in Fig. 5.30.

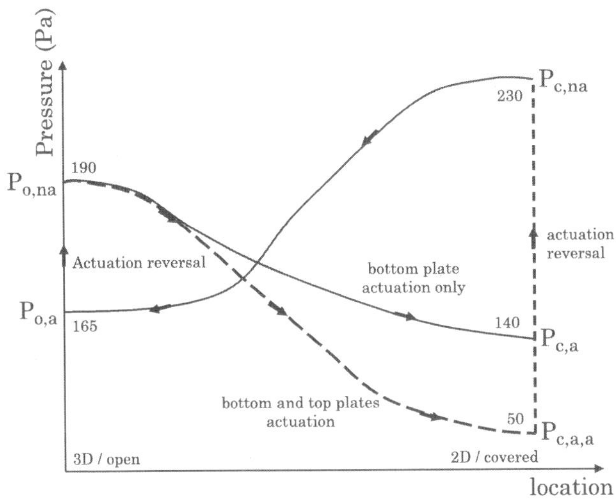

Figure 5.31 Droplet internal pressure during a cycle: the motion of the droplet corresponds to a decrease in pressure. Increase in pressure is obtained by suppressing the electric actuation when the droplet is on either side of the device. The two different curves correspond to the two configurations of covered EWOD devices, with or without electrodes in the upper cover.

Fig. 5.41 shows the general pressure evolution in the droplet during a cycle. A cycle is defined by motion from the open region (noted 3D/open) to the covered region (noted 2D/covered) and back. Suppose the droplet starts from the 3D/open region (top left, Fig. 5.31). Electrodes in the 3D/open region are not actuated whereas the electrodes in the 2D/covered region are actuated. The droplet then moves towards the covered region. When it has crossed the boundary, and is located in the 2D/covered region (bottom right), the actuation is switched off and the droplet internal pressure suddenly increases (top right). The actuation in the 3D/open region is then switched on and the droplet moves back to this latter region. When the droplet is entirely located in the 3D/open region (bottom right) the actuation is switched off and the droplet recovers its initial conditions.

This analysis leads to the conclusion that a motion from one region to the other is accompanied by a monotonous decrease in pressure. The condition for the motion from open to covered region is

$$P_{\text{open,not-actuated}} > P_{\text{covered,actuated}} \tag{5.43}$$

and conversely, for the motion from covered to open region

$$P_{\text{covered,not-actuated}} > P_{\text{open,actuated}}. \tag{5.44}$$

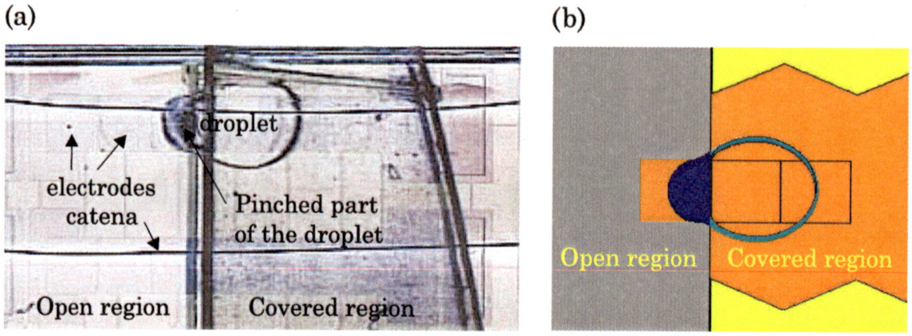

Figure 5.32 Left: experimental view of a droplet blocked on the covered/open boundary by pinching effect on the open region electrode. Right: same situation predicted by Surface Evolver calculation. From [6]. © 2006 NSTI; reprinted with permission.

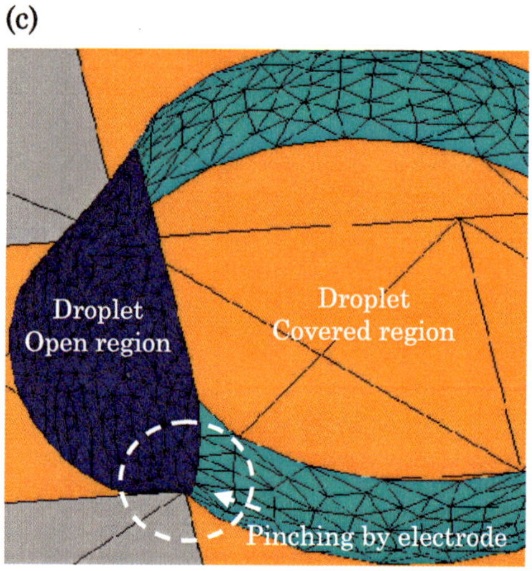

Figure 5.33 Close up view of the droplet blocked the covered/open boundary due to a pinching effect provoked by a narrow electrode (the upper plate of the covered region has been dematerialized for clarity). From [6]. © 2006 NSTI; reprinted with permission.

Note that if the covered region has electrodes embedded in its top and bottom plates, i.e., actuation can be switched on in both plates, the crossing towards the covered region is facilitated (dashed line in Fig. 5.41).

However, there exists an exception to the rule stated by inequalities (5.43) and (5.44). If the drop volume is too large compared to the electrode size, i.e., the droplet overfills the corresponding electrode volume, blockage of the drop due to a pinching effect may occur during the motion from covered to open configuration (Figs. 5.32 and 5.33). This case corresponds to a non-monotonous pressure curve. This drawback can easily be avoided by increasing the width of the electrodes at the transition region.

Droplet pressure in each region can be calculated by using the Laplace law. First, for a drop of volume V, confined between two horizontal plates separated by a distance δ, internal pressure is given by

$$P_C = \gamma_c \left(\frac{-\cos\theta_t - \cos\theta_b}{\delta} + \sqrt{\frac{\pi\delta}{V}} \right) \tag{5.45}$$

where θ_t and θ_b are the contact angles with the top and bottom plates and γ_c the surface tension in the covered region—which can be different from that of the open region γ_o depending on the surrounding fluid (oil or air). In (5.45), the first term of the right hand side corresponds to the vertical curvature, and the second term to the horizontal curvature. For a sessile drop of same volume (3D/open configuration), we obtain, again using the Laplace law

$$P_O = 2\gamma_o \left(\frac{3V}{\pi\left(2 - 3\cos\theta + \cos^3\theta\right)} \right) \tag{5.46}$$

where θ is the contact angle with the substrate. We are then left with two conditions derived from (5.43) and (5.44); the first one is for the motion from the open to covered region

$$P_O\left(\theta = \theta_0\right) - P_C\left(\theta_b = \theta_a\right) \geq 0.$$

So that

$$2\gamma_O \left(\frac{3V}{\pi(2 - 3\cos\theta_0 + \cos^3\theta_0)} \right)^{-\frac{1}{3}} - \gamma_c \left(\frac{-\cos\theta_a - \cos\theta_t}{\delta} + \sqrt{\frac{\pi\delta}{V}} \right) \geq 0. \tag{5.47}$$

And, the second one is for the opposite motion

$$P_C\left(\theta_b = \theta_0\right) - P_O\left(\theta = \theta_a\right) \geq 0.$$

So that

$$2\gamma_o \left(\frac{3V}{\pi\left(2 - 3\cos\theta_a + \cos^3\theta_a\right)} \right)^{-\frac{1}{3}} - \gamma_c \left(\frac{-\cos\theta_0 - \cos\theta_t}{\delta} + \sqrt{\frac{\pi\delta}{V}} \right) \leq 0 \tag{5.48}$$

where θ_0 is the non-actuated contact angle with the solid substrate and θ_α the actuated contact angle.

Experiments have been performed on standard EWOD chips (Figs. 5.29 and 5.32). The electrodes are 800 × 800 μm. Different conditions have been tested:

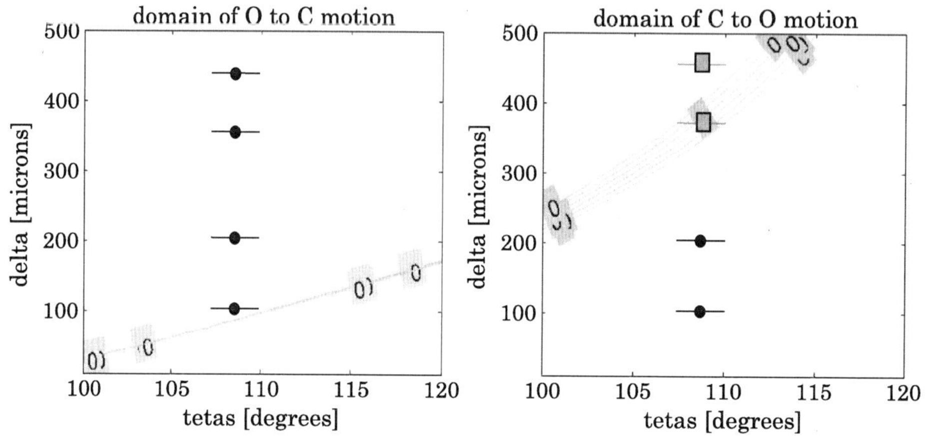

Figure 5.34 Left: conditions for water droplet motion (in air) from open to covered EWOD configuration. Right: conditions for droplet motion from covered to open EWOD configuration. Vertical scale "Delta" is the vertical gap in the covered region and the horizontal scale is the contact angle θ_t on the upper plate in the covered region. Continuous lines correspond to the analytical model—equations (5.47) and (5.48); they separate the domain where motion is possible from the domain where motion is impossible. Dots are the experimental results. Circle dots correspond to observed drop motion from one configuration to the other, while square dots signify that the motion was not obtained. The different continuous lines correspond to different non-actuated contact angle values θ_0 (bottom plate).

the water droplet can be surrounded by air ($\gamma = 72$ mN/m) or by silicone oil ($\gamma = 33$ mN/m), and the height of the vertical gap in the covered region can take the values 110, 200, 360, or 450 μm. The zero potential electrode is provided by a catena that runs all along the electrode row, just below the level of the upper plate. The non-actuated contact angle is close to 113° for water/air, and electric potential was chosen so that the actuated contact angle is 80° (potential value close to 60 V). It has been found that the contact angle on the upper plate of the covered region is a very important parameter of the problem, as is shown in Fig. 5.34.

Fig. 5.34 (left) shows that the motion from open to covered region is not possible if the vertical gap is too small: the counter-pressure in the covered region is then too large and equation (5.47) is not satisfied. Motion in the reverse direction is not possible if the gap is too large, i.e., equation (5.48) is not satisfied because the driving pressure in the covered region is too small. There is a domain where both motions are always possible. This domain is shown in Fig. 5.35. It appears immediately that this domain is sufficiently large to leave room to build a component where back and forth motions are possible.

From this analysis, it has been shown that the motion from an open to a covered region of an EWOD micro-device—and conversely—is closely related to the difference of drop internal pressure between the departing and arriving

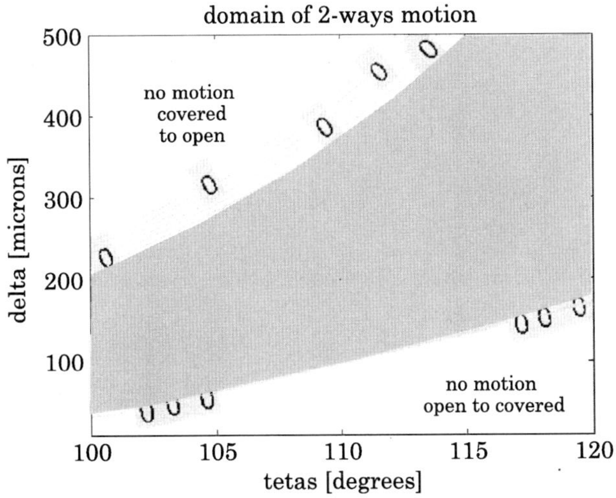

Figure 5.35 Superposition of the two previous figures: the domain for which the motion is possible both ways is the yellow region comprised between the green lines.

regions. Electrodes at the covered/open region boundaries must be adequately designed in function of the liquid droplet volume; they should not be too narrow in order to avoid a pinching effect resulting in stopping the motion.

For a droplet of water in air, motion from open to covered region is possible provided that the vertical gap in the covered region is not too small, and motion from covered to open region is possible if the vertical gap is not too large. Contact angle with the upper plate θ_t is an important parameter of the motion; below 90° the droplet will have difficulties exiting the covered region towards the open region due to hydrophilic grip on the upper plate; above 120°, the motion towards the covered region will be increasingly difficult due to hydrophobic repulsion on the upper plate. There is an important dimension for a component where both motions are easily possible. A vertical distance of about 200 microns seems to work well for buffer liquid drops of 0.1 to 0.5 μl on Teflon substrate and electrodes of 800 × 800 μm.

5.3.5 Effect of a Catena

The effect of a catena has been treated in Section 3.2.9. We just recall here that the catena is used as a zero potential contact for some open EWOD systems (Fig. 5.36).

It has been observed that the catena modifies the behavior of the drop. In the case of a hydrophobic contact the drop adopts a position tangential to the catena, and in the case of a hydrophilic contact, the drop wants to be positioned symmetrically on the catena. The explanation of this phenomenon is presented in Chapter 2. Fig. 5.37 shows a droplet having a lyophobic contact with the

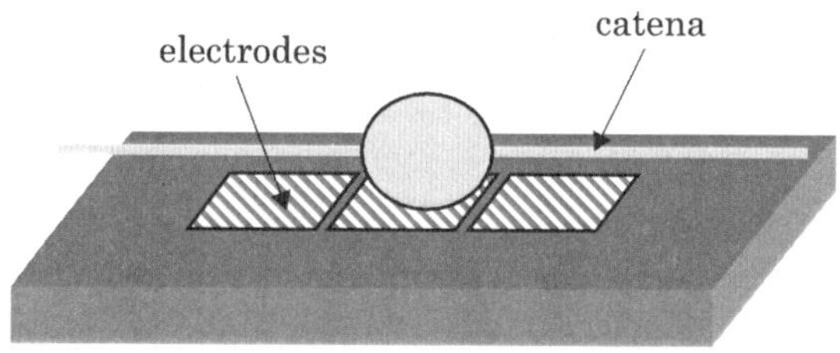

Figure 5.36 Schematic view of an open EWOD microsystem with catena.

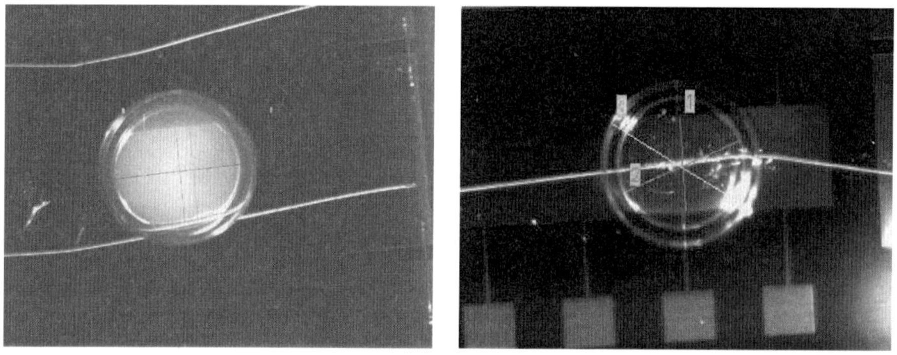

Figure 5.37 Left: the micro-drop is tangent to the catena due to an hydrophobic contact. Right : catena intersects the drop in its middle when the contact is hydrophilic due to the addition of surfactants (Photo Ph. Clementz).

catena. It is positioned tangentially to the wire. After adding surfactants, the contact becomes lyophilic and the drop moves to a position centered on the wire.

In conclusion, the position of the catena above the electrodes has to be carefully chosen depending on its wettability. A hydrophilic catena should be located above the median of the electrode row, whereas a hydrophobic catena is best located above the edge of the electrode row. A catena at a non-optimized position can block the drop motion.

5.4 Division of Droplets

A very important operation for the digital manipulation of droplets in EWOD microsystems is the splitting, or cutting of droplets, i.e., the division of a droplet into two "daughter droplets". This operation is needed because merging of droplets increases by two times the volume of liquid that the system must treat. Division in two half-volumes restores the possibilities of the system.

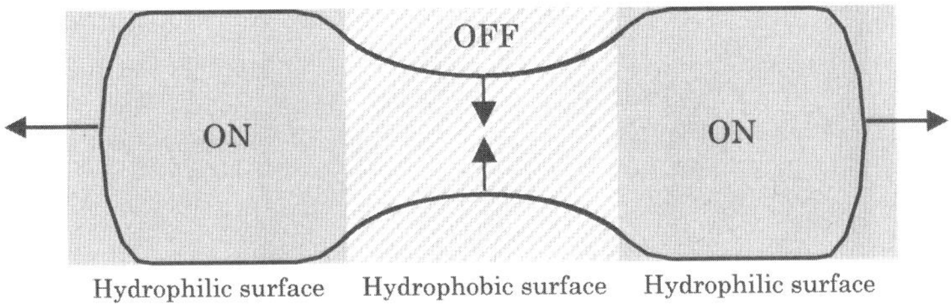

Figure 5.38 Drop division: scheme of the forces on the contact line; the principle is to apply a stretching force in one direction combined with a pinching force in the other direction.

5.4.1 Theoretical Approach

The principle of droplet cutting/splitting/division is shown in Fig. 5.38. The liquid is first stretched or elongated by two actuated electrodes at both ends of the droplet, creating two lyophilic (hydrophilic) areas with electro-capillary forces pulling in opposite directions. Second, the non-actuated lyophobic (hydrophobic) electrode exerts a pinching force on the triple contact line [4,5]. Depending on the force balance and the elasticity of the interface, the droplet can be cut in two.

Splitting a drop into two same daughter drops require an energy input which is a function of the increase in free surface. The total drop energy is the sum of the surface energies

$$E = \gamma_{\mathrm{LG}} S_{\mathrm{LG}} + \gamma_{\mathrm{SL}} S_{\mathrm{SL}}. \tag{5.49}$$

It can be shown that the division of a spherical drop (open system) corresponds to the relative energy input of

$$\frac{\Delta E}{E} = 2^{\frac{1}{3}} - 1 \approx 0.26 \tag{5.50}$$

whereas the division of a drop of radius a, confined between two plates separated by a distance δ, requires a relative energy increase of

$$\frac{\Delta E}{E} \approx \frac{(\sqrt{2} - 1)}{1 + \frac{\gamma_{\mathrm{SL}} a}{\gamma_{\mathrm{LG}} \delta}}. \tag{5.51}$$

Since $a \gg \delta$ and $\gamma_{\mathrm{SL}} > \gamma_{\mathrm{LG}}$, the denominator of (5.51) is much larger than 1, and we find that it requires much less energy to cut a droplet into the covered EWOD configuration

$$\left.\frac{\Delta E}{E}\right|_{\text{covered}} \ll \left.\frac{\Delta E}{E}\right|_{\text{open}}. \tag{5.52}$$

Figure 5.39 A spherical drop cannot be split; it just escapes to any one of the two hydrophilic regions.

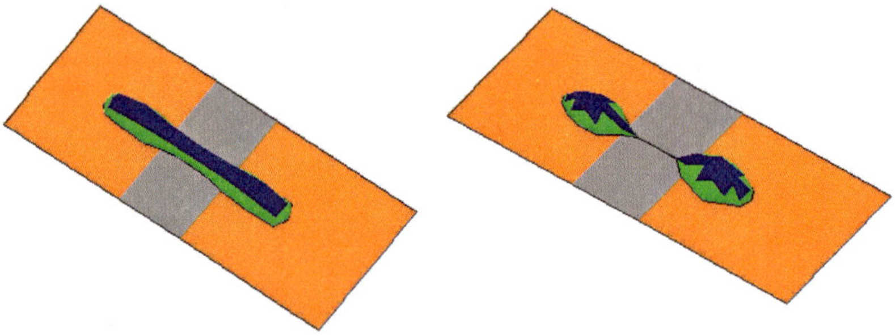

Figure 5.40 Division of an initially stretched droplet predicted by the Evolver numerical software.

Indeed, it is experimentally checked that the division of a sessile water droplet is impossible by electrowetting actuation, whereas this operation is feasible in covered EWOD microsystems.

5.4.2 Droplet Division in Open EWOD Systems

Numerical results confirm the experimental impossibility of splitting sessile water droplets with EWOD. The electrowetting forces are not large enough to elongate the droplet sufficiently. Fig. 5.39 shows that the droplet does not elongate significantly and squirts randomly to the side.

Furthermore, numerical simulations show that, in an open EWOD configuration, a drop can be split if it has been previously elongated by other means (Fig. 5.40).

From all the liquid that we have tested, it appears that only ionic liquids may be split in an open system EWOD, because they have very small contact angles with the actuated electrodes and their elasticity is larger than that of aqueous liquids.

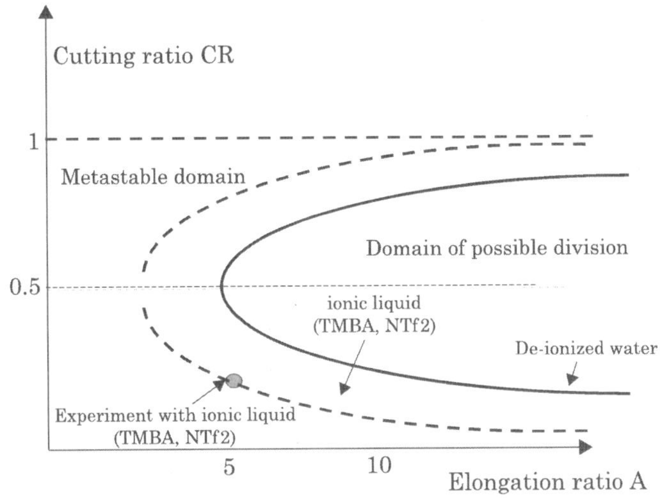

Figure 5.41 Domain of possible drop division for open EWOD systems obtained by a series of Evolver calculations. Continuous line: de-ionized water with contact angles 70 and 115° and surface tension 70 mN/m. Dotted line, ionic liquid with contact angles 60° and 93°, and surface tension 40 mM/m.

Drop division potentially depends on two dimensionless parameters: the initial elongation ratio A between drop length and drop width, and the "cutting" ratio CR between pinching length and initial drop length. The domain of possible splitting is plotted in Fig. 5.51 for two very different liquids. For the usual buffer liquid (continuous line), with a surface tension $\gamma_{LG} = 70$ mN/m and hydrophilic and hydrophobic contact angles of 70° and 115°, the optimal "cutting" ratio is $CR = 1/2$ and the elongation required is $A > 5$. An "open" water droplet has to be stretched at least on five electrodes to be split by electrowetting. The domain where splitting is possible depends on the values of the contact angles and on the elasticity of the interface. In Fig. 5.41, the dotted line corresponds to an ionic liquid with 60° and 93° contact angles.

5.4.3 Droplet Division in Covered EWOD Systems

Numerical simulation confirms the possibility of splitting a droplet in a covered system provided that the vertical gap is sufficiently small. Fig. 5.42 shows how a droplet confined between two parallel plates is easily cut in two by electrowetting forces.

We have seen that in open EWOD systems, division is very difficult for most liquids. So, it is understandable that for a closed EWOD system, there exists a limit value for the vertical gap $\delta_{\lim}$ above which division is not possible. For a square electrode of dimension e, an approximate criterion for the limit vertical

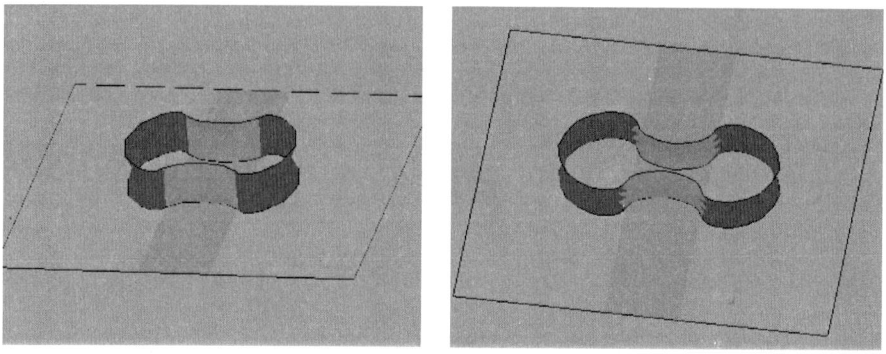

Figure 5.42 Simulation of splitting of a droplet confined between two horizontal plates (the upper plate has been dematerialized for visualization). The hydrophobic contact angle is 115°, the hydrophilic contact angle is 75°, and the liquid/gas surface tension is 70 mN/m.

gap is

$$\frac{\delta_{\lim}}{e} = -\cos\theta_0 \tag{5.53}$$

where θ_0 is the non-actuated contact angle (note that θ_0 is larger than $\pi/2$, so that $\cos\theta_0$ is negative). This relation may be derived by using the Laplace law in the pinching region. The pressure inside the liquid is related to the two curvature radii by

$$\Delta P = \gamma_{\mathrm{LG}}\left(\frac{1}{R_1} - \frac{1}{R_2}\right) \tag{5.54}$$

where the minus sign takes into account the concavity of the drop surface. As we have seen in Chapter 2, the vertical curvature radius R_1 is

$$R_1 = -\frac{\delta}{2\cos\theta_0} \tag{5.55}$$

and, we have approximately, when the width of the pinching region goes to zero

$$R_2 \approx \frac{e}{2}. \tag{5.56}$$

After substitution of (5.55) and (5.56) into (5.54), we obtain a pressure difference between the drop interior and exterior of

$$\Delta P = 2\gamma_{\mathrm{LG}}\left(-\frac{\cos\theta_0}{\delta} - \frac{1}{e}\right). \tag{5.57}$$

Relation (5.57) indicates that the pressure inside the drop decreases when the vertical gap δ increases. The lowest possible pressure difference is zero, so that we find the maximum vertical gap δ_{lim} defined by relation (5.53).

Relation (5.53) produces a rule for scaling up or down covered EWOD devices: if the ratio δ/e is kept constant, and using the same materials, drop division will still be possible. Relation (5.53) also confirms that a very hydrophobic contact angle is best for the efficiency of drop splitting. In our case $e = 800$ μm, and $\theta_0 = 115°$, then the vertical gap δ should not exceed a value of about 340 μm.

5.4.4 Experimental Results

The simulation results of the preceding sections have been verified experimentally. First, for a covered EWOD microsystem (Fig. 5.53) with 800 × 800 μm electrodes and a vertical gap of 100 μm, the drop is made of water and is immersed in silicone oil. At the beginning the drop has a rounded shape and it evolves to the shape of Fig. 5.43 as soon as the electrodes to the left and to the right have been actuated; then the pinching process acts on the middle part of the drop until it is split into two identical "daughter drops". Note the circular profiles in the "cutting" region due to surface energy minimization.

Second, we have used ionic liquids to show the principle of drop division after elongation in an open EWOD system. Ionic liquids have good wettability, do not evaporate, and are used in microchemistry (see Chapter 6). The drop spreads easily on the five actuated electrodes and takes a very rectangular shape, because the elasticity of the surface is large. Switching off the central electrode is sufficient to obtain the splitting of the drop (Fig. 5.44).

5.4.5 Droplet Dispensing

At the beginning of any EWOD process, micro-drops have to be extracted from a reservoir. This step is called droplet dispensing [36–38]. In the following section we investigate the conditions for satisfactory droplet dispensing [39].

5.4.5.1 Analysis of the Dispensing Mechanism

First, we observe that experimental and numerical simulations show that drop dispensing in an open EWOD system is not possible for the usual buffer fluids (aqueous solutions and biological buffers). As a consequence, we analyze the dispensing in a covered EWOD microsystem. As shown in Fig. 5.45, to be effective, dispensing is constituted by three steps:

(1) Liquid extrusion from the reservoir onto the electrode row by applying an electric potential on the electrode row and by switching off the reservoir electrodes. Extrusion occurs because there is an electrowetting force driving

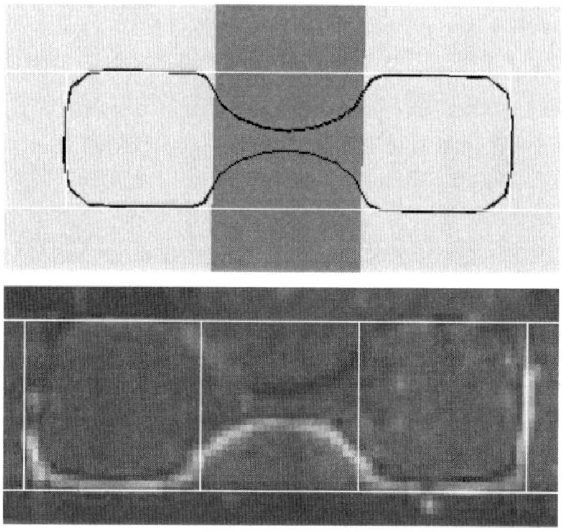

Figure 5.43 Shape of drop comparison between experiment and Evolver calculation during an attempt of division of a confined drop (water in oil). In this case, the electric potential is 20 V and not sufficient to obtain drop splitting. From [5]. © 2006 Elsevier; reprinted with permission.

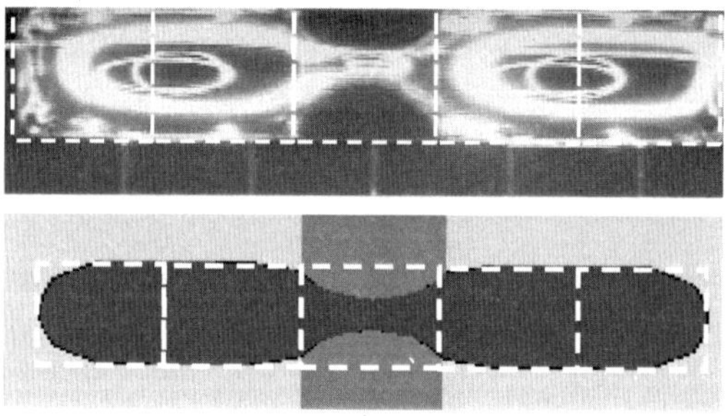

Figure 5.44 Comparison between experiment and Evolver calculation during the division of an ionic liquid droplet: the droplet can be sufficiently elongated and the splitting is effective. From [5]. © 2006 Elsevier; reprinted with permission.

the liquid onto the electrode row and a hydrophobic force pushing the liquid out of the reservoir.

(2) A pinching effect shrinks the liquid filament at the level of the cutting electrode when the latter is switched off. This pinching effect has been analyzed in the preceding section. This pinching step is sometimes enough to separate a droplet from the reservoir, but it has been observed that a third step, called the "back pumping" step, was useful to easily extract well calibrated droplets.

(3) Final dispensing is obtained by "back pumping" the liquid into the reservoir after re-actuation of the reservoir electrodes. The role of back-pumping is to decrease the droplet pressure so that pinching becomes more effective.

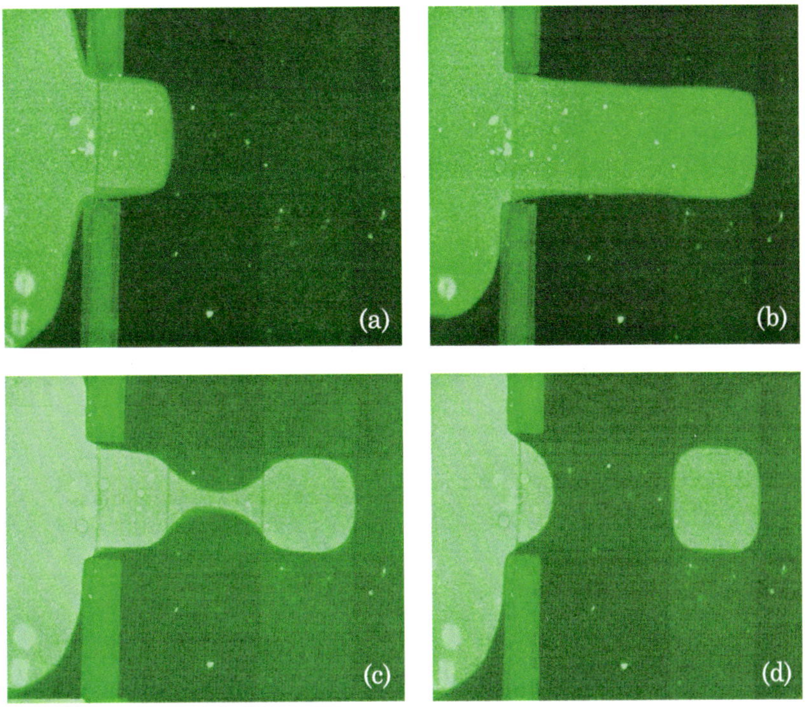

Figure 5.45 (a) and (b): liquid extrusion from reservoir, (c): pinching of the liquid extrusion; (d) separation by back pumping. From [5]. © 2006 Elsevier; reprinted with permission.

It has been checked that the dispensing process is facilitated if the reservoir is separated from the processing electrodes by a solid wall made of plastic (for example Ordyl). Simulation results obtained with the Evolver software are very close to the experimental results (Fig. 5.46) and confirm the key role of back pumping for drop dispensing. In fact, back pumping consists of switching the contact angle from a hydrophobic value—which results in pressurizing the drop—to a hydrophilic (<90°) or neutral (90–95°) value in order to reduce the internal pressure. This decrease in pressure facilitates the pinching effect on the cutting electrode.

Fig. 5.47 shows the time evolution of the drop internal pressure during the different phases of the extraction process. If the decrease in pressure due to back-pumping is sufficient, the pinching effect on the cutting electrode becomes efficient and separation occurs.

5.4.5.2 Insensitivity to the Surface Tension

It is observed numerically and experimentally that the value of the surface tension has little influence on drop dispensing. Table 5.4 indicates the limit

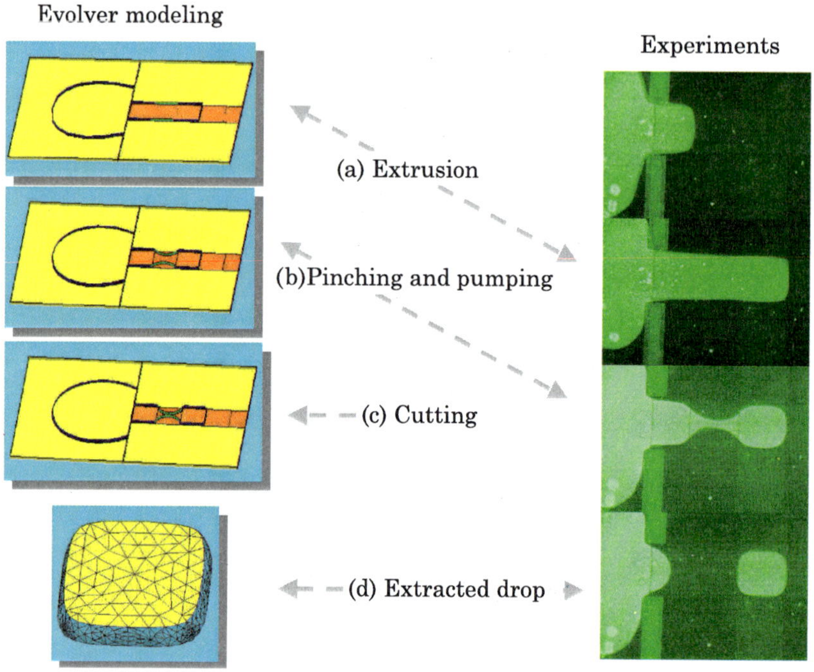

Figure 5.46 Comparison between the numerical and experimental results for drop dispensing in a closed EWOD microsystem. From [5]. © 2006 Elsevier; reprinted with permission.

values of contact angle in the reservoir during back pumping to obtain drop dispense for different values of the surface tension, in the case of 800×800 µm electrodes and a vertical gap of 100 µm. The non-actuated contact angle on the cutting electrode is 120°, the value of the capacitance is $C = 20$ µFarads/m^2, and the reservoir volume is 0.9 µl. The drop does not separate from the liquid of the reservoir if the contact angle is larger than the listed contact angle. The results of Table 5.4 show that, in this geometry, it is not always necessary to have hydrophilic contact angle for back-pumping. Contact angles less than 94° are sufficient. However, it is best to have the smallest possible actuated contact angles if we want to dispense many drops until there remains only a very small volume of liquid in the reservoir.

The present results agree with the dispensing equation obtained in [39] linking the actuated contact angle θ_a to the dimensions and characteristics of the microsystem, and showing that the surface tension does not affect the dispense possibility. We recall this equation because it bears the physics of the dispensing

$$\cos\theta_a - \cos\theta_0 = \frac{2\delta}{e} + \frac{\delta\sqrt{\left(\theta_w - \frac{\sin 2\theta_w}{2}\right)}}{\sqrt{S_R}} \tag{5.58}$$

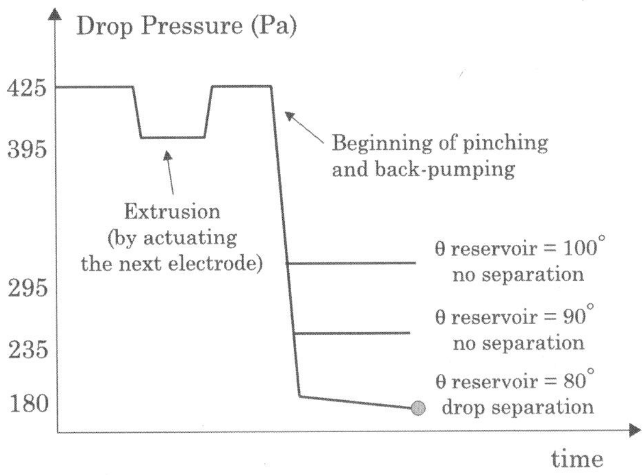

Figure 5.47 Pressure evolution during drop dispense. Each time an electrode in the electrode row is actuated, internal pressure decreases and the drop spreads on the new electrode. When the reservoir electrode is actuated for back pumping with a contact angle of 80°, the pressure decreases to a level where the pinching effect becomes effective and drop separates ($\gamma_{LG} = 40$ mN/m).

Table 5.4 Values of Contact Angles in the Reservoir to Obtain Dispense, as a Function of the Surface Tension. Value of surface tension has nearly no effect on the dispensing

Liquids	Water + surfactant/oil	Water/oil	Water/air
Surface tension (mN/m)	20	36	72
Maximum contact angle in reservoir during back pumping (°C)	95	94	93

where θ_w is the contact angle with the separation wall (Ordyl), and S_R is the contact surface of the drop in the reservoir. To the right hand side of (5.58), the first term corresponds to the "pinching" effect and the second term to the "back-pumping". Equation (5.58) shows that the smaller the vertical gap δ, the easier the drop dispensing. It shows also that the dispensing becomes more difficult when the volume of liquid in the reservoir decreases.

5.4.5.3 Velocity Field in the Liquid During Drop Dispensing

Liquid motion during dispensing can be visualized by following the motion of particles dispersed in the liquid. Fig. 5.48 shows the progression of the reservoir liquid (de-ionized water) towards the electrodes. In photograph (a) the first electrode is actuated and the liquid moves to cover the electrode surface; we notice that the velocity of the liquid is larger on the interface. After the liquid has invaded the first electrode, the next electrode is actuated (b). A similar

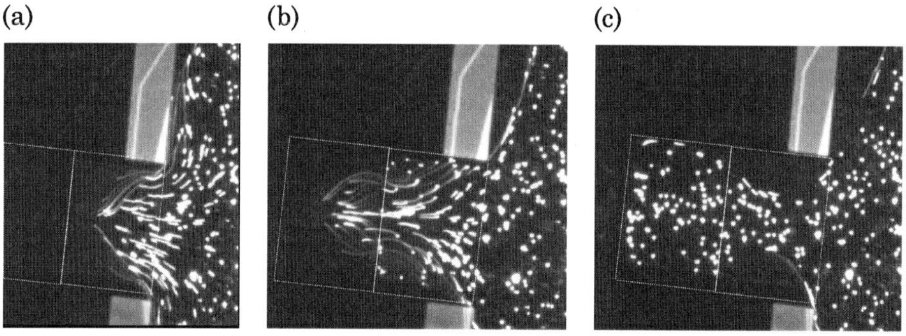

Figure 5.48 Visualization of the hydrodynamic motion during dispensing: the molecules in the vicinity of the interface move much faster than those in the bulk (images a and b). The motion stops immediately when the extracted droplet reaches the boundaries of the actuated electrode, confirming that surface tension forces prevail on inertia (c). The dotted lines outline the electrode boundaries. Photographs courtesy D. Jary (CEA-LETI).

motion starts again with higher velocities at the interface. It seems that the liquid invades the volume above the actuated electrode in the form of a thin layer near the droplet interface, the bulk of the reservoir being little affected. Note the shape of the interface in the reservoir; it does not contact the Ordyl wall, because of the curvature of the interface at the inlet corner. Motion stops abruptly when the electrode is invaded by the liquid, confirming that inertia is small compared to surface tension forces.

5.4.6 Influence of the Cutting Electrode on Drop Volume and Reproducibility

In biology and biotechnology, the extracted drops from the reservoir are used as biological units for the biological operations that take place in the biochip. It is then extremely important to control precisely the volume of the drops and to ensure their reproducibility.

It has been observed that the dispensing process described in the preceding section produces very reproducible droplets. The model shows that the size of the cutting electrode closely controls the size of the final extracted drop. The reason is purely morphological as shown by the comparison between experiments and numerical simulations. It is the tail of the drop that retracts at the very moment of separation and adds to the drop volume (Fig. 5.49). A limitation of the tail size results in a better controllability of the extracted drop volume. Hence, there is a definite advantage to reducing the size of the cutting electrode.

As we have seen in this chapter, depending on the surface tension, the shape of the drop follows more or less the shape of the underlying electrode (Fig. 5.50). However, when retracting after separation, the tail forces a supplementary

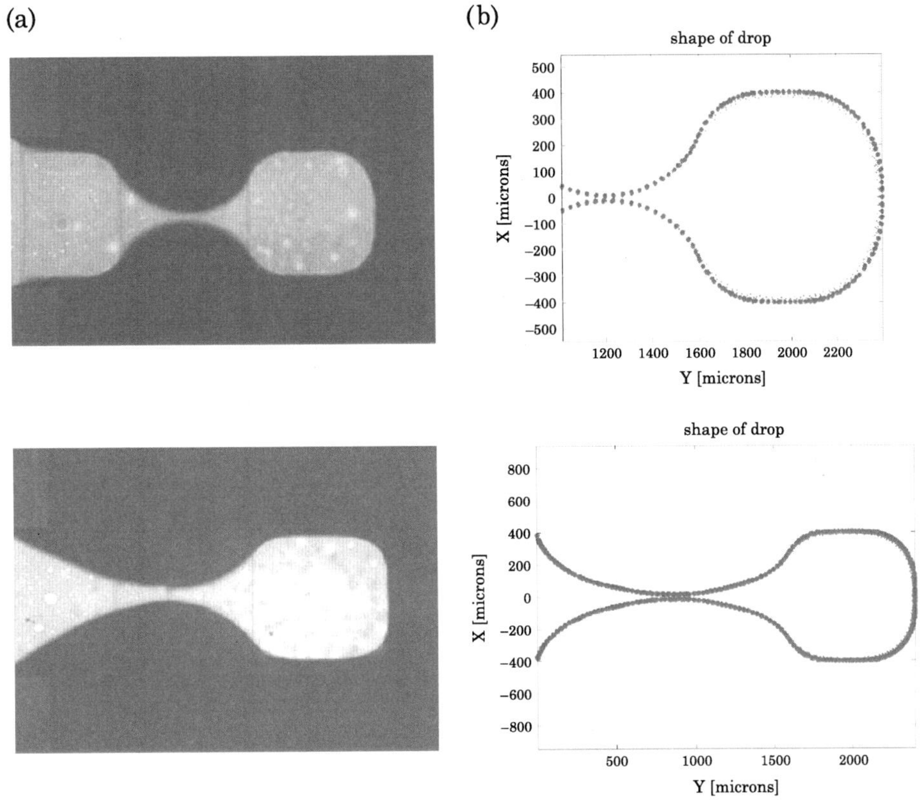

Figure 5.49 Extraction of a micro-drop from the reservoir. Left: experimental view of the extraction obtained by using one or two cutting electrodes; right: numerical simulations with the numerical software Surface Evolver in the same conditions. Using a single cutting electrode produces more controllable volumes of droplets.

volume of liquid into the already formed droplet. It is this "tail" volume that may produce variability in the droplet volume. Fig. 5.51 shows two different configurations where different volumes of liquid are stocked above the electrode.

Table 5.5 shows the final volume of the extracted drop depending on the size of the cutting electrode. Note that for a cutting electrode length less or equal to that of the standard electrode (here 800 μm), the drop volume is less than the volume of $800 \times 800 \times 100\ \mu m^3 = 0.064\ \mu l$, and that for an electrode length of twice the electrode size, the liquid volume bulges out of the electrode (its volume is much larger than 0.064 μl).

The calculated values of Table 5.2 are close, but a little below the experimental values. Using a cutting electrode of the same size as that of the electrodes, we have obtained a very good reproducibility of the drop volume which fluctuates between 64.5 and 66.0 nl (Fig. 5.52). It is a general rule that the reproducibility is satisfactory when the size of the cutting electrode is less than the size of the electrodes in the electrode row.

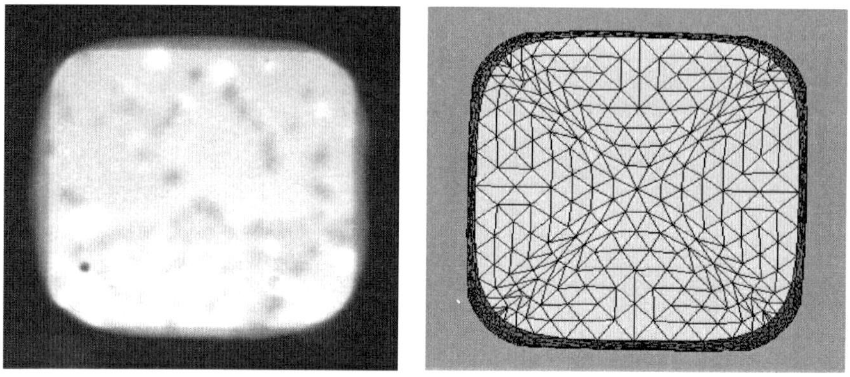

Figure 5.50 Left: fluorescent micro-drop observed with a binocular, blurred contour corresponds to the upper and lower contact line. Right: micro-drop on electrode calculated with Surface Evolver. The lower electrode being actuated, the contact angle is in the range 70 to 85°; the upper plate being at zero potential, the contact angle is hydrophobic (110 to 115°).

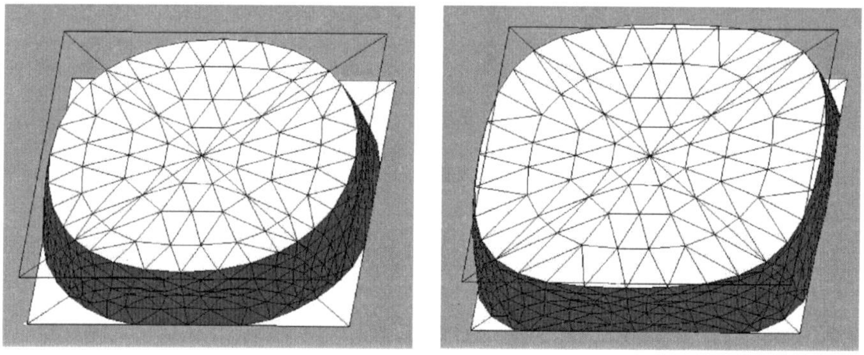

Figure 5.51 Two different shapes of a drop confined between two plates in a covered EWOD microsystem depending on the liquid volume "forced" into the droplet.

Table 5.5 Calculated Values of Extracted Drop Volume in Nanoliters vs. Length of Cutting Electrode and Surface Tension. Note that below 600 μm, dispensing does not occur anymore, because pinching becomes inefficient

Length of cutting electrode [μm] →	600	800	1600
Surface tension [mN/m]			
$\gamma = 40$	56	60	69
$\gamma = 20$	56	60	71
$\gamma = 8$	56	61	77

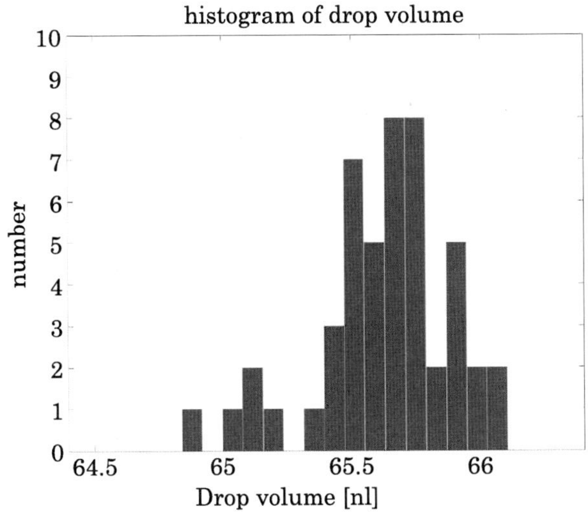

Figure 5.52 Volume reproducibility of extracted droplet: histogram of drop volume. The average volume is $< \mathrm{Vol} >= 65.626$ nl—very close to the volume comprised between the lower and upper electrode (64 nl)—and the standard deviation is $\sigma = 0.26$ nl ($CV = 0.4\%$).

5.4.7 Centering Electrode and Shape of Reservoir Electrode

5.4.7.1 Centering Electrode

Solid substrates of EWOD microsystems are micro-fabricated using extra care to make them as smooth as possible. We have seen in Chapter 3 that surface defects can lead to unwanted pinning, resulting in the malfunctioning of the micro-chip. But a consequence of the smoothness of the surface is that micro-drops, if not anchored by a boundary line, may not always be positioned at the same location on the surface. They show an unstably positioned drop (Fig. 5.53) that drifts until it finds an anchored position by pinning to a singular point or to a boundary line. To maintain a micro-drop at a given location, star-shaped electrodes are used.

The principle of drop centering by star-shaped electrodes is intuitive; it stems from the analysis of the preceding section where opposite electro-capillary forces are created by using crenellated electrodes. When the star-shaped electrode is actuated, the resultant of the electro-capillary forces is zero provided that the droplet is positioned at the center of the electrode (Fig. 5.54).

A numerical simulation using the Surface Evolver software confirms this analysis (Fig. 5.55). A sessile droplet located on the plane is displaced towards the center by the actuation of the star-shaped electrode. After release of the actuation, it regains the same spherical shape, but is now positioned at the center of the star-shaped electrode.

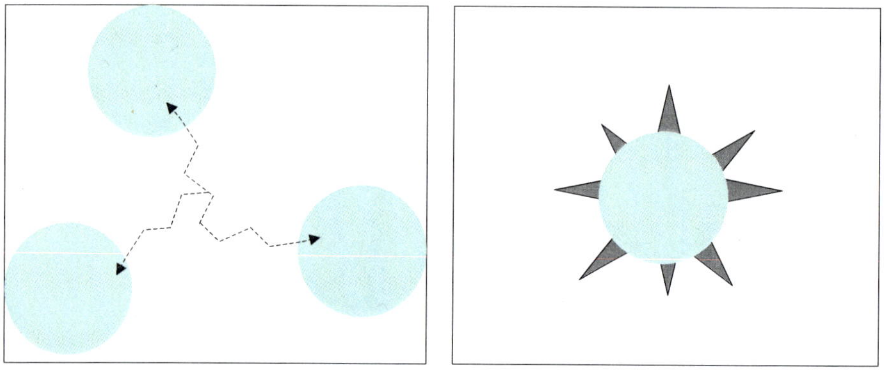

Figure 5.53 Left: on a smooth surface droplets drift until they are pinned on a boundary; the free energy of the droplet is the same for any of the shown locations; right: a star-shaped electrode centers the droplet.

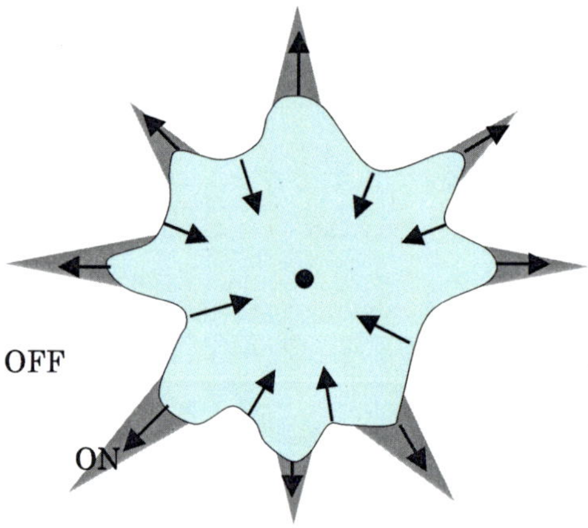

Figure 5.54 Droplet artificially pinned on a central position by the use of a star-shaped electrode. Any deformation of the triple contact line increases the free energy of the system.

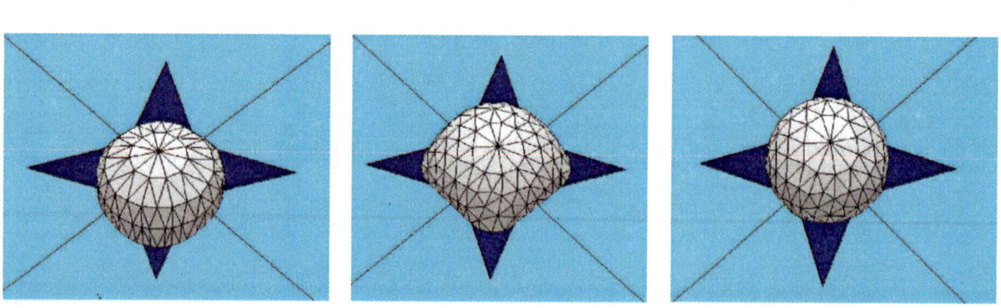

Figure 5.55 Surface Evolver simulation of a droplet on a star-shaped electrode. Left: the droplet is initially positioned arbitrarily on the plane; middle: when the star-shaped electrode is actuated, the droplet moves in such a way that it maximizes its contact surface on the actuated electrode; right: after the actuation is turned off, the droplet regains its spherical shape and is positioned at the center of the star-shaped electrode.

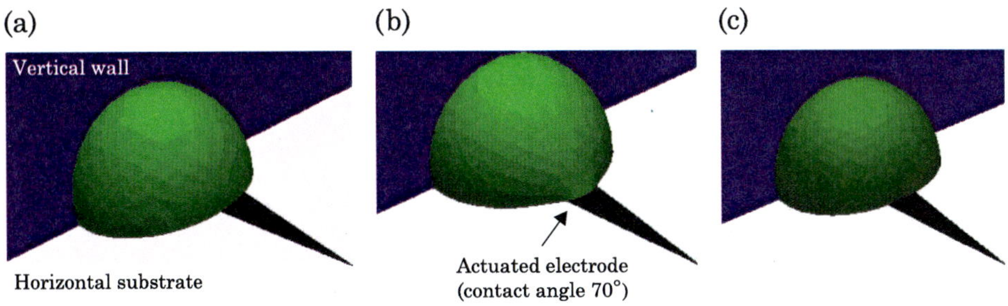

Figure 5.56 Simulation of a droplet being moved to the desired location by the successive actuation/de-actuation of a spike-shape electrode: (a) Droplet "de-centered", (b) actuation of the spike-electrode brings the droplet in the center, (c) after de-actuation of the electrode, the droplet regains its original shape. Operation (a), (b), (c) are equivalent to a translation of the droplet along the vertical wall.

5.4.7.2 Shape of Reservoir Electrode

In the reservoir of EWOD microsystems, the volume of liquid decreases each time a micro-drop is dispensed into the system. Thus, the large droplet of the reservoir progressively decreases. It is essential that this large droplet be positioned right in front of the first "working" electrode of the microsystem, or else dispensing liquid may be stalled.

5.4.7.2.1 Centering electrode

We show first that, in the geometry of droplet dispensing, a spike-shaped electrode can be used to "center" the droplet in front of the "opening gate" in the vertical wall separating the reservoir from the "working" electrodes. In Fig. 5.56, we have numerically simulated how a droplet at first not positioned at the right location can be re-centered by actuation followed by de-actuation of a spike electrode.

From a physical standpoint, the effect of the spike-shaped electrode is schematized in Fig. 5.57. When the droplet has a dissymmetrical position relative to the electrode, there is a resultant of the capillary forces on the triple line parallel to the vertical wall. This force translates the droplet until its position becomes symmetrical and the resultant vanishes.

5.4.7.2.2 Star-shaped electrode

We have shown in the preceding section that a spike-shaped electrode centers the droplet in front of the opening gate. However, this is not the only constraint we want to put on the drop in the reservoir. We have seen in Section 5.4.5.3 (Fig. 5.48) that the liquid in the reservoir can move away from the opening gate after the dispensing of a droplet because of the curvature of the contact line imposed by the corner of the opening gate. In Fig. 5.58 we show that a half star-shaped electrode, when actuated, exerts a force that brings the droplet

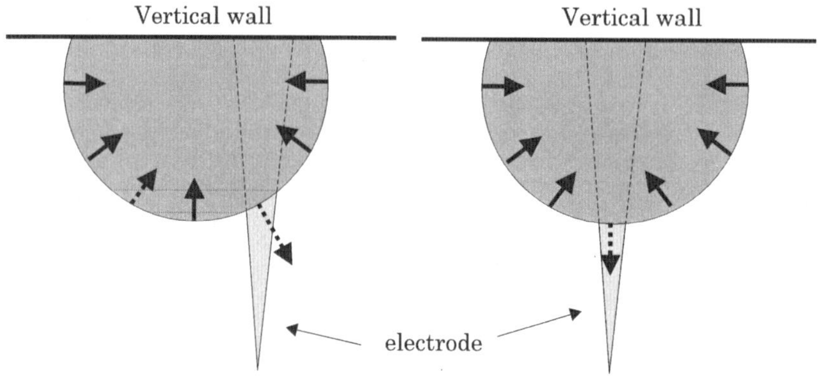

Figure 5.57 Sketch of the forces on the contact line: in a dissymmetrical position relatively to the electrode, there is a resultant capillary force parallel to the vertical wall. In the symmetrical position, this resultant vanishes and the droplet is at equilibrium.

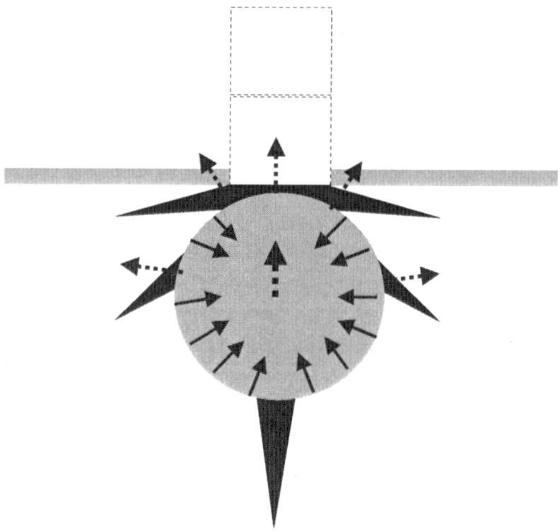

Figure 5.58 Sketch of the resultant of the forces when the drop in the reservoir has moved away from the opening gate. The star-shaped electrode brings back the liquid towards the opening gate.

back towards the "opening gate". First, the long spike-shaped electrode centers the droplet; second, a resultant electrocapillary force is exerted on the liquid, due to the force on the triple contact line on the star-shaped electrode close to the "opening gate". The liquid on the reservoir then is pushed towards the Ordyl wall and ready for a new droplet dispensing.

Fig. 5.59 shows the liquid in the reservoir positioned on the star-shaped electrode (after it has been de-actuated). The liquid is correctly placed for the dispensing of the next droplet through the opening gate at the top.

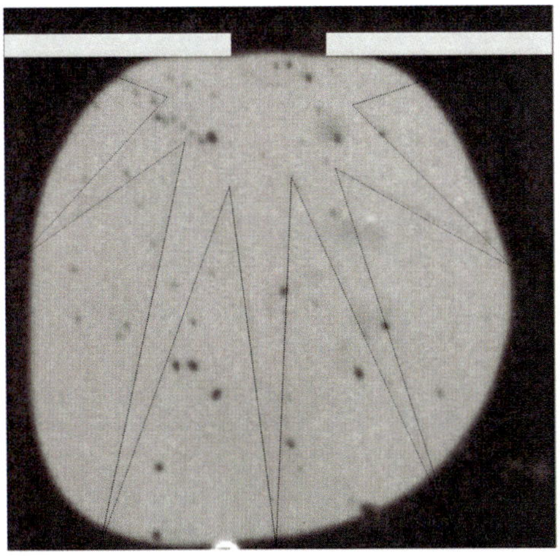

Figure 5.59 Verification of the role of the star-shaped electrode: the droplet has been pulled back towards the Ordyl wall by actuation of the star-shaped electrode; then the star-shaped electrode is switch off and the system is ready for a new droplet dispensing. Photograph Y. Fouillet (CEA-LETI).

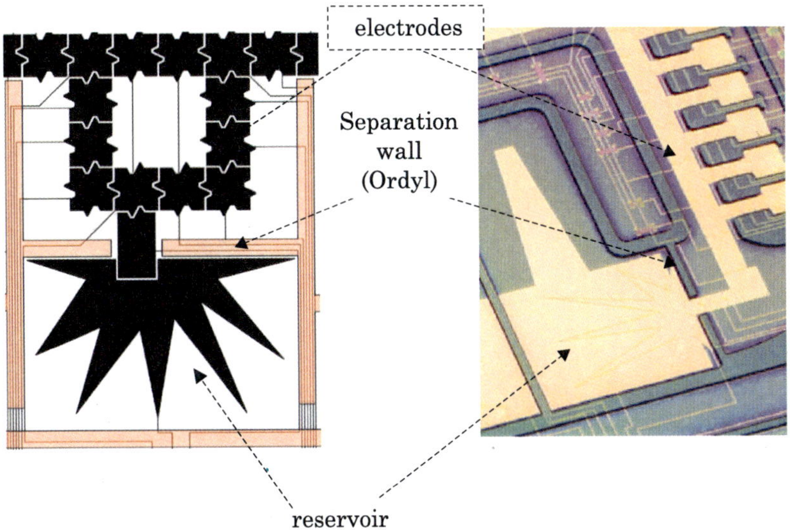

Figure 5.60 Left: schematic view of the reservoir electrodes. Right: EWOD microsystem with reservoir and star-shaped electrode for dispensing (photo Y. Fouillet, CEA-LETI).

5.4.7.2.3 Examples of reservoir designs

From the technological standpoint, it is easy to construct the reservoir with two electrodes: a primary electrode covering most of the surface of the reservoir and a secondary half star-shaped electrode (Fig. 5.60). During the first phase

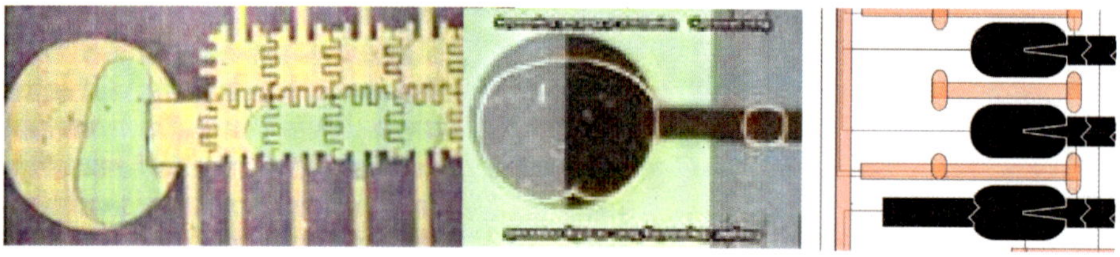

Figure 5.61 Other types of reservoir electrodes. Left: circular electrode with an entering "working" electrode [40]; middle: circular reservoir with semi-circular centering electrode from Duke University; right: one spike star electrode [LETI].

of liquid extrusion onto the "working" electrode row, the main electrode is not actuated (OFF), resulting in a hydrophobic pushing force on the triple contact line, while the secondary star-shaped electrode is at first actuated (ON), resulting in a translation of the reservoir liquid towards the injection electrode. When the liquid is engaged at the working electrodes, the star-shaped electrode is switched off, increasing the pushing force towards the electrode row. When the liquid is sufficiently pulled by the capillary forces exerted by the "working" electrodes, the back-pumping step is engaged, and the two reservoir electrodes are actuated again (ON), leading to droplet separation.

Other reservoir electrodes, less sophisticated, but somewhat easier to fabricate are described in the literature; they are shown in Fig. 5.61.

5.5 Droplet Merging and Mixing

Merging of two droplets is a straightforward operation. It suffices that the two droplets be displaced towards the same electrode (or the same couple of electrodes). Usually, the two drops contain aqueous liquids and their coalescence is immediate (except if the viscous effects characterized by the Ohnesorge number are dominating). However, mixing of the liquids of the two droplets is a complex process. As a matter of fact, mixing in fluids in their laminar state is a very complicated phenomenon. In this case, mixing is promoted by diffusion (which can be a very slow process, since the diffusion coefficients can be very small for macromolecules) and by stretching and folding of the streamlines [41] (Fig. 5.62). For more insights, consult Ottino [42] on the concepts of chaotic mixing with laminar flow. Mixing inside droplets is even more complicated, because of "skin effects" that we will point out below. From a theoretical standpoint, mixing inside droplets has yet to be fully delineated, and we will only tackle the problem by an experimental approach.

Figure 5.62 The different processes leading to mixing. Striation results from successive stretching and folding of the streamlines [42].

5.5.1 The Difficulty of Mixing in Covered EWOD Microsystems

For a covered EWOD system, different actuation patterns can be thought of in order to increase the striation process. In the following, we present the results of three processes: (1) back and forth motion, (2) successive division and merging operations, and (3) loop motion. These operations are sketched in Fig. 5.63.

5.5.1.1 Back and Forth Motion

Mixing by a back and forth motion on a row of electrodes is shown in Fig. 5.64. One droplet is marked with fluorescent dye, while the other is not marked. Efficiency of mixing is monitored by the diffusion of the fluorescent dye. It is interesting that the mixing is induced by two filaments of the liquid—which is located at the tail during the motion—moving along the free (liquid–air) interface (Fig. 5.65). This pattern seems to be general, and has also been observed by Paik et al. and Fowler et al. [43–45]. We notice the similarity with Fig. 5.48 where the particles close to the interface are moving much faster than those in the bulk. We interpret this result as an electric Marangoni effect (see Chapter 1) caused by the electric forces acting on the liquid–air interface. Intuitively, it is as if a liquid "skin" was pulled towards the newly created interface by the droplet motion.

The mixing is sketched in Fig. 5.66. We see that the back and forth motion of the liquid skin reduces the efficiency of mixing, the bulk of the two liquids staying separated for a long time.

5.5.1.2 Successive Merging and Divisions

Another way to mix liquids from two droplets is to induce successive merging and divisions, as shown in Fig. 5.67. We observe the same pattern as in the preceding section. The two droplets exchange their "liquid skin", as is seen

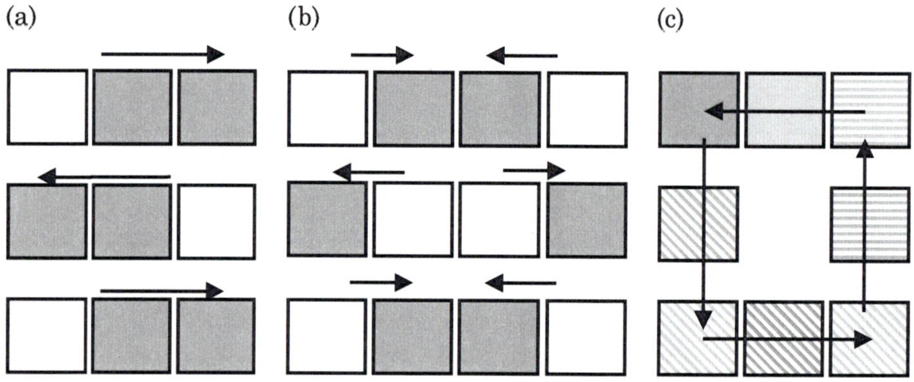

Figure 5.63 Three sequences of actuation for mixing droplets in EWOD microsystems: (a) back and forth motion; (b) successive merging and divisions; (c) loop motion.

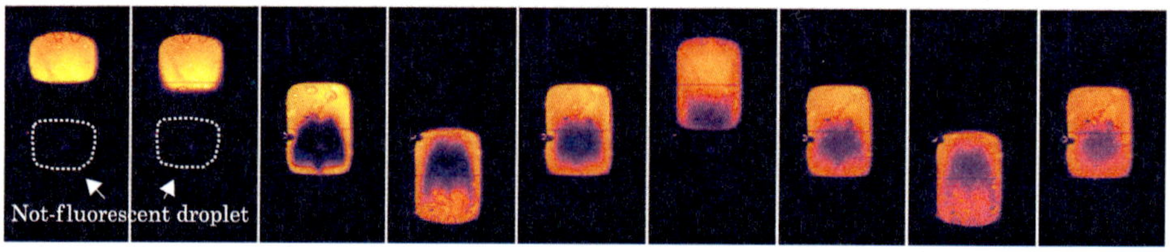

Figure 5.64 Images of the mixing of two droplets of water in a covered EWOD microsystem. One of the droplets is painted with fluorescent dye. Remark how the mixing is induced by a flow pattern alongside the liquid–air interface. Photograph courtesy D. Jary (CEA-LETI).

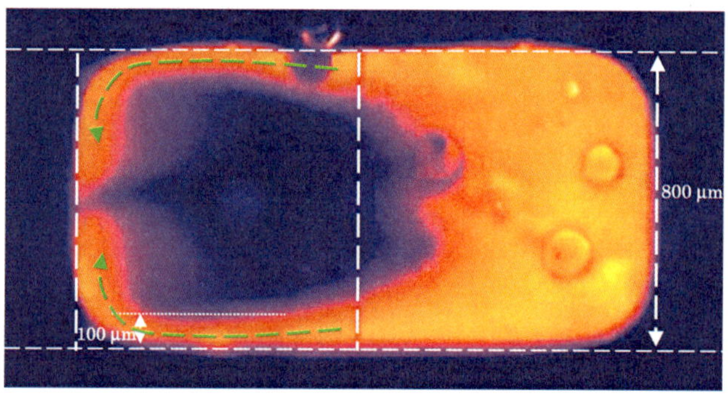

Figure 5.65 Detail of mixing pattern: the liquid in the tail of the merged droplet progresses very quickly along the edges of the electrodes, like a skin effect, probably under the action of electric forces. Photograph courtesy D. Jary (CEA-LETI).

on the top right photograph, where a skin of bright liquid surrounds the dark droplet and conversely a skin of the black droplet surrounds the bright droplet. Again, mixing occurs slowly by diffusion of the liquid in the "skin".

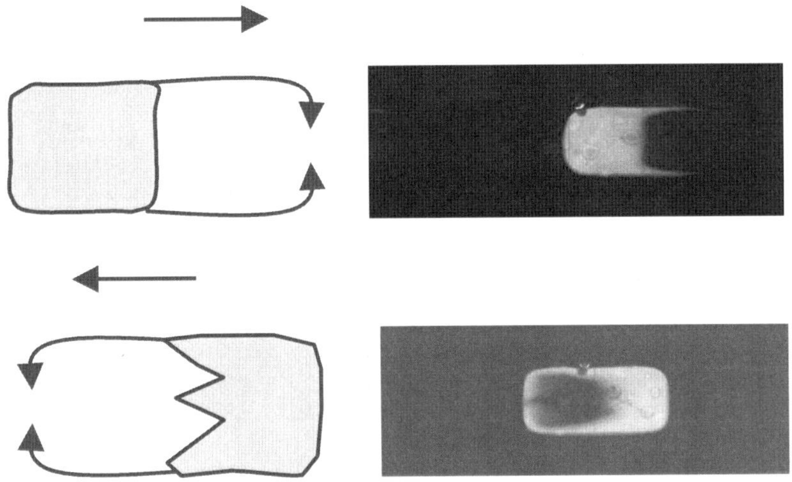

Figure 5.66 Sketch of the mixing by back and forth motion.

5.5.1.3 Loop Motion

The two preceding mixing processes do not induce the stretching-folding pattern which is the best promoter for mixing [42]. In order to induce such a stretching-folding pattern, loop motions have been investigated. Fig. 5.68 shows the mixing pattern obtained by a square loop motion.

We still observe the "skin" effect, but this time the bulk is stretched and folded at each corner leading to a much better mixing efficiency.

5.5.2 Mixing in Open EWOD Systems

In this section, we investigate the mixing patterns in sessile droplets in open EWOD microsystems. We will see that mixing liquids in open EWOD systems is easier, at least for aqueous droplets because of their small viscosity.

5.5.2.1 Mixing by Marangoni Effects

Marangoni convection is a ubiquitous feature in droplet microfluidics. In Chapters 2 and 3, we discuss the different types of Marangoni convection: thermal, concentration, electric. Note that some authors do not favor the name of "electric Marangoni convection," like Jones [11], because it is not formally rigorous; the term electro-convection is more appropriate; however, following Colin [46], it is convenient to use Gibbs' approach and to use the notion of effective surface tension when electric charges are located close to the interface.

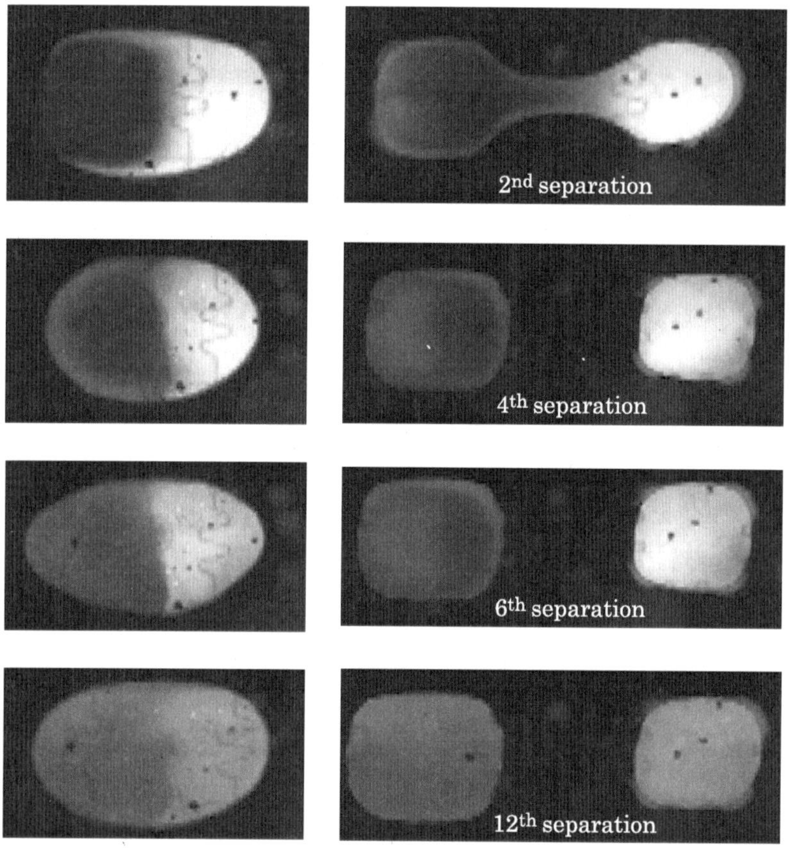

Figure 5.67 Mixing of droplets is a slow process: if the droplet follows cycles of division and merging, it takes many cycles before mixing is achieved. Mixing is due to diffusion of the "skin" towards the central regions, not by hydrodynamic mixing of the central regions. Photograph Y. Fouillet (CEA-LETI).

Fig. 5.69 shows two types of Marangoni flow in a sessile droplet: first, the well-known random motion due to evaporation (no actuation of electrodes); second, typical electric Marangoni—or electroconvective—patterns in a droplet whose interface is submitted to electric stresses [47].

The convective motions can be very active in water-based droplets and induce a very strong mixing.

5.5.2.2 The Difficulty of Mixing Viscous Fluids

So far we have investigated the mixing of aqueous droplets, which have a small dynamic viscosity. The situation is somewhat different for highly viscous fluids, like ionic liquids. For such liquids, the Marangoni effect seems ineffective. In order to induce some mixing, back and forth motions on electrodes have been investigated. Markers in the liquid have been monitored (Fig. 5.70)

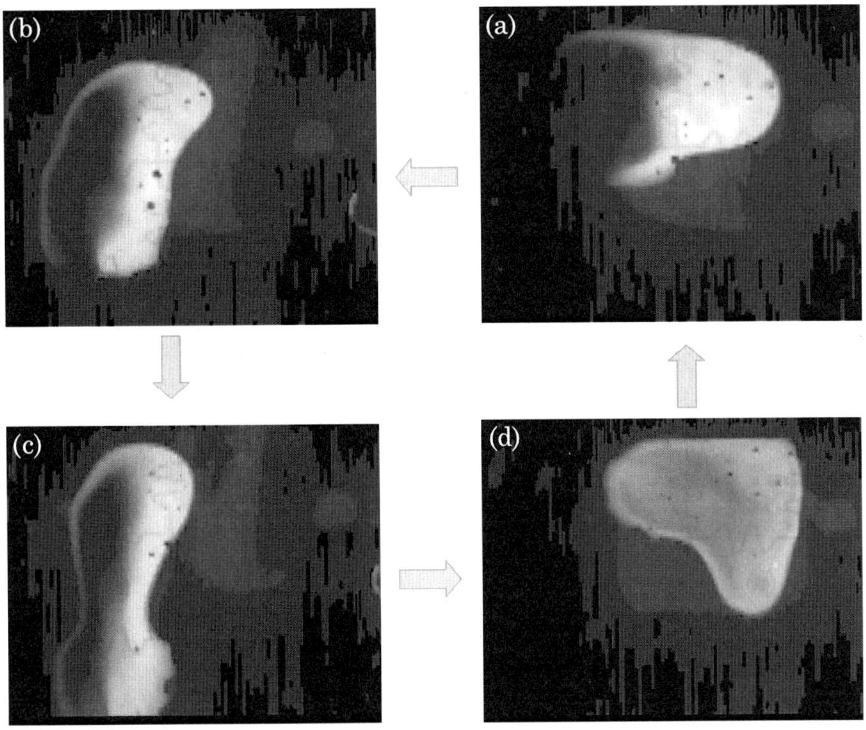

Figure 5.68 Mixing droplets by a square loop motion. The motion starts at the upper right (a). Mixing is more efficient because of the stretching and folding of the droplet in the corners (image c in the bottom left corner). A single cycle on the 8 electrodes loop is nearly enough to obtain the mixing of the two liquids (photo Y. Fouillet, CEA-LETI).

showing an extremely linear pattern. Hence, the mixing efficiency is very low in very viscous fluids. Besides, these fluids behave like "gels": their motion from one electrode to the next is approximately a translation of the whole droplet.

5.6 Dilution

5.6.1 Dilution of Solutes

Controlling and monitoring the dilution of reactants is a very important operation in biology. Concentrations must be very accurately dosed to follow the biologic protocols. The difficulty of diluting reactants is often overlooked; traditionally in biology, it is done using pipettes, but the control of the diluted concentrations is not very accurate. We have seen that EWOD can handle with precision tiny volumes of liquids. In this respect, EWOD is well adapted to dilution operations. As mentioned earlier, the size and shape of the electrodes

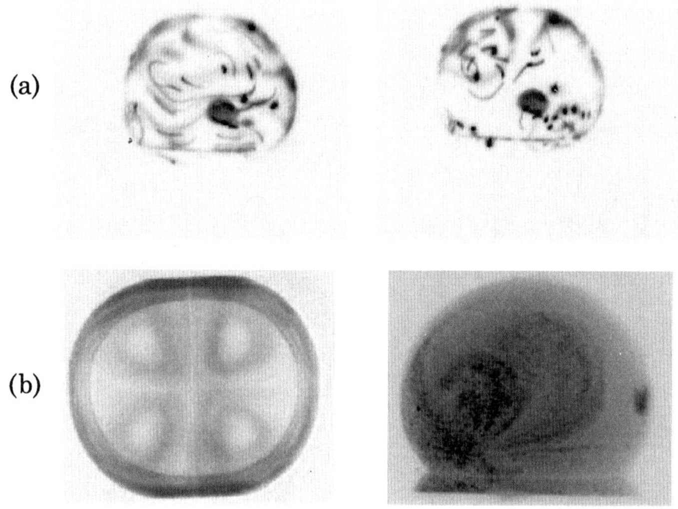

Figure 5.69 Marangoni flow in a sessile droplet: (a) thermal Marangoni convection: effect of evaporation; (b) electric Marangoni convection: effect of the electric field; left: recirculation rolls caused by the presence of two actuated electrodes; right: typical electric Marangoni flow in a droplet on an actuated electrode. (Photographs courtesy Ph. Dubois, Y. Ishida and O. Raccurt).

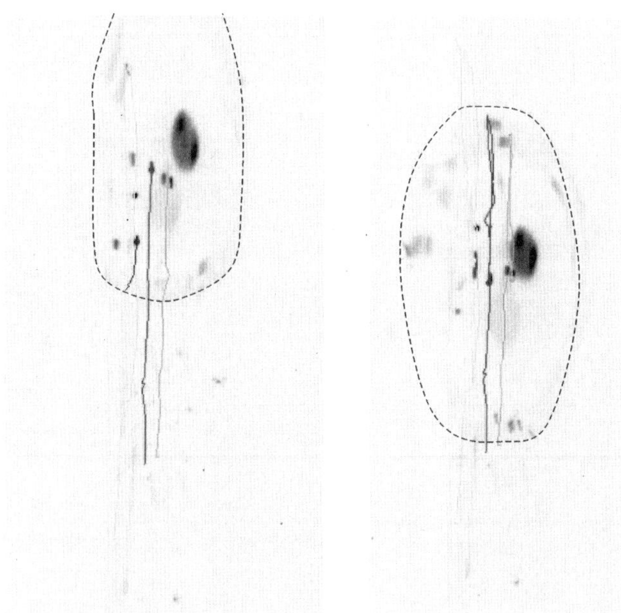

Figure 5.70 Straight line motion of markers in an ionic liquid sessile droplet moved back and forth in an open EWOD device. Photograph courtesy Ph. Dubois (CEA-LETI).

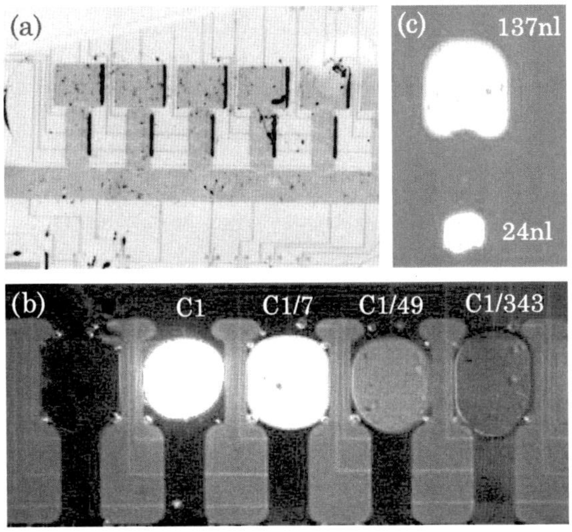

Figure 5.71 (a) View of the EWOD architecture to perform dilution operations; (b) a liquid at a concentration c_1 in a biochemical species can be diluted 343 times in 3 successive dilution operations; (c) dilution is based on extraction of 1/7 of the initial droplet volume (courtesy Y. Fouillet, CEA-LETI).

determines precisely the volume of the droplet. In Fig. 5.71, we show how precise dilution can be realized on EWOD microchips.

The principle of EWOD dilution is the extraction of a fixed, well determined volume of a liquid and the merging of this volume in a fresh droplet with a zero concentration, as shown in Fig. 5.72. Each step divides the concentration by a factor n equal to the ratio of the volume of the reservoir electrode and the working electrode

$$n = \frac{V_{\text{reservoir}}}{V_{\text{elec}}} \cong \frac{S_{\text{reservoir}}}{S_{\text{elec}}}. \tag{5.59}$$

Note that at each step a mixing operation must be performed in order to homogenize the solution. Homogenization is not always possible if the particles in the solution are large, discrete entities like polymers, macromolecules, etc., as we will show next.

5.6.2 Dilution of Discrete Particles

In the preceding section, the dilute species diffuse in the available volume, and the concentration homogenizes, given sufficient time, or efficient mixing. The situation is not quite the same if we consider discrete entities like particles, magnetic beads, cells, etc. Concentration of such particles is not homogeneous: the particles are submitted to body forces like gravity and to interaction forces

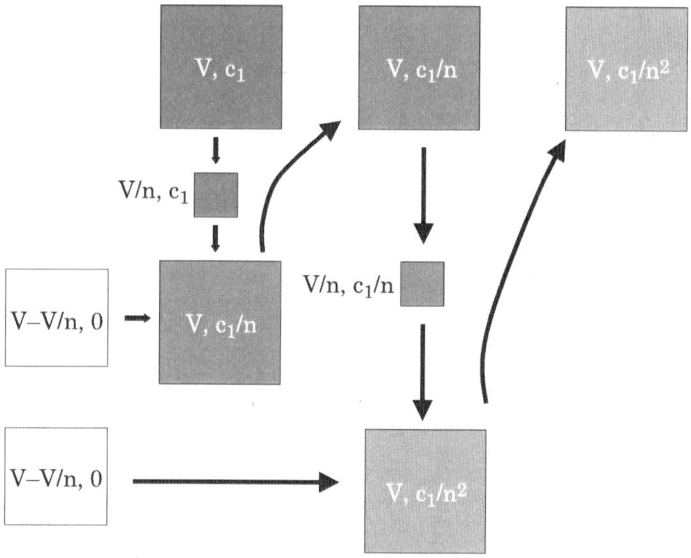

Figure 5.72 Principle of dilution on EWOD microchips.

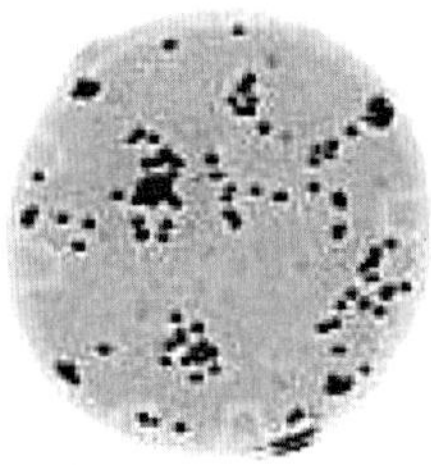

Figure 5.73 Photograph of an aqueous droplet containing cells (photo D. Jary, CEA-LETI).

(electrostatic, van der Waals, etc.) and Brownian motion, or even mixing, is not sufficiently efficient to disperse the particles and cannot homogenize the liquid. This is, for example, the case of cells in an aqueous droplet (Fig. 5.73).

Fig. 5.74 shows the dilution process of solid spherical particles in EWOD reservoirs: obviously the concentration is not homogeneous in the reservoirs; hence the dilution rate between two steps of dilution is not constant. This problem is fundamental in biology and biotechnology, when one wants to study "single cells". This problem is presently being investigated in many labs in the world.

5.7 Magnetic Beads in EWOD Microsystems

In biotechnology, magnetic beads are often used for the transport of macromolecules [41,48]: super-paramagnetic beads are functionalized or labeled to

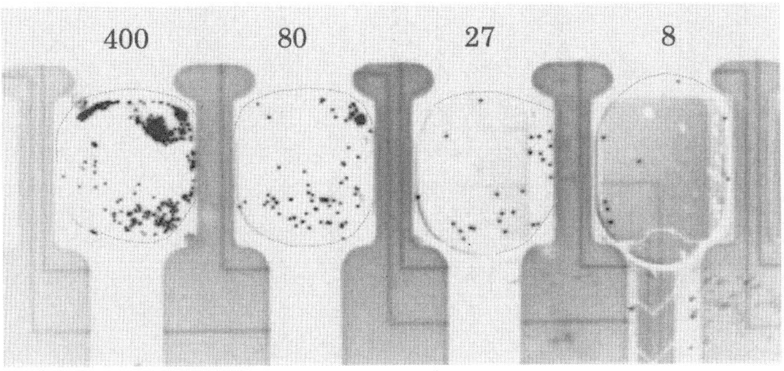

Figure 5.74 Dilution of concentration of particles on an EWOD-based microchip. The number on top of each reservoir is the number of particles (courtesy D. Jary, CEA-LETI).

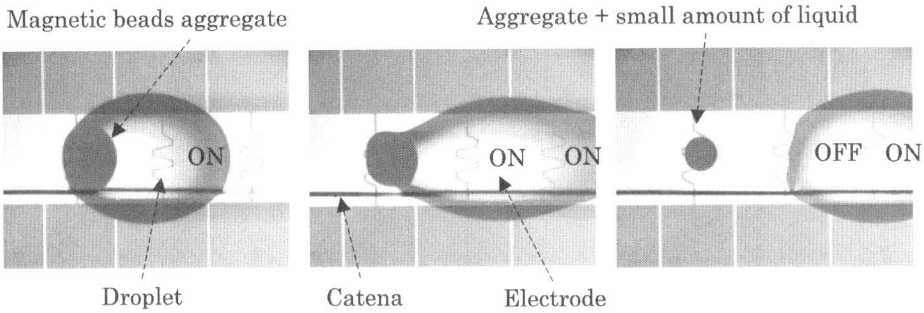

Figure 5.75 Combination of magnetic forces exerted on an aggregate of magnetic beads and electrowetting forces exerted on a conductive liquid droplet. In this case, the magnetic forces are sufficiently important to pin the aggregate and the electrowetting forces are sufficiently large to move the droplet, leaving behind the aggregate with a small amount of liquid (photo courtesy N. Chiaruttini, CEA-LETI).

recognize and bind to a specific molecule. Functionalization is achieved by coating the magnetic bead with molecules having a chemical affinity for the target molecules. Hence, when dispersed in a liquid, the labeled beads capture or bind to the target molecules, forming a composite macromolecule. In the absence of a magnetic field, the particles disperse in the liquid phase. In the presence of a magnetic field, they are attracted towards the magnetic pole; besides they aggregate together because of the formation of chains of magnetic dipoles [41].

In biotechnology, it is thought that processes based on magnetic capture and concentration, similar to those using microflows, can be developed in digital microfluidic devices, with all the advantages of smaller liquid volumes, allowing more precise and sensitive recognition and/or bioanalysis. Fig. 5.75 shows an open EWOD device (with catena) and a droplet containing magnetic beads aggregated by a mini-magnet placed below the substrate. If the electrowetting forces are sufficient and if the magnetic forces are sufficient, the magnetic aggregate separates from the droplet.

Fig. 5.76 demonstrates the principle of concentration of target molecules by paramagnetic labeled micro-beads. First, the beads are dispersed in the droplet and some capture the target molecules (a); then the beads are aggregated by using a mini-magnet (b); the droplet is then motioned by electrowetting actuation (c and d); if the magnetic and electrowetting forces are sufficient, the droplet continues its motion, leaving behind the aggregate with a small amount of liquid (e). Target molecules are now concentrated in "nano" droplet.

Fig. 5.77 shows the balance of forces applied on the receding contact line. In the absence of magnetic and electrowetting forces, the balance of forces is given by the Young law (a). The aggregate is blocked near the tip of the magnet; the shift in the horizontal direction of the magnetic aggregate produces a horizontal magnetic force $f_{x,\mathrm{mag}}$; the equilibrium position is found to give the maximum force $f_{x,\mathrm{mag}}$ (b). The limit angle α of the droplet is given by the diagram (c). If the electrowetting force is larger than the equilibrium force of diagram (c), the contact angle decreases under the value α, and the droplet separates from the magnetic aggregate (d).

5.8 Architecture of EWOD Microsystems

5.8.1 General Architecture

In the preceding sections we have dealt with the basic manipulations of droplets in EWOD systems. These manipulations involved only a few electrodes. In this section, we give some insights into the more complex architecture of complete EWOD micro-devices. Usually EWOD biochips include a few reservoirs containing the different reagents and buffers, and a sufficient number of electrodes to perform the required biological and chemical operations.

Fig. 5.78 shows a typical biologic EWOD chip with its main and secondary, dump reservoirs, and the electric wiring. The figure especially shows the complexity of the wiring. The electrodes are most of the time individually addressed. However, some electrodes are often difficult to connect to an electric line; besides it is very costly in terms of microfabrication to have two levels of wiring in the substrate. A technique called "multiplexing" is then used to address groups of electrodes without losing their individual functionality. We discuss this technique in the following section.

5.8.2 Multiplexing

We have seen that EWOD chips require a paving of the substrate with electrodes. Some EWOD applications require a large number of electrodes and reservoirs. Addressing each one of the electrodes individually is not always

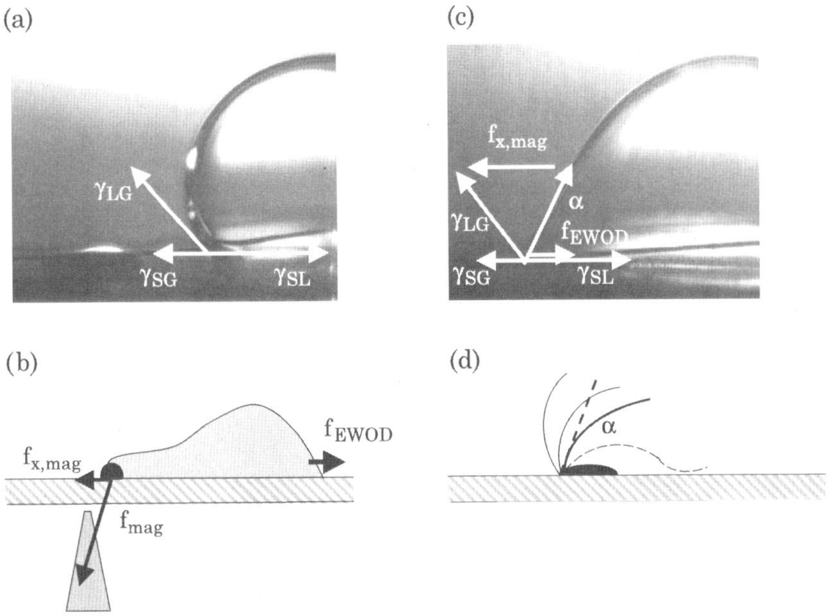

Figure 5.76 Principle of extraction/concentration using a combination of electrowetting and magnetic beads. At the end of the process (f), the concentrated aggregate of magnetic beads attached to target molecules can be removed with a pipette.

possible if the wires are all in the same plane inside the substrate. However, depending on the travel scheme of the droplets on the chip, some electrodes can be put on the same electric line [49]. Consider the design of Fig. 5.79 [7]. The system comprises $M = 5$ electrodes in a row and $N = 3$ rows. Droplets can be individually handled if it is possible to switch the catenaries to the voltage 0 or V and the electrodes to 0 or V. Hence, the $N * M = 15$ electrodes require only $M + N = 8$ electric switches.

Multiplexing is very efficient for systems doing droplet manipulations in parallel. An example of such microsystems is that of Moon et al. [50]. It realizes the generation of multiple droplets from a reservoir and parallel in-line sample purification of a carrier fluid containing proteins.

5.9 Other EWOD Microsystems

EWOD microsystems can be used to perform functions other than those presented above. A first functionality is a "conveyor system"; a second functionality is the "electro-capillary filaments".

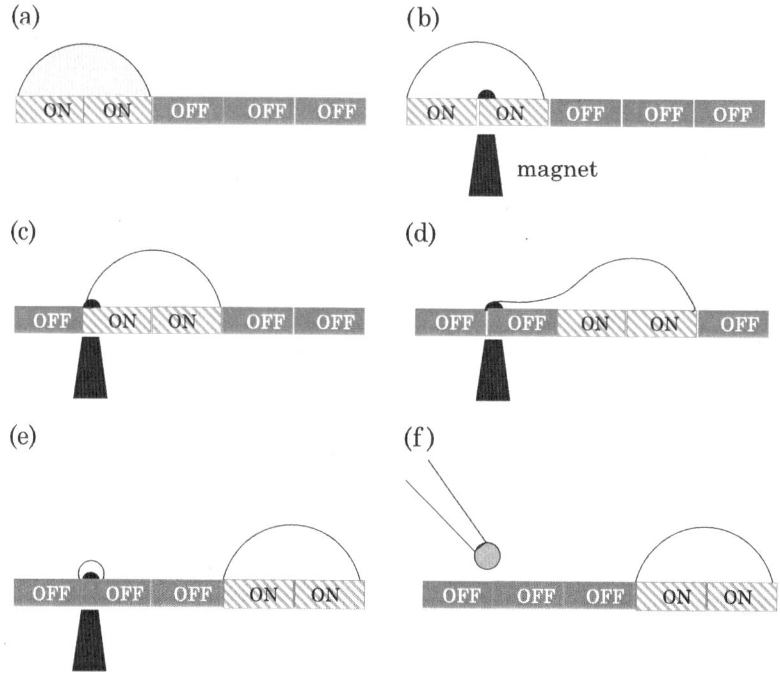

Figure 5.77 Balance of forces on the receding contact line. (a) Young's law in absence of magnetic and electrowetting forces; (b) deformation of the droplet under the two actions of the magnetic and electrowetting forces; (c) balance of the forces at the rupture limit; (d) if the electrowetting forces are sufficient for pulling the interface and decreasing the contact angle below its limit value, rupture inevitably occurs.

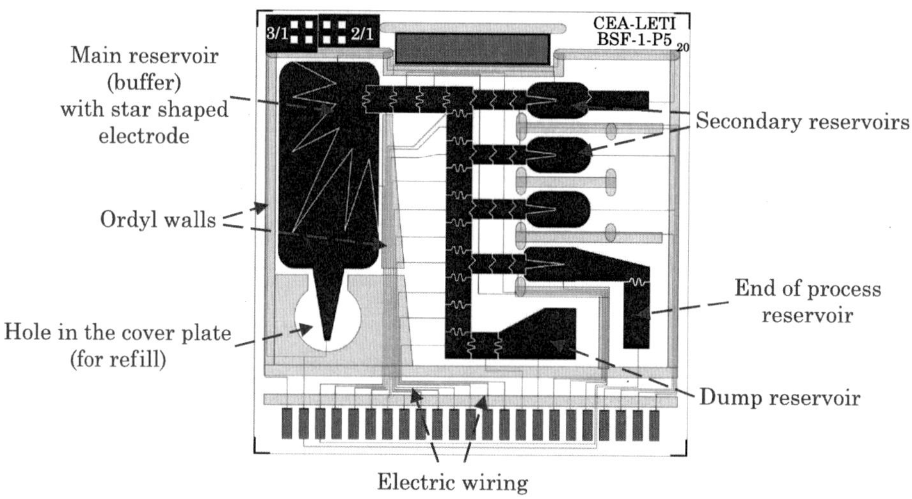

Figure 5.78 Covered EWOD biochip from LETI; courtesy Y. Fouillet (CEA-LETI).

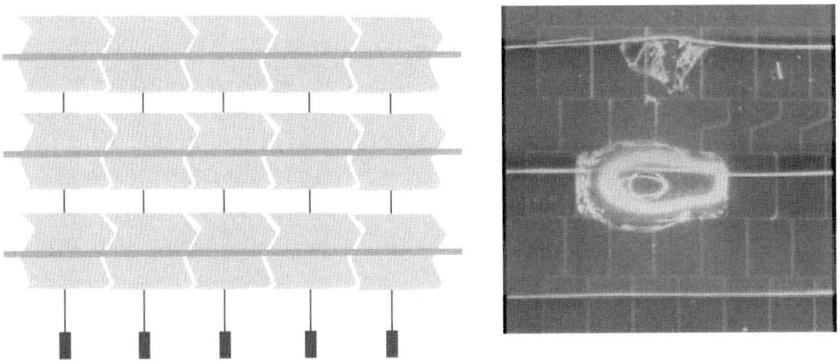

Figure 5.79 Left: multiplexing scheme for an open EWOD system with catenae: number of electrodes in a row is $M = 5$, number of rows $N = 3$, number of electrical lines $M + N = 8$, total number of electrodes $N * M = 15$. Right: EWOD system with multiplexing. Photograph Ph. Dubois CEA-LETI.

5.9.1 Electrowetting Conveyor System

A particularly interesting use of EWOD has been proposed by Moon et al [2]. The concept is that of a conveyor system moved by electrowetting actuation. The system is an open EWOD system without catena. The "carriage" is composed of a thin flat plate placed on four droplets (Fig. 5.80). These four droplets are the "wheels" of the "carriage". A first requirement is that the four droplets move in parallel and at the same speed on two electrode rows, constituting the "road". A second requirement is that the plate placed upon the four droplets be hydrophobic, or else the droplets will merge together.

It has been observed that the moving plate can carry an important load, and can be used as a moving frame for biological/chemical applications.

5.9.2 Extension of the Electrowetting Concept: Electrowetting Filament and Open Microfluidics

A very interesting use of electrowetting actuation is the ability to work with "liquid filaments" or "fingers" [51]. The principle is derived from that of space applications for micro-engines: a filament of liquid may move even if it is not totally enclosed between solid walls. It suffices that capillary forces maintain the free boundaries of the liquid. Such type of flow is sometimes called "open microfluidics". It is clear that such flows cannot be motioned by pressure; however, if the filament is not too long, it can be moved by electrocapillary forces. Thus Satoh and coworkers have developed a microsystem based on electrowetting, which can move conductive liquid filaments. The cross section of such a system is shown in Fig. 5.81.

Fig. 5.82 shows the extension of water filaments. The system does not need valves to orient the flow; it is done by the actuation of the proper electrodes.

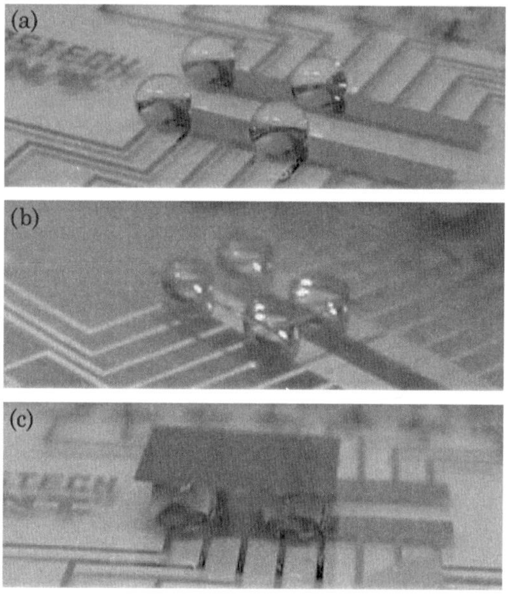

Figure 5.80 Conveyor belt constituted of 4 droplets on EWOD substrate (a) and a top pad made of glass (b) or silicon (c). Reused with permission from [2]. © 2002 American Institute of Physics.

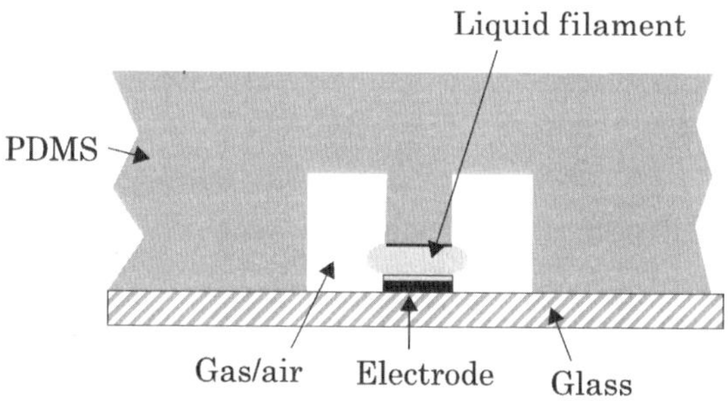

Figure 5.81 Cross section of the device: the lateral sides of the liquid filament are maintained by capillary forces.

An important advantage of such devices is the large interfacial area between the liquid and the gas. Typically it could be very effective for chemical micro-reactions or to extract particles and molecules from the gas phase.

5.10 Summary

In this chapter, we have presented the physical principles of open and covered EWOD micro-devices. We have shown how and under what conditions droplets can be moved, merged, divided, mixed, and diluted in such devices. The basic

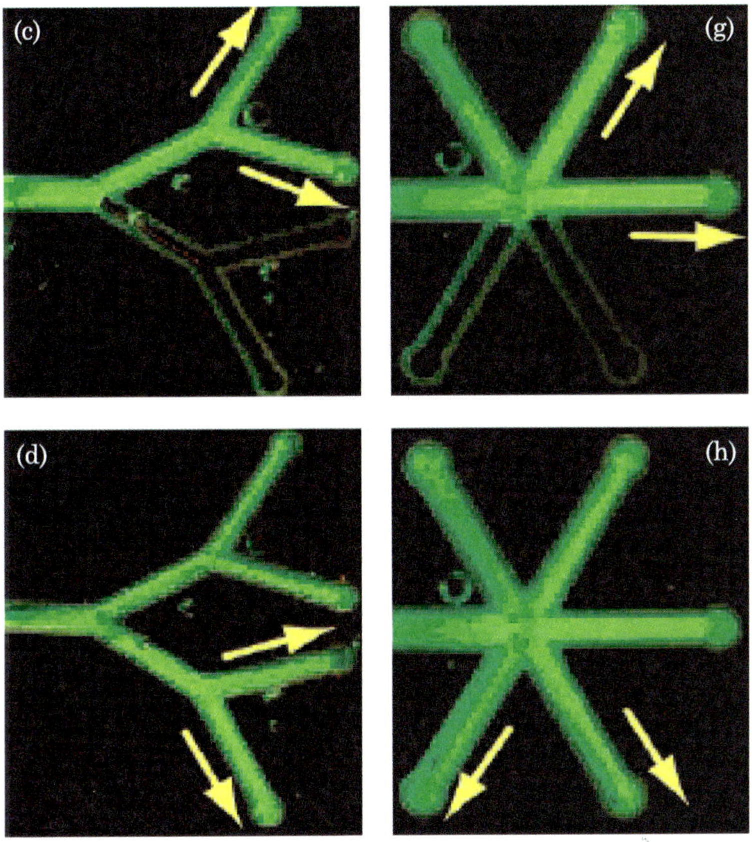

Figure 5.82 Extension of liquid "fingers" by electrowetting actuation. Reprinted with permission from [51]. © 2005 American Chemical Society.

operations combined with digital manipulation of droplets constitute the heart of EWOD micro-devices. Assembling a complete biochip requires a complex architecture with micro-fabricated electrodes, whose complexity is even more daunting if multiplexing is needed.

In Chapters 6 and 7, EWOD systems dedicated to biology, biochemistry, and chemistry are presented.

References

1. M.G. Pollack, R.B. Fair and D. Shenderov, "Electrowetting-based actuation of liquid droplets for microfluidic applications," *Applied Physics Letters*, Vol. 77, no. 11, pp. 1725–1726, 2000.
2. H. Moon, S.K. Cho, R.L. Garrell and C.-J. Kim, "Low voltage electrowetting-on-dielectric," *Journal of Applied Physics*, Vol. 92, no. 7, pp. 4080–4087, 2002.

3. B. Shapiro, H. Moon, R.L. Garrell and C.-J. Kim, "Equilibrium behavior of sessile drops under surface tension, applied external fields, and material variations," *Journal of Applied Physics*, Vol. 93, no. 9, pp. 5794–5811, May 2003.
4. S.K. Cho, H. Moon and C.-J. Kim, "Creating, transporting, and merging liquid droplets by electrowetting-based actuation for digital microfluidics circuits," *Journal of Microelectromechanical systems*, Vol. 12, no. 1, pp. 70–80, 2003.
5. J. Berthier, Ph. Clementz, O. Raccurt, D. Jary, P. Claustre, C. Peponnet and Y. Fouillet, "Computer aided design of an EWOD microdevice," *Sensors and Actuators A: Physical*, Vol. 127, pp. 283–294, 2006.
6. J. Berthier, Ph. Clementz, J.-M. Roux, Y. Fouillet and C. Peponnet, Modeling microdrop motion between covered and open regions of EWOD microsystems, Proceedings of the 2006 Nanotech Conference, Boston, USA, 7–11 May, 2006.
7. Y. Fouillet, H. Jeanson, D. Jary, O. Constantin and C. Vauchier, Moving droplets with microcatenaries, Proceedings of the 7th International Conference on Miniaturized Chemical and Biochemical Analysis Systems (μTAS), Squaw-Valley, Ca, USA, October 5–9, 2003.
8. K. Brakke, "The surface evolver," *Exp. Math.*, Vol. 1, p. 141, 1992.
9. M.K. Chaudhury and G.M. Whitesides, "How to make water run uphill," *Science*, 256, pp. 1539–1541, 1992.
10. N. Moumen, R.S. Subramanian and J. McLMaughlin, The motion of a drop on a solid surface due to a wettability gradient, Proceedings of the AIChe 2003 Annual Meeting, 2003.
11. T.B. Jones, "An electromechanical interpretation of electrowetting," *J. Micromech. Microeng.*, Vol. 15, pp. 1184–1187, 2005.
12. M. Washizu, "Electrostatic actuation of liquid droplets for microreactor applications," *IEEE Trans. On Industry Applications*, Vol. 34, no. 4, p. 732, 1998.
13. Ilju Moon and Joonwon Kim, "Using EWOD (electrowetting-on-dielectric) actuation in a micro conveyor system," *Sensors and Actuators A: Physical*, Vol. 130–131, pp. 537–544, 2006.
14. A.A. Darhuber and S.M. Troian, "Principles of microfluidic actuation by modulation of surface stresses," *Annual Review of Fluid Mechanics*, Vol. 37, pp. 425–455, 2005.
15. R.-Q. Duan, S. Koshizuka and Y. Oka, "Two-dimensional simulation of drop deformation and break-up around the critical Weber number," *Nuclear Engineering and Design*, Vol. 225, pp. 37–48, 2003.
16. F. Blanchette and T.P. Bigioni, "Partial coalescence of drops at liquid interfaces," *Nature Physics*, Vol. 2, pp. 254–257, 2006.
17. Chao-Yi Chen, E.F. Fabrizio, A. Nadim and J.D. Sterling, Electrowetting-based microfluidics devices: design issues, 2003 Summer Bioengineering Conference, Sonesta Beach Resort in Key Biscayne, Florida, USA, 25–29 June, 2003.
18. A. Dolatabadi, K. Mosheni and A. Arzpeyma, "Behavior of a moving droplet under electrowetting actuation: numerical simulation," *Canadian Journal of Chemical Engineering*, Vol. 84, pp. 17–21, February 2006.
19. M.G. Pollack, A.D. Shenderov and R.B. Fair, "Electrowetting-based actuation of droplets for integrated microfluidics," *Lab. Chip*, Vol. 2, pp. 96–101, 2002.
20. T.P. Lodge, "Reconciliation of the molecular weight dependence of diffusion and viscosity in entangled polymers," *Physical Review Letters*, Vol. 83, no. 16, 1999.
21. P.-G. de Gennes, F. Brochard-Wyart and D. Quéré, *Drops, Bubbles, Pearls, Waves*, Springer, New-York, 2004.
22. T. Ondarçuhu, "Total or partial pinning of a droplet on a surface with chemical discontinuity," *J. Phys. II France*, Vol. 5, pp. 227–241, 1995.
23. L.Y. Yeo and Hsueh-Chia Chang, "Electrowetting films on parallel line electrodes," *Phys. Rev. E*, Vol. 73, p. 011605, 2006.

24. T.B. Jones, J.D. Fowler, Young Soo Chang and Chang-Jin Kim, "Frequency-based relationship of electrowetting and dielectrophoretic liquid microactuation," *Langmuir*, Vol. 19, pp. 7646–7651, 2003.
25. Jun Zeng and T. Korsmeyer, "Principles of droplet electrohydrodynamics for lab-on-a-chip," *Lab. Chip.*, Vol. 4, pp. 265–277, 2004.
26. J. Berthier, Ph. Dubois, Ph. Clementz, P. Claustre, C. Peponnet and Y. Fouillet, "Actuation potentials and capillary forces in electrowetting based microsystems, sensors and actuators," *A. Physical*, Vol. 134, no. 2, pp. 471–479, 2007.
27. R. Aveyard, J.H. Clint and D. Nees, "Theory for the determination of line tension from capillary condensation," *J. Chem. Soc., Faraday Trans.*, Vol. 93, no. 24, pp. 4409–4411, 1997.
28. J. Berthier and C. Peponnet, "A model for the determination of the dimensions of dents for jagged electrodes in EWOD microsystems," *BioMicrofluidics*, Vol. 1, pp. 014104-1–014104-10, 2007.
29. Fei Su, Krishnendu Chakrabarty and R.B. Fair, "Microfluidics-based biochips: Technology issues, implementation platforms and design-automation challenges," *IEEE Transactions on Computer-Aided Design and Integrated Circuits and Systems*, Vol. 25, no. 2, February 2006.
30. M. Armani, S. Walker and B. Shapiro, Modeling and control of electrically actuated surface tension driven microfluidic systems, Proceedings of the 13th Mediterranean Conference on Control and Automation, Limassol, Cyprus, 27–29 June, 2005.
31. J. Lienemann, A. Greiner and J.G. Korvink, "Modeling, simulation, and optimization of electrowetting," *IEEE Transactions on Computer-Aided Design and Integrated Circuits and Systems*, Vol. 25, no. 2, February 2006.
32. F. Caron, J.C. Fourrier, J. Carlier, S. Arscott, V. Thomy, J.C. Camart, C. Druon and P. Tabourier, 4th International Discussion Meeting on Electrowetting, Blauberen, Germany, 6–8 September 2004.
33. D. Jary, A. Chollat-Namy, Y. Fouillet, J. Boutet, C. Chabrol, G. Castellan, D. Gasparutto and C. Peponnet, DNA repair enzyme analysis on EWOD fluidic microprocessor, Proceedings of the 2006 Nanotech conference, Boston, 7–11 May 2006.
34. P. Clementz, J. Berthier, J.M. Roux, R. Blanc, G. Castellan, C. Chabrol, P. Claustre, O. Constantin, D. Jary, D. Lauro, O. Raccurt, Y. Fouillet and C. Peponnet, EWOD contribution to sample preconcentration: droplet geometrical transformation from drop formation to evaporation step, NanoBio Europe Conference, Münster, 22–24 September 2005.
35. N.A. Patankar and Y. Chen, Numerical simulation of droplets shapes on rough surfaces, Nanotech 2002; Technical Proceedings of the 5th International Conference on Modeling and Simulation of Microsystems, pp. 116–119, 2002.
36. R. Ahmed and T.B. Jones, "Dispensing picoliter droplets on substrates using dielectrophoresis," *J. Electrostatics*, Vol. 64, pp. 543–549, 2006.
37. K.-L. Wang, T.B. Jones and A. Raisanen, "Dynamic control of DEP actuation and droplet dispensing," *J. Micromech. Microeng.*, Vol. 17, pp. 76–80, 2007.
38. C. Peponnet, EWOD actuated droplet microfluidics: from a fluidic function toolbox to biological validations, 8th LETI Annual Meeting, Grenoble, France, 29–31 May, 2006.
39. J. Berthier, O. Raccurt, Ph. Clementz, D. Jary, P. Claustre, C. Peponnet and Y. Fouillet, An analytical model for the prediction of microdrop extraction and splitting in digital microfluidics systems, 8th annual NSTI Nanotech 2005, Anaheim, California, USA, 8–12 May, 2005.
40. M.W.L. Watson, M. Abdelgawad, G. Ye, N. Yonson, J. Trottier and A.R. Wheeler, "Microcontact printing-based fabrication of digital microfluidic devices," *Anal. Chem.*, Vol. 78, pp. 7877–7885, 2006.
41. J. Berthier and P. Silberzan, *Microfluidics for Biotechnology*, Artech House Publishers, 2005.

42. J.M. Ottino, *The Kinematics of Mixing: Strectching, Chaos, and Transport*, Cambridge University press, 1989.
43. P. Paik, V.K. Pamula and R.B. Fair, "Rapid droplet mixers for digital microfluidic systems," *Lab Chip.*, Vol. 3, pp. 253–259, 2003.
44. P. Paik, V.K. Pamula, M.G. Pollack and R.B. Fair, "Electrowetting-based droplet mixers formicrofluidic systems," *Lab Chip*, Vol. 3, pp. 28–33, 2003.
45. J. Fowler, Hyejin Moon and C.J. Kim, Enhancement of mixing by droplet-based microfluidics, Proceedings, IEEE micro electro mechanical systems, pp. 97–100, 2002.
46. S. Colin, *Microfluidique*, Hermès Science Publications, 2004.
47. Olivier Raccurt, Jean Berthier, Philippe Clementz, Mathias Borella and Marc Plissonnier, "On the influence of surfactants in electrowetting systems," *J. Micromech. Microeng.*, Vol. 17, pp. 1–7, 2007.
48. U. Häfeli, W. Schütt, J. Teller and M. Zborowski, *Scientific and Clinical Applications of Magnetic Carriers*, Plenum Press, 1997.
49. Shih-Kang Fan, C. Hashi and Chang-Jim "CJ" Kim, Manipulation of multiple droplets on NxM grid by cross-reference Ewod driving scheme and pressure-contact packaging, 17th IEEE International Conference on Micro Electro Mechanical Systems MEMS 2004, Maastricht, The Netherlands, 25–29 January, 2004.
50. Hyejin Moon, A.R. Wheeler, R.L. Garrell, J.A. Loob and Chang-Jin "CJ" Kim, "An integrated digital microfluidic chip for multiplexed proteomic sample preparation and analysis by MALDI-MS," *Lab. Chip*, Vol. 6, pp. 1213–1219, 2006.
51. Wataru Satoh, Hiroki Hosono and Hiroaki Suzuki, "On-chip microfluidic transport and mixing using electrowetting and incorporation of sensing functions," *Anal. Chem.*, Vol. 77, pp. 6857–6863, 2005.

6 Biological Applications of EWOD☆

6.1 Introduction

The preceding chapters were dedicated to the physics of the digital microfluidics chip (DMC). In the following chapters, concrete applications of such microsystems are presented. At the present time, applications are found in the domains of biology and biotechnology, chemistry, and optics. This chapter is devoted to biological and biotechnological applications. In these fields, applications are principally related to the development of biosensors (lab-on-a-chip, immunoassays), and, to a lesser extent, of cell-on-a-chip.

First of all, let us ask the question: what would be an ideal on site biochemical analysis system? An ideal sensor is required to be inexpensive, fast, robust, sensitive, and reliable, to have low power consumption, a high degree of automation and integration, a high throughput, to use small reagent volumes, and to cover a wide spectrum of samples and biological parameters. With all these requirements, it is clear that the ultimate biochemical sensor is still a long way ahead. The complexity of such systems and the technological problems to solve are huge. However, as we shall see in this chapter, considerable progress is being made in the domain of sample collection and preparation, DNA analysis, DNA repair, protein recognition, and cell sorting. In this chapter, we present examples illustrating these applications. A preliminary step is to determine which biological and physiological liquids are compatible with EWOD-based systems, i.e., can successfully be displaced and maneuvered on EWOD-based microsystems.

6.2 Biological Liquids and Physiological Samples Compatible with EWOD Devices

We have seen in the preceding chapters that a condition for a liquid to be maneuvered on EWOD-based microsystems is that the liquid is electrically conductive. However, we do not know if this condition is sufficient. Examples corresponding to de-ionized water, buffers, and ionic liquids have been shown. By extrapolation of these results, it is logical to presume that human physiological liquids, which are electrically conductive due to their salt content, might be compatible with EWOD-based microsystems. In this section, we investigate the compatibility of usual biological liquids with EWOD-based microsystems.

☆This Chapter was written in collaboration with Christine Peponnet (Director of the Biochip Laboratory at the CEA-LETI).

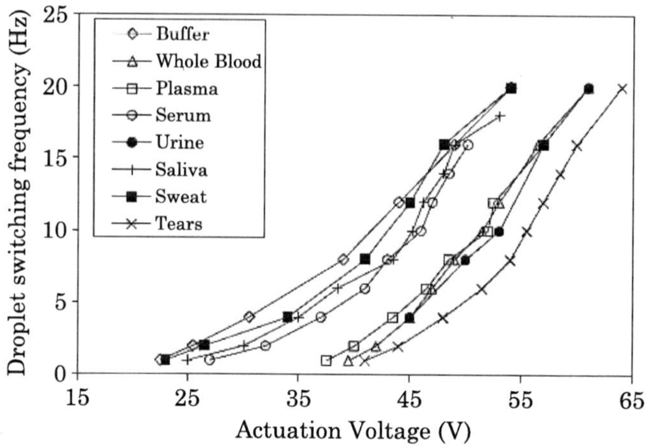

Figure 6.1 Transportability of different human physiological liquids: maximum droplet switching frequency as a function of the applied voltage; the maximum switching frequency is the maximum rate at which a droplet can be moved across adjacent electrodes, it is a measure of the transportability of a droplet in an EWOD-based microsystem. The droplet volume is 1.12 μl and the surrounding media is silicone oil. (From [2]; reproduced by permission of The Royal Society of Chemistry).

6.2.1 Physiological Liquids

Srinivasan et al. [1,2] and Lei Li et al. [3] have shown that physiological liquids, like blood, sweat, tears, serum, milk, etc. can be moved on EWOD-based microsystems, although not always so easily as de-ionized water or buffers. A convenient method to assess the maneuverability of a liquid on a EWOD chip is to measure the maximum switching frequency as a function of the applied voltage. The method consists in observing if a back and forth motion of a droplet between two neighboring electrodes can be obtained by switching the actuation of these two electrodes. If the switching of actuation between the two electrodes is too fast, motion does not occur. Hence, the maximum switching frequency is the maximum rate at which a droplet can be moved across adjacent electrodes (Fig. 6.1). This is a measure of the aptitude of a droplet to move on a digital microfluidic device. It takes into account at the same time the electric force exerted on the droplet and the resistance to motion due to the viscous forces (friction on the substrate).

An interesting observation stemming from Fig. 6.1 is that serum (blood plasma) is more easily transportable than whole blood. This is not surprising since whole blood is constituted of serum plus red and white blood cells which make it much more viscous, and give it a non-Newtonian rheological behavior. This property is general: carrier fluids are moved at lower voltages than the same carrier fluids transporting biological macromolecules, proteins, and cells. We will come back to the motion of proteins and cells in the next section. It is especially interesting to know that whole blood is compatible with EWOD-based systems, since many applications are linked to blood analysis. Fig. 6.2

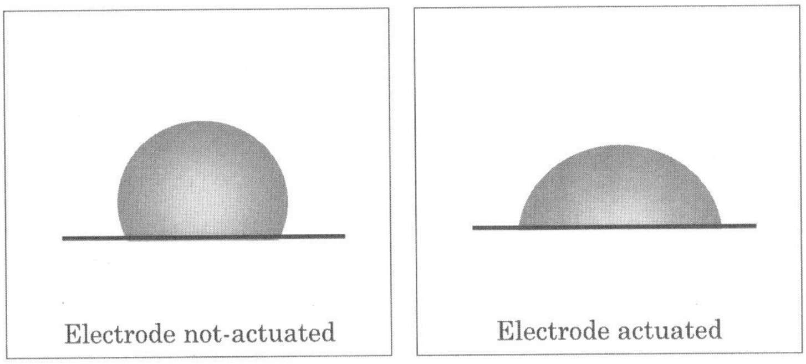

Figure 6.2 Droplet of blood on an electrode; the contact angle varies with the applied voltage according to the Lippmann–Young law.

shows that whole blood obeys the general Lippmann–Young law, and can be displaced on EWOD electrodes.

6.2.2 Proteins and Biofouling

The situation is more complex with fluids containing enzyme-laden reagents, proteins, and cells. These liquids pose several challenges. By necessity, EWOD surfaces are hydrophobic, and most proteins adsorb irreversibly to hydrophobic surfaces. This phenomenon is called biofouling. Proteins adsorbed to the surface prevent the correct functioning of the chip by rendering the surface hydrophilic [4], and by inducing droplet pinning and even dielectric breakdown (see Chapter 4). In order to avoid—or reduce—biofouling, silicone oil is widely used. The effect of silicone oil is to maintain a film between the droplet and the solid surface. The effect of an oil film regarding electrowetting has been studied in Chapter 4.

Often a protein called bovine serum albumin (BSA)—or other proteins like casein—are added to biological reagents to reduce adhesion of enzymes and proteins to the solid surfaces [5]. Advantages of BSA—or casein—are their biocompatibility with most biological reactions and their low cost. However, they adsorb to the solid surfaces. The aptitude of a droplet containing BSA to be maneuvered on EWOD devices depends on the initial concentration in BSA. Fig. 6.3 shows that a droplet containing a buffer with BSA (surrounded by oil) is maneuverable on an EWOD-based device if the concentration in BSA is less than 0.01 mg/ml [6]. Above this limit, the required voltages are large, and the saturation limit or the breakdown level are quickly reached.

6.2.3 Cells

One of the applications expected from digital microfluidic systems is the precise handling of cells. The first step is to verify that cells can be moved by electrowetting forces on EWOD-based microchips. Evidently there is a size

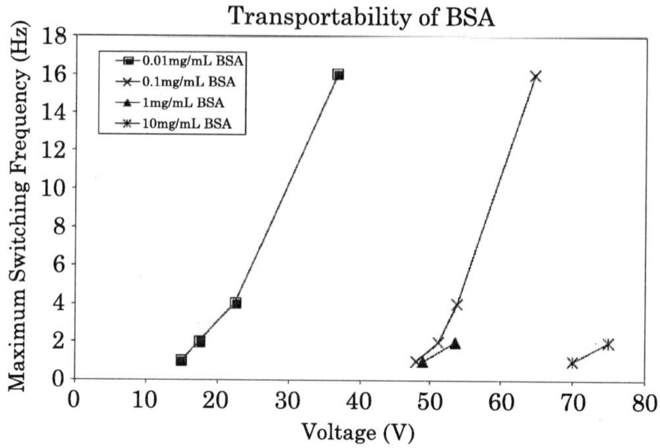

Figure 6.3 Transportability of droplets on EWOD-based microsystems decreases quickly when the concentration of BSA increases. From [6].

Figure 6.4 Jurkat cells transported by a droplet on an EWOD-based microsystem. The cell size is approximately 50 μm. (Photograph D. Jary, CEA-LETI).

limit above which the cells cannot be moved. For small to medium sized cells (less than 70 μm), the challenge is that the cells involved must be maintained alive and that, with the cells, a nutrient solution must also be moved. To prevent adherence on the solid surfaces, surfactants must be added to the carrier liquid. However, the usual surfactants, like Tween, are not bio-compatible; they usually kill the cells. We have seen in the preceding section that BSA can be used only at very low concentration. Usually it is not enough for cells. In the literature, specific surfactants have been studied and found to be biocompatible. This is the case of Pluronic [7,8]. Let us mention that Pluronic is a poloxamer, i.e., an amphiphilic copolymer composed of a central hydrophobic chain of polyoxypropylene (poly-propylene oxide) flanked by two hydrophilic chains of polyoxyethylene (polyethylene oxide). Fig. 6.4 shows a cell laden droplet on an EWOD-actuated chip.

6.2.4 Conclusion

Many physiological liquids are compatible with the EWOD mode of transportation, opening a wide range of biological applications to these systems.

Some care must be taken with enzymes, proteins, and cells to prevent them from adsorbing to the solid surfaces. To this extent the use of biologically adapted surfactants is a major improvement. In the following sections, we present some applications of EWOD-based microsystems to sample collection, DNA analysis, DNA repair, protein recognition, and cell concentration.

6.3 Sample Collection and Analysis on a Lab-on-a-chip Platform

The following example [9,10] describes a system to collect and analyze airborne particles using microdroplets circulating in an EWOD chip. This microdevice prefigured the digital microfluidic systems that are being developed to address environmental problems with the detection of the chemical and biochemical composition of aerosol; it is also a good illustration of the integration in the same device of different physical phenomena already encountered and analyzed in this book.

The collection, pre-concentration, and analysis of the target particles are realized in four steps: (1) the first step consists of collecting airborne particles by impinging a jet on the surface of an open EWOD chip. During this phase, the particles adsorb on the surface. (2) The air stream is blanked off and a scanning droplet is moved as fast as possible on the surface (velocities of 10 cm/s can be reached in open EWOD-based systems). In order to increase the velocity of motion and to reduce evaporation, the water droplet is clad in a thin oil film. The surface is scanned in a few seconds and the droplet does not have time to evaporate. Fig. 6.5 shows the scanning phase and particle collection by the droplet. During this phase, the collected particles are concentrated in the moving droplet. (3) The droplet is then transferred to a covered EWOD region filled with silicone oil. The conditions of transfer from open to covered EWOD-based systems have been analyzed in Chapter 5. Fig. 6.6 illustrates such a transfer. In the new covered environment, the droplet cannot evaporate. In this region, the droplet is mixed with specific reagents that produce colored reaction products. (4) The concentration of the analyte is quantitatively determined by measuring adsorption at specific wavelengths using a simple light emitting diode and photodiode setup.

Applications include sampling of airborne contaminants—such as sulfates, nitrates and ammonium—bioagents, and also of nitroaromatic explosive particles such as TNT. Very low detection limits are achieved due to the efficiency of the concentration in the droplet. If the absorption of the impinging particles is total, following [9], the concentration in the droplet c_w is related to that in the air c_a by the relation

$$c_w = \frac{c_a Q_a t}{V_w} \tag{6.1}$$

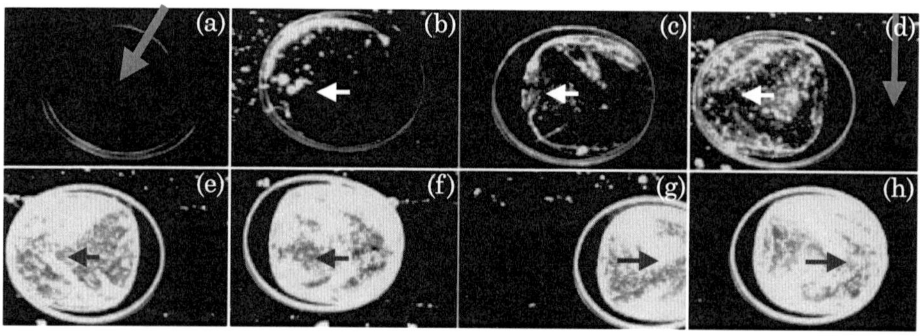

Figure 6.5 Photographs of a droplet sweeping a solid surface and collecting the particles initially adsorbed on the surface. (a) Initial droplet; (b) droplet scans over beads; (c) beads collected in drop; (d) the surface is cleaned along the droplet trajectory; (e) and (f) more beads are collected; (g) and (h) the droplet returns to its initial location over a cleaned surface. From [9]. © 2004 SPIE; reprinted with permission.

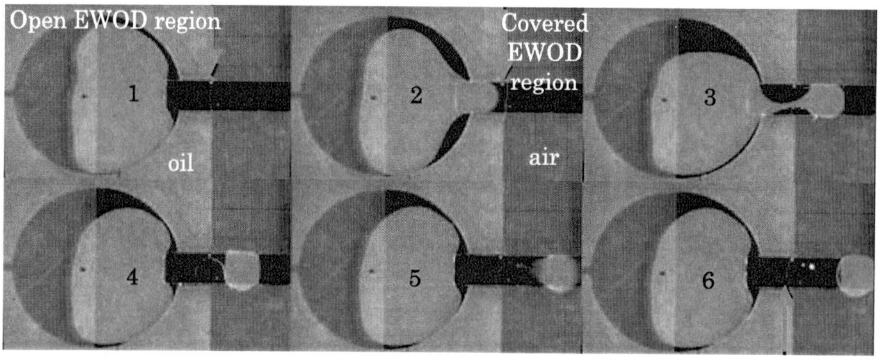

Figure 6.6 Droplet motion from an open EWOD region to a covered EWOD region. In the open region, droplet is surrounded by silicone oil, whereas it is surrounded by air in the covered region. From [9]. © 2004 SPIE; reprinted with permission.

where V_w is the volume of the water droplet, Q_a the air flow rate, and t the collection time. In terms of limit of detection (LOD), equation (6.1) yields

$$LOD_a = \frac{LOD_w V_w}{Q_a t}. \tag{6.2}$$

Formula (6.2) shows that the detection level is proportional to the detection level on the chip times the extraction volume and divided by the sampling flow rate. Due to the very small value of the drop volume, a very low detection limit in air can be achieved in a short time (1 mn). Of course, the reasoning assumes that the capture of the impinging particles is total and that the scanning by the droplet is extremely efficient.

6.4 Polymerase Chain Reaction (PCR) on Digital Microfluidics Systems

6.4.1 Introduction

Most of the time, biologic samples only contain a very small number of DNA strands; bioanalysis of DNA for the detection of viruses, hereditary diseases, or genotyping requires a sufficiently large number of strands. Hence, the initial DNA strands must be replicated—or amplified—in a sufficiently large number. Polymerase chain reaction is the most used in-vitro method to amplify DNA sequences, because it amplifies exponentially the sequences via enzymatic replication. It is not the purpose of this book to go into the details of biological processes like PCR [11,12]. We shall just mention here the information needed for the comprehension of the design and functioning of a PCR digital system. A PCR requires a PCR mix containing several basic components, such as the primers, which are complementary to the DNA regions at the 5' and 3' ends of the DNA region that is to be amplified, a DNA polymerase used to synthesize a DNA copy, deoxynucleotide triphosphates from which the DNA polymerase builds the new DNA, and a buffer solution, which provides suitable chemical environment for optimum activity of the DNA polymerase.

Basically, a PCR requires 20 to 30 cycles, each cycle doubling the number of DNA sequences. Each cycle comprises three steps: the first step is a denaturation step during which the double-stranded DNA is separated in two single DNA strands. Denaturation requires a temperature of 94–96 °C. Denaturation is followed by the annealing step. In this step the reaction temperature is lowered to 55–64 °C: the primers and the DNA strands bind together if their sequence is complementary. A relatively low temperature (lower than the fusion temperature) is needed for attachment. The annealing step is followed by an extension/elongation step during which the DNA polymerase synthesizes new DNA strands complementary to the DNA template strands. The temperature at this step depends on the DNA polymerase used. Usually, a temperature of 70–74 °C is optimum.

Hence, the reaction mix must be submitted to multiple temperature cycles. The duration of the cycles depends on the uniformity of the temperature and on the volume of liquid. In microsystems, the time required for a temperature cycle is much shorter than for conventional systems. Assuming perfect conditions and perfect mixing, the denaturation step requires a stabilization time of a few seconds, approximately 10 seconds is needed for the annealing step, and elongation process is nearly instantaneous. One of the advantages of microsystems is the ability to work with very small volumes of liquid, for which temperature homogeneity and diffusion efficiency are higher than that of macrosystems. However, the thermal cycles induce constraints on the system, especially regarding evaporation and bubble formation.

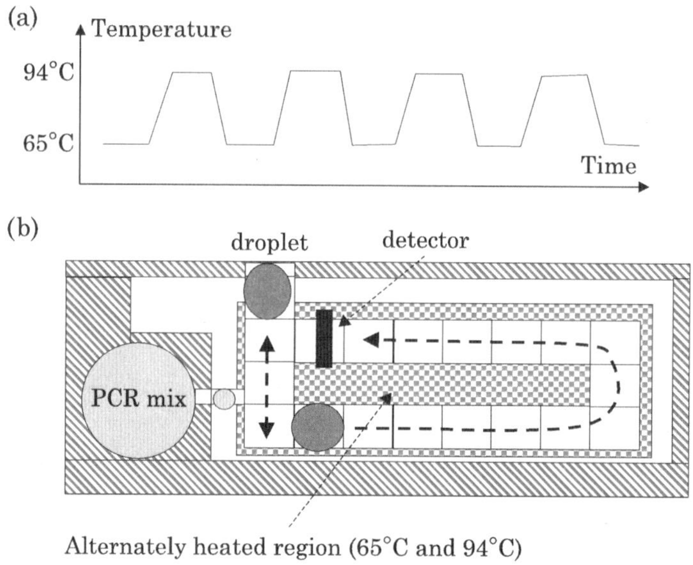

Figure 6.7 (a) Temperature cycles (between 65 °C and 94 °C) that the liquid must observe for the PCR to be effective; the temperature is monitored by a Peltier device located below the substrate; (b) schematic view of a PCR chip (one zone of thermal regulation). The incoming droplet is merged with a droplet of PCR mix, then is mixed by back and forth motion; the droplet is then moved step by step towards the detector. At each step a thermal cycle is performed to achieve the duplication of the DNA strands. (Sketch Y. Fouillet CEA-LETI). The liquid is progressively enriched in target DNA and eventually reaches the detector.

In the next section we first present the architecture of a PCR-EWOD microsystem; then we analyze the thermics of a PCR digital microsystem, and finally we show PCR results in digital microfluidic systems.

6.4.2 Architecture of a PCR-EWOD Microsystem

In order to achieve the PCR, the following operations must be successively performed: (1) dispensing of the sample mix, (2) merging of the droplet with another droplet containing the PCR-mix (primers, polymerase, triphosphates), (3) mixing the components, (4) submitting the droplet to the temperature cycles, (5) detecting the reaction results. The three first steps have already been detailed in the preceding chapter. We now focus on steps (4) and (5).

Conceptually, the thermal regulation can be performed using two different approaches: either there is only one region of thermal heating/cooling and the droplet is at rest in this region (Fig. 6.7), or the droplet alternately moves from a region maintained at 64 °C to a region maintained at 95 °C (Fig. 6.8). These two categories are just the translation to digital microfluidics of the "stationary-chamber micro PCR chips" and the "continuous-flow micro PCR chips" that

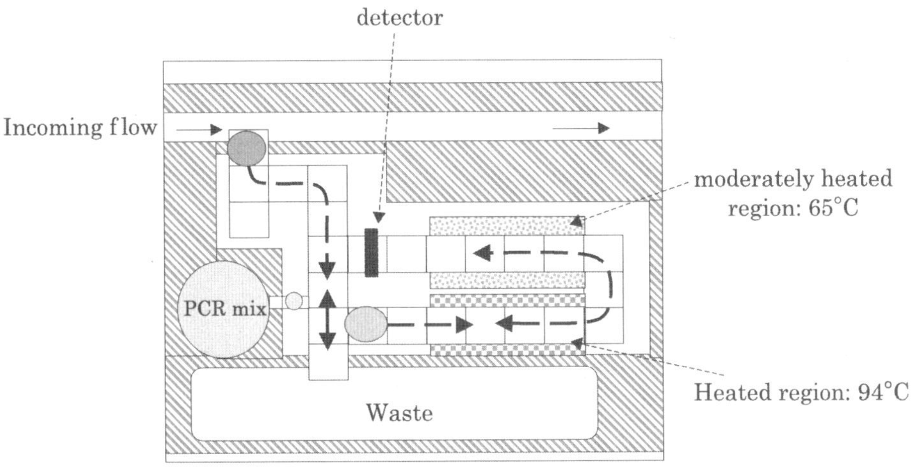

Figure 6.8 Schematic of a PCR EWOD microsystem with two zones of thermal regulation: the droplet is extracted from a flowing liquid then transported on the EWOD chip; there, it merges with a droplet of PCR-mix. After mixing by back and forth motion, the droplet is moved alternately from the heated to moderately heated regions. After approximately 20 cycles, the droplet is directed towards the detector. (Sketch Y. Fouillet CEA-LETI).

have been developed for conventional microfluidic systems, depending if the samples are stationary or moving inside the channels.

In the following section, we analyze the temperature control of PCR-EWOD devices.

6.4.3 Thermal Analysis

Mini-Peltier devices are often used in microsystems for temperature regulation. Peltier devices are based on the thermoelectric effect [13]. In 1821, J.T. Seebeck discovered that dissimilar metals that are connected at two different locations (junctions) will develop a micro-voltage if the two junctions are held at different temperatures. This effect is known as the "Seebeck effect"; it is the basis for thermocouple thermometers. A little later, in 1834, J. Peltier discovered the inverse of the Seebeck effect, now known as the "Peltier effect": a temperature difference builds up at the junction between two dissimilar metals or semiconductors when a voltage is applied. This results in a small heat pump, also known as a thermo-electric cooler (TEC). The advantages of Peltier devices are numerous: they can be miniaturized (their size varies from a few millimeters thick by a few millimeters to a few centimeters square), they can be used as well as cooling or heating devices (one junction cools off while the other heats up), they have no moving parts, they need no Freon refrigerant, they induce no vibrations, they have a long lifetime and a good precision on the temperature control; besides, they do not require large voltages for miniaturized systems.

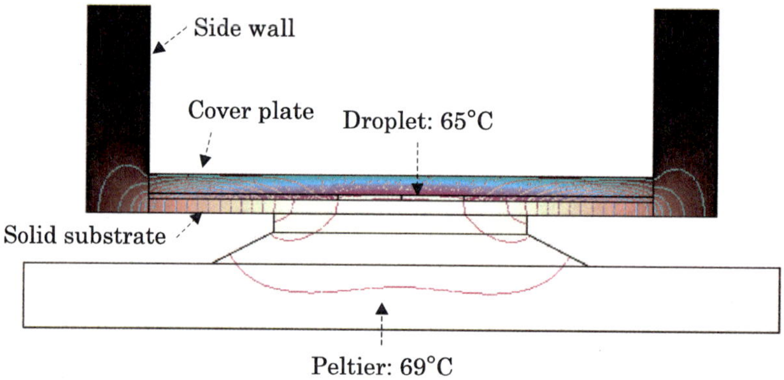

Figure 6.9 Covered EWOD system: the temperature field in the droplet is homogeneous within a 0.5 °C margin. The side walls are made of plastic to reduce lateral heat loss. Calculation made with the COMSOL numerical software [14].

For a system with a unique thermal region regulating the temperature cycles, a unique mini-Peltier device is used. For two regions of different temperatures, two mini-Peltier devices are needed.

6.4.3.1 Single Thermal Region—Single Peltier Device

The conception and design of the PCR-EWOD chip necessitates the knowledge of the transfer function between the Peltier and the EWOD system, and the thermal response of the system. In the following, we use a numerical approach [14] to determine the temperature field in the system. The equation governing the thermal behavior of the chip is simply the heat conduction equation

$$\frac{\partial T}{\partial t} = \frac{k}{\rho C_p} \nabla^2 T \tag{6.3}$$

with the adequate boundary conditions. In (6.3) T is the temperature in kelvin, k the heat conductivity (W/m/K), ρ the density (kg/m^3), and C_p the specific heat capacity (J/K/kg). We investigate the cases of a covered EWOD system with an insulating cover and of an open EWOD system with an oil cover. Figs. 6.9 and 6.10 show the temperature field in the system during the annealing step (liquid temperature 65 °C), with the two different cover conditions. In both cases, there is approximately a 3 °C temperature difference between the Peltier device and the droplet. The droplet temperature is extremely uniform in the case of a covered system—and sufficiently uniform in the case of an oil cover—to ensure an adequate annealing. Similar results are obtained for the denaturation phase.

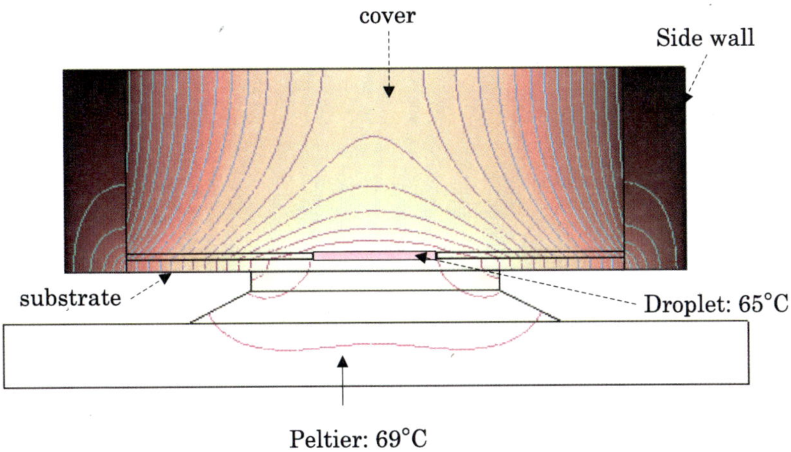

Figure 6.10 Open EWOD system with a droplet immersed in oil: the temperature field in the droplet is homogeneous within a 1 °C margin. The relatively thick layer of oil reinforces the thermal insulation.

Concerning temperature transient, the response of the system is very fast (Fig. 6.11). The droplet temperature follows the temperature transient of the Peltier device without any time shift.

6.4.3.2 Two Thermal Regions—Two Peltier Devices

A two-region EWOD-PCR device is sketched in Fig. 6.12. In order to function correctly, the two regions must be maintained at a right, stable temperature. But, if the substrate is thermally conductive, like silicone or glass, there is a temperature leakage through the substrate. This temperature leakage prevents a uniform temperature in the droplet. The solution is to insert an insulating separation in the substrate (Fig. 6.13). This separation blocks the heat flux through the substrate and the two regions are thermally homogeneous.

6.4.3.3 Conclusions

In conclusion, droplet microsystems present many advantages over conventional systems for thermal heating/cooling and thermal cycling of liquids. On one hand, the temperature of the liquid is highly homogeneous and can be very precisely tuned, under the condition that the oil layer—or the cover layer—and the side walls are sufficiently insulated. On the other hand, thermal transients are very fast, due to the small amount of liquid. The only drawback of miniaturization is the confinement of the temperature to a very small region around the droplet. A separation wall constituting a thermally insulating barrier can be incorporated into the substrate to enhance the confinement but requires a more complicated fabrication process.

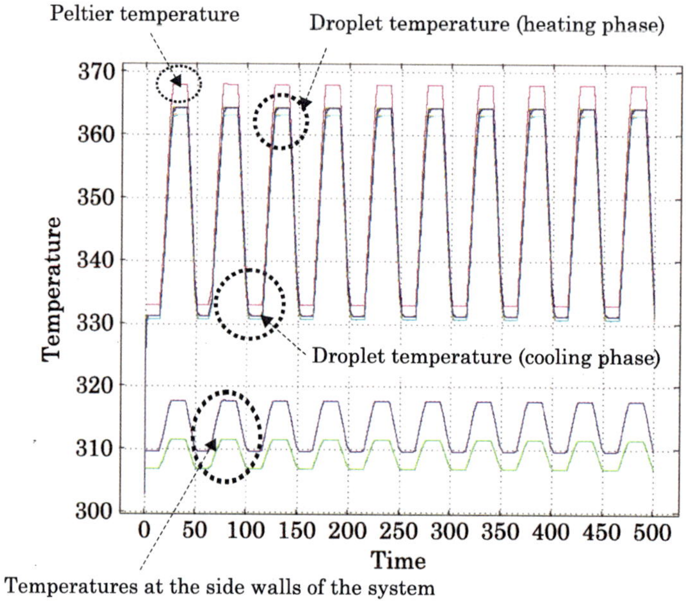

Figure 6.11 Heating phase: even for an open EWOD system, temperatures are very homogeneous inside the droplet. Temperature transient inside the droplet follows perfectly that of the Peltier, showing no time shift.

6.4.4 Experimental Realization of PCR and Results

6.4.4.1 Experimental Conditions

The demonstration of the feasibility of PCRs in digital microfluidic systems has already been made by Yi-Hsien Chang and coworkers [15] for the dengue II virus, and by Pollack and coworkers [16] for single nucleotide polymorphisms or SNPs. We present here the results of SNP-PCR amplification performed in the LETI laboratories. First, let us recall that SNPs are DNA sequence variations that occur when a single nucleotide (A, T, C, or G) in the genome sequence is variable. For example, an SNP is the possibility of finding in the population a position in the genome where two different cases can be found such as the DNA sequences ACGGCTAA and ATGGCTAA. For a variation to be considered an SNP, it must occur in at least 1% of the population. SNPs, which make up about 90% of all human genetic variation, occur every 100 to 300 bases along the 3-billion-base human genome. Two of every three SNPs involve the replacement of cytosine (C) with thymine (T). SNPs can occur in both coding (gene) and noncoding regions of the genome. Many SNPs have no effect on cell function, but scientists believe others could predispose people to disease [17]. Hence, identification of SNPs is a major issue in biology.

As we mentioned earlier, the biologic protocol for a PCR is complex, and we only mention here its main features. As the aim is to reproduce conventional

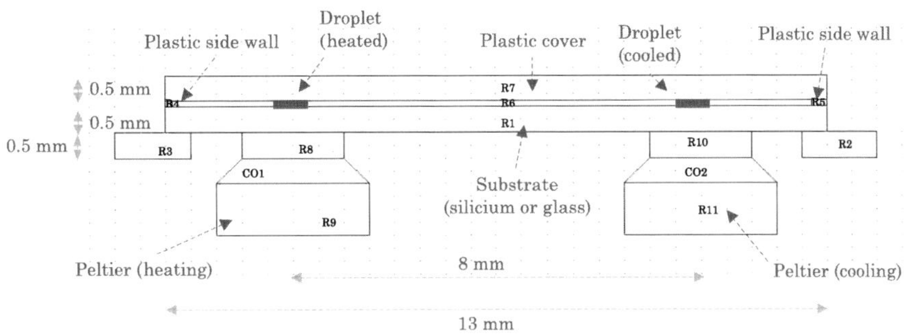

Figure 6.12 Schematic of a PCR-EWOD system with two temperature control regions: the two Peltier devices are separated by approximately 8 mm. The calculation was performed with the COMSOL numerical software.

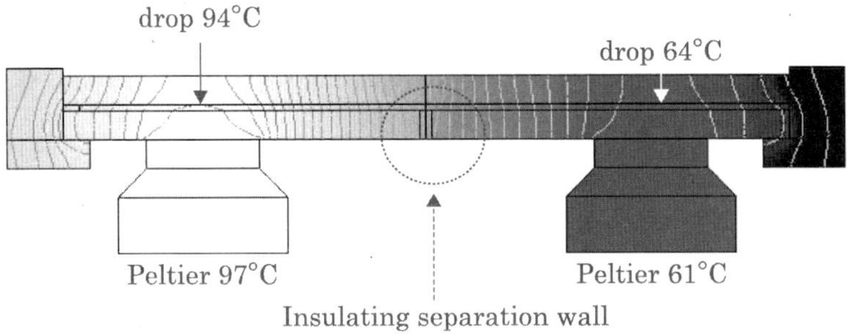

Figure 6.13 Temperature control is achieved by two Peltier devices, one heated to 97 °C, one cooled to 61 °C. The droplet temperatures are very homogeneous at 94 °C in the heated zone and at 64 °C in the cooled zone. Calculation made with the COMSOL numerical software [14].

PCR reactions performed in tubes, the protocol is directly derived from that of a tube PCR. The primers are the same, and the polymerase is also a Taq-polymerase [12]. A small concentration (0.08 μg/μl) of BSA is added to the mix to prevent polymerase adsorption on the solid substrate; this is needed since the surface/volume ratio is very high. The human placenta DNA to be analyzed is introduced in a reservoir on the chip, and a negative control is realized by the use of a water droplet in another reservoir. The DNA concentration is the same as that of the conventional test of 0.01 mg/μl corresponding to only 25 copies (i.e., initial DNA strands) for a 64 nl droplet, and to 4200 copies for a 10 μl tube reaction.

The EWOD chip used for this PCR test is covered (covered EWOD), and initially filled with mineral oil (Fig. 6.14). Just after oil filling, the chip is placed under vacuum for 10 mn to avoid potential air trapping (air/gas bubbles prevent the correct functioning of the reaction). Thermal cycling occurs in a temperature plateau of 10 mn at 94 °C, followed by 40 cycles of 30 seconds at 65 °C and 30 seconds at 94 °C. A dedicated two-color fluorescent detector is

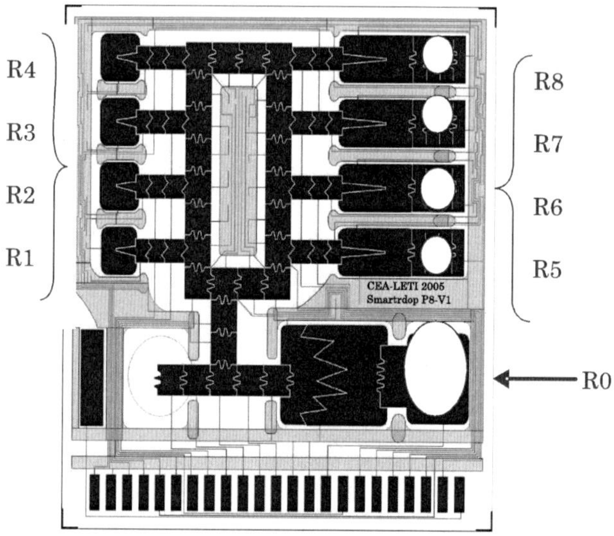

Figure 6.14 EWOD microchip for the realization of SNP polymerase chain reaction. The white dots are the holes drilled into the upper cover used to introduce the liquids. R0 is the main reservoir (for DNA sample), and R1 to R8 are 8 auxiliary reservoirs; R5 to R8 are directly accessible for loading with the different PCR-mix. A length scale is given by the small electrodes which are 800 μm × 800 μm. (Photo Y. Fouillet CEA-LETI).

used to measure the fluorescence level of the PCR reactions at each PCR cycle. Fluoresence detection is adapted to the FAM and VIC fluorophores—one label for each allele—with excitation wavelengths of 488 and 520 nm, respectively, and emission wavelengths of 510 and 555 nm.

6.4.4.2 On-chip PCR Results

Experimental results are very satisfactory: Fig. 6.15 shows the comparison between in-tube and on-chip PCR results. The fluorescent time profiles are very similar. In both cases the fluorescent level of the FAM marker rises quickly after approximately 27 cycles. At the end of the reaction (after 40 cycles), the PCR product is quantified by capillary electrophoresis (Bioanalyzer, Agilent^{TM}). Comparable yields were obtained for droplets and tube reactions, as shown in Fig. 6.16.

6.4.4.3 Conclusions

On-chip EWOD-PCRs present some very interesting advantages. First, on-chip PCRs can be really fast. Thermal-cycling was reduced to cycles of 2 seconds at 94 °C and 10 seconds at 60 °C, and were found to give satisfactory results. The total time for 40 cycles is then only 8 mn. Second, only a very

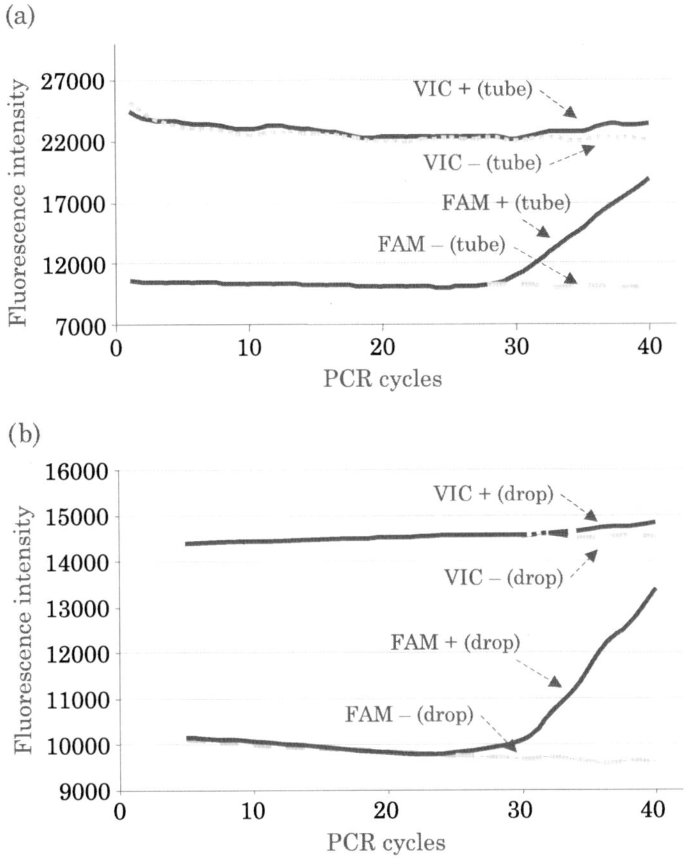

Figure 6.15 SNP-PCR results: (a) conventional test in tube; (b) test on EWOD-based microsystem. Both cases have similar results and FAM signal shows a sharp turn after 27 cycles. (Courtesy A-G Brachet, CEA-LETI).

small quantity of PCR mix is needed compared to the same reaction in-tube. Third, the signal-to-noise ratio is much better in a droplet PCR than that in a tube PCR, resulting in an increased sensitivity of the assay. Indeed, the number of PCR cycles required to amplify a certain number of copies depends on the concentration. Thus, in a smaller volume, PCR amplification is detected more readily than in a large tube reaction. Finally, PCRs can be easily run in parallel on the same chip. Fig. 6.17 shows the results of three DNA sequence amplifications (corresponding to PhiX, S-pneumoniae and lambda DNA sequences) obtained in parallel on the same chip, compared with in-tube conventional analysis realized in three different tubes.

However, digital PCRs present some difficulties. First, we must be assured that there is no cross contamination between the different drops. Cross contamination could happen through the oil phase or by biofouling, i.e., adsorption/desorption on the solid surface. On their EWOD-based system,

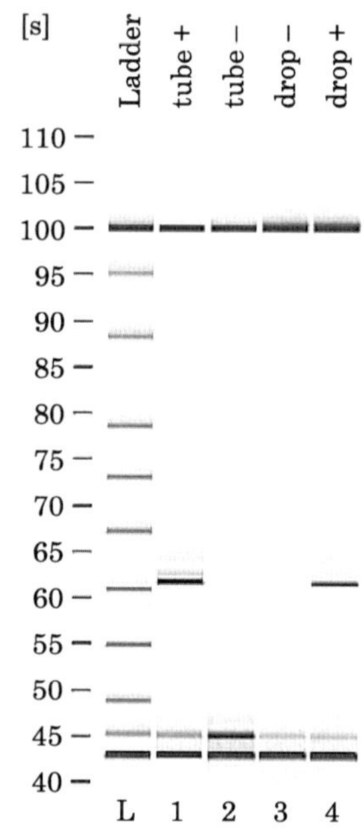

Figure 6.16 Comparison of PCR yieldsand specificity: the same DNA sequence is amplified by the conventional in-tube method and by the EWOD method. (Courtesy A-G Brachet, CEA-LETI).

Pollack et al. found no substantial cross contamination on 15,000 transfer opportunities between a droplet with and without DNA (Fig. 6.18) [16]. In our case, we did not find any contamination through the oil phase and, like Pollack, no contamination of sample DNA; however, PCR products were found to cause some contamination due to adsorption/desorption on the solid surfaces. This could result in a problem if the PCR chips are used several times for PCR amplification. An important remark is that one should be cautious with the new superhydrophobic substrates that are being developed to increase the electrowetting force (see Chapter 4), when these substrates present rugosities; cross contamination and biofouling could become an important issue in such a case.

Second, gas bubbles are prone to form due to the relatively high temperatures (94 °C). Bubbles are a serious caveat in the functioning of EWOD-based chips. They might lead to dielectric breakup and chip deterioration. De-gassing precautions should be used to avoid bubble formation.

6.5 DNA Repair Microprocessor

DNA repair activity tests have a large field of diagnostic applications: diseases associated with DNA repair deficiencies, identification of cancer predisposition,

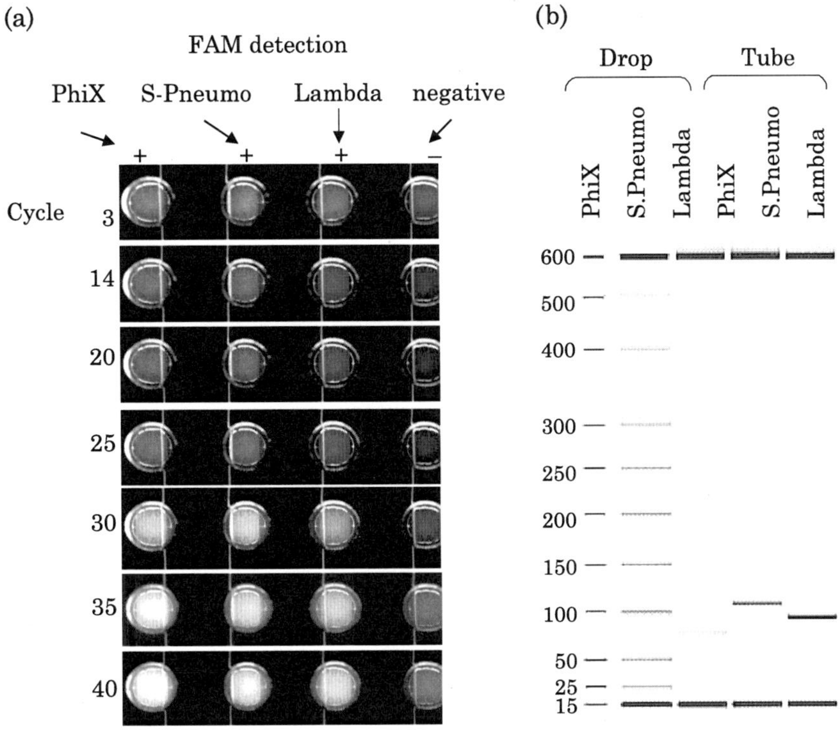

Figure 6.17 (a) Fluorescence control of the four droplets at the same cycle number (the fourth droplet is a droplet without DNA for differential imaging); (b) analysis by gel electrophoresis after the 40th cycle. Again the in-tube and on-chip results are similar. (Courtesy A-G Brachet, CEA-LETI).

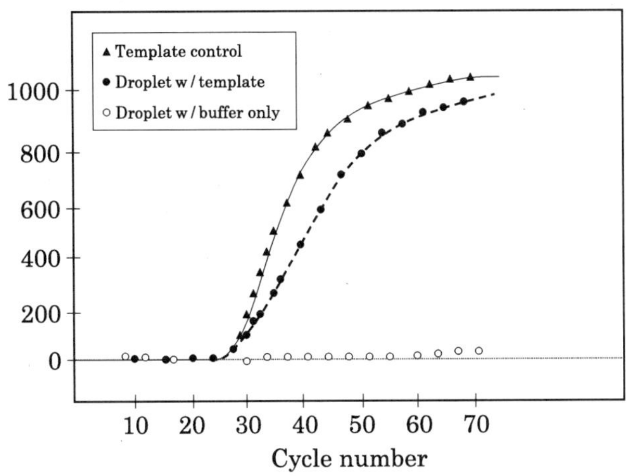

Figure 6.18 Cross-contamination results: droplets both with or without DNA were collected after 15 000 intersecting transfers, added as the template component to a PCR mixture and amplified in a commercial real-time PCR system [17].

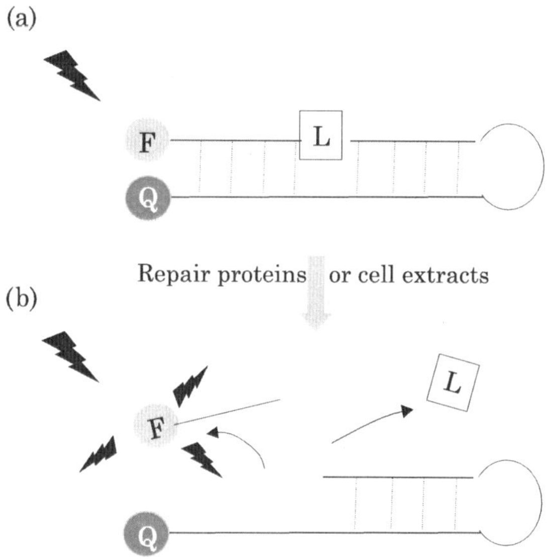

Figure 6.19 Principle of FRET-based DNA repair assay using DNA molecular beacon (F = fluorescent marker, Q = quencher) with a lesion (L). The presence of the quencher next to the fluorescent marker prevents the emission of light by the fluorophore (a); once the DNA lesion L is cut out of the DNA strand, fluorescence is restored (b). (From: [18]. © 2006 NSTI. Reprinted with permission).

prediction of individual response to chemo or radiation therapy, and finally high throughput screening of DNA repair enzymes. A generic challenge is to test in parallel various enzymatic activities with a very limited amount of cellular extracts. A promising way is to run biological tests allowing the DNA repair activity to be monitored by real time fluorescence measurement on an EWOD-based microchip [18].

Quantification of enzymatic repair activities is made by using a DNA molecular beacon in which a specific lesion has been inserted, as shown in Fig. 6.19. The test consists of monitoring the DNA repair enzyme activity contained in the cell extracts; these enzymes excise the DNA lesion by fragmenting the DNA strand at the location of the lesion. As soon as the excision is performed, the fluorescent energy transfer (FRET) between the fluorophore and the quencher is cancelled and there is an emission of fluorescence due to DNA breakage.

In the assay described here, the molecular beacon contains an 8-oxo-Guanine lesion, and the DNA repair enzyme is a fapy glycosylase (Fpg). Incubation buffer is constituted by a solution of 25 mM Tris (pH 7.6), 2 mM EDTA, and 50 mM NaCl. All experiments were simultaneously performed on a chip using a 120 nl droplet and in an Eppendorf tube containing a 6 μl volume, on a real time PCR instrument (Stratagene MX 3500P). The on-chip experiment consists in dispensing two 64 nl droplets of enzyme and substrate reagents, mixing them,

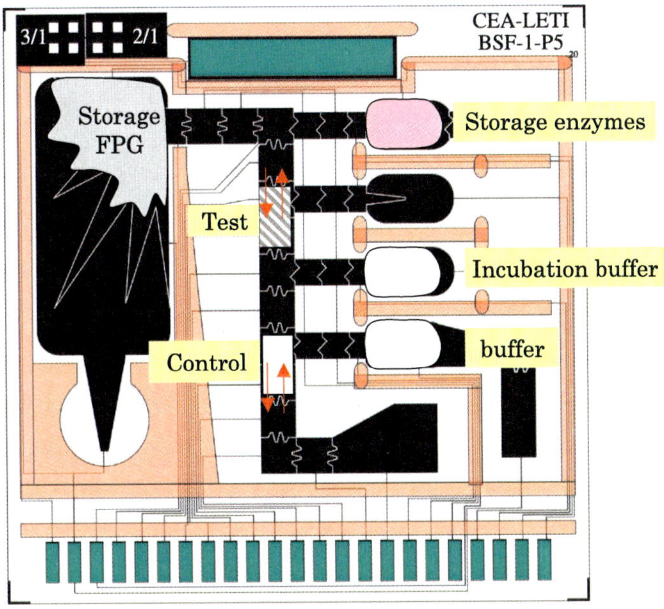

Figure 6.20 Covered EWOD chip for DNA repair activity test. Light emission from the test droplet is compared to a negative control droplet. (From: [18]. © 2006 NSTI. Reprinted with permission).

and incubating at 37 °C for 30 mn. A negative control droplet with no enzyme is located next to the incubating droplet in order to make a differential measure of the fluorescence (Fig. 6.20).

Two different comparisons with in-tube reference were made: first, with a fixed repair enzyme concentration [Fpg] = 1 μM and various concentrations of substrate ranging from 0.25 to 1.2 μM; second, with a fixed substrate concentration [8oxoGua] = 1 μM and an enzyme concentration ranging from 200 nM to 1 μM. The results are shown in Fig. 6.21 where the initial velocity of the reaction is plotted versus concentrations of enzymes or substrate. The results show that the on-chip tests compare very well with the in-tube conventional tests.

On the other hand, reaction kinetics are compared in Fig. 6.22 for a fixed concentration of GFP and three different concentrations of 8-oxo-guanine. Here again, the kinetics of reactions are similar.

In conclusion, the EWOD-based assay performs quite well, a sensitivity of 5 nM of GFP was found, and no loss of activity due to volume size reduction was detected. A clear advantage is the use of tiny volumes of costly enzymes, compared to in-tube assays. However, cell extracts have not yet been used; only an enzyme has been tested. More systematic tests need to be developed to assess the validity of the approach on EWOD microprocessors.

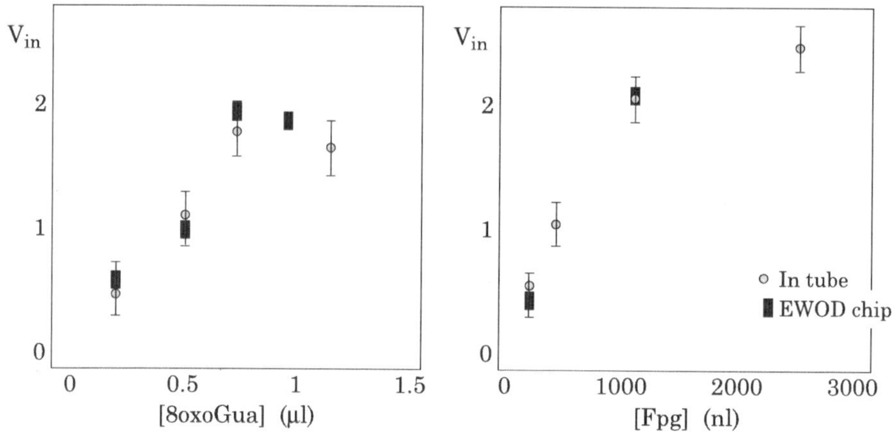

Figure 6.21 Left: experimental results for a 1 µl concentration of Fpg; right: same for a 1.2 µl concentration of 8-oxo-Guanine. (From: [18]. © 2006 NSTI. Reprinted with permission).

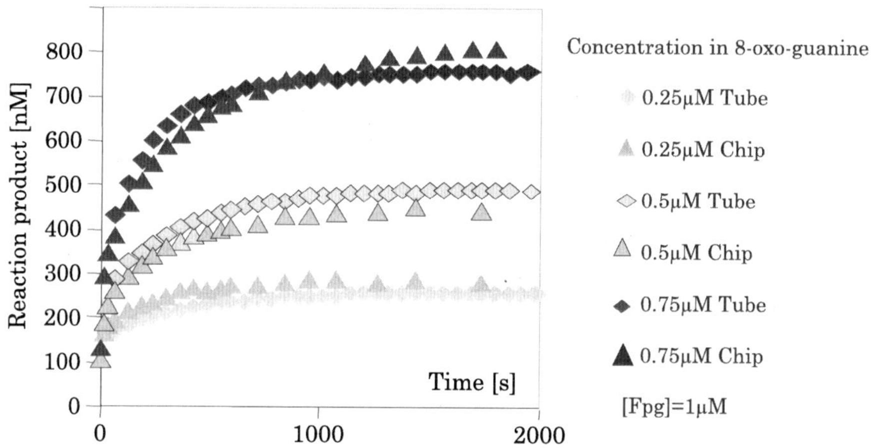

Figure 6.22 Comparison of kinetics between in-tube and on-chip reactions.

6.6 Protein Analysis Coupled with Mass Spectrometry

Protein analysis has become a major issue in biology. Proteins are linked to gene expression; they are essential parts of all living organisms and participate in every process within cells. Many proteins are enzymes that catalyze biochemical reactions, and are vital to metabolism. The analysis and classification of proteins is a major challenge. Microsystems now contribute to this search. The first integrated systems are being developed using microflow technology [19,20]. However, because there is a huge amount of a protein in liquid samples, it is advantageous to reduce as much as possible the analyzed

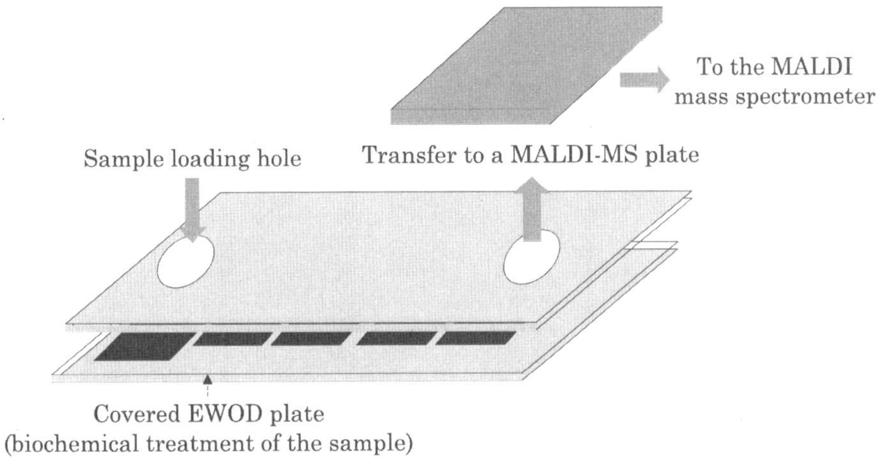

Figure 6.23 Schematic of a covered-EWOD system coupled to a MALDI mass spectrometer [21].

volume, in order to be more specific. To this extent, digital microfluidics can bring a solution. However, an integrated digital microfluidic system is yet to be developed. At this time, the first systems being developed combine a digital microfluidic system and a conventional analysis system. They use a digital microfluidic system to transport the carrier fluid droplets, extract the proteins, and perform the enzymatic reactions; then the liquid is transferred onto a special plate which is incorporated in a conventional MALDI-MS device. Let us recall briefly that MALDI-MS or MALDI-TOF-MS stands for matrix assisted laser desorption ionization time-of-flight mass spectrometry. This complicated device can identify large biomolecules, over 300,000 daltons, in the time they take to traverse a region submitted to an electric field. First, the biomolecules are accelerated by an electric field; the kinetic energy acquired by the biomolecule is given by

$$E_c = \frac{1}{2}mv^2 = ze\nabla Es \tag{6.4}$$

where m is the mass of the ion, v the velocity of the ion, z the number of charges, e the charge on an electron in coulombs, ∇E the electric field gradient, and s the distance of the ion source region. The time taken by the biomolecule to cross a region of constant electric field is recorded. This time depends on the ratio m/z of the biomolecule and identifies this molecule.

Fig. 6.23 shows the principle of protein characterization micro-device, realized by coupling an EWOD-based microsystem with a MALDI-MS conventional apparatus.

Now, let us come back to the digital microfluidic system itself. The principle is based on soft-printing and is sketched in Fig. 6.24.

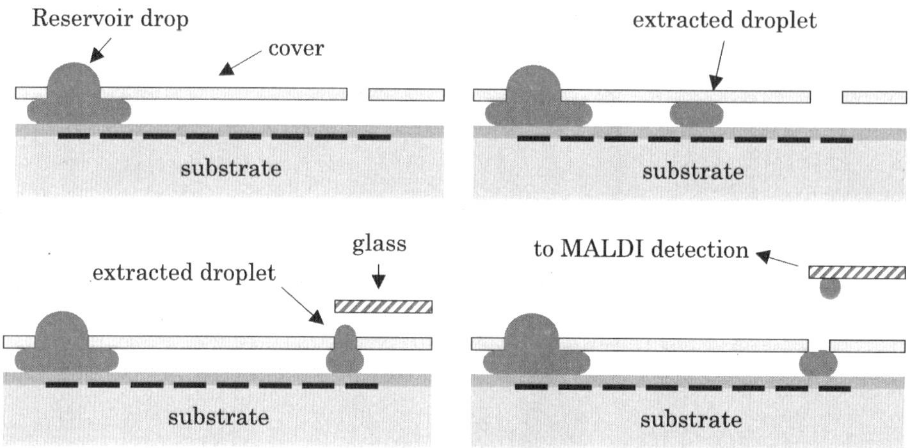

Figure 6.24 Soft printing process to a glass plate: (a) sample loading; (b) creating and transporting discrete droplets; (c) droplet bulging in the outlet port, approach of a glass plate; (d) contact with the glass plate; moving away the glass plate with the droplet.

Let us examine here the conditions for optimal droplet bulging through the hole. We showed in Chapter 3 that the maximum bulging is obtained when the droplet shape is a half-sphere. The pressure of the droplet is given by the Laplace law

$$P = \frac{4\gamma}{D} \tag{6.5}$$

where γ is the surface tension of the interface and D the diameter of the hole (Fig. 6.25). The droplet pressure can also be derived from the droplet shape between the two plates

$$P = \gamma\left(\frac{-2\cos\theta}{d} + \frac{1}{R}\right). \tag{6.6}$$

Relations (6.5) and (6.6) determine the value of R and consequently the size of the droplet in the covered-EWOD system

$$\frac{R}{d} = \frac{1}{2}\frac{1}{\left(\frac{2d}{D} + \cos\theta\right)}. \tag{6.7}$$

The orifice diameter D should not be too large, because the interface would spread over onto the upper cover (because the pressure in the dome would be small compared to the pressure of the liquid between the two plates, and there would be no equilibrium possible), nor too small, because the volume of liquid bulging out would be too small ($h = D/2$).

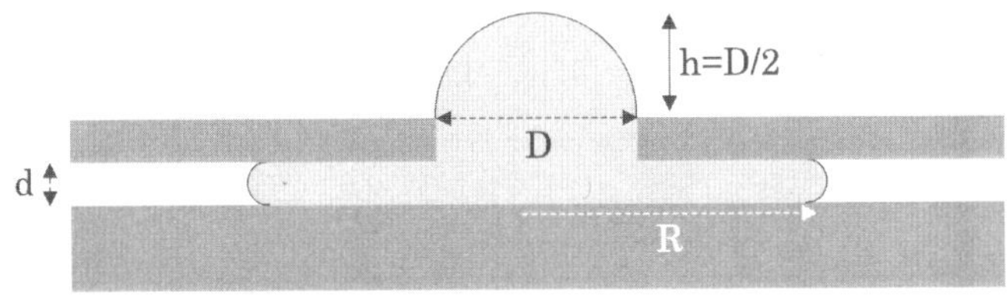

Figure 6.25 Sketch of a droplet bulging in the outlet port.

In conclusion, the principle of a microsystem for protein analysis using digital microfluidics has been proposed. However, there is still a long way to go to obtain a prototype, and an integrated micro-device is still to be developed.

6.7 Cell-on-a-chip Microsystems: Example of a Cell Concentrator

We have seen that cells can be moved on EWOD-based chips—with the conditions that they are sufficiently small and that adequate surfactants are used. The presence of electrodes in the chip makes it convenient to combine the electrowetting effect—for transporting droplets containing the cells—to a dielectrophoretic effect—for the manipulation of the cells [22]. It is noteworthy to mention that electrowetting is primarily a contact line phenomenon, and refers to electric field-induced interfacial tension changes between a liquid and a solid conductor—as was shown in Chapter 4. On the other hand, dielectrophoresis (DEP) is a bulk phenomenon caused as a result of polarization induced in a dielectric liquid by a non-uniform electric field. Combining the two effects leads to very interesting applications, like that of the cell concentrator microsystem developed by Po-Weng Huang and co-workers [23]. Their device is schematized in Fig. 6.26.

First, the EWOD electrodes are used to move the droplet containing the cells on the DEP electrodes. Then the DEP electrodes are successively actuated in order to progressively confine—by negative dielectrophoresis—the cells to an extremity of the droplet (Fig. 6.27). Note that dielectrophoresis can produce an attraction or repulsion effect on cells and particles depending on the actuation frequency; negative DEP corresponds to repulsive forces [22]. Finally the droplet is cut in two by EWOD actuation. As a result, the cells are concentrated in a smaller droplet.

We note that the mechanism presented here is the opposite of that presented in Chapter 5 for the dilution of particulate solutions. Fig. 6.28 shows the concentration of neuroblastoma cells contained in a solution of 3% PBS with a concentration of 280 nM of isotonic sucrose. In this case, a high frequency positive dielectrophoresis was used because the low frequency

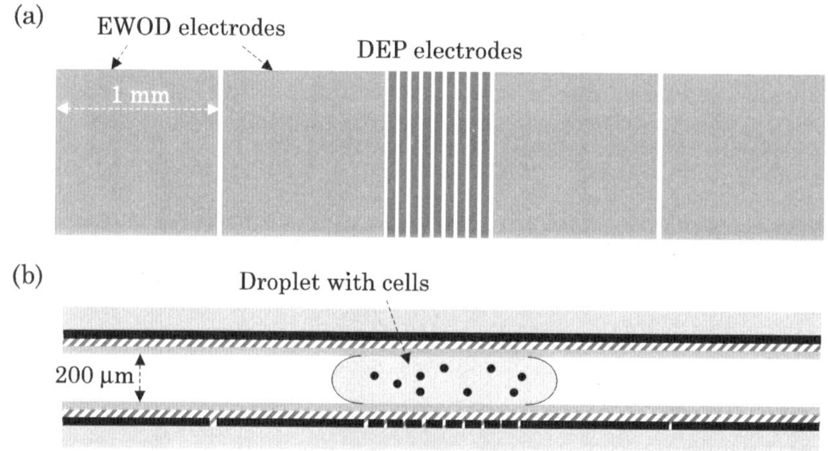

Figure 6.26 (a) View from top of the device with the EWOD electrodes—for droplet motion—and the DEP electrodes—for cells manipulation. (b) side view of the device showing the top cover and the droplet.

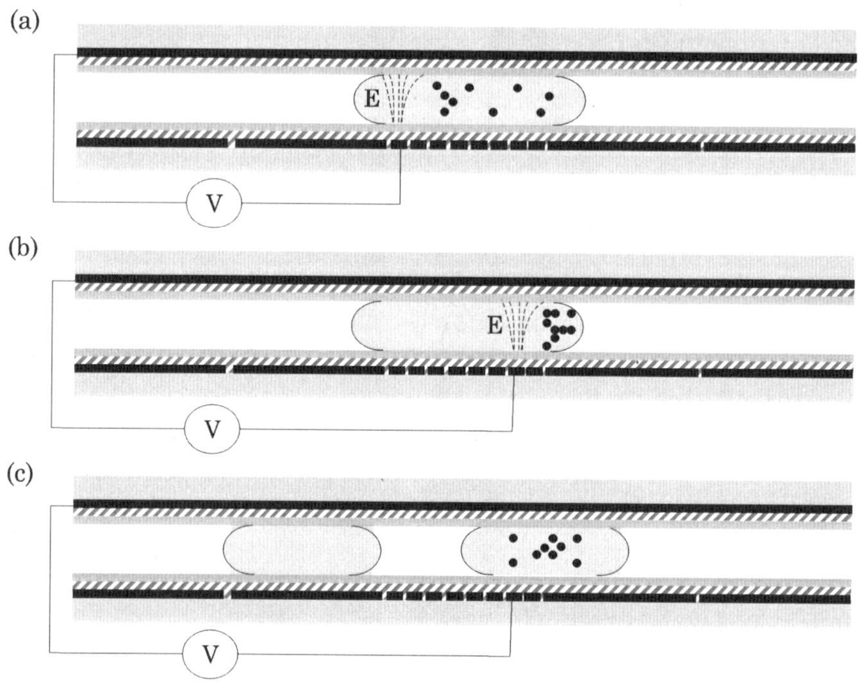

Figure 6.27 Principle of cell concentration by EWOD and DEP effects. (a) Negative dielectrophoresis repels the cells to the right. (b) By successively displacing the actuation to the right the cells are progressively confined in the right corner of the droplet. (c) The droplet is cut in two by EWOD actuation; the cells are concentrated in the right droplet.

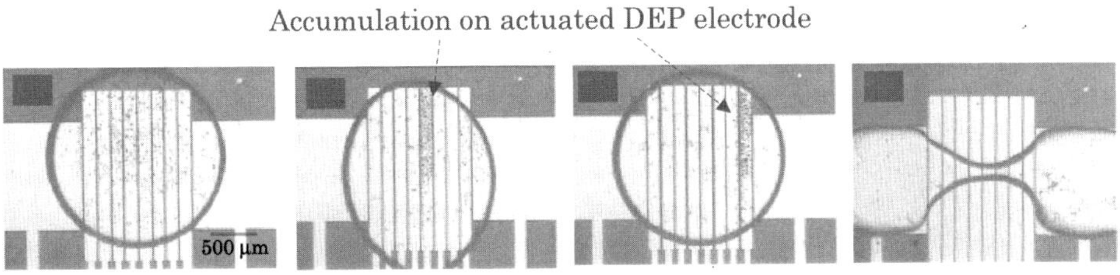

Figure 6.28 Photograph of progressive concentration of Neuroblastoma cells by positive DEP, followed by an EWOD splitting of the droplet. DEP was produced by a 60 V_{rms} voltage at a frequency of 2 MHz, whereas electrowetting splitting was obtained with a voltage of 84 V_{rms} at a frequency of 1 kHz. From [23]. © 2006 IEEE; reprinted with permission.

negative dielectrophoresis is not sufficient to displace the cells due to the screening of the dielectric layer.

This example shows that combining EWOD and DEP effects could lead to interesting technological solutions for cell manipulation. It is noteworthy that the voltages used for each effect do not alter the viability of the cells.

6.8 Summary

The development of digital microfluidics for biotechnological and biological applications is just at its beginning. These microsystems are not yet at the industrial stage, but rather are laboratory prototypes. The first results are promising and the potential of digital microfluidics seems important: necessary volumes of biologic samples are extremely reduced, volumes of costly reagents are also minute, sensitivity is expected to be very high, time of reaction/detection is short, and efficient parallelization can be achieved.

It is expected that digital microfluidic systems will follow the same evolution as that of conventional microflow systems. However, there are considerable remaining difficulties to overcome, like fabrication, integration, and portability.

References

1. Vijay Srinivasan, Vamsee K. Pamula and Richard B. Fair, "An integrated digital microfluidic lab-on-a-chip for clinical diagnostics on human physiological fluids," *Lab-on-a-Chip*, Vol. 4, pp. 310–315, 2004.
2. Lei Li, Huan Hu, Hong Lin and Da-Tian Ye, Electrowetting of the blood droplet on the hydrophobic film of the EWOD chips, Proceedings of the 2005 IEEE Engineering in Medecine and Biology 27th Annual Conference, Shanghai, China, 1–4 September, 2005.
3. Jeong-Yeol Yoon and R.L. Garrell, "Preventing biomolecular adsorption in electrowetting-based biofluidic chips," *Anal. Chem.*, Vol. 75, pp. 5097–5102, 2003.
4. http://en.wikipedia.org/wiki/Bovine_serum_albumin.

5. V. Srinivasan, K. Pamula, M.G. Pollack and R.B. Fair, A digital microfluidic biosensor for multianalyte detection, Technical Digest IEEE MEMS, pp. 327–330, 2003.
6. Vijay Srinivasan, Vamsee Pamula, Phil Paik and Richard Fair, Protein stamping for MALDI mass spectrometry using an electrowetting-based microfluidic platform, Lab-on-a-Chip: Platforms, Devices, and Applications, Conf. 5591, SPIE Optics East, Philadelphia, 25–28 October, 2004.
7. L. Gan, X. Xiong, R. Wang, J.B. Power and M.R. Davey, Plant regeneration from cell suspension protoplasts of actinidia deliciosa, ISHS Acta Horticulturae, V International Symposium on Kiwifruit, Vol. 610, 2003.
8. J.H. Jeong, Y.M. Moon, S.O. Kim, S.S. Yun and H.I. Shin, "Human cartilage tissue engineering with pluronic and cultured chondocyte sheet," *Key Engineering Materials*, Vol. 342–343, pp. 89–92, 2007.
9. R.B. Fair, A. Khlystov, V. Srinivasan, V.K. Pamula and K.N. Weaver, Integrated chemical/biochemical sample collection, pre-concentration, and analysis on a digital microfluidic lab-on-a-chip platform, Lab-on-a-Chip: Platforms, Devices, and Applications, Conf. 5591, SPIE Optics East, Philadelphia, 25–28 October, 2004.
10. Yuejun Zhao and Sung Kwon Cho, "Microparticle sampling by electrowetting-actuated droplet sweeping," *Lab. Chip*, Vol. 6, pp. 137–144, 2006.
11. K.B. Mullis, F. Ferré and R.A. Gibbs, *The Polymerase Chain Reaction*, Birkhäuser, Boston, 1994.
12. PCR: http://www.britannica.com/eb/article-9002524/polymerase-chain-reaction.
13. http://en.wikipedia.org/wiki/Peltier-Seebeck_effect.
14. COMSOL Reference Manual. Stocholm: COMSOL AB. http://www.comsol.com.
15. Yi-Hsien Chang, Gwo-Bin Lee, Fu-Chun Huang, Yi-Yu Chen and Jr-Lung Lin, "Integrated polymerase chain reaction chips utilizing digital microfluidics," *Biomed. Microdevices*, Vol. 8, pp. 215–225, 2006.
16. M.G. Pollack, P.Y. Paik, A.D. Shenderov, V.K. Pamula, F.S. Dietrich and R.B. Fair, Investigation of electrowetting-based microfluidics for real-time PCR applications, Proceedings of the 2003 MicroTas Conference, Squaw-Valley, Ca, USA, 5–9 October, 2003.
17. http://www.ornl.gov/sci/techresources/Human_Genome/faq/snps.shtml.
18. D. Jary, A. Chollat-Namy, Y. Fouillet, J. Boutet, C. Chabrol, G. Castellan, D. Gasparutto and C. Peponnet, DNA repair enzyme analysis on EWOD fluidic microprocessor. Proceedings of the NSTI-Nanotech Conference, Vol. 2, 2006.
19. N. Sarrut, S. Bouffet, O. Constantin, J. Garin, F. Mittler, J. Sudor and F. Vinet, Comparison of hydrodynamic versus electroosmotic driven flows for enzymatic protein digestion in a microreactor. Proceedings of the MicroTas 2004 Conference, 26–30 September, 2004.
20. Fabien Sauter-Starace, C. Pudda, C. Delattre, H. Jeanson, C. Gillot, N. Sarrut, O. Constantin and R. Blanc, *Lab-on-a-chip: Advances in Packaging for MEMS and Lab-on-a-chip. Multi-Material Micro Manufacture*, W. Menz, S. Dimov and B. Fillon, Eds., Elsevier, 2006.
21. U.-C. Yi and C.-J. Kim, "Soft Printing of Droplets Pre-Metered by Electrowetting," *Sensors and Actuators A*, Vol. 114, nos. 2–3, pp. 347–354, 2004.
22. J. Berthier and P. Silberzan, *Microfluidics for Biotechnology*, Artech House Publishers, 2005.
23. Po-Weng Huang, Tsu-Te Wang, Sheou-Wei Lin, Yu-Chi Chang and Shih-Kang Fan, Dielectrophoretic cell concentrator on EWOD-based chips, Proceedings of the 1st IEEE International Conference on Nano/Micro Engineered and Molecular Systems," Zhuhai, China, 18–21 January, 2006.

7 Chemical Applications☆

7.1 Introduction

In the last decade, lab-on-a-chips (LOC) dedicated to chemical applications have seen a growing interest [1–5]. Many reasons are similar to those already mentioned in the preceding chapters, namely, reduction of the volume of reagents, optimization of the reactions leading to a faster production of desired products with a greater yield and purity, portability of the systems, reduction of waste production; in the case of chemical applications, another advantage of miniaturization is related to the increase of security during the dangerous synthesis by an efficient dissipation of heat generated by exothermic reaction owing to the high surface-area-to-volume ratios. Furthermore, microsystems offer a unique approach to the spatial and temporal control of chemical reactions [6].

7.1.1 Continuous Flow Systems

Most microsystems described in the literature are continuous flow systems. These consist of a network of micrometer-sized channels embedded into a solid substrate such as glass [6,7], silicon [8], polymers [9], steel or stainless steel [10], and ceramics [11]. In most of these microsystems, liquids and suspended solids are transported in channels from one location to another using generally hydrodynamic (pressure) pumping (Fig. 7.1 (a)). However, in a pressure-driven flow, the velocity profile in a cross section is quadratic (approximately a paraboloid if the channel is rectangular), and the mean flow rate depends on the channel cross-section area. This shape generates unwanted diffusion; Watts and Haswell reported a new kind of micro-reactor in borosilicate glass [12], as shown in Fig. 7.1 (b), based on electro-osmotic flow (EOF) [13]. Electro-osmotic flow offers significantly less unwanted dispersive effects than pressure-driven flow and, with appropriate application of electric potentials, its valve-less control of fluid flow is a convenient high-performance sample separation technique. In an ideal EOF, the velocity profile will be flat—except very close to the walls—and the mean velocity is independent of the cross-sectional area due to the very small thickness of the electric double-layer. However, EOF in microfluidic devices can be difficult to control due to many complex factors such as solid surface material, buffer characteristics, and hydrodynamic effects.

☆This chapter was written by **Gilles Marchand** (CEA/Leti-Minatec, Department of Technology for Biology and Health, 17 rue des Martyrs 38054 Grenoble, France), **Philippe Dubois** (CEA/Leti-Minatec, Department of Technology for Biology and Health, 17 rue des Martyrs 38054 Grenoble, France; University of Rennes 1, UMR CNRS 6510, 263 avenue du Général Leclerc, 35042 Rennes France) and **Michel Vaultier** (University of Rennes 1, UMR CNRS 6510, 263 avenue du Général Leclerc, 35042 Rennes France).

(a)

(b)

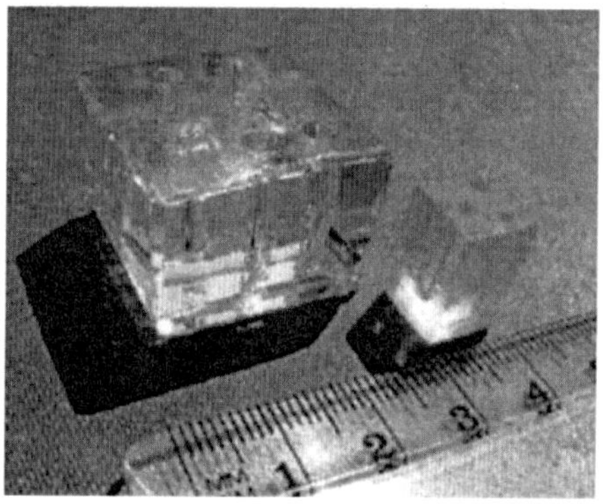

Figure 7.1 (a) View of the device etched in borosilicate glass. The on-chip reactions were carried out in a 197 mm long, 200 μm wide, 100 μm deep microchannel under pressure-driven flow [7]. (b) The device consists of a borosilicate glass base plate containing an etched channel network and a top block with 3 mm drilled holes to introduce the reagent into the reservoirs. Thermal bonding of the two layers results in a sealed microreactor with typical dimensions in the range of 2.5 cm × 2.5 cm × 2.0 cm. Transport under EOF. A. from [7], reproduced by permission of The Royal Society of Chemistry. B. From [6]. © 2002 Elsevier; reprinted with permission.

A range of chemical reactions such as Suzuki coupling [14], Wittig reaction [15], Michael addition [16], nitration of benzene, or tripeptide synthesis (Fig. 7.2) have been carried out to demonstrate the advantages that micro-reactors offer in terms of stereo-selective product control [17–19], multistep synthesis [20], reaction kinetics, or conversion rate [6]. For example, the time necessary for Suzuki coupling [6] has been reduced from 8 hours to 6 seconds with the same conversion rate.

In spite of many advantages, micro-reactors suffer from several drawbacks such as the presence of large dead volume, the frequent obstruction of the channels, the difficulty of maintaining a constant hydrodynamic pressure, especially for viscous liquids, and the control of surface charges when electro-osmotic pumping is used. For this last actuation mode, the requirement of electric charges on the solid surface limits the choice of materials.

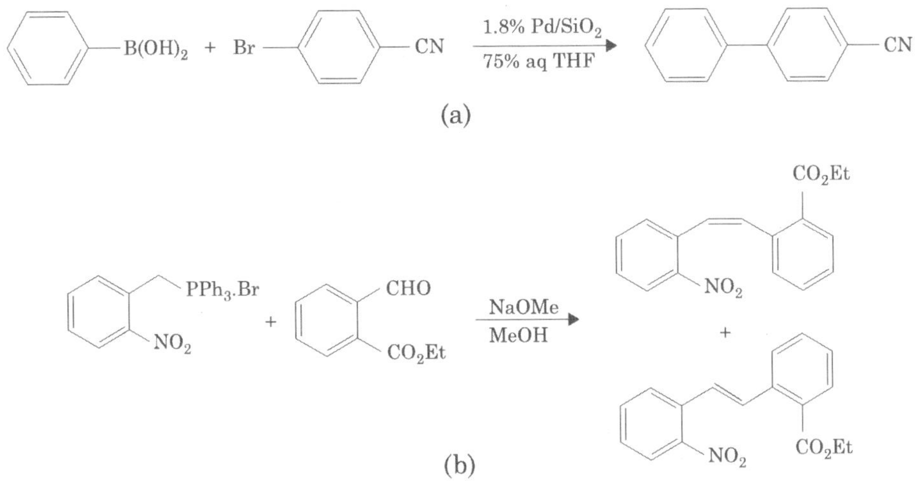

Figure 7.2 Examples of reaction performed in a microreactor. Schematic representation of the Suzuki coupling (a) and Wittig reaction (b) described by Watts et al. (P. Watts, S.J. Haswell, Chemical Society Reviews, Vol. 3, pp. 235–246, 2005).

In an attempt to avoid these problems, a continuous flow microfluidic system in which droplets are separated by immiscible liquid spacers (Fig. 7.3) has been proposed [21] (more details on droplet microfluidics are given in Chapter 9). The system consists of a series of nano-liter plugs of many different aqueous reagents, separated and surrounded by a fluorinated carrier fluid and spaced by gas bubbles inside a hydrophobic glass or plastic capillary. Reactant filled droplets may be issued from the emulsification process of a multiple fluid co-flow immersed in an immiscible carrier fluid [22]. This three-phase liquid/liquid/gas system was preferred to a two-phase system, since it enhances reliability in the manipulation and transport of these plugs. Indeed, in a two-phase system, plugs may coalesce during subsequent use if the viscosity of the carrier fluid is very different from that of the plugs or if the viscosities within plugs vary significantly. Air bubbles can be used for actuation of steady microfluidic flow [23].

However, although the droplet format resolves many problems existing with liquids in flow, complex protocols are relatively difficult to perform with this system, and for the moment, only a few validations [24,25] have been reported for aqueous solvents which are often not suitable for chemical synthesis, and one for octadecene [26].

7.1.2 Digital Microfluidic

Digital microfluidic is an alternative approach to droplet microfluidic systems (see Chapter 1). In such microsystems, droplets containing reagents, samples, etc., are manipulated on two-dimensional planar chips—contrary to microfluidic

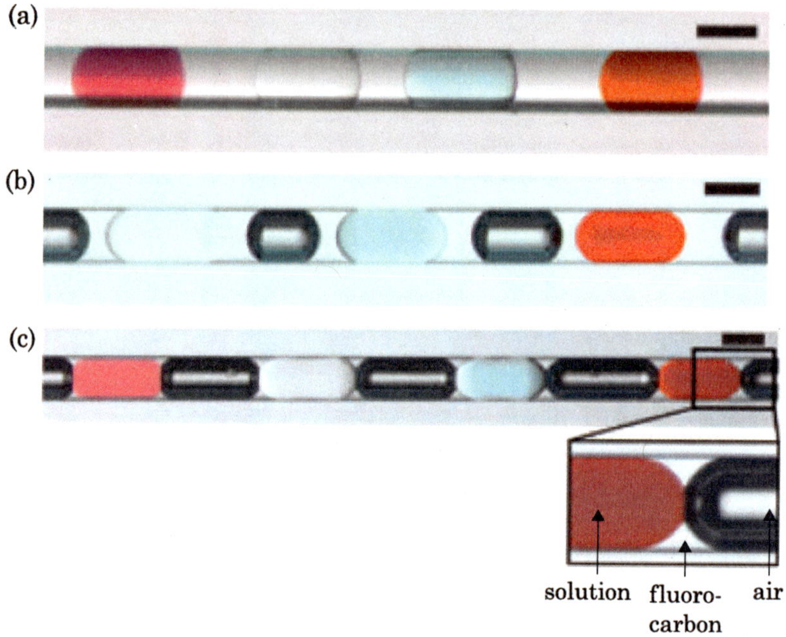

Figure 7.3 (a) Array of plugs of four different reagents in a capillary. The flow contains $KMnO_4$ (purple), NaCl (colorless), $CuSO_4$ (blue) and $Fe(SCN)_3$ (red) plugs respectively. The carrier fluid is fluorocarbon. (b) and (c) An array of plugs of different reagents transported by a fluorocarbon flow and separated by air bubbles (dark) in a capillary. In (b) the aqueous plugs are separated from the air bubbles by a layer of fluorocarbon, thereby preventing cross-contamination between plugs. From [21]. © 2005 Wiley; reprinted with permission.

channels. Such devices have several advantages, such as no dispersion due to diffusion—the droplet interface limits the extension of diffusion, no or negligible cross-mixing, and simplified fluidic manipulation. This very interesting concept allows easy programming of displacements and merging of droplets containing different reagents ($A_1 \ldots A_x, B_1 \ldots B_x$) on a chip. Thus, the concept of digital microfluidics opens the way to multistep reactions and parallel or combinatorial synthesis (Fig. 7.4).

Several fluidic actuators have been reported such as chemical [27–29] or thermal gradient, acoustic waves [30–32], and electrical methods. Dielectrophoresis (DEP) [33] and electrowetting on dielectric (EWOD) [34] are the most common electrical methods (Fig. 7.5). These electrical methods are based upon the micro-manipulation of droplets by an individually addressable electrode array via a programmed external electric field. They offer the possibility of performing elementary fluidic functions without any moving parts or pressure actuation. Hence, these electrical methods open the way to digital lab-on-a-chips requiring elementary operations like dispensing of droplets containing reagents from reservoirs, displacement of these droplets in the lab-on-a-chip, mixing, cutting and merging of the droplets leading

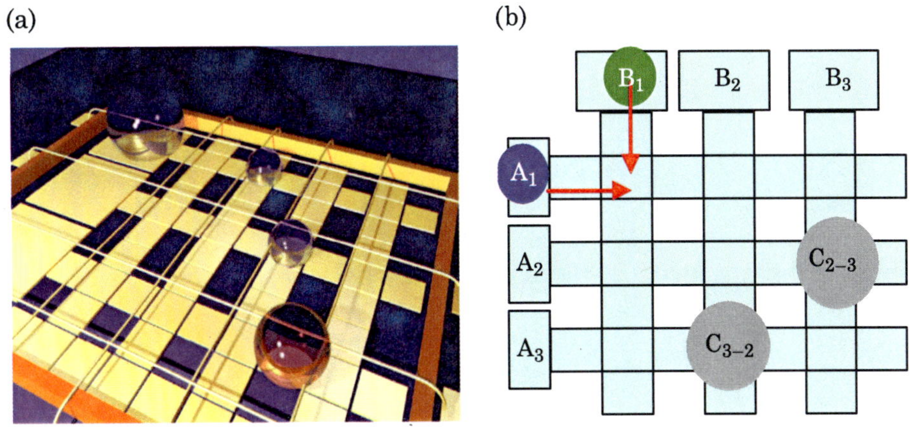

Figure 7.4 Principle of a digital microfluidic system: (a) droplets on a planar chip; (b) Design of a multiplexed chip allowing the performing of combinatorial synthesis.

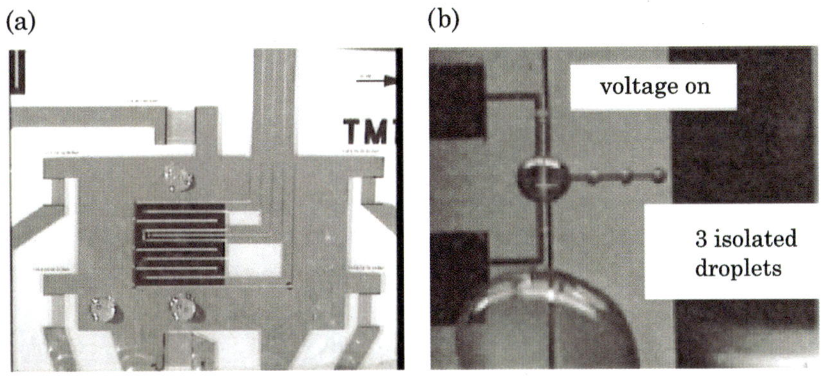

Figure 7.5 (a) Photo of aqueous droplets on a planar lab-on-a-chip displaced by acoustic waves. From A. Wixforth, Superlattices and Microstructures, Vol. 33, pp. 389–396, 2003. © 2003 Elsevier; reprinted with permission. (b) View of water droplet formation with DEP. Reused with permission from [33]. © 2001 American Institute of Physics.

to a biological, biochemical, or chemical reaction. Some biological [35] and chemical [36] applications using DEP were reported. However, DEP requires high frequencies to displace droplets (AC voltage >50 kHz) raising the problem of electrical connection and heating of the solution. Conversely, EWOD requires only low frequencies (1 to 5 kHz), and, even better, can be used with DC voltage. Moreover, EWOD heating of droplets is negligible even with high ionic strength solutions [37]. Therefore, electrowetting on dielectric appears to be applicable for a wider matrix of samples than DEP.

This chapter deals with droplets displaced by EWOD actuation and used as micro-reactors for chemical applications. First, the validation of elementary fluidic functions with different solvents is reported. Then examples of chemical synthesis performed in water, organic solvents, and ionic liquids are described. Finally, the possible perspectives of droplet micro-reactors are investigated.

7.2 Non Aqueous Solvents on EWOD Chips

Most applications of EWOD-based microsystems are found in biotechnology, with reactions such as PCR (polymerase chain reaction) [38,39] or DNA reparation [40], performed in aqueous solvents (see preceding chapter). Consequently, the displacement, merging, mixing, and dispensing of aqueous solutions have been widely studied. However, even if some specific organic synthesis can be performed in water, chemical synthesis usually requires non aqueous solvents. The use of non aqueous solvents—more adapted to organic synthesis—in EWOD-based microsystems [41–43] has been developed recently.

7.2.1 Volatile Organic Solvents (VOS)

7.2.1.1 Displacement in Closed Systems

We have seen in the preceding chapters (4 and 5) that closed EWOD systems were more favorable for moving droplets than open EWOD systems, and that the motion of aqueous liquids was easier in oil. However, not all liquids are immiscible with oil. The possibility of using digital microfluidics to move and split droplets of pure organic solvents and solutions *under air* in a closed microsystem has been reported by Chatterjee et al. [41]. The EWOD microsystems used for these experiments are constituted of a microfabricated substrate including an array of conductive gold electrodes, a dielectric layer (SiO_2 or parylene-C), hydrophobic Teflon-AF coating, and a top plate with a single ITO electrode laminated on a glass substrate and coated with a Teflon-AF layer. The gap between the electrode array and the top plate is in the range 100 to 300 μm. Solvent droplets (0.1–0.6 μL for moderate to high boiling point liquids and up to 2 μL for low boiling point liquids) are sandwiched between the two plates (Fig. 7.6).

Operations on droplets are performed by applying 90 V_{rms} AC potentials at 10 Hz–8 kHz between the electrode in the top plate and the appropriate electrodes in the bottom plate. It was demonstrated that a wide range of organic solvents can be manipulated *under air* in such a system. Table 7.1 shows the voltage—induced contact angle changes and the movabilities of the organic solvents and solutions in comparison with water.

Many solvents can be actuated, and motion is not always correlated with either a high liquid surface tension or an electrically induced change in the liquid–solid contact angle. For example, ethanol and chloroform can be actuated

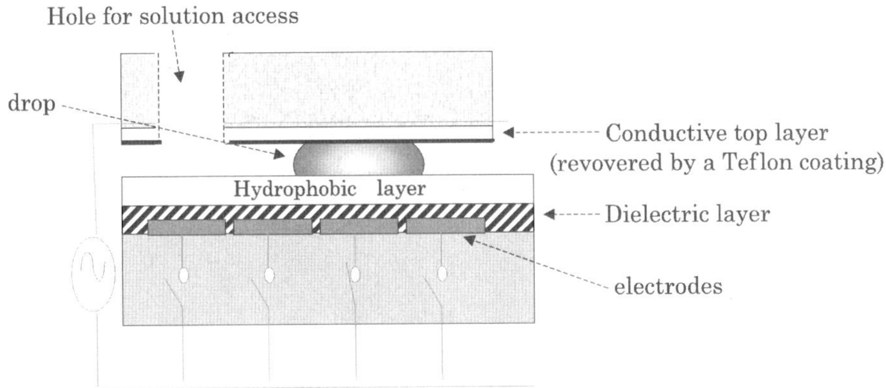

Figure 7.6 Scheme of a closed digital microsystem.

Table 7.1 Movability of Organic Solvents and Solutions in EWOD Systems; d is the Vertical Gap. From Chatterjee et al. [41]

Liquid	ε_r	σ[S/m]	γ[mN/m]	$\Delta\Theta$ [degrees]	Movable
Fornamide	111	3.5×10^{-3}	57	16.1	Y
Water	80	8.7×10^{-4}	72	30.0	Y
Formic acid	51.1	7.0×10^{-3}	37	26.3	Y
DMSO	47.2	3.0×10^{-5}	43	15.3	Y
DMF	38.3	3.2×10^{-5}	37	6.9	Y
Acetonitrile	36.6	1.9×10^{-5}	29	9.8	Y
Methanol	33.0	1.7×10^{-4}	22	9.8	Y
Ethanol	25.3	7.4×10^{-5}	22	10.5	Y
Acetone	21.0	5.0×10^{-7}	23	6.4	Y
Piperidine	4.3	1.0×10^{-5}	29	8.9	Y
1-Pentanol	15.1	8.0×10^{-7}	25	12.8	Y ($d < 100$ µm)
1-Hexanol	13.0	1.6×10^{-5}	26	14.6	Y ($d < 100$ µm)
Dichloromethane	8.9	1.0×10^{-7}	27	3.7	Y ($d < 100$ µm)
Dibromomethane	7.8	2.6×10^{-6}	39	7.3	Y ($d < 100$ µm)
THF	7.5	5.0×10^{-8}	26	4.9	Y ($d < 100$ µm)
Chloroform	4.8	7.0×10^{-8}	27	0.5	Y ($d < 100$ µm)
65% Toluene, 35% 1-Hexanol	3.7	3.0×10^{-8}	28	6.1	Y ($d < 100$ µm)
4.7 mM TBATFB in toluene	2.3	1.8×10^{-7}	28	5.6	Y ($d < 100$ µm)
Toluene	2.4	8×10^{-14}	28	0.2	N
Carbon tetrachloride	2.2	4×10^{-16}	26	1.0	N
Cyclohexane	2.0	7×10^{-16}	25	0.2	N

and moved. Ethanol presents a low surface tension (22×10^{-3} N/m) and a contact angle change of 10.5 degrees, while chloroform has a surface tension of

27×10^{-3} N/m and nearly no voltage-induced contact angle change. Remember that even if there is no change of apparent contact, an electrowetting force can exist (Chapter 4). Generally, polarizable liquids with dielectric constants greater than 3 and/or conductivities greater than 10^{-9} S m^{-1} are movable. The mixture of unmovable liquids and movable liquids can be moved as long as the mixture has a dipole moment, a dielectric constant, and/or conductivity above the threshold. Furthermore, liquids having conductivities ranging from 10^{-8} to 1 S m^{-1} can be actuated in the same device. Hence, most of hydrocarbon and apolar solvents of which the conductivities were near 10^{-14}–10^{-16} S m^{-1} cannot be actuated or displaced on EWOD devices. This impossibility will limit the use of some essential solvents, and then, it will reduce the scope of the possible chemical reactions in droplets.

Additionally, it is observed that solvents become more movable if the AC voltage frequency is decreased (the limit being a DC actuation). Because the properties of Table 7.1 do not depend on the frequency, it was suggested that the frequency-dependent permittivityε^* be used to predict if a droplet can be moved.

$$\varepsilon^* = \varepsilon_0 \left(\varepsilon_r - j\frac{\sigma}{\omega\varepsilon_0} \right) \tag{7.1}$$

where ε_0 is the permittivity of vacuum, ε_r is the dielectric constant (relative permittivity), σ is conductivity, and ω the angular frequency. Fig. 7.7 shows this correlation. For instance, on the microsystem used in [41], authors noted that solutions with $|\varepsilon^*|$ greater than 2×10^{-9} F/m can always be moved, whereas solutions with $|\varepsilon^*|$ lower than 8×10^{-11} F/m were never movable. Solutions with an intermediate value of $|\varepsilon *|$ require a smaller vertical gap d between the array of electrodes and the top-plate to be displaced by electrowetting.

7.2.1.2 Dispensing of VOS

The potential to move a droplet on an EWOD-based chip is a necessary criterion, but it is not sufficient in itself to ensure a satisfactory functioning of the chip. An important point in chemistry, essentially in combinatorial chemistry or in parallel synthesis, is the possibility to split the solvent droplets containing different reagents or to generate droplets from a reservoir. The mechanisms of droplet dispensing and splitting have been investigated in Chapter 5 [44]. Droplet volume is approximately defined by the dispensing electrode surface multiplied by the gap value between the micro-structured substrate and the cover plate (with a bias due to the shape of the free interfaces).

Garrell and coworkers [41] have reported EWOD dispensing of ethanol (Fig. 7.8) but no information was given for the other organic solvents.

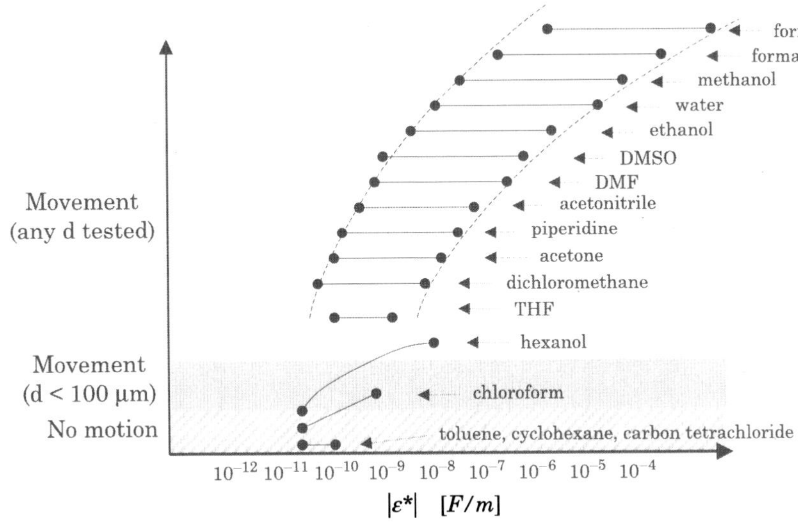

Figure 7.7 Manoeuverability of different chemical species on covered EWOD chip. The left and right points correspond respectively to an actuation frequency of 8 and 0.01 kHz.

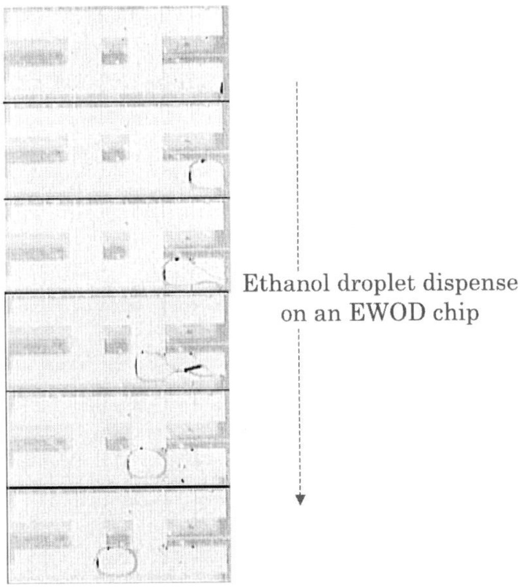

Figure 7.8 Dispense of ethanol droplet on a covered EWOD device. From [41], reproduced by permission of The Royal Society of Chemistry.

7.2.1.3 Mixing in Closed Systems

Due to the very small dimension of a droplet, internal recirculating flow inside a droplet is laminar. Mixing is then a diffusional phenomenon, enhanced by the process of stretching and folding (see Chapter 3). Stretching and folding is much

Table 7.2 Viscosities of Common Volatile Organic Solvents and Aqueous Solutions [45]

Solvents	Temperature [°C]	$10^3 * \eta$ Viscosity [Pa.s]
Acetone	0	0.399
	15	0.316
	30	0.295
Acetonitrile	0	0.442
	15	0.375
	25	0.345
Ethanol	0	1.773
	10	1.466
	30	1.003
Ethylacetate	0	0.582
	10	0.512
	30	0.400
Water	0	1.787
	10	1.307
	30	0.7975
NaCl solution (1% wt)	20	1.004

easier in low viscosity fluids, so mixing in droplets is then closely associated with the viscosity of the liquid. As shown in Table 7.2, viscosities of organic solvents (VOS) are comparable to that of aqueous solutions.

Therefore, the mixing in VOS droplets is similar to that of aqueous solutions. EWOD-actuated mixing in aqueous droplets has been detailed in Chapter 5. Let us recall that back and forth linear motions are not the most effective way to enhance mixing in droplets [46], and that loop motions are more effective because they create more successive stretching and folding of the fluid filaments [47].

7.2.1.4 The Problem of Evaporation

VOS displacement, dispensing, and mixing are currently investigated in order to assess their adaptability to EWOD-based microsystems. However, tiny droplets of volatile organic solvents evaporate very rapidly, even if some of these solvents have high boiling points. Evaporation constitutes a serious drawback for most chemical synthesis. Indeed, many organic reactions necessitate heating and/or pressurization during several hours of incubation. A solution consists in manipulating the VOS droplets in covered microsystems and under oil [48] or orthogonal solvent. However, this way of proceeding introduces other problems such as complexity of the fluidic connections—which are similar to that of microchannel systems, cross-contaminations, and eventual miscibility between oil or orthogonal solvent and droplet solvent, and expulsion of smaller droplets due to the instability of the liquid contour [49]. Furthermore, in the case of covered microsystems in air, the evaporation of low boiling solvents (e.g., chloroform, dichloromethane, and acetone) cannot be avoided. When the size of the droplet becomes smaller than an electrode, maneuverability is lost. This difficulty explains why chemical synthesis examples in the literature in volatile organic solvents are scarce. Non-volatile organic solvents such as ionic liquids are now preferred for chemical synthesis on EWOD-based microsystems. In the following section, we present the characteristics of such liquids and their properties on EWOD-based micro-devices.

7.2.2 Displacement of Ionic Liquids

7.2.2.1 Definition of Ionic Liquids

The term "ionic liquid" is usually used to refer to an organic salt with a melting point less than 100 degrees Celsius. Often, this melting point is even lower than room temperature, therefore ionic liquids are free flowing at room temperature, hence the generic term, room temperature ionic liquids (RTILs) [50,51]. An ionic liquid usually consists of an organic cation (e.g., ammonium, imidazolium, phosphonium, pyrrolidinium, guanidinium, etc.) and an inorganic anion (e.g., tetrafluoroborate, hexafluorophosphate, bis(trifluoromethylsulfonyimide), nitrate, etc.). Fig. 7.9 represents typical ionic liquids like [tmba] [NTf_2] and [bmin] [NTf_2]. Some ionic liquids are in a dynamic equilibrium where at any time more than 99.99% of the liquid is constituted of ionic rather than molecular species.

The main characteristics of ionic liquids are

- negligible vapor pressure which makes them very easy to contain, manipulate, transfer, and allows their use under high vacuum or heating conditions; thus they are considered "environment friendly" alternatives to VOSs,
- intrinsic conductivity (Coulombic fluids),
- non-flammability associated with the low vapor tension,
- thermal stability (some of them can be heated up to 400 °C without degradation which is much greater than most classical organic solvents),
- wide liquid range (by simply adjusting the substitution pattern around a central core structure for the cation or varying the anion),
- solvating properties for a wide range of organic [52], inorganic, and organo-metallic compounds [50].

The drawback to the use of ionic liquids is their high viscosities (from 30 to 400 cP compared to 0.89 cP for water at 25 °C) [53], which limits their continuous microfluidic use.

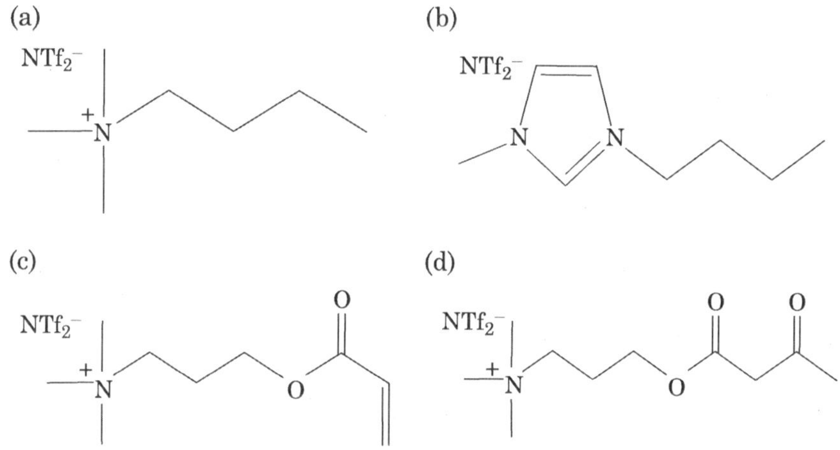

Figure 7.9 Examples of ionic liquids: room temperature ionic liquids RTILs ((a) and (b)) and task-specific ionic liquids TSILs ((c) and (d)). (a) N-trimethyl-N-butylammonium bis(trifluoromethylsulfonyl)imide [tmba][NTf_2]; (b) 1-butyl-1-methylimidazolium bis(trifluoromethylsulfonyl)imide [bmim][NTf_2]; (c)(3-acryloyloxy-propyl)-trimethylammonium bis(trifluoromethylsulfonyl)imide; (d) (3-acetoacetoxy-propyl)- trimethylammonium bis(trifluoromethylsulfonyl)imide.

Because of their properties, ionic liquids attract great attention in many fields, including organic chemistry, electrochemistry, catalysis, physical chemistry, and engineering. Moreover, a subclass of ionic liquids called task-specific ionic liquids (TSILs), has widened the utilization scope of ionic liquids [54–56]. Their functional groups covalently linked to the cation, the anion, or both, confer additional properties to these salts, thereby expanding their potential applications far beyond those of conventional ionic liquids. Moreover, TSILs and RTILs can be combined in solutions. Hence, supported synthesis in a homogeneous solution can be achieved which represents a major advantage [57,58]. Indeed, the reaction can thus occur on a supported phase (easy purification) in solution (no limitation of the kinetics and yields) and the final product can be easily purified since starting reagents and byproducts can be eliminated by a simple washing with diethyl ether.

7.2.2.2 Static Behavior

Several authors [41–43] have dealt with the static behavior of ionic liquids under EWOD, more specifically to their ability to spread on a surface under electric actuation [59,60]. The authors found that the electrowetting curves are very similar to the electrowetting curves recorded for aqueous solutions (water or buffer) on a hydrophobic surface. They observed that the Lippmann–Young relation is valid at low voltage (linear relation between the cosine of the contact angle and the square of the voltage, with a slope equal to $C/2\gamma$), and, at high

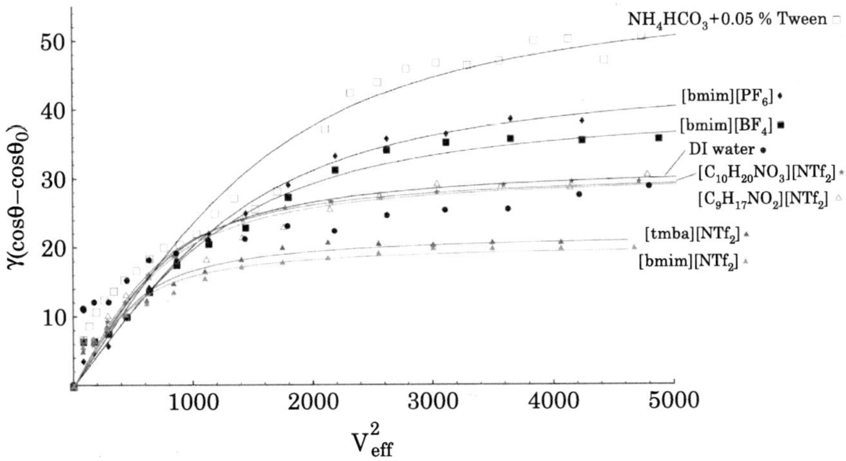

Figure 7.10 Representation of the relation $\gamma(\cos\theta - \cos\theta_0) = C/2V^2$ and Langevin's function for various ionic liquids, deionized water and NH_4HCO_3 25 mM + 0.05% Tween 20 buffer. From [62]. © 2006 NSTI, reprinted with permission.

voltage, the saturation threshold was observed [61] whatever the sign of the potential [26]. According to the developments detailed in Chapter 4, we have represented in Fig. 7.10 the Lippmann–Young relation for ionic liquids under the form

$$\gamma(\cos\theta - \cos\theta_0) = \frac{C}{2}V^2. \tag{7.2}$$

We observe the saturation phenomenon for voltages larger than 60 V_{rms} approximately. In Chapter 4, we have shown that the Langevin function [62] given by equations (7.3) and (7.4) could be used to fit the electrowetting curves of water and biological buffers with surfactants. We observe that this same function also fits the ionic liquid curves; our conclusion is that ionic liquids and aqueous solutions are ruled by the same constitutive law.

$$\frac{\cos\theta - \cos\theta_0}{\cos\theta_S - \cos\theta_0} = L\left(\frac{CV^2}{2\gamma(\cos\theta_S - \cos\theta_0)}\right) \tag{7.3}$$

where L is the Langevin function defined by

$$L(X) = \coth(3X) - 1/3X \tag{7.4}$$

and $X = CV^2/2\gamma(\cos\theta_S - \cos\theta_0)$

However, despite the same constitutive law, the efficiency of electrowetting for ionic liquids is lower than aqueous salt solutions. On Fig. 7.10, we observe that the electrowetting force$\gamma(\cos\theta - \cos\theta_0)$ is lower for ionic liquids. On the other

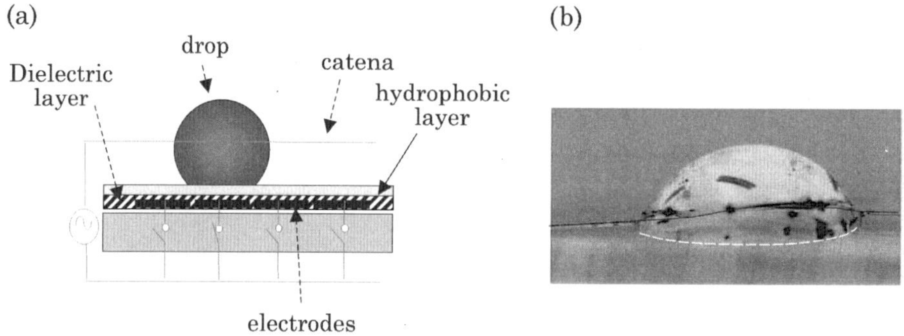

Figure 7.11 (a) Sketch of an open digital microsystem with catena; (b) photo of a drop of ionic liquid pierced by a gold catena. (Photo Ph. Dubois, CEA/LETI).

hand, contrary to Millifiorini who concludes that the nature of the anion is not of such a crucial importance with respect to electrowetting, it has been shown [43] that both anion and cation have an influence on electrowetting curves. Indeed, the same cation bmim$^+$ with three different anions (PF_6^-, BF_4^-, NTf_2^-) leads to three different behaviors in terms of saturation level. Thus, electrowetting forces of 40, 36, and 18 mN were observed for [bmim][PF_6], [bmim][BF_4], and [bmim][NTf_2] respectively. On the other hand, for the same anion NTf_2^- with four different cations, only two distinct behaviors were observed. This could be explained by, first, the relatively similar structures of the cations bmim$^+$ and tmba$^+$ (Fig. 7.9 (a) and (b)) and, second, the strong similarity of structure between the cations $C_{10}H_{20}NO_3^+$ and $C_9H_{17}NO_2^+$ (Fig. 7.9 (c) and (d)).

7.2.2.3 Motion of Ionic Liquids

As ionic liquids respond to electrowetting actuation, their displacement on a matrix of electrodes has been tested. Garrell et al. [41] demonstrated that [bmim][BF_4] and [bmim][PF_6] can be moved in a covered system under an oil layer. P. Dubois et al. [43,63] explored the possibility of moving ionic liquids on an open system under air. A gold wire (25 μm diameter) was used as a catena (Fig. 7.11). This design presents the advantage of simplicity as it allows a direct access to droplets (Chapter 4).

The motion of deionized water (DI water) droplets was observed on such systems; a maximum motion speed of 125 mm/s was obtained [43]. Typical velocities of ionic liquids on EWOD-based microsystems are comprised between 1 and 10 mm/s. However, the comparison of the nature of motion of DI water droplet with a quick camera and ionic liquids does not show a significant difference. It is reasonable to attribute this difference between the maximum motion speed of DI water and ionic liquids to the difference in viscosity between the two species. Indeed, as previously described, ionic liquids present viscosities 30 to 400 times more important than water. Nonetheless they are movable

Table 7.3 Ohnesorge and Reynolds Non-dimensionnal Numbers for Room Temperature Ionic Liquid and Water

Liquid	ρ^{a}	u^{b}	D^{c}	γ^{d}	ν^{e}	η^{f}	On	Re
[tmba][NTf_2]	1.43	1.3	1.3	34.5	58.0	82.9	1×10^3	3×10^{-2}
[bmim][BF_4]	1.17	4.0	1.1	46.3	71.8	84.0	1×10^3	6×10^{-2}
[bmim][PF_6]	1.38	1.3	1.1	43.7	210.3	290.2	4×10^3	7×10^{-3}
DI water	1.00	120	1.2	72.5	0.9	0.9	1×10^1	2×10^2

[a] density (g/cm^3).
[b] motion speed (mm/s).
[c] droplet diameter (mm).
[d] surface tension (mN/m).
[e] cinematic viscosity (mm^2/s).
[f] dynamic viscosity (cP).

on EWOD-based microsystems. Two non-dimensional numbers illustrate the importance of viscous forces in the dynamic behavior of ionic liquids in EWOD systems. The Ohnesorge (On) number illustrates the ratio between the viscous forces and surface tension and the Reynolds (Re) number represents the ratio between the inertia forces and the viscous forces. They are given by

$$On = \frac{\eta}{\sqrt{\rho D \gamma}} \tag{7.5}$$

and

$$Re = \frac{uD}{\nu} \tag{7.6}$$

where ρ is the density of the liquids, u the maximum motion speed of the droplet, D the diameter of the droplet, γ the liquid/air surface tension of the liquid, ν the cinematic viscosity of the liquid, and η the dynamic viscosity of the liquid.

Both numbers show large differences between DI water and ionic liquid, as is shown in Table 7.3. Ionic liquids have a large Ohnesorge and a small Reynolds number, whereas it is the opposite for water. Thus we can establish the following rules:

- for ionic liquids: *viscous forces* > *surface tension forces* > *inertia*
- for water: *surface tension forces* > *inertia* > *viscous forces.*

The velocity of the digital displacement of droplets on EWOD devices is related to the electrowetting force (driving force) and to the viscosity of the liquid (resistive force) [64]. A high electrowetting force and a low viscosity correspond to the larger velocities. To this extent, water and ionic liquids are at the two extremes of the scale. Note that the electrowetting force is governed by the saturation limit, and for both types of liquids, the velocity of motion saturates when the electric saturation is reached.

7.2.2.4 Working Potentials

It is observed that a droplet of conductive liquid does not move from one electrode to the next as soon as an electric actuation is applied. A minimum voltage threshold is required in order to obtain the motion of the drop. This minimum electric potential—noted $V_{\min}$—depends on the nature of the liquid/fluid/solid triplet. On the other hand, a maximum actuation voltage—noted $V_{\max}$—exists above which the electrocapillary force on a drop does not increase any more, only the risk of dielectric breakdown increases [65–69]. Moreover above $V_{\max}$ there is no more gain in the motion speed of a droplet. It has been shown in Chapter 4 that $V_{\min}$ is proportional to $2\sqrt{\alpha\gamma \sin\theta_0}$ and $V_{\max}$ to $\sqrt{\gamma_{SL} - \gamma_{LG}\cos\theta_0}$.

Consequently, for defining the functioning range of the microsystem, it is necessary to know the two limits given by the potentials $V_{\min}$ and $V_{\max}$. The working potential was determined for several ionic liquids and water based solutions [43]. Table 7.4 shows the working potentials for different ionic liquids in air on a Teflon covered substrate. In air, ionic liquids show a wider working interval than water. Indeed, the minimum potential to move ionic liquid droplets (18–30 V) is lower than that of water (40 V), whereas the saturation threshold is similar for both. In Chapter 4, an estimate of the working potentials, i.e., of the values of the minimum and maximum potentials $V_{\min}$ and $V_{\max}$ have been derived based respectively on an analysis of the hysteresis phenomenon for $V_{\min}$ and on a saturation correlation for $V_{\max}$ (PQRS model) [69]. Fig. 7.12 shows that the estimates of the minimum and maximum potentials are still valid for ionic liquids. Note that the minimum actuation potential V_{min} for ionic liquids and that of a biological buffer containing 1% of Tween 20 are similar.

7.2.2.5 Ionic Liquid Dispensing

An important point in the design of EWOD chips is related to the dispensing of liquid from a reservoir located on the chip. If liquids can be directly dispensed from an on-chip reservoir, it will avoid the use of a syringe pump or spotting robot. Such a feature renders the chip portable and allows for chemical reactions of combinatorial chemistry.

Generation of ionic droplets from a reservoir in covered EWOD-based microsystems is similar to that described in Section 7.2.1.2 for VOS [41]. In such covered microsystems, it has been shown that ionic liquids can be distributed from a reservoir with a 100 μm gap while applying a $55V_{\mathrm{rms}}$ voltage [70]. Thus, ionic liquid droplets with volumes less than 100 nl may be delivered. However, addition of perfluorous solvent was needed to reduce the wettability of ionic

Table 7.4 Physical Characteristics of Ionic Liquids in Comparison with Water. All measurements were made at room temperature. Surface tensions (mN.m) were measured with the pendant drop method on a Drop Shape Analysis System G10/DSA10. All contact angles were measured with a commercial Digidrop GBX system, with droplets of 1 µL). V_{min}:minimum actuation voltage (V_{rms}) to displace a droplet of 0, 4 µL on a row of 800 × 800 µm electrodes. V_{max}: saturation voltage (V_{rms}). For buffer +1% Tween 20, evaporation and very long equilibration time makes it impossible to obtain meaningful data

Liquids	γ[mN/m]	θ_0	θ_s	V_{min}	V_{max}
[tmba] [NTf_2]	34.5	77.0	30–45	18	48
[bmin] [BF_4]	46.3	94.2	40–45	22	58
[bmin] [PF_6]	43.6	94.8	30–40	26	58
[bmin] [NTf_2]	33.9	76.5	30–35	18	53
[teba] [NTf_2]	–	76.5	30–32	18	48
[emin] [NTf_2]	–	81.1	35–40	23	58
[$C_8H_{19}BrN$][NTf_2]	–	85.7	30–35	27	48
[$C_7H_{16}NO_2$][NTf_2]	–	89.5	31–36	30	58
[$C_{25}H_{54}$] [OTf]	–	79.5	30–35	22	53
[$C_6H_{16}NO$][NTf_2]	–	89.1	30–35	28	58
[$C_9H_{17}NO_2$][NTf_2]	37.4	85.1	28–30	21	63
[$C_{10}H_{20}NO_3$][NTf_2]	39.4	86.6	33–40	22	58
De-ionized water	72.5	110.0	85–90	40	63
NH_4HCO_3 25 mM + 0.05% Tween 20	65.7	98.0	50–55	23	65
NH_4HCO_3 25 mM + 1% Tween 20	38.0	81.0	–	20	–

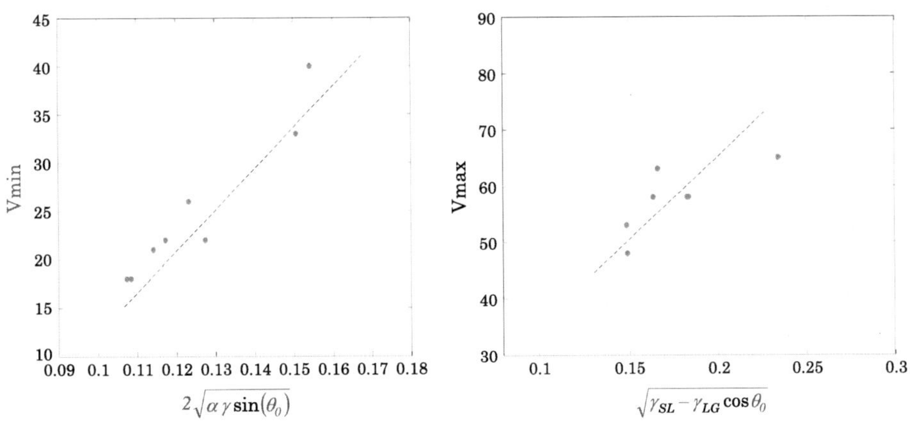

Figure 7.12 Left: plot of V_{min} as a function of $2\sqrt{\alpha\gamma \sin(\theta_0)}$ based on the values of Table 7.4. Right: plot of V_{max} as a function of $\sqrt{\gamma_{SL} - \gamma_{LG}\cos\theta_0}$, based on the values of Table 7.4.

liquids on the polymer walls of the reservoir. Moreover, it has been checked that drops of ionic liquid can be transported easily from a closed system to an open one (see Section 5.3).

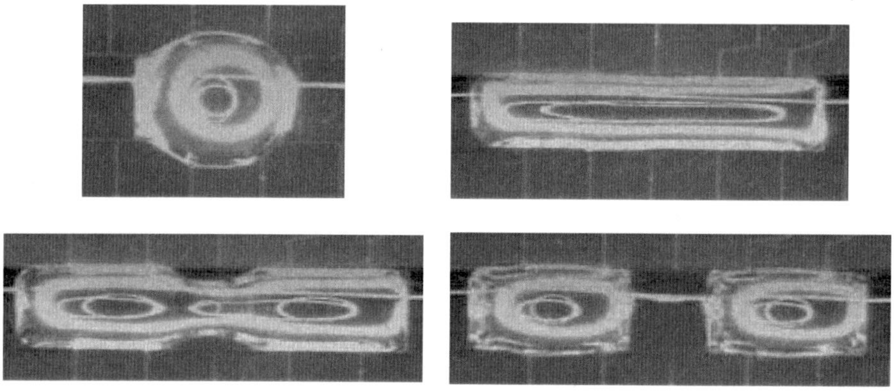

Figure 7.13 Splitting [tmba][NTf_2] droplet in an open EWOD system. A 0.8 μl droplet was deposited on a chip. By successive actuations of electrodes ($V = 55\ V_{rms}$), the droplet was stretched on five electrodes; cutting is then obtained by switching off the central electrode. (Photo Ph. Dubois).

7.2.2.6 Cutting Ionic Liquid Droplets

Cutting a droplet in two has been recognized as a very important operation. Most studies in this area deal with water [71] showing the necessity of performing the splitting in a covered system. Experiments with ionic liquids have shown that splitting ionic liquids by EWOD is possible, even in open systems, as shown in Fig. 7.13. A drop of ionic liquid [tmba][NTf_2] is stretched on five electrodes by successive actuation of the electrodes; the switch off of the central electrode leads to the generation of two droplets of the same size. At the present time, it is not clearly established why ionic liquid droplets are much easier to cut than aqueous droplets. The most striking aspect is the possibility of stretching ionic liquid droplets. We tend to associate this phenomenon with the large elasticity of ionic liquid interfaces that we analyzed in Chapter 5 [70,72].

7.2.2.7 Mixing of Ionic Liquids on an Opened System

Homogeneous mixing of components—biomolecules, chemical species—inside a droplet is essential for obtaining an optimal bio or chemical reaction. Insufficiently mixed liquids have poor reaction rates. We have already analyzed the mixing of aqueous solutions on EWOD substrates (Chapter 5) [73]. In this section we focus on ionic liquids. In order to follow the convective movement inside a droplet, fluorescent beads are added to the liquid (Fig. 7.14).

Two different types of motion can be induced inside a droplet: first convective motions triggered by external conditions: a thermal gradient induces a Marangoni convective effect, mass transfer to the air by evaporation induces convection [74], and an electric field imposes tangential constraints resulting in

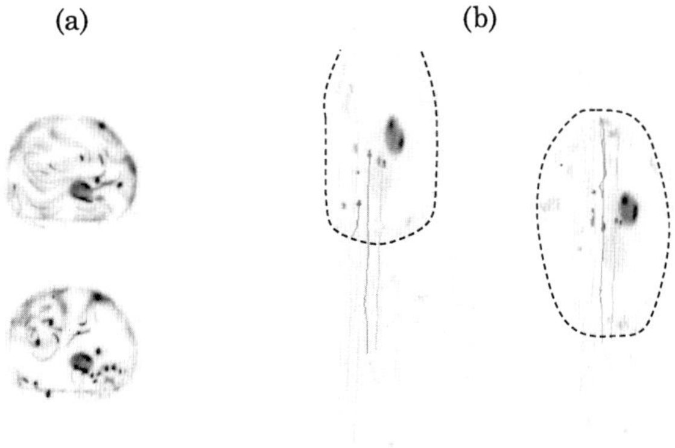

Figure 7.14 Internal motion tracked by fluorescent beads inside droplet; (a) de-ionized water (0, 8 μL–70 V_{rms}) during evaporation; (b) ionic liquid [bmim][BF_4] (0, 4 μL–70 V_{rms}) during back and forth motion on a row of electrodes; edges are indicated by a doted line. (Photo Ph. Dubois).

an electroconvective effect [75]. The second type of motion is that triggered by the motion (displacement) of the droplet itself on the solid substrate. This case has been illustrated in Chapter 5, where we showed what pattern of motion was the most effective to induce mixing inside the droplet.

Whatever the cause of the internal motion, an important difference of behavior between water and ionic liquids was observed. Fluorescent beads draw circular trajectories showing strong convection inside a water droplet. This is not the case for ionic liquids where beads draw linear trajectories coming back near their initial position after back and forth motion on a row of aligned electrodes. It appears that ionic liquids behave in some way like a gel. Ionic liquid high values of viscosity (up to 300 times more viscous than water) can shut off the electroconvection effects observed in a water droplet. Thus, in the absence of additional mixing factors, mixing occurs only by diffusion in ionic liquids. No local recirculation occurs inside droplets during motion, resulting in a very slow mixing of two ionic components in a droplet. This fact has been confirmed by another experiment with fluorescent molecules. After the merging of fluorescent droplets, and back and forth displacement on the chip, the fluorescent droplet recovers its original position (Fig. 7.15); homogenization of the solution required many back and forth motions, taking a total time of 6 minutes. This poor mixing can be predicted by the Reynolds number (Re) as shown in Table 7.3. The Reynolds number obtained with [tmba][NTf_2] is around 3×10^{-2}. This very low Reynolds indicates a very laminar flow which is disadvantageous for a quick mixing.

The preceding analysis shows that mixing in ionic liquid droplets is not easy and mixing enhancement solutions must be developed. Solutions using thermal heating to decrease the viscosity or acoustic actuation to increase the convective forces are currently being investigated.

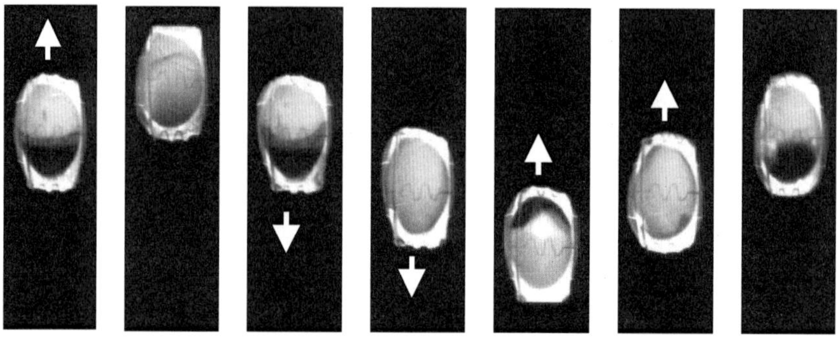

Figure 7.15 Pictures of the mixing of a fluorescent droplet of [tmba][NTf_2] by back and forth displacement on a row of aligned electrodes. See how the last photo is alike the first one. (Photo Ph. Dubois).

The high value of viscosity in ionic liquids renders them very cumbersome to use in microchannel-based systems. The driving pressure would be too important for the microfabricated components. Again, thermal heating of the liquid is a solution [76].

7.3 Chemical Synthesis in Droplets in EWOD-based Systems

In the following section, examples of synthesis performed in droplets handled by EWOD-based systems are reported. The section is divided into three parts, one for each media: first water or aqueous solutions, second VOSs, both types of liquid requiring a covered EWOD-based systems, and finally ionic liquids, allowing the use of an opened EWOD-based system. Advantages and drawbacks of all these systems are discussed.

7.3.1 Synthesis in Aqueous Media

Aqueous solutions are generally reserved for biological reactions, and only some very specific chemical reactions can be performed in such media. On the other hand, water, more precisely buffer, is an ideal medium to perform biochemical synthesis. For example, a biocatalytic reaction is described in references [77,78] which demonstrates the feasibility of a bioassay in a covered system. A calorimetric enzyme-kinetic method based on the Trinder's reaction [79] has been used for the determination of glucose concentrations (Fig. 7.16). In this system two droplets of aqueous solutions containing respectively glucose oxidase, peroxidase, 4-aminoantipyridin, and *N*-ethyl-*N*-sulfopropyl-*m*-toluidine in one droplet and glucose in the second one, were

$$\text{Glucose} + H_2O + O_2 \xrightarrow{\text{glucose oxidase}} \text{Gluconic acid} + H_2O_2$$

$$2\,H_2O_2 + \text{4-AAP} + \text{TOPS} \xrightarrow{\text{peroxidase}} \text{Quinoneimine} + 4\,H_2O$$

Figure 7.16 Schematic representation of the Trinder's reaction.

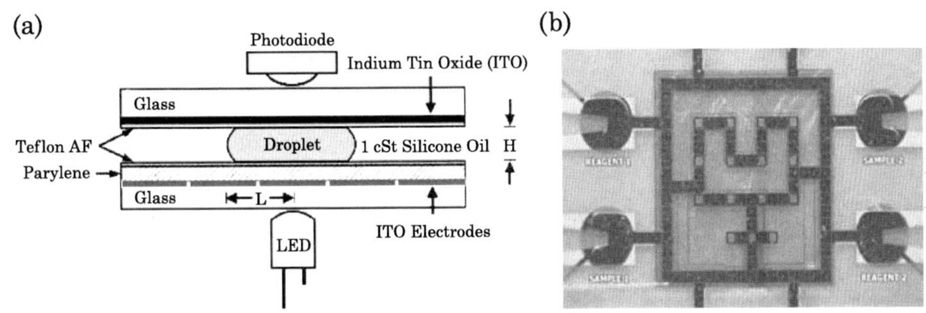

Figure 7.17 (a) Vertical cross-section of the electrowetting chip showing the detection device. (b) Microfabricated EWOD chip used for multiplexed bioassay (gluocose detection). From [78], reproduced by permission of The Royal Society of Chemistry.

sandwiched between two plates, in air or in any other immiscible liquid (Fig. 7.17 (a)). After merging the two droplets together, the glucose is enzymatically oxidized to gluconic acid in the presence of glucose oxidase. The hydrogen peroxide reacts then with 4-amino antipyrine and *N*-ethyl-*N*-sulfopropyl-*m*-toluidine in the presence of peroxidase to form violet colored quinoneimine which exhibits an absorbance maximum at 545 nm.

The lab-on-a-chip used for the reaction of a multiplexed bioassay is shown in Fig. 7.17 (b). Sample injection holes, reservoirs, fluidic pathways, mixing areas, and optical detection sites are all located on the same substrate.

No chemical reaction in water droplet displaced under EWOD actuation has been reported in the literature. However, aqueous media were used with DEP (dielectrophoresis) actuation, which is the other possible electrical method for displacement of droplets in LOC. A first demonstration of a chemical reaction performed in a droplet is the alkalization of phenolphthalein [80] which has been realized in an oil environment to prevent evaporation. This reaction was performed by coalescence of two water droplets containing respectively NaOH (1 mM) and phenolphthalein under DEP actuation. The red color taken by the liquid is an indicator of thc reaction (Fig. 7.18). In this reaction, phenolphthalein, a pH indicator, is deprotonated by NaOH. The two

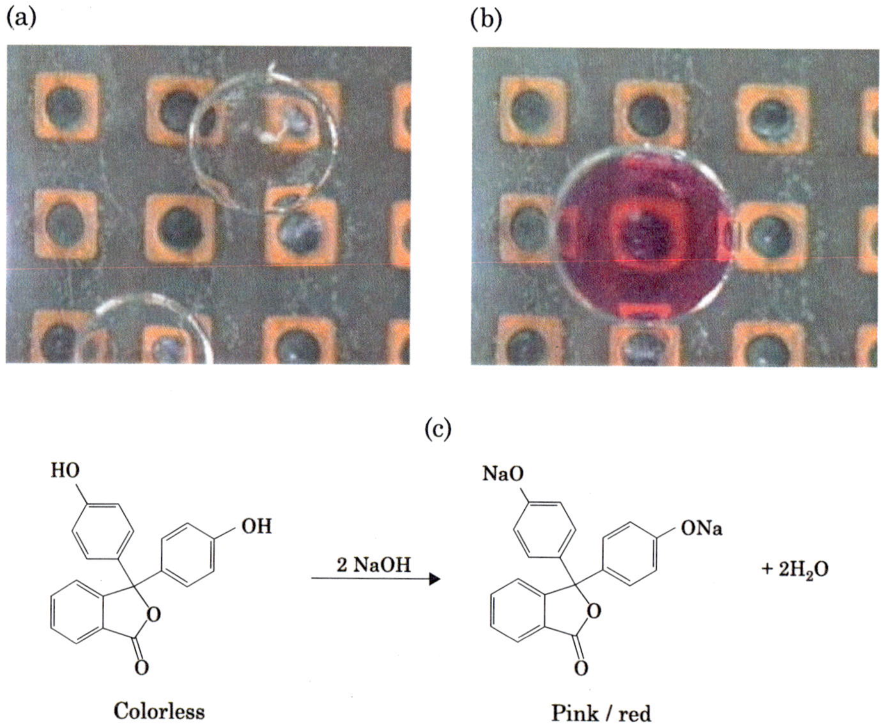

Figure 7.18 (a) Picture of the droplets containing NaOH (right) and phenolphthalein (left); (b) after merging, the droplet takes a red color indicating that the reaction has taken place. (c) Schematic representation of the reaction between phenolphthalein and soda. From [80], reproduced by permission of The Royal Society of Chemistry.

phenol functions are then transformed in two phenolate functions favoring the electronic delocalization on the conjugated backbone.

Even if some reactions can be performed in aqueous solutions [81], these examples are not very numerous and limited to some types of chemical reactions. Indeed, despite the fact that water is the cheapest, safest, and most nontoxic solvent known, most of the time the presence of water is not compatible with the reaction and is eliminated by drying of substrates and solvents. The use of water as a medium for organic reactions is, therefore, one of the latest challenges for modern organic chemists.

7.3.2 Synthesis in Volatile Organic Solvents

Chemical syntheses in VOS are not very widespread due to their relatively high volatility necessitating a cover plate or an oil layer that has to be immiscible with the VOS. Among the few examples reported in the literature, the very simple reaction between nitroaromatic compounds and potassium

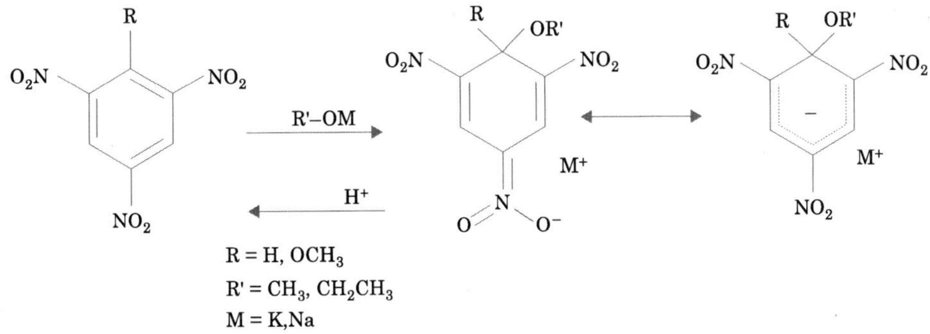

Figure 7.19 Schematic representation of the reaction between an arene and a nucleophile leading to a Jackson–Meisenheimer complex.

hydroxide (KOH) can be mentioned. This reaction allows the detection of trinitroaromatic compounds, the most commonly used class of explosives. To perform this reaction in droplet format, Pamula and coworkers [82] have dissolved in DMSO 2,4,6-trinitrotoluene (TNT) or 2,4-dinitrotoluene (DNT) and KOH in a water/DMSO (1:2000) mixture. Droplets of TNT and KOH were dispensed manually by a pipette under a silicone oil layer on an electrowetting chip and the two droplets are coalesced by applying voltages (50 V) between appropriate electrodes. The droplet is then further mixed by shuttling it across four electrodes for 30 seconds. The mixing should be complete in less than 30 seconds for this pattern of mixing [83]. The reaction between TNT/DNT and KOH leads to colored Jackson–Meisenheimer complexes [84] which can easily be detected on chip by UV-visible measurements. These complexes are produced by reaction between an arene carrying electron withdrawing groups and a nucleophile (Fig. 7.19). These salts have been identified as reactive intermediates in nucleophilic aromatic substitution but stable and isolated.

As shown above, chemistry in droplets of water or volatile organic solvents is possible in EWOD covered systems. However, evaporation of tiny droplets is a major drawback for most of chemical synthesis. Moreover, many reactions necessitate heating, pressure, and/or several hours in incubation. To avoid evaporation, some authors suggest using a surrounding oil environment. Certainly it stops evaporation, but, as previously noted, this approach brings other problems such as fluidic connections and liquid/liquid cross contamination between droplets through the surrounding solvent.

7.3.3 Synthesis in Ionic Liquids

Dubois and coworkers [43,63] have recently reported experiments using droplets of RTILs and TSILs as stable microreactors displaced by EWOD in chemical applications. The approach consists of using an open digital microfluidic chip coated with Teflon, ionic liquid droplets containing different

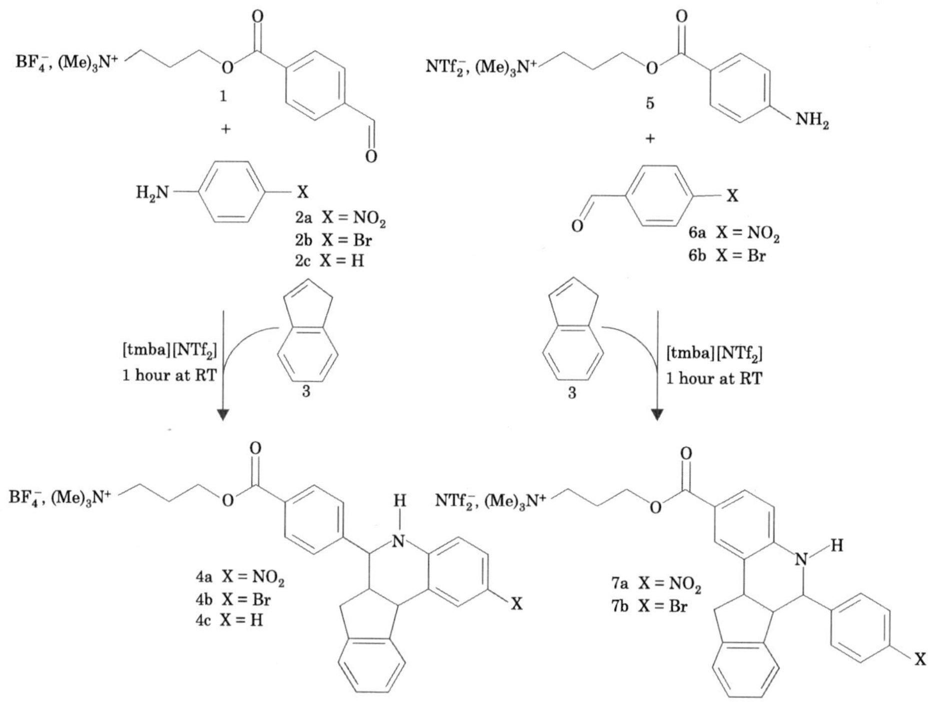

Figure 7.20 Schematic representation of Grieco's reaction in a droplet microreactor.

reagents, and EWOD actuation for moving and merging the droplets. This concept takes advantage of the negligible vapor pressure of ionic liquids allowing the handling of very small liquid volumes (<1 μL) on open microsystems, thus simplifying the experimental approach, facilitating direct access to the droplets and avoiding cross contamination induced by the presence of oil. Chemical syntheses with minute amounts of reagents (especially useful for potentially explosive or expensive reactions) can be completely achieved in solution using ionic liquid matrices such as RTILs or on soluble supported supports using TSILs or task specific onium salts in solution in RTILs. In this last case, the excess of reagents and byproducts could be eliminated at the end of the reaction by simply washing or heating the droplets.

The demonstration of ionic liquid droplets displacement in EWOD microsystems has opened the way to performing on-chip multicomponent chemical reactions. A tetrahydroquinoline synthesis via a well-known three-component reaction described by Grieco and coworkers [85–87] has been chosen for several reasons: (i) tetrahydroquinolines are an important class of biologically active compounds [88]; (ii) the reaction is a relatively complex reaction (a good model for proof of feasibility), (iii) this reaction occurs rapidly at room temperature with a quantitative yield and has been demonstrated on solid supports [89]. In this reaction, aniline reacts with an aldehyde in the presence of an electron-rich olefin and an acidic catalyst (trifluoroacteic acid

-TFA) to produce tetrahydroquinolines at room temperature (Fig. 7.20). The first step of this reaction is the imine formation followed by an Aza Diels-Alder reaction ([4 + 2]-cycloaddition) [90]. The electron-withdrawing effect of the nitrogen atom and the activation of the dienophile by the acidic catalysts allows for the reaction to occur very rapidly at room temperature contrary to the classical Diels-Alder reactions.

A similar reaction was run in an EWOD microsystem, using task specific ionic liquids (TSIL) in solution in [tmba][NTf_2]. In a more detailed way, a 0.2 μL droplet of [tmba][NTf_2] containing a TSIL containing either an aldehyde function (1) or an aniline function (5) and 10 equivalent of TFA (trifluoroacetic acid) was deposited manually on the chip. The droplet was approached from a second droplet of [tmba][NTf_2] containing the derivate 2 or 6. After coalescence, the first step of the reaction takes place and the activated imine starts to be generated. Next, the resulting droplet was merged with a third droplet containing an excess of indene. The reaction was then incubated within 1 hour at room temperature. At the end of the reaction, the final product was analyzed off line and on line (Fig. 7.21).

This approach seems to be very promising for low-volume organic synthesis in solution. Indeed, ionic liquids enable the use of open systems, simple to use and avoiding the cross contamination induced by the use of a secondary fluid. Moreover, the combination of RTILs and TSILs (or task specific onium salts) allows a fast separation of the expected final product and the starting reagents or byproducts. Thus, complex protocols can be achieved on this system due to the easy displacement of droplets by EWOD. However, some difficulties occur in such an approach which can limit its significance.

The first difficulty is the evaporation of the non-supported species through the droplet of ionic liquids. Indeed, even if the ionic liquids have a negligible volatility, because of the high surface/volume ratio, volatile reagents can diffuse through the droplet and evaporate at its surface. For instance, to perform fully the Grieco reaction described in Fig. 7.20, 10 equivalents of TFA are needed in the droplet format, whereas one equivalent is enough in a flask. Solutions to this problem exist. First, TFA can be replaced by a TSIL containing an acid function. Second, in the wide range of components available for organic chemistry, a solid reagent, with the same physical and chemical properties, can be used instead of the volatile compound. For instance, if TFA is replaced by trichloroacetic acid (TCA), which is solid at ambient temperature, a 100% conversion rate is obtained with only two equivalents. Third, a saturated atmosphere can prevent the full evaporation of the volatile compound, which should also be a possible way to introduce volatile reagents inside droplets.

The second difficulty arises from the large surface/volume ratio offered by a digital microsystem. This large surface/volume ratio can boost unwanted reactions, such as oxygen quenching or hydrolysis. In the particular case of RTIL, we have to keep in mind the high hydroscopic capacity of ionic liquids. Therefore, high hydrophobic RTIL, e.g., the ones with NTf_2^- or fluorinated anions, have to be preferred. A dry atmosphere which surrounds the chip can

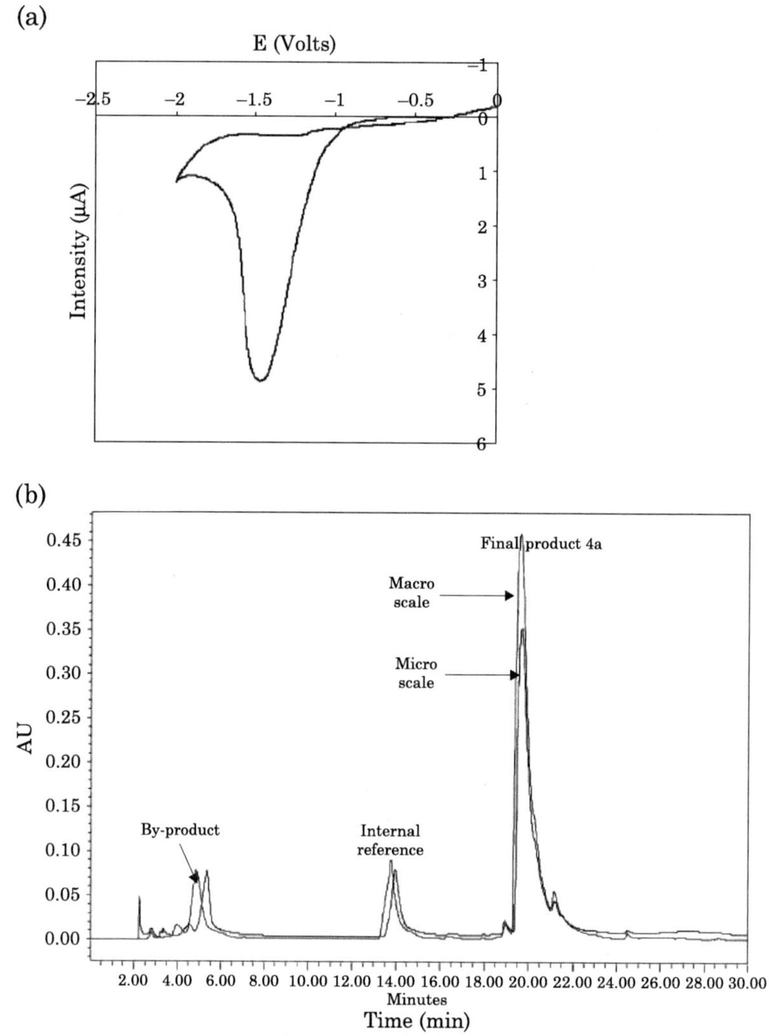

Figure 7.21 (a) On line detection: cyclic voltammogram of 7a in [tmba][NTf_2] at a gold electrode ($\phi = 25$ μm and V = 50 mV/s. (b) Off line detection: comparison of the chromatograms of the macro- and microscale reactions after the reaction of 1 with 2a and 3. The observed byproduct in this chromatogram corresponds to a residual protonated p-nitroaniline. The numbers 7a, 1, 2a and 3 correspond to Fig. 7.20.

protect from such unwanted reactions. One should note that covered EWOD systems also suffer from this drawback.

Finally, reactions in open digital microsystems suffer from the poor mixing of ionic liquids. Indeed, the high viscosities of ionic liquids prevent reagents from a quick diffusion through the droplets [91]. Thus, at room temperature, after the merging of two droplets, around 40 minutes are needed to obtain a total conversion (whereas 5 minutes were needed in a stirred beaker). In this case, reagents are only mixed by a passive way, i.e, by diffusion. Therefore, an

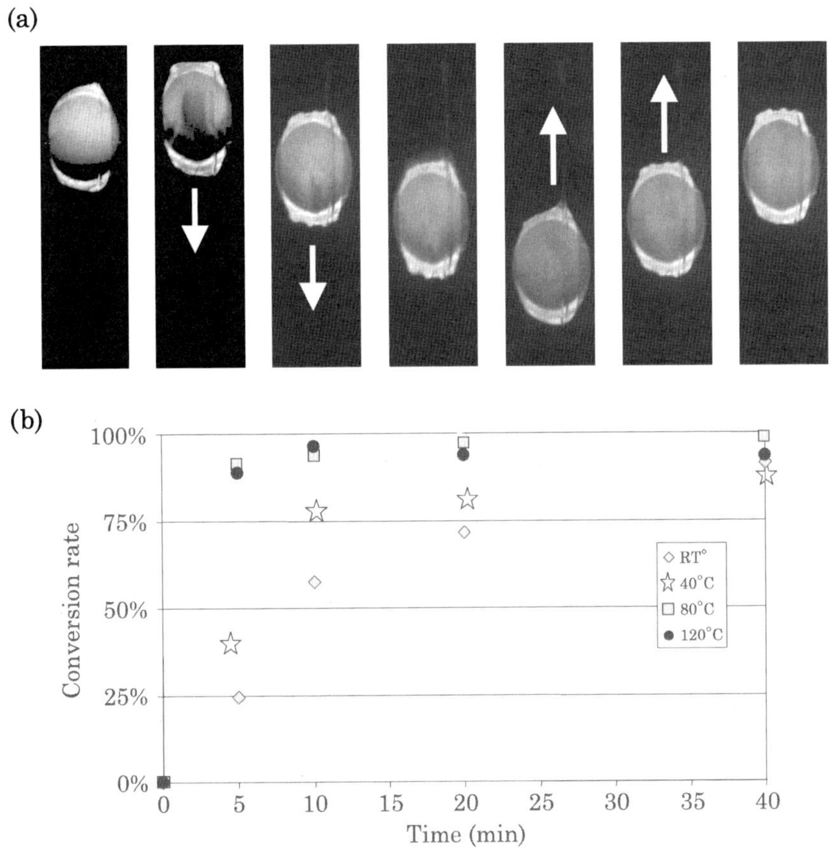

Figure 7.22 (a) Photographs of the mixing of a fluorescent droplet of [tmba][NTf_2] with another non marked droplet with heating from below (120 °C) after one forth and back motion; (b) improvement of kinetic rate with the temperature increase.

active, faster process is required. Back and forth displacement of droplets on the chip has been investigated, but as previously noted, it was found to be not very efficient for ionic liquids. In order to enhance internal convective motions, droplets have been heated from below. In this case, mixing is promoted by two effects. First, the viscosity of the ionic liquids decreases which facilitates internal motion. Second, internal convection appears due to a Marangoni effect (see Section 2.8). Heat related mixing of droplets of RTIL was investigated and a strong effect was observed (Fig. 7.22 (a)). Consequently, the Grieco reaction was performed at a higher temperature and it was observed that the rate enhancement was clearly improved. One may object that the improvement of the kinetic rate is due to the favorable thermodynamic effect of temperature on Grieco reaction and not to the improvement of the droplets mixing. However, the Grieco reaction is already really fast at ambient temperature in the bulk scale, which indicates a low thermodynamic barrier. Besides, at 80° and 120 °C, the time needed for full mixing is identical; and we have observed only little difference in the kinetic rate (Fig. 7.22 (b)).

7.4 Conclusions and Perspectives

Compared to traditional batch reactions, digital chemical micro-reactor technology is showing definite advantages. These types of micro-reactors produce outputs with higher yield and purity, in a shorter time, with safer conditions, and using smaller amounts of reagents. As mentioned above, microflow devices exhibit a number of drawbacks associated with the nature of the technology itself. The development of droplet-based micro reactors, although still in its infancy, is bringing in promising new possibilities in this area, especially because it avoids the problems associated with continuous microflow systems.

We have shown in this chapter that droplets of a number of molecular solvents or ionic solvents, e.g., room temperature ionic liquids (RTILs) can be moved on planar microsystems by EWOD technique. It is interesting to note that the walls of the micro-reactors have become the interface between the liquid and the ambient atmosphere. In order for the micro-reactor to have a sufficient lifetime to be useful, it is necessary that the liquid constituent of the droplet does not evaporate too fast. Molecular solvents, even the ones with a high boiling point, do evaporate, introducing a severe limitation as to their use for such devices. Solutions to these problems have been proposed, but they introduce important limitations. These limitations disappear when using RTILs which have extremely low vapor pressures and therefore do not evaporate. The developments of the chemistry of and in ionic liquids are becoming so numerous that it opens up a very large number of possibilities for digital microfluidic reactors. This is even more the case since the discovery that ionic liquids can be functionalized, thus conferring to the liquid and consequently to the droplet specific chemical, physical, and spectroscopic properties. It is important to keep in mind that ionic liquid droplet micro-reactors can be functionalized by solubilization of onium salts bearing one or several functional groups in a non-functionalized ionic liquid matrix. This opens up the field of homogenous catalysis via onium salt supported catalysts, and the field of solution phase chemistry via soluble supports immobilized in ionic liquids matrixes, offering new possibilities for combinatorial chemistry. Multiplexing has been shown to make possible the handling of a large number of droplets (Chapter 5). Therefore, digital microfluidic reactors are a promising tool for future developments in the combinatorial chemistry area. For example, they can be used for the search of new active molecules or scaffolds. One can also envision the use of this technology for process intensification. Since droplets can be merged and also can be dispensed from a three-dimensional reservoir, it is reasonable to predict that these systems will be used for automated multistep synthesis. A proper functionalization of the onium salt should also open the way to intelligent droplet micro-reactors able to act as micro-sensors and/or micro-probes. The capacity of online detection must be added to these devices in order to make the digital micro-reactors an even more attractive technology.

Amongst the foreseen applications, the ones using enzymes are especially interesting: enzymes can be dissolved in ionic liquids; it has been shown that their catalytic efficiency and thermal stability are increased in such an environment. This could lead to the design of digital micro-reactors with enzyme activity aimed at organic synthesis, analytical purposes, and molecular biology. It may be safe to say that applications of this technology are only limited by the capacity to imagine new applications, keeping in mind that there are still a number of technical problems waiting to be solved.

References

1. Y. Kikutani, H. Hisamoto, M. Tokeshi and T. Kitamori, *Lab chip*, Vol. 4, pp. 328–332, 2004.
2. D.M. Ratner, E.R. Murphy, M. Jhunjhunwala, D.A. Snyder, K.F. Jensen and P.H. Seeberger, *Chem. Commun.*, pp. 578–580, 2005.
3. G.N. Doku, W. Verboom, D.N. Reinhoudt and A. van den Berg, *Tetrahedron*, Vol. 61, pp. 2733–2742, 2005.
4. S. Taghavi-Mogadam, A. Kleemann and K.G. Golbig, *Org. Proc. Res. & Develop*, Vol. 5, pp. 652–658, 2001.
5. D.A. Snyder, C. Noti, P.H. Seeberger, F. Schael, T. Bieber, G. Rimmel and W. Ehrfeld, *Helv. Chim. Acta*, Vol. 88, pp. 1–9, 2005.
6. P.D.I. Fletcher, S.J. Haswell, E. Pombo-Villar, B.H. Warrington, P. Watts, S.Y.F. Wong and X. Zhang, Micro reactors: principles and applications in organic synthesis, *Tetrahedron*, p. 4735, 2002.
7. M. Brivio, R.E. Oosterbroek, W. Verboom, M.H. Goedbloed, A. van den Berg and D.N. Reinhoudt, *Chem. Comm.*, pp. 1924–1925, 2003.
8. K.F. Jensen, *Chem. Eng. Sci.*, Vol. 56, pp. 293–303, 2001.
9. D.C. Duffy, J.C. McDonald, O.J.A. Schueller and G.M. Whitesides, *Anal. Chem.*, Vol. 70, p. 4974, 1998.
10. K. Jähnish, V. Hessel, H. Löwe and M. Baerns, *Angew. Chem. Int. Ed.*, Vol. 43, pp. 406–446, 2004.
11. C.M. Mitchell, D.-P. Kim and P.J.A. Kenis, *Journal of Catalysis*, Vol. 241, pp. 235–242, 2006.
12. P. Watts and S.J. Haswell, *Chemical society reviews*, Vol. 3, pp. 235–246, 2005.
13. P.D. Fletcher, S.J. Haswell and V.N. Paunov, *Analyst*, Vol. 124, p. 1273, 1999.
14. G.M. Greenway, S.J. Haswell, D.O. Morgan, V. Skelton and P. Styring, *Sensors & Actuators B*, Vol. 63, p. 153, 2000.
15. V. Skelton, G. Greenway, S.J. Haswell, P. Styring, D.O. Morgan, B. Warrington and S.Y.F. Wong, *Analyst*, Vol. 126, p. 7, 2001.
16. C. Wiles, P. Watts, S.J. Haswell and E. Pombo-Villar, *Lab Chip*, Vol. 2, pp. 62–64, 2002.
17. V. Skelton, G.M. Greenway, S.J. Haswell, P. Styring, D.O. Morgan, B.H. Warrington and S.Y.F. Wong, *Analyst*, Vol. 126, p. 7, 2001.
18. V. Skelton, G.M. Greenway, S.J. Haswell, P. Styring, D.O. Morgan, B.H. Warrington and S.Y.F. Wong, *Analyst*, Vol. 126, p. 11, 2001.

19. C. Wiles, P. Watts, S.J. Haswell and E. Pombo-Villar, *Lab Chip*, Vol. 4, pp. 171–173, 2004.
20. P. Watts, C. Wiles, S.J. Haswell, E. Pombo-Villar and P. Styring, "The synthesis of peptides using micro reactors," *Chem. Commun.*, p. 990, 2001.
21. B. Zheng and R.F. Ismagilov, *Angew. Chem., Int. Ed.*, Vol. 44, pp. 2520–2523, 2005.
22. G. Cristobal, L. Arbouet, F. Sarrazin, D. Talage, J.-L. Bruneel, M. Joanicot and L. Servant, *Lab Chip*, Vol. 6, pp. 1140–1146, 2006.
23. P. Marmottant and S. Hilgenfeldt, *Proc. Natl. Acad. Sci. USA*, Vol. 101, p. 9523, 2004.
24. I. Shestopalov, J.D. Tice and R.F. Ismagilov, *Lab Chip*, Vol. 4, pp. 316–321, 2004.
25. Z. Nie, S. Xu, M. Seo, P.C. Lewis and E. Kumacheva, *J. Am. Chem. Soc.*, Vol. 127, pp. 8058–8063, 2005.
26. E. Chan, A. Alivisatos and R.A. Mathies, *J. Am. Chem. Soc.*, Vol. 127, pp. 13854–13861, 2005.
27. R. Yamada and H. Tada, *Langmuir*, Vol. 21, pp. 4254–4256, 2005.
28. E. Raphaël, *C.R. Acad. Sci., Ser. II*, Vol. 306, p. 751, 1998.
29. M.K. Chaudhury and G.M. Whitesides, *Science*, Vol. 256, p. 1539, 1992.
30. F. Brochard, *Langmuir*, Vol. 5, p. 432, 1989.
31. A. Wixforth, C. Strobl, Ch. Gauer, A. Toegl, J. Scriba and Z. Guttenberg, *Analytical and Bioanalytical Chemistry*, Vol. 379, nos. 7–8, pp. 982–991, 2004.
32. Z. Guttenberg, H. Müller, H. Habermüller, A. Geisberg, J. Pipper, J. Felbel, M. Kielpinski, J. Scriba and A. Wixforth, *Lab Chip*, Vol. 5, pp. 308–317, 2005.
33. T.B. Jones, M. Gunji, M. Washizu and M.J. Feldman, *J. Applied Phys*, Vol. 89, pp. 1441–1448, 2001.
34. J. Gong, S.-K. Fan and C.-J. Kim, Tech. Dig. IEEE Int. Conf. Micro Electro Mech. Syst., 17th, Maastricht, The Netherlands, 25–29 January, p. 355, 2004.
35. J. Schwartz, J.V. Vykoukal and P. Gascoyne, *Lab Chip*, Vol. 7, pp. 11–17, 2004.
36. J. Millman, K. Bhatt, B. Prevo and O.D. Velev, *Nature Mater.*, Vol. 4, p. 98, 2005.
37. M.G. Pollack, A.D. Shenderov and R.B. Fair, *Lab Chip*, Vol. 2, pp. 96–101, 2002.
38. V. Srinivasan, V.K. Pamula and R.B. Fair, *Lab Chip*, Vol. 4, pp. 310–315, 2004.
39. Y. Fouillet, D. Jary, A.G. Brachet, C. Chabrol, J. Boutet, P. Clementz, D. Lauro, R. Charles and C. Péponnet, *Micro Total Anal. Syst.*, Boston, MA, 2005.
40. D. Jary, A. Chollat-Namy, Y. Fouillet, J. Boutet, C. Chabrol, G. Castellan and D. Gasparutto, Péponnet, Nanotech 2006, Boston, 7–11 May, 2006.
41. D. Chatterjee, B. Hetayothin, A.R. Wheeler, D.J. King and R.L. Garrell, *Lab Chip*, Vol. 6, pp. 199–206, 2006.
42. S. Millifiorini, A.H. Tkaczyk, R. Sedev, J. Efthimiadis and J. Ralston, *J. Am. Chem. Soc.*, Vol. 128, pp. 3098–3101, 2006.
43. P. Dubois, G. Marchand, Y. Fouillet, J. Berthier, T. Douki, F. Hassine, S. Gmouh and M. Vaultier, *Anal. Chem.*, Vol. 78, pp. 4909–4917, 2006.
44. Y. Fouillet, D. Jary, A.G. Brachet, J. Berthier, R. Blervaque, L. Davous, J.M. Roux, J.L. Achard and C. Peponnet, 4th International Conference on Nanochannels, Microchannels and Minichannels, 19–21 June, Limerick, Ireland, 2006.
45. *Handbook of Chemistry & Physics*, 70th Edition, 1989–1990.
46. P. Paik, V.K. Pamula, M.G. Pollack and R.B. Fair, *Lab Chip*, Vol. 3, pp. 28–33, 2003.
47. P. Paik, V.K. Pamula and R.B. Fair, *Lab Chip*, Vol. 3, pp. 253–259, 2003.
48. P.R.C. Gascoyne, J.V. Vykoukal, J.A. Scwartz, T.J. Anderson, D.M. Vykoukal, K.W. Current, C. McConaghy, F.F. Becker and C. Andrews, *Lab Chip*, Vol. 4, pp. 299–309, 2004.
49. M. Vallet, B. Berge and L. Vovelle, *Polymer*, Vol. 37, pp. 2465–2470, 1996.
50. P. Wasserscheid and T. Welton, *Ionic Liquids in Synthesis*, Wiley CH, Weinheim Germany, 2003.
51. T. Welton, *Chem. Reviews (Washington, D.C.)*, Vol. 99, pp. 2071–2083, 1999.
52. R.D. Rogers and K.R. Seddon, *Science*, Vol. 302, pp. 792–793, 2003.

53. A. Bagno, C. Butts, C. Chiappe, F. D'Amico, J.C.D. Lord, D. Pieraccini and F. Rastrelli, *Org. Biomol. Chem.*, Vol. 3, pp. 1624–1630, 2005.
54. J. Fraga-Dubreuil and J.-P. Bazureau, *Tetrahedron Lett.*, Vol. 42, pp. 6097–6100, 2001.
55. S.T. Handy and M. Okello, *J. Org. Chem.*, Vol. 70, pp. 2874–2877, 2005.
56. J.H. Davies, *Chem. Lett.*, Vol. 33, pp. 1072–1077, 2004.
57. M. Vaultier and S. Gmouh, French patent, FR 2845084 application no. 2002-11910, WO 2004 029004.
58. M. Vaultier, S. Gmouh and F. Hassine, French patent, FR 2857360 application 2003-8413, WO 2005 005345.
59. E. Colgate and H. Matsumoto, *J. Vac. Sci. Technol., A*, Vol. 8, p. 3625, 1990.
60. M. Washizu, *Proc. IEEE Trans. Ind. Appl.*, Vol. 34, p. 732, 1998.
61. F. Mugele and J.-C. Baret, *J. Phys.: Condens. Matter.*, Vol. 17, p. R705, 2005.
62. J. Berthier, O. Raccurt, P. Clementz, D. Jary, C. Peponnet and Y. Fouillet, Nanotechnology Conference and Trade Show, Anaheim, CA, 8–12 May, 2005.
63. P. Dubois, G. Marchand, Y. Fouillet, C. Peponnet, C. Chabrol, J. Berthier and M. Vaultier, *Proc. Micro Total Anal. Syst., Boston*, Vol. 1, pp. 412–414, 2005.
64. R. Bavière, J. Boutet and Y. Fouillet, SHF-Microfluidics, 12–14 December 2006, Toulouse, France.
65. V. Peykov, A. Quinn and J. Ralston, *Colloid Polymer Science*, Vol. 278, pp. 789–793, 2000.
66. A. Quinn, R. Sedev and J. Ralston, *Journal of Physical Chemistry B.*, Vol. 109, pp. 6268–6275, 2005.
67. M. Vallet, M. Vallade and B. Berge, *The European Physical Journal B.*, Vol. 11, pp. 583–591, 1999.
68. H.J.J. Verheijen and M.W.J. Prins, *Langmuir*, Vol. 15, no. 20, pp. 6616–6620, 1999.
69. J. Berthier, Ph. Dubois, Ph. Clementz, P. Claustre, C. Peponnet and Y. Fouillet, *Sensors and Actuators A: Physical*, Vol. 134, no. 2, pp. 471–479, 2007.
70. J. Berthier, Ph. Clementz, O. Raccurt, D. Jary, P. Claustre, C. Peponnet and Y. Fouillet, *Sensors and Actuators A*, Vol. 127, pp. 283–294, 2006.
71. P.K. Wong, T.H. Wang, J.H. Deval and C.M. Ho, *Transactions on Mechanotronics*, Vol. 9, pp. 366–376, 2004.
72. S.K. Fan, C. Hashi and C.J. Kim, IEEE Conf. MEMS, Kyoto, Japan, p. 694, January 2003.
73. E. Dussan, *Annual review of Fluidic Mechanical*, Vol. 11, p. 371, 1979.
74. J.J. Hegseth, N. Rashidnia and A. Chai, *Phys. Rev. E*, Vol. 54, pp. 1640–1644, 1996.
75. G.I. Taylor, *Proc. Roy. Soc.*, Vol. CCLXXX, pp. 501–509, 1964.
76. A.J. de Mello, M. Habgood, N.L. Lancaster, T. Welton and R.C. Wooton, *Lab Chip*, Vol. 4, pp. 417–419, 2004.
77. V. Srinivasan, V.K. Pamula and R.B. Fair, *Analytica Chimica Acta*, Vol. 507, pp. 145–150, 2004.
78. V. Srinivasan, V.K. Pamula and R.B. Fair, *Lab Chip*, Vol. 4, pp. 310–315, 2004.
79. P. Trinder, *Ann. Clin. Biochem.*, Vol. 6, p. 24, 1969.
80. T. Taniguchi, T. Torii and T. Higuchi, *Lab Chip*, Vol. 2, pp. 19–23, 2002.
81. U.M. Lindström, *Chem. Rev.*, Vol. 102, pp. 2751–2772, 2002.
82. V.K. Pamula, V. Srinivasan, H. Chakrapani, R.B. Fair and E.J. Toone, Proc. IEEE 2005, Int. Conf. Micro Electro Mech. Syst., 18th, January 30–February 03, p. 722–725, 2005.
83. P. Paik, V.K. Pamula and R.B. Fair, *Lab Chip*, Vol. 3, pp. 28–33, 2003.
84. G.A. Artamkina, M.P. Egorov and I.P. Beletskaya, *Chem. Rev.*, Vol. 82, pp. 427–459, 1982.
85. P. Grieco and A. Bahas, *Tetrahedron Lett.*, Vol. 29, pp. 5855–5858, 1988.
86. P. Grieco and S.D. Larsen, *Organic Syntheses*, Vol. 68, p. 206, 1990.
87. P. Gricco and S.D. Larsen, *Organic Syntheses, Coll.*, Vol. 8, p. 31, 1993.
88. A.R. Katritzky, S. Rachwal and B. Rachwal, *Tetrahedron*, Vol. 52, pp. 15031–15070, 1996.

89. A. Kiselyov and R.W. Armstrong, *Tetrahedron Lett.*, Vol. 38, pp. 6163–6166, 1997.
90. V. Jurcik and R. Wilhelm, *Org. Biomol. Chem.*, Vol. 3, pp. 239–244, 2005.
91. P. Dubois, G. Marchand, S. Gmouh and M. Vaultier, *Chemistry—A European Journal*, Vol. 13, no. 19, pp. 5642–5648, 2007.

8 Acoustic Methods for Manipulating Droplets☆

8.1 Introduction

The recent development of microfluidic systems in some sense marks a rebirth of the old discipline of hydrodynamics, however, on a much smaller scale than the one we are familiar with in our daily lives. Hence, the laws of hydrodynamics still apply, but need to be adjusted to the changed physical environment of the microscale. The regime for microfluidics is governed by strong interactions between the fluid and the outer world, be it containers, tubes, or trenches. Here, surface effects like wettability, surface tension, adhesion and cohesion, and others lead to a completely different fluid behavior than the one we are used to. Usually, this new behavior is physically described by the ratio of inertial terms and viscous terms in the equations of motion of a fluid volume element. This ratio is usually referred to as Reynold's number. Flow at low Reynolds numbers in the macro world is usually restricted to very viscous fluids, like ice in a glacier, for example. Low Reynolds numbers are related to laminar flow and the absence of turbulence which has enormous consequences for chemical and biochemical reactions. Interestingly, the microscale behavior of fluids and particles is not that unusual even in our macroscopic world, taking the fact that the vast majority of living species on our planet has a size of only a few micrometers, bacteria, namely. They live in a world where the classical laws of hydrodynamics are not really applicable. In a famous paper, Purcell describes the fate of such microbes in a microfluidic environment [1], describing all the obstacles these creatures experience as compared to an average size fish, but also outlining the strategies life developed to still be comfortable in the micro world.

Microfluidic systems miniaturize chemical and biological laboratory processes to a sub-millimeter scale. Reducing the dimensions of macroscopic biological or chemical laboratories is advantageous, as the small scale allows for the integration of various processes on one chip analogous to integrated microelectronic circuitry. Also, the required reagent volumes are reduced which saves material costs and allows the reactions to be carried out at high sample concentrations. These high concentrations drive the reactions towards the products' side and accelerate the kinetics. Finally, miniaturization results in enhanced precision by providing homogenous reaction conditions.

☆This chapter was written by Thomas Frommelt, Matthias Schneider (Chair for Experimental Physics I, University of Augsburg, D-86519 Augsburg, Germany) and Achim Wixforth (Advalytix AG, D-85649 Brunnthal, Germany).

Several approaches to realize microfluidic systems have been reported in the literature. In this context, the term lab-on-a-chip has emerged, already outlining the use and typical applications of such systems. In most cases the reagents are transported in channels or capillaries with typical diameters ranging from 10 to 500 μm. These channels can be fabricated by deep etching processes on appropriate substrates such as glass, quartz, or silicon [2]. Alternatively, hot embossing is used to structure polymer substrates. The channels are capped by anodic bonding or glue processes. Generally, these systems do not allow the reagents to be handled separately, as the channels need to be completely filled in order for the fluidics to work properly. Therefore, the application of these systems is restricted to continuous flow processes rather than batch processes as found in macroscopic laboratories.

The number of different pumping mechanisms is even greater than the number of substrate materials employed in microfluidics. Some pumping units are not an integral part of the chip and must be linked with appropriate tubes or pipes. They use, e.g., piezoelectric actuation or mechanically moving parts to drive the reagents through the channels. Others take advantage of the small dimensions of the microfluidic channel itself [3]. The chemical potentials of the channel walls and that of the liquid inside differ considerably, and a space charge region forms at the interface. A voltage applied along the channel induces a flow at the space charge region which drags along the liquid closer to the center of the channel. This electrokinetic effect works only for narrow channels and relatively high voltages. Fluidic motion can also be induced by spatially modulating the wetting properties of a substrate. For aqueous solutions, this can be achieved by patterning the substrate with hydrophobic and hydrophilic regions. The techniques used to realize such a modulation of the wetting properties include micro-contact printing [4], vapor deposition, and photolithography [5,6]. Aqueous solutions tend to cover the hydrophilic regions and avoid the hydrophobic areas. A guided flow can be achieved [7] by changing the wetting properties with time. For example, illumination can induce a guided motion of liquids as the free energy of the surface changes locally under illumination [8]. Other pumping mechanisms include peristaltic pumps based on thin membrane [9], or polymer films [10] with a controlled deformation creating a guided flow along microchannels [11].

In this chapter, we present a novel approach of miniaturized liquid handling which does not displace the reagents in channels but rather on the surface of piezoelectric substrates. In our case interdigital transducers excite surface acoustic waves (SAW) which transfer momentum to liquids placed on the chip. The reagents can be manipulated either as discrete droplets or by streaming patterns induced in macroscopic volumes. The technology allows both batch and continuous processes to be carried out at high speed. The most important feature, however, is the programmability of the chip as different assay protocols can be realized with the same chip layout.

8.2 Digital Microfluidics

Watching a "typical" chemist or biologist do laboratory work, the first thing that strikes the observer is that they usually operate discrete fluid volumes. They pipette a small amount of a fluid reagent from a reservoir, load a test tube or a beaker with it, pipette a second reagent volume, add it to the first one, stir it, heat it, etc. Unlike the fluid handling in a chemical plant, nearly all the fluid handling is done in a "digital" fashion. The reason is obvious. Closed volumes connected by tubes and hoses ("analog fluidics") are much more difficult to meter, and cross contamination, intermixing, inter-diffusion, and similar difficulties are easily avoided. Moreover, the flow resistance in tubes strongly depends on the liquids actually used, making the precise pumping and dispersion of the liquid a cumbersome task in many cases. A multiple connected network of tubes and hoses represents a complex, very nonlinear system, where changes at one end of the network have a notable impact and crosstalk to the fluid/pressure behavior at the other end. For this reason, many lab-on-a-chip microfluidic applications have emerged, where the small amounts of fluids are defined as small, independent quantities.

At this point, droplet based fluidic systems enter the scene. Once we have a platform at hand that is able to control the movement and the handling of small fluid volumes in the form of droplets, and we find a way to deliberately actuate these droplets, we have a system that can be truly regarded as a miniaturized version of a macroscopic laboratory. Small droplets serve as micro test tubes and reactor vessels, being completely independent from each other and hence reducing the risk of cross contamination and intermixing to a minimum. These droplets, if small enough, do not need any containers, as surface tension acts as a potential well holding the fluid in place.

The next task on the route towards a laboratory on a chip is the fluid actuation. To be able to perform complicated experiments involving different reagents, some patterned surfaces are needed on the chip. This includes the flat world analogue of channels, tubes, reservoirs, mixing chambers, and similar building blocks usually employed to guide, contain, or process liquids in a fluidic network. Hence, we need areas on the chip along which liquids can be moved in a predetermined manner (Fig. 8.1). By a chemical modification of parts of the chip surface (i.e., silanization employing an octa-decyl-trichlorsilane, OTS based surface chemistry) we are able to create patterns of preferred wettability (hydrophilic regions), being separated by regions of surface chemistry, where fluids are repelled (hydrophobic areas). Employing lithographic techniques borrowed from semiconductor microelectronics, we thus can create completely flat, two-dimensional fluidic networks, where liquids are confined to virtual tracks, reservoirs, and reaction chambers by surface tension alone.

In Fig. 8.2, we depict some of such self-assembled virtual potential wells for fluids on a surface. Photolithographic techniques have been employed to create "containers" for the smallest amounts of liquid, having shapes like a hexagon (left) or a "tube" with a narrowing (acting as a pressure driven valve). Given

Figure 8.1 Droplets on a patterned chip surface: if the droplets are small enough and the surface consists of areas of preferred wettability (hydrophilic anchors), surrounded by hydrophobic regions, the droplets are stabilized at a predetermined volume. They act as individual "virtual test tubes" for microscale laboratories. (Courtesy Advalytix).

Figure 8.2 Photolithographically defined surface modification to modulate its wettability. This way, virtual fluidic tracks are created to confine small amount of liquid to predetermined geometries or to guide a SAW driven droplet along a predetermined path on the chip surface. From [23]. © 2003 Elsevier, reprinted with permission.

this surface functionalization, either closed fluid volumes or single droplets may be acoustically guided along predetermined pathways along the surface of the chip.

8.2.1 Acoustic Streaming

In this chapter, we introduce the basics that are necessary to understand the acoustic actuation scheme for fluid manipulation on a chip. We start with a short and comprehensive description of surface acoustic waves on a solid, as we employ SAW to interact with a fluid, induce internal acoustic streaming, or supply enough force to actuate a droplet as a whole. This SAW-fluid interaction

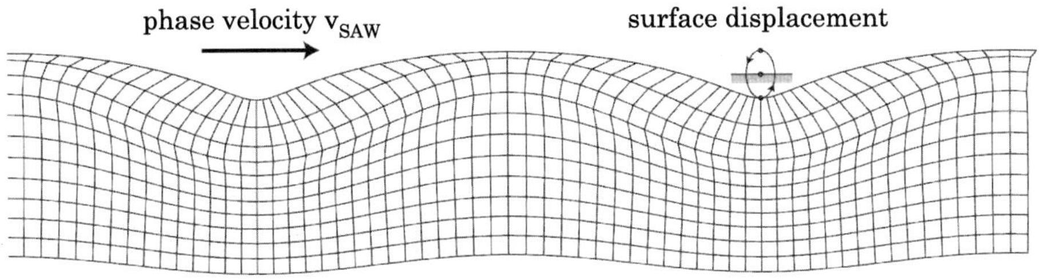

Figure 8.3 Sketch of a surface acoustic wave propagating on a piezoelectric substrate. Typical wavelengths are in the micrometer range, typical amplitudes less than a nanometer. Note the elliptically polarized displacement of the surface. (Courtesy T. Frommelt).

is presented in the next sections; with special emphasis given on the internal acoustic streaming in a small amount of fluid and droplet actuation, on the other hand.

8.2.1.1 Surface Acoustic Waves on a Solid

In a crystal, or more generally spoken in a solid, optical and acoustic lattice vibrations are observable. Both types of lattice vibrations in solids are distinguished from each other according to their vibrational modes. Optical vibrations show a dispersed motion of individual atoms and, in case each individual atom carries a charge of opposite sign, such vibrations can couple to electromagnetic waves, i.e., light. In acoustic modes, atoms move commonly as in the acoustic sound waves of long wavelength in air. Another major difference between optical and acoustic modes is their dispersion. Acoustic phonons exhibit a linear dispersion close to the zone center. At the zone boundary the group velocity vanishes such that lattice vibrations do not propagate and standing waves occur. Propagating lattice vibrations generate waves, which can be typically bulk waves or surface waves, depending on the boundary conditions [12]. The special case of surface acoustic waves was theoretically described in 1885 by Lord Rayleigh from the perspective of seismology [13].

Meanwhile, reduced to the significantly smaller nano-scale, acoustic waves found their way into much friendlier fields: SAW devices are widely used for RF signal processing and filter applications, and became a huge industry in mobile communication [14]. SAW devices have been around for years in communication circuitry—every cell phone has filters using the effect. An electrical signal fed into so-called transducers on the surface of a piezoelectric chip is converted into a deformation of the crystal underneath. Given the right frequency of the signal, a mechanical wave is launched across the chip. In Fig. 8.3, we sketch a snapshot of a SAW propagating on a solid.

In the recent past, SAWs have also been used to act in a completely different way than for filtering and signal processing, just by converting electrical signals into mechanical vibrations and vice versa. Excited on piezoelectric substrates, they are accompanied by large electric fields. Those electric fields are traveling at the speed of sound of the substrate (approx. 3000 m/s) having the same spatial and temporal periodicity as their mechanical companions. Charges at or close to the surface are coupling to these electric fields, and currents are induced within a conducting layer. Nearly 20 years ago, SAW was introduced to study the dynamic conductivity $\sigma(\omega, k)$ of low-dimensional electron systems in high magnetic fields and at low temperatures. It turns out that the interaction between a SAW and the mobile charges in a semiconductor is strongest for very low sheet conductivities as they are observed, e.g., in the regime of the quantum Hall effect [15]. However, SAW can not only be used to probe the properties of quantum systems, but also to deliberately alter some of them as SAW represent a spatially modulated strain and stress field accompanied by strong electric fields in a solid, and propagating at the speed of sound. Such an interaction between SAW and the optical properties of a semiconductor quantum well led us to the discovery that photo-generated electron hole pairs in a semiconductor quantum well can be spatially separated under the influence of a SAW mediated electric field. This, in turn has an enormous impact on the photoluminescence (PL) of the semiconductor. We were able to show that the PL not only is quenched under the influence of a SAW, but also can be reestablished at a remote location on the sample and after a certain delay time [16]. Further studies include the acoustic charge transport and the creation of dynamically induced electron wires, as well as the study of nonlinear acoustic interaction with low dimensional electron systems in semiconductors [17].

However, as will be outlined below, the piezoelectric effect is usually only a small contribution to the elastic properties of a solid: Most of the energy propagating in a SAW (usually more than 95%) is of mechanical nature. Hence, not only electrical interactions as described above, but also mechanical interactions are a possible scope for experimental investigations. Having wavelengths of a few micrometers and amplitudes of about only a nanometer, however, the forces and electric fields within the nano-quake are sufficient to have a macroscopic effect. Any piece of matter at the surface along the way of a SAW experiences its vibrating force: viscous materials like liquids absorb a lot of their energy. It turns out that the interaction between a SAW and a liquid on top of the substrate surface induces an internal streaming, and, as we will point out below, at large SAW amplitudes this can even lead to a movement of the liquid as a whole.

8.2.1.2 Elastic Waves in a Solid

Imagine an isotropic and homogenous medium (density ρ) without external forces and piezoelectricity. An infinitesimal volume element $dV = dx_1\, dx_2 dx_3$ experiences only the forces exerted by neighboring volume elements. Those forces can be described in terms of the symmetrical strain tensorT_{ij} [18]. Here, the T_{ii} are describing strain or stress forces, and $T_{ij,i \neq j}$ represents the shear forces in the system. Neglecting gravity, the equation of motion for such an infinitesimal volume element can be written as

$$\rho \frac{\partial^2 u_i}{\partial t^2} = \frac{\partial T_{ij}}{\partial x_j}. \tag{8.1}$$

Here, as throughout this chapter, the so-called Einstein sum convention has been used, indicating summation over double indices. In (8.1) u_i is the mechanical deformation in the direction x_i. The actual deformation of an infinitesimal volume element dV is described by the symmetrical deformation tensor S_{kl} [14]

$$S_{kl} = \frac{1}{2}\left(\frac{\partial u_k}{\partial x_l} + \frac{\partial u_l}{\partial x_k}\right). \tag{8.2}$$

Strain and deformation are as usual connected via the elastic constants c_{ijkl}:

$$T_{ij} = c_{ijkl} S_{kl}. \tag{8.3}$$

This is the three-dimensional version of Hooke's law, which for the case of a piezoelectric crystal has to be extended by an additional term in the strain, expressing the piezoelectricity. Together with the equation describing the dielectric displacement$D$, this results in the equations of constitution for a piezoelectric crystal

$$\begin{aligned} T_{ij} &= c_{ijkl} S_{kl} - p_{kij} E_k \\ D_i &= p_{ijk} S_{jk} + \varepsilon_{ij} E_j. \end{aligned} \tag{8.4}$$

Here, E denotes the electric field strength, p the piezoelectric tensor, and ε the dielectric tensor for the material under consideration.

The symmetric nature of T_{ij} and S_{kl} requires that $c_{ijkl} = c_{jikl}$ and $c_{ijkl} = c_{ijlk}$. From energy conservation it can moreover be shown that $c_{ijkl} = c_{klij}$ [14]. Given these symmetry relations, it turns out that there are 21 independent elastic constants remaining for an arbitrary crystal structure. As can be seen from the above, the piezoelectric properties of a given material are also influenced by electromagnetic quantities as described by Maxwell's equations. However, it turns out that this purely electromagnetic influence is only weak; hence it is in most cases justified to approximate the electric field by the gradient of a scalar potential $E_i = -\partial\varphi/\partial x_i$. Using this approximation, the constituting equations (8.4), the definition of the deformation tensor (8.2), the equation of motion (8.1), and Maxwell's equationdiv$D = 0$, we finally end up with a system of four wave equations for the three components of the mechanical elongations

u_i and the electrical potential φ

$$\begin{aligned} \rho\frac{\partial^2 u_j}{\partial t^2} - c_{ijkl}\frac{\partial^2 u_k}{\partial x_i \partial x_l} - p_{kij}\frac{\partial^2 \varphi}{\partial x_i \partial x_k} &= 0 \\ p_{ikl}\frac{\partial^2 u_k}{\partial x_i \partial x_l} - \varepsilon_{ik}\frac{\partial^2 \varphi}{\partial x_i \partial x_k} &= 0. \end{aligned} \tag{8.5}$$

In general, this results in three different wave modes for an infinite crystal: one longitudinal mode and two orthogonally polarized transversal modes (shear waves). For the special case of a plane bulk wave in an infinitely large yz-cut $LiNbO_3$ crystal, the materials being discussed here, the equations further simplify. If the propagation direction is denoted by $x_3 = z$, one obtains for the longitudinal mode

$$\begin{aligned} \rho\frac{\partial^2 u_z}{\partial t^2} - c_{33}\frac{\partial^2 u_z}{\partial z^2} - p_{z3}\frac{\partial^2 \varphi}{\partial z^2} &= 0 \\ p_{z3}\frac{\partial^2 u_z}{\partial z^2} - \varepsilon_{zz}\frac{\partial^2 \varphi}{\partial z^2} &= 0. \end{aligned} \tag{8.6}$$

Elimination of φ results in the following wave equation

$$c_{33}\left(1 + \frac{p_{z3}^2}{\varepsilon_{zz}c_{33}}\right)\frac{\partial^2 u_z}{\partial z^2} = \rho\frac{\partial^2 u_z}{\partial t^2}. \tag{8.7}$$

For the simpler case of a non-piezoelectric material, equation (8.7) reduces to

$$c_{33}\frac{\partial^2 u_z}{\partial z^2} = \rho\frac{\partial^2 u_z}{\partial t^2} \tag{8.8}$$

which makes it obvious to express the effect of the piezoelectricity in terms of an *effective* elastic constant

$$c_{33}^* = c_{33}\left(1 + \frac{p_{z3}^2}{\varepsilon_{zz}c_{33}}\right) = c_{33}\left(1 + K^2\right). \tag{8.9}$$

K^2 in Eq. (8.9) is the so-called electromechanical coupling coefficient, in general a tensor of fourth rank. From the above it becomes clear that the effect of piezoelectricity "stiffens" a material of given elasticity by the amount of K^2. The reason for this "piezoelectric stiffening" is additional restoring forces apart from the elastic forces, caused by a charge separation under deformation of a piezoelectric material. These additional restoring forces lead to an increase of the sound velocity for a piezoelectric medium, if compared to the corresponding non-piezoelectric substrate

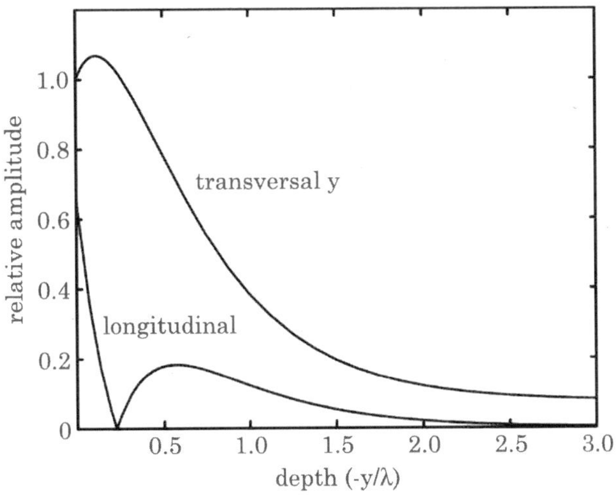

Figure 8.4 Normalized surface elongations for a Rayleigh SAW on y-cut $LiNbO_3$ [19]. (Courtesy T. Frommelt).

$$v^* = \sqrt{\frac{c_{33}^*}{\rho}} = \sqrt{\frac{c_{33}\,(1+K^2)}{\rho}} \approx v\left(1+\frac{K^2}{2}\right) \tag{8.10}$$

where $v = \sqrt{c_{33}/\rho}$. This characterizes the role and the definition of the electromechanical coupling coefficient K^2

$$K^2 \approx 2\frac{\Delta v}{v}. \tag{8.11}$$

Here, $\Delta v = v^* - v$ denotes the renormalization of the sound velocity if piezoelectricity is taken into account. The electric field related terms in equation (8.4) can be strongly modified if the substrate is, for example, covered by a thin conducting metal film. A varying conductivity at or close to the surface on which a SAW is propagating strongly influences both the amplitude of the SAW and its sound velocity. The difference in sound velocities of a "free" and a "shortened" surface is then given by $K^2/2$. This leads to the interactions between a SAW and quantum systems shortly described above.

8.2.1.3 Rayleigh Waves

So far, we have only discussed the case of volume or bulk waves. The special case of a surface acoustic wave can be obtained from the above calculations if the boundary conditions of a stress-free surface are considered: $T_{31} = T_{32} = T_{33} = 0$ for $x_3 = 0$. Here, the components of the stress normal to the surface vanish. The important case of the Rayleigh wave is visualized in Fig. 8.4, where we depict the surface elongations along a cut through the substrate (y-cut $LiNbO_3$ in this case). As can be seen, both elongation components decay in a specific manner with depth into the material. The exact shape of the decay, however, depends on the actual material, crystal cut, and anisotropy of the substrate under consideration.

A Rayleigh SAW thus exhibits an elliptically polarized particle movement directly at the surface of the crystal (see Fig. 8.3). The motion of a volume element is *retrograde*, i.e., moving against the direction of SAW propagation, and within the sagittal plane. As it turns out, the interaction between the SAW and a fluid on top of the substrate surface is governed by the transversal elongation of the surface, whereas the interaction between a SAW and an electrically conducting layer at the surface is mainly governed by the lateral fields [15].

8.2.1.4 SAW Excitation

Rayleigh SAW on a piezoelectric solid can be excited in a particularly simple way: a lithographically designed electrode system in the shape of an interdigitated grating is deposited on top of the surface, using semiconductor technology. The period of the fingers determines the frequency of the SAW, which depends on the SAW velocity of the given crystal orientation and cut, respectively. There are many different designs for such interdigital transducers (IDT) for special purposes in high frequency signal processing or sensor applications. We only want here to give an idea of their operation and restrict ourselves to the simplest cases. For a fixed frequency operation, IDT consist of a comb-like metal structure where every second metal finger is connected. Application of a radio frequency (RF) signal to the transducer causes a periodic crystal deformation underneath, via the inverse piezo effect. In first order, the resonant frequency of such an IDT is then given by:

$$f = \frac{v_{\mathrm{SAW}}}{p} \tag{8.12}$$

where p denotes the periodicity of the IDT, as sketched in Fig. 8.5 (left). If two such transducers are deposited onto the same substrate, a so called delay line is formed. Here, an RF signal fed into one of the transducers is converted into a SAW, propagating across the substrate and, at the second transducer, is re-converted into an RF signal via the piezoelectric effect. As the IDT and the delay line represent a resonant device, it exhibits a specific frequency response which is heavily exploited for RF signal processing. The bandwidth Δf of a delay line is basically given by the IDT layout and the number of finger pairs.

As can be suspected from the sketch presented in Fig. 8.5, a SAW delay line is extremely sensitive to surface contamination. Both the transmitted SAW amplitude and the frequency response of the device are easily disturbed by any change of the boundary conditions at the free surface. This is true for all

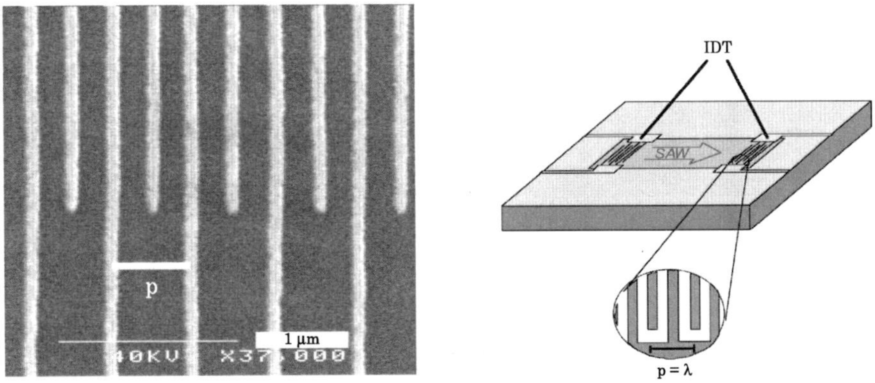

Figure 8.5 Left: SEM micrograph of an IDT on a piezoelectric substrate. The surface acoustic wavelength λ is given by the periodicity p of the IDT and the SAW velocity v, respectively. Right: a delay line formed by two IDT on a piezoelectric substrate.

disturbances able to interact with the SAW. Most prominent, of course, is a mechanical load of the surface by contamination with a fluid, dust, or the like.

This is why the SAW device industry puts a lot of effort into proper sealing of SAW devices to avoid time unstableness and deterioration. Any piece of matter at the surface along the way of a SAW experiences its vibrating force: viscous materials like liquids absorb a lot of energy, which indicates a strong interaction between the fluid and the SAW. This interaction induces internal streaming and eventually droplet actuation.

8.2.2 Internal Streaming

The interaction between a SAW and a fluid on the surface of the substrate on which the SAW is propagating represents a delicate, though extremely interesting hydrodynamical problem. To approach it, we start from the Navier–Stokes-equation, describing the flow in a hydrodynamic system. It is a nonlinear equation in the velocity components, reading

$$\rho\frac{\partial\vec{v}}{\partial t} + \rho\left(\vec{v}.\text{grad}\right)\vec{v} = -\text{grad}p + \eta\Delta\vec{v} + \vec{f}. \tag{8.13}$$

Here, $\vec{v}$ is the velocity field of the flow, η the viscosity, and ρ the mass density of the fluid. p denotes the pressure that a fluid element experiences from its surroundings and $\vec{f}$ is an externally applied body force driving the flow. The term $\rho\left(\vec{v}.\text{grad}\right)\vec{v}$ describes the inertia of the fluid element and $\eta\Delta\vec{v}$ marks the viscous term. In microfluidics, the nonlinear term in the Navier–Stokes equation is often much smaller than the viscous term and can therefore be neglected. The validity of this approximation depends on the ratio of inertial forces and viscous forces, which is known as the dimensionless Reynolds number Re

$$Re = \frac{\rho v^2}{l}\frac{l^2}{\eta v} = \frac{\rho v l}{\eta}. \tag{8.14}$$

In this definition, l denotes a typical length scale in the system under consideration, for example, the channel diameter. For a microfluidic system, Re is usually a small number, indicating the little importance of inertia in the problem. The most prominent consequence thereof and hence the most important difference to a macroscopic fluid volume is the lack of turbulent flows in a microfluidic system. The transition between turbulent and laminar flow usually occurs at a threshold Reynold's number $Re \approx 2000$. For a typical microfluidic system, given $l \approx 10$ μm, $v \approx 100$ μm/s, and the material parameters of water ($\rho = 10^3$ kg/m^3, $\eta = 10^{-3}$ kg/m/s), we end up with $Re \approx 0.001$. Solutions of the Stokes equations exhibit several remarkable properties. First, the flow is laminar and reversible. The streamlines of a laminar flow are locally straight and parallel. The global streamline pattern, however, may be complicated. Material lines of the fluid, when evolved together with the flow, may fold into even more complicated structures. If the sign of the flow is reversed, however, the complicated material lines will be restored into their initial shape. This behavior is due to the linearity of the Stokes equations and has grave implications for the ability of small living organisms trying to propel themselves through the fluid [1].

For a lab-on-a-chip application, this usually causes severe problems. For instance, mixing of two fluids or stirring a liquid to enhance homogeneity or to speed up a chemical reaction is a very difficult task for a purely laminar flow system. Also, pumping of a low Reynold's fluid is difficult, as the interaction of the fluid with the vessel walls is predominant. We will show that many of these obstacles can be overcome by employing acoustic methods.

To model the interaction between a SAW and a fluid, we treat the SAW as an array of point sources f_i of force, acting on a volume element of the fluid above the surface. This view is again justified, as it turns out [1] that the fluid flow for the given length and velocity scales is found to be over-damped. If the driving force f_i is switched off, then the flow ceases nearly immediately.

Acoustic streaming in classical, macroscopic systems has been treated in detail by several authors in the past, and for a detailed description of it, we wish to refer the reader to the works by Eckart [20] and by Nyborg [11] as a starting point. The general framework of their theories on acoustic streaming is the following: The effect takes place on two separate time-scales. On the short time-scale, which in our case is the time-scale of the SAW, the motion of the interface between the substrate and the fluid is resolved. On this scale, water does not perform an incompressible flow but has to be described by the equations of compressible fluid dynamics which consider the mass-density of the fluid not as a constant but as part of the problem. Compared to the stress tensor in the previously formulated Navier–Stokes equations, the stress then contains an additional term accounting for the compressibility. It is this compressibility, which gives rise to higher order terms in the pressure and velocity, finally leading to a DC term in the pressure distribution and the acoustic streaming. They further rely on conservation of mass, and relate density and pressure. The SAW, which is a wave on the substrate surface only, then causes a longitudinal sound wave propagating from the substrate into the fluid. This sound wave is subject to viscous damping along its path in the fluid. Furthermore, the nonlinear terms in the compressible Navier–Stokes equation give rise to higher harmonic pressure and velocity fields, which oscillate with all multiples of the initial frequency. In a simple, one-dimensional representation, the sound induced pressure is given by

$$p(x,t) = p_{\max} \cos\left[\omega\left(t - x/v\right)\right] + 2\omega \cos^2\left[\omega\left(t - x/v\right)\right] + p_0. \tag{8.15}$$

Here, ω denotes the frequency of the SAW, v the sound velocity in the fluid, and p_0 a static pressure. The time average of the first and last terms in equation (8.15) equals zero, but the $\cos^2$ term results in a time independent pressure. If attenuation is taken into account, this leads to a pressure gradient and the streaming velocity field which is observed in the experiment.

It turns out that due to phase matching arguments, the longitudinal sound wave in the fluid is obeying a diffraction law similar to the ones known from ray optics, leading to a diffraction of the wave from the parallel motion along the surface (SAW) to an oblique entrance into the fluid above the surface. The angle of diffraction turns out to be given by the different sound velocities in the substrate and the fluid, respectively

$$\sin\left(\theta_R\right) = \frac{v_{\text{fluid}}}{v_{\text{SAW}}}. \tag{8.16}$$

The diffraction effect can be observed in an experiment, where a SAW generating IDT is placed directly underneath a fluid reservoir, containing water and a small amount of ink to visualize the acoustic streaming in the container. In Fig. 8.6, we depict the result of such an experiment.

Fig. 8.7 shows a similar experiment performed in a small droplet. Here, a small amount of fluorescent dye was deposited on the surface of the chip, then a small droplet (approx. 50 nl) was pipetted onto this dye spot. A SAW impinging from the left quickly dissolves the dye and distributes it throughout the droplet. Apart from the beauty of these pictures, the experiment has an enormous impact for future biochips, as it indicates the possibility of disolving chemical reagents in a small virtual test tube by virtues of a SAW. This would enable the manufacturer of the chip to ship it preloaded with reagents. Only a suitable buffer needs to be applied to the chip and the SAW driven fluidics do the rest.

This typical fingerprint of the SAW driven acoustic streaming can also be theoretically understood and visualized when we look at a finite element simulation of the pressure distribution in a SAW driven fluid. If the SAW is modeled as an array of point sources for the forces f_i, being damped in

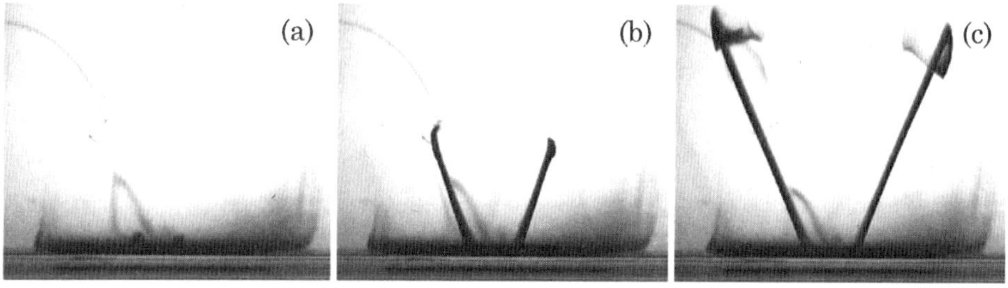

Figure 8.6 Fluid jets are caused by the actuation of an IDT at the bottom of a well covered by a small amount of dye. The fluid is contained in a cusp; the part shown in the pictures is approximately 18 mm wide. Between the snapshots, time spans of 0.16 s and 0.48 s have elapsed. The long and thin shape of the colored fluid fingers indicates the direction in which the SAW acts on the fluid (Courtesy Advalytix).

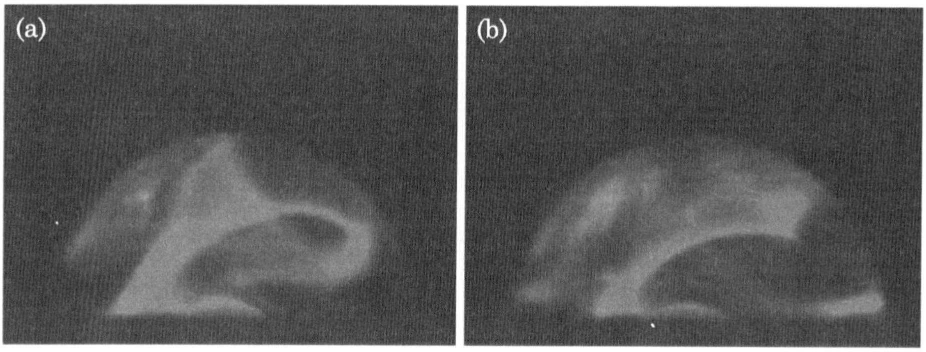

Figure 8.7 SAW induced internal acoustic streaming in a small droplet. To visualize the streaming, prior of the droplet deposition, a small amount of fluorescent dye had been deposited on the chip surface. The SAW impinging from the left quickly dissolves this dye. From [23]. © 2003 Elsevier, reprinted with permission.

the direction of SAW propagation into the fluid reservoir, a Huygens like superposition leads to plane waves in the fluid (Fig. 8.8). Phase matching results in a Rayleigh angle θ_R as given in equation (8.16).

SAW induced acoustic streaming can be exploited for efficient pumping of closed fluid volumes on microfluidic chips, as well as for efficient stirring. The pumps in this case are integrated to the chip, have no moving parts, and are electrically addressable.

8.2.3 Droplet Actuation

For high SAW amplitudes, not only internal acoustic streaming within a small fluid volume is observed, but an actuation of a droplet as a whole can be achieved. This SAW driven droplet actuation if combined with a surface

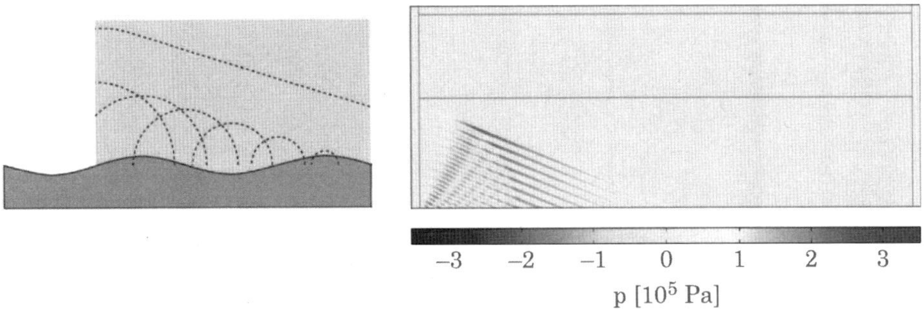

Figure 8.8 Finite element calculations of the pressure distribution in a substrate-fluid system (right) and the simple Huygens model (left) (courtesy T. Frommelt).

modification resulting in "virtual tracks" or lateral confinement of the droplet by an appropriate shaping of the SAW wavefront opens a wealth of applications for a lab-on-a-chip system. A typical example for SAW driven droplet actuation is shown in Fig. 8.9. In this figure, we show a series of snapshots, where a small (approx. 50 nl) droplet of water sitting on a chip surface is viewed from the side. An intense SAW is impinging from the left, being pulsed at a certain time sequence. Once the SAW pulse enters the droplet, violent internal streaming within the droplet is observed (cf. Fig. 8.7; not visible in Fig. 8.9) which is accompanied by a dynamic deformation of the droplet. Here, too, a relic of the Rayleigh diffraction angle can be observed.

The theoretical description of this droplet actuation represents a particular challenge, as we obviously have to deal with extremely complex boundary conditions in this case. The fluid volume is finite, and is only confined by surface tension effects, except for the base on which it is resting. Very recently, however, the problem has been solved leading to a beautiful agreement between theory and experimental observations [21]. In this last reference, a finite-element algorithm for computing free-surface flows driven by arbitrary body forces is presented.

Schindler et al. represent the free surface shape by the boundaries of finite elements that move according to the stress applied by the adjacent fluid. Additionally, the surface is modeled to minimize its free energy and by adapting its curvature to balance the normal stress at the surface. The numerical approach consists of the iteration of two alternating steps: the solution of a fluidic problem in a prescribed domain with slip boundary conditions at the free surface and a consecutive update of the domain driven by the previously determined pressure and velocity fields. For a Stokes problem the first step is linear, whereas the second step involves the nonlinear free-surface boundary condition. This algorithm is justified both by physical and mathematical arguments. It is tested in two dimensions for two cases that can be solved analytically. For the details of this nice piece of calculations, we again refer the reader to the original publication [21]. However, the results are worth being summarized in Fig. 8.10.

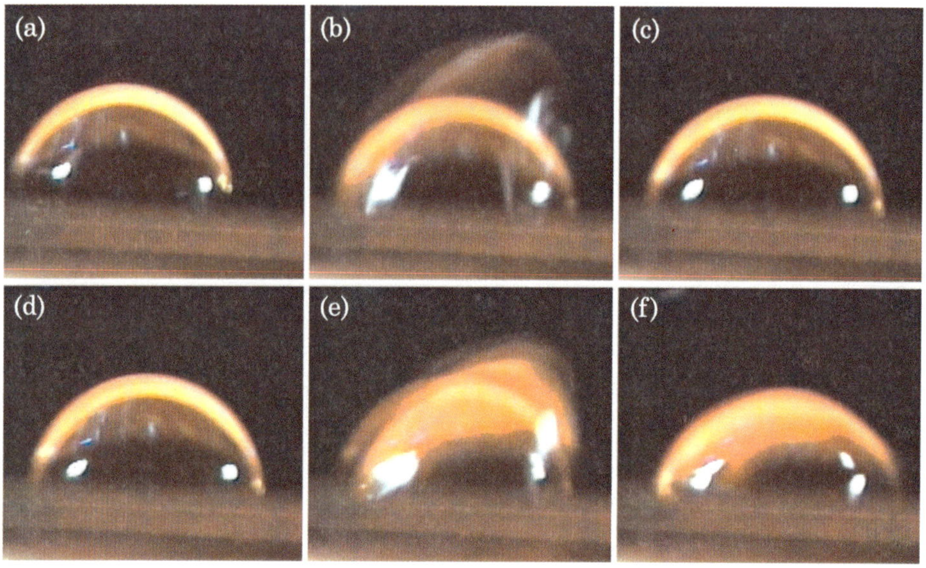

Figure 8.9 A "jumping" droplet containing 50 nl water, on a flat substrate. The snapshots are taken at consecutive times. In the center pictures (b) and (e), a powerful pulsed SAW hits the droplet from the left, leading to a severe deformation. When the SAW is turned off, the droplet relaxes into its initial spherical shape at a shifted downwind position to the right. From [23]. © 2003 Elsevier, reprinted with permission.

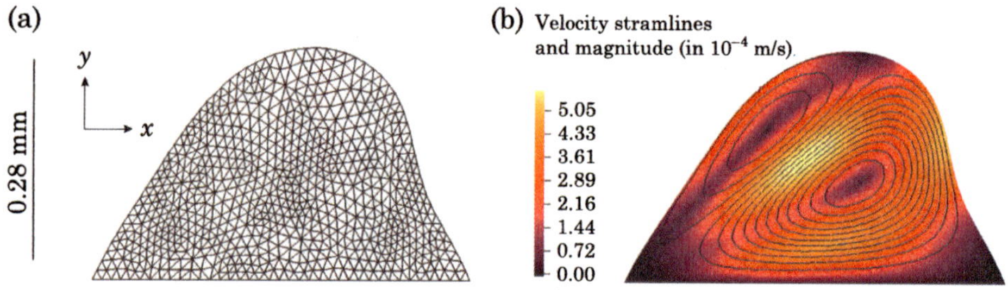

Figure 8.10 Finite element calculation of a SAW driven droplet deformation and acoustic streaming within [21]. Note the similarity of the resulting streaming pattern to the one experimentally observed in Fig. 8.7. Reused with permission from [21]. © 2006 American Institute of Physics.

8.3 Example 1: Acoustic Mixing

Efficient mixing of small amounts of fluid is probably the most challenging task for microfluidic devices. For liquids confined to small volumes, interaction between the fluid and the walls of the container become dominant, as the surface-to-volume ratio becomes very large. In general, this increased liquid–wall interaction is parameterized by Reynold's number, usually being a small quantity in microfluidic systems. A low Reynold's number, however, is equivalent to an increase of the apparent viscosity. As turbulent streaming only occurs at high Reynold's numbers, this in turn leads to slow mixing in microfluidic systems, in which only laminar flow processes are possible. Hence, the only way for small fluid volumes to mix effectively is by diffusion. Here, the smallness of the system is in favor of the diffusion limited time scales, as the respective length scales are equally small. However, for many applications, especially micro-array based assays [22–24], a deliberate and controlled agitation of the fluid under investigation would be of great importance. This is because diffusion driven mixing is effective only on a short range. Micro-array hybridization, however, requires transport of molecules in the fluid over large distances, and agitation is mandatory.

There are many different approaches for mixing in microfluidics. Many rely on external means like pumping through connections to the microfluidic chips. Other approaches rely on the flow geometry. Duct flows have recently received particular interest and many concepts for mixing the channel throughput exist [25]. Either the diffusion path is decreased by hydrodynamic focusing [26,27], or the contact surface is increased by multilamination [28,29]. Parabolic velocity profiles are used to mix by Taylor-Aris dispersion in ring channels [30,31]. Even chaotic advection can be induced within a channel, e.g., by patterning the walls [32], resulting in a meandering channel [33] or active flow disturbances [34–36]. Many or most of the above approaches, however, require either a sophisticated channel layout or external means which make the actual application difficult. Acoustically induced internal streaming seems to offer a path out of this dilemma. The same SAW that is used to actuate the fluid on a chip or in a closed volume can be employed to induce complex streaming patterns. Moreover, SAW can be modulated in frequency as well as in time, this being electrically activated.

A recent approach is related to the folding of material flow lines by modulated SAW induced streaming. Here, a SAW of a given frequency and direction with respect to the fluid volume is modulated either in space (trajectory) or in time. Each SAW induces a (stable) streaming pattern like the one shown in Figs. 8.6 or 8.7. The streaming pattern changes according to the SAW direction, frequency, or amplitude. Such SAW modulations can easily be achieved by switching the excitation parameters of the IDT.

One example for SAW mixing in a thin capillary gap formed between the SAW chip and a cover slip on top of it is shown in Fig. 8.11. The plot shows an experimentally obtained micrograph of a SAW spread dye in a capillary gap together with a theoretical calculation of the flow profile for the same geometry [37]. The experimental picture represents a snapshot some time after the SAW has been turned on; the inset of the right panel shows the time evolution of the shape as extracted from our model calculation. The lower part of the right panel exhibits a snapshot of the material folding or flow line after some time. Similar but different flow lines are obtained if, for example, the direction of the SAW is changed. The resulting material (dye in this case) distribution then represents a superposition of the single quasi static cases. If

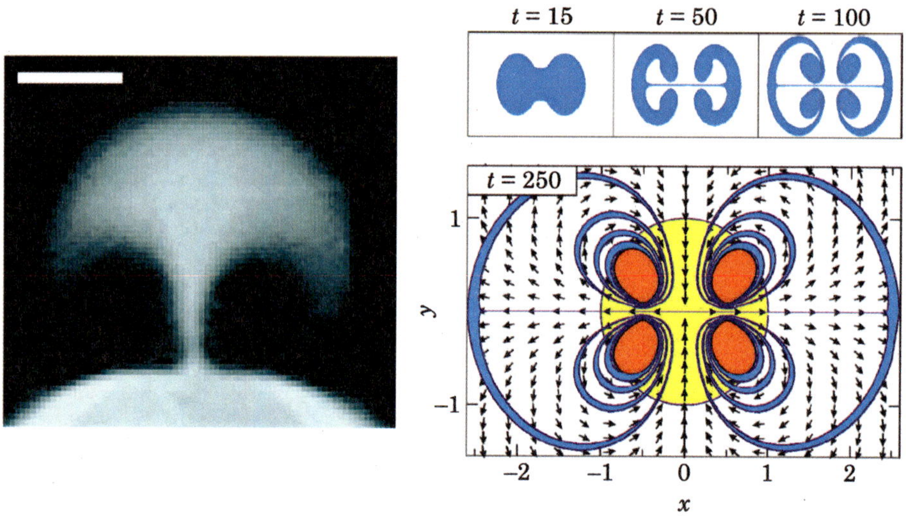

Figure 8.11 Left: a dye jet pumped by a SAW transducer, shortly after switching the RF power. In the center of the transducer, the fluid is pumped upwards; on both sides the fluid is sucked in. The white scale bar is 0.5 mm. Right: results of the theoretical modeling of this capillary gap flow profile (view from above). The inset shows the temporal evolution of the dye distribution.

this procedure is repeated several times, a homogenous distribution of the dye is obtained.

More recently, SAW induced mixing in closed channels has been observed [38], triggering chaotic advection in a microfluidic system. The optimum SAW propagation parameters have been determined to achieve perfect mixing after shortest possible times.

8.4 Example 2: Acoustic Droplet Actuation

The possibility that we can deliberately steer very small fluid volumes across a microfluidic chip, dissolve pre-deposited reagents, and mix and stir together with additional functional elements being integrated on the same chip, offers a wealth of applications for a miniaturized laboratory. The fact that the experiments or assays are operated in a batch mode, meaning that each step is independent from the other, makes the SAW driven microfluidic system an ideal candidate to provide a programmable microfluidic processor. The same hardware can be used for completely different protocols, with computer software monitoring the different steps of the process. The pumps, mixers, and dissolvers are SAW driven and can easily be combined with heaters, sensors, etc. on the same chip. The technology to produce the chips is well established, and borrowed from semiconductor technology. The chips can be produced in a few mask steps on whole wafers, making them so cheap that they can be used as consumables.

One example of such an acoustically driven, programmable microfluidic processor is given in Fig. 8.12. The figure shows a series of four snapshots of a chip, where three different chemicals in the form of little droplets are actuated, merged, and mixed. The chemicals in this case have been chosen such that each merger results in a color change of the output. The chip in this case is equipped with six different transducers, arranged around the circumference of the chip. Each transducer can be operated independently, two of each facing each other forming a delay line. The corresponding sound paths are patterned with a surface functionalization, leaving a hydrophillic virtual track behind, along which the droplets are guided (not visible, as monolayer chemistry). In the center of the chip, additional elements like heaters and a thermometer can be seen, all being produced by planar technology.

The whole area of the chip is about 1×1 cm; the droplets used in this experiment are about 100 nl of volume. Application of a suitable RF signal to one of the transducers induces SAW driven actuation of the desired droplet in its sound path. Interestingly, a second droplet being located in the lee side of the actuated one is not affected by the SAW (Fig. 8.12 (a) and (d)). The reason is that the SAW is strongly attenuated by the first droplet, leaving not enough amplitude behind it to actuate a second one. This remarkable feature makes individual actuation of individual droplets possible. Once two droplets merge, they are mixed instantaneously, leading to a nearly immediate homogenous chemical reaction as visualized by the respective color change. Control experiments, where the two droplets were carefully brought into contact by slightly pushing them with a pipette tip towards each other, revealed a very slowly moving reaction front within the resulting droplet. Without being able to estimate the time difference between the SAW mixed reaction and the diffusion driven one, the SAW driven reaction is definitely at least two orders of magnitude faster.

A very similar chip has been used to test a completely computer controlled feedback system, enabling the software to actually determine the position of a droplet on the chip. As the SAW actuation depends on many parameters like wetting angle, precise droplet size, quality of the surface functionalization, defects or particle contaminations along the fluidic track, there is no reliable way to predict the position of a droplet from the excitation parameters alone. Instead, we make use of the fact that a SAW is strongly attenuated by a liquid droplet in its way. For this purpose, we employ one of the temporarily unused delay lines on the chip to monitor the SAW transmission between them. Once a droplet is pushed into the sound path of this monitoring SAW, it is strongly attenuated. The corresponding SAW transmission signal can then be used to determine the position of the droplet on the chip.

The result of such a position measurement is shown in Fig. 8.13. Here, we show a photograph of the chip and the droplet together with a transmission measurement of a perpendicularly oriented SAW delay line employing a network

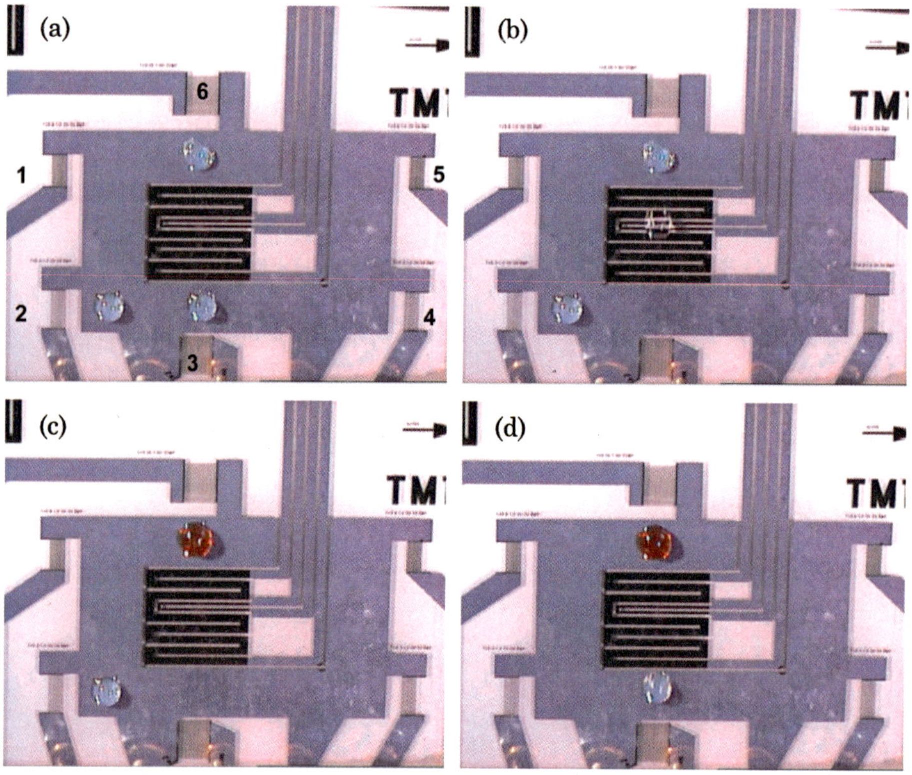

Figure 8.12 A SAW driven microfluidic processor: three droplets (approx. 100 nl each) are moved independently and remotely controlled by 6 acoustic nano-pumps. (a) to (d) represent a series of subsequently taken snapshots to visualize the movement, and the chemical reactions occurring when the droplets are merged and mixed by the action of the surface wave. The chip not only contains the acoustic nano-pumps and the fluidic environment but also additional systems like sensors and heaters. From [23]. © 2003 Elsevier, reprinted with permission.

analyzer. In this case, so called tapered IDT (TIDT) have been used [39] to increase the spatial resolution of the experiment. A droplet in the sound path of the monitoring SAW leads to a significantly reduced transmitted intensity, as can be seen on the display of the network analyzer in the right panel. The attenuation of a SAW caused by a small droplet can exceed several dB, hence making it very simple to collect the data for a position analysis on the chip.

8.5 Applications

8.5.1 Micro-array Hybridization (Biological Application)

A first and very important application for the SAW induced acoustic streaming as outlined above is its use for micro-array hybridization. Small

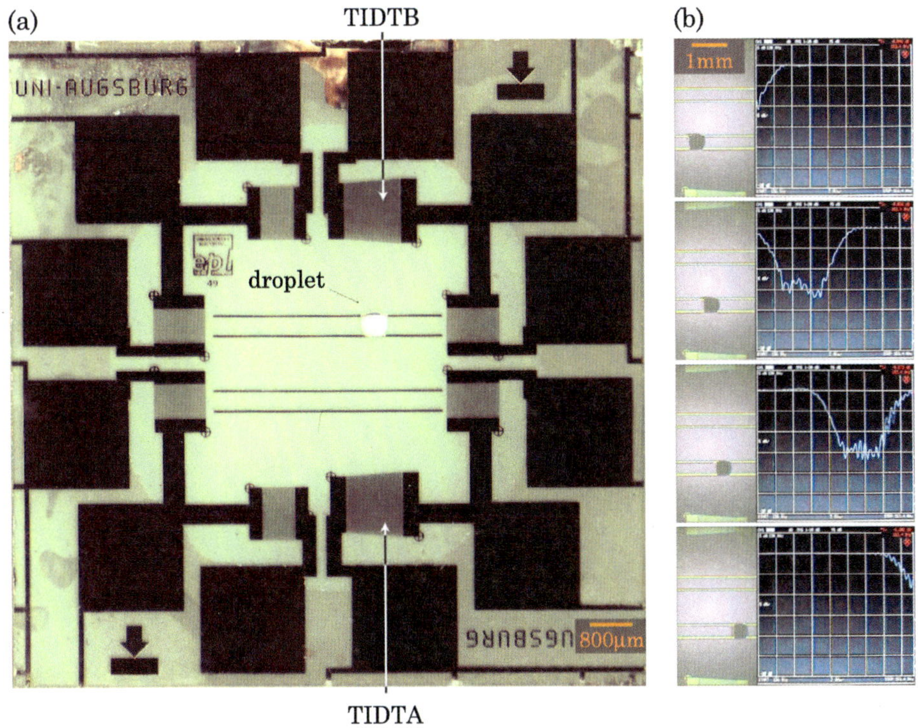

Figure 8.13 SAW driven microfluidic chip with means to determine the actual droplet position on the chip. One set of IDT (left, right) is used to actuate the droplet; a second delay line (TIDTA, TIDTB) is employed to measure its position. Here, the SAW transmission between TIDTA and TIDTB is monitored as a function of the droplets position. Strong attenuation is observed, once the droplet resides within the sound path of this delay line (seen right panel). To increase spatial resolution, a special tapered IDT has been used [39] (Courtesy T. Frommelt).

amounts of fluid containing an unknown constituent of DNA fragments is spread over a more or less densely packed array of "spots" containing different known constituents, acting as capture molecules for the unknown ones. The latter are being tagged with a fluorescent dye, in most cases. Upon successful hybridization between the target molecule in solution and a corresponding one on a spot, the local fluorescence intensity for this spot raises as compared to the areas without capture molecules or unspecific binding events somewhere else. Employing a sophisticated micro-array layout and multi-color tagging schemes, a wealth of information about the unknown sample can be gained from such experiments. The probably best known example would be the "genetic fingerprint."

Protein and DNA micro-arrays have thus become a standard tool in proteomics/genomics research. Up to several thousand spots of oligonucleotides or cDNA [40] probes with known identity are implemented on the chip in a checkerboard pattern. In gene expression profiling assays, the ratio of binding of complementary nucleic acids from test and control samples is determined. This allows a parallel, semi-quantitative analysis of transcription levels in a single experiment. In order to guarantee fast and reproducible hybridization results, however, the diffusion limit must be overcome. Hybridization experiments are performed at either body or slightly elevated temperatures, and it turns out that diffusion—especially for large macromolecules like oligonucleotides, DNA fragments, or the like—is an extremely slow process. On the other hand, typical micro-array areas are quite large, and the fluid film containing the precious sample is only a few tens of micrometers high, forming a capillary gap being spread over the micro-array. Especially in case of low concentrated cDNA molecules corresponding to low expressed genes, the immediate vicinity of the corresponding probe spot will be quickly depleted. Without active agitation, diffusion is the only mechanism for the DNA strands to be transported to their complementary spots. It has been estimated that it would require weeks for the hybridization reaction to reach equilibrium [41,42].

To achieve reasonable hybridization time while saving precious sample and dye, stirring or mixing is an important issue. SAW micro-agitation chips efficiently agitate the smallest sample volumes (down to 10 μl and below) without introducing any dead volume. The advantages are reduced reaction time, increased signal-to-noise ratio, improved homogeneity across the micro-array, and better slide-to-slide reproducibility. Such SAW micro-mixer chips are the heart of the Advalytix *ArrayBooster*, or the Advalytix *SlideBooster* [44], being compatible with all micro-arrays based on the microscope slide format.

Three different methods to apply the SAW agitation technique to micro-array hybridization assays have been successfully demonstrated. First, in the so called *ArrayBooster* geometry: a SAW mixer chip, embedded into a glass or epoxy slide, is placed on top of the sample fluid from above. The SAW chip and the supporting glass or epoxy slide thus form a "mixer card," and at the same time serve as the cover slip. Spacers between the mixer card and the slide containing the micro-array define a well controllable fluid layer, if the sandwich is filled with sample liquid. To visualize the mixing process, in the experiment shown in Fig. 8.14, we drilled four holes into the mixer card, through which we pipetted small amounts of ink into the buffer solution confined between the slide and the mixer card. At the left of the figure (a) through (c), the progress of the mixing can be observed, once the SAW mediated acoustic streaming takes place. In the right parts, the SAW had not been switched on. Hence, diffusion is the only source to act upon the ink, and nearly no spreading of the dye is observed.

A second method to apply SAW induced acoustic streaming to a micro-array assay is to couple the ultrasonic SAW energy through an index matching fluid and through the slide hosting the micro-array into the sample fluid on top of it. Even without direct contact between the sample hybridization solution and the mixer chip, the SAW mediated agitation can be applied to the hybridization assay. This *SlideBooster* geometry [43] has proven to be particularly successful for protein and tissue micro-arrays for immuno-histochemistry applications. In

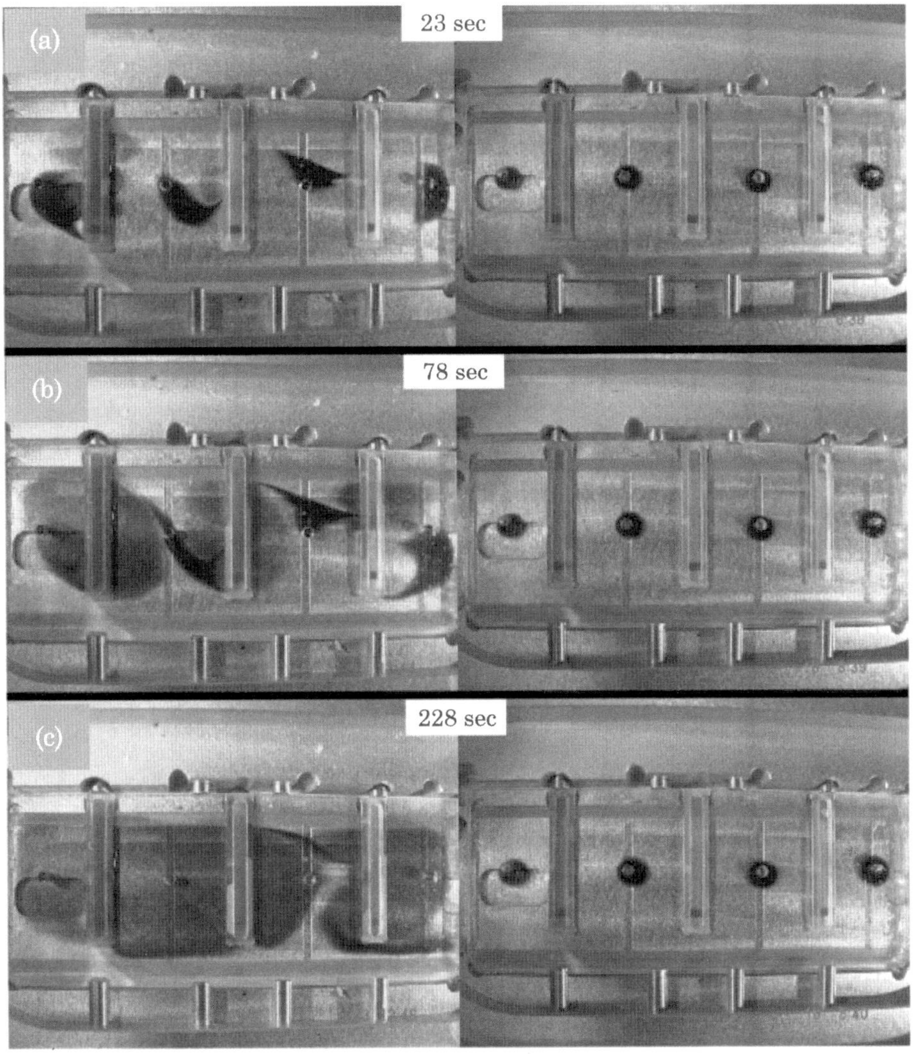

Figure 8.14 Visualization of the SAW mediated agitation of a micro-array sandwich. This top view into the hybridization chamber shows a mixer card with three SAW chips, which is used to confine a fluid layer of $d = 60\ \mu m$ on top of a glass slide containing an oligonucleotide micro-array. Ink is pipetted through four holes to visualize the agitation process. In the left part of the figure, the mixer chips are SAW activated. To the right, the SAW had been switched off, thus showing the effect of diffusion alone. After a several minutes, the SAW induced acoustic streaming led to a complete homogenization of the ink, indicating a very efficient mixing effect, whereas the ink in the "diffusion only" case has not spread at all. From [44]. © 2005 JAI; reprinted with permission.

both cases, the signal intensity as well as the homogeneity has been remarkably enhanced as compared to conventional methods. The third technique relies on the fact that mixing of a sample fluid in thin liquid layers is particularly

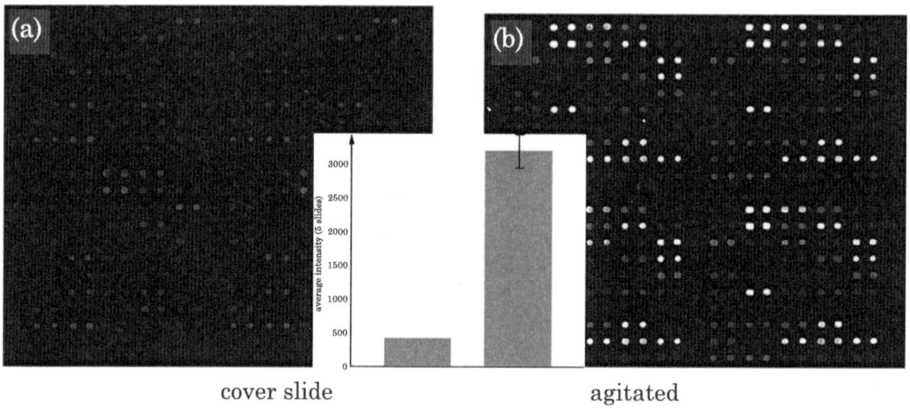

Figure 8.15 Fluorescence image of a typical micro-array without (a) and with (b) micro-agitation. Clearly, a higher intensity and homogeneity due to agitation is observed. The inset shows the averaged intensities for 5 different slides. From [45]. © 2003 JBT; reprinted with permission.

effective if the sources of the ultrasonic energy are distributed randomly across the hybridization area on the micro-array [44]. Although both the *ArrayBooster* and the *SlideBooster* geometries make use of a distributed source mixing by using multiple mixer chips containing multiple SAW transducers being operated at different frequencies (multiplexing), the distribution of SAW sources in these cases is spatially fixed and given by the layout of the mixer chips used. The frequency multiplexing distributes the sources of ultrasonic energy in time, following a predetermined protocol. In other words, a set of transducers is powered at a given SAW intensity over a certain period of time and hence induces a specific streaming pattern. After a certain dwell time, a second set of transducers with a second streaming pattern is powered, and so on.

Fig. 8.15 is a typical example for such a fluorescently labeled micro-array experiment. In (a), we show the fluorescence of a series of spots on the micro-array, which has been hybridized employing a conventional cover slip technique [45], whereas panel (b) shows the same micro-array after agitation during hybridization. Clearly, in (b), the fluorescence intensities and the homogeneity are much higher than in (a). Details for the hybridization parameters are found in [45]; many more examples for a successful application of agitation in the micro-array field including protein micro-arrays and immuno-histochemistry (IHC) applications can be found in reference [43].

8.5.2 Polymerase Chain Reaction

Many modern experiments and analytical techniques in genetics rely on the amplification of genetic material prior micro-array hybridization assays. The reason is that in some cases (for example for the genetic fingerprinting), not enough material is available for further investigation. In other cases, the amplification of genetic material can also lead to a purification of the existing sample. The technique of choice for genetic amplification is the polymerase chain reaction (PCR), the invention of which was rewarded by the Nobel price for chemistry in 1993 [46]. PCR is a biochemistry and molecular biology technique for enzymatically replicating DNA without using a living organism. Like amplification using living organisms, the technique allows a small amount of DNA to be amplified exponentially. As PCR is an in vitro technique, it can be performed without restrictions on the form of DNA, and it can be extensively modified to perform a wide array of genetic manipulations. It is commonly used in medical and biological research labs for a variety of tasks, such as the detection of hereditary diseases, the identification of genetic fingerprints, the diagnosis of infectious diseases, the cloning of genes, paternity testing, and DNA computing [47].

In this section, we present a complete PCR lab on a SAW driven chip, comprising many of the building blocks as described and outlined above. The whole chip has a size of about 2 by 2 cm, and SAWs are generated by IDT structures being patterned by photolithography on a planar $LiNbO_3$ substrate. Employing the same process, other metal structures can be processed as well. This allows for building up a chip with different functional units. The transport of liquid between the units can then be achieved by SAW. Chemical surface functionalization is employed to dispense and guide droplets. The first functional structure on the chips is a metal thin film resistance heater with an integrated temperature sensor. The heater is 2×2 mm^2 in size with a window in its center. This window allows for an optical monitoring through the transparent piezoelectric substrate. The temperature is computer controlled with 0.1 K accuracy. The chip heater is used for two processes in DNA analysis. The first is the amplification of a small amount of template DNA in a polymerase chain reaction where repeated cycles of three different temperatures lead to amplification of the DNA strand [48]. To finally analyze the amplified DNA, hybridization assays are commonly used, as has been described in the previous section. Employing this method, the binding of DNA in solution to surface-coupled complementary strands is observed, while the temperature is kept on a value characteristic for the reactants. This provides information on the corresponding nucleotide sequence.

With the two processes performed in series on one chip, a tool for complete DNA analysis is built suitable for various applications like the study of gene function, gene expression, and single nucleotide polymorphism (SNP) [49]. A picture of such a SAW driven PCR chip is shown in Fig. 8.16. The chip shown in Fig. 8.16 allows for so-called "multi spot PCR," where several samples are amplified simultaneously [43,48].

A severe drawback to the otherwise "open" geometry of SAW driven programmable microfluidic chips is the sensitivity to elevated temperature and the related evaporation of the open droplets. To prevent this evaporation, the aqueous solution containing the PCR mix is covered with mineral oil, forming

Figure 8.16 Programmable microfluidic biochip for multi-spot PCR. The sample solution in red is covered by a mineral oil layer to prevent evaporation at elevated temperatures. Apart from SAW driven nanopumps for the fluid actuation, the chip also hosts a heater and a thermometer. The chip is connected to the outside via pogo contacts. Underneath the transparent piezoelectric substrate, a microscope objective is mounted for fluorescence monitoring of the PCR progress employing an intercalating dye. Microsystem realized by Avalytix and the Institute for Physical High Technology in Jena, Germany. From A. Wixforth, Flat fluidics: programmable on-chip networks for biological and chemical applications. © 2005 NSTI; reprinted with permission.

a semispherical cap due to the chemical modification of the surface. We recall here that the PCR mix consists of two oligonucleotide primers which hybridize to either side of the DNA region to be amplified, deoxynucleotide triphosphates, or dNTPs, the substrates for DNA synthesis, and the heat resistant Taq polymerase (an enzyme that comes from the bacterium *thermophillus aquaticus* native to hot springs). This mix can either be applied as a liquid or be preloaded to the chip in dry form. In this case, sample buffer needs to be added, and SAW agitation is used to dissolve the PCR mix into solution. The application of the oil cap can also be achieved with the SAW by pushing the oil onto the water droplet, while the different surface tensions of the two liquids account for the spatial arrangement. Therefore, encapsulation of the PCR solution against evaporation can be reached much easier than with valves in a microfluidic channel system. The contact area of the oil/water droplet to the chip plane has about the size of the heater structure. This whole fluidic micro-chamber can be moved across the chip with the SAW.

PCR cycling has been described in detail in Chapter 6. Let us mention that to gain a better control over progress and efficiency of the PCR, an optical online monitoring system can be integrated. This is done with different fluorescent dye systems that exhibit varying emission intensities correlated with the amount of double stranded DNA or the concentration of a special DNA product with their emission intensity. To investigate the possibilities for a direct PCR process control on a planar chip, a small optical unit that can collect the light emitted

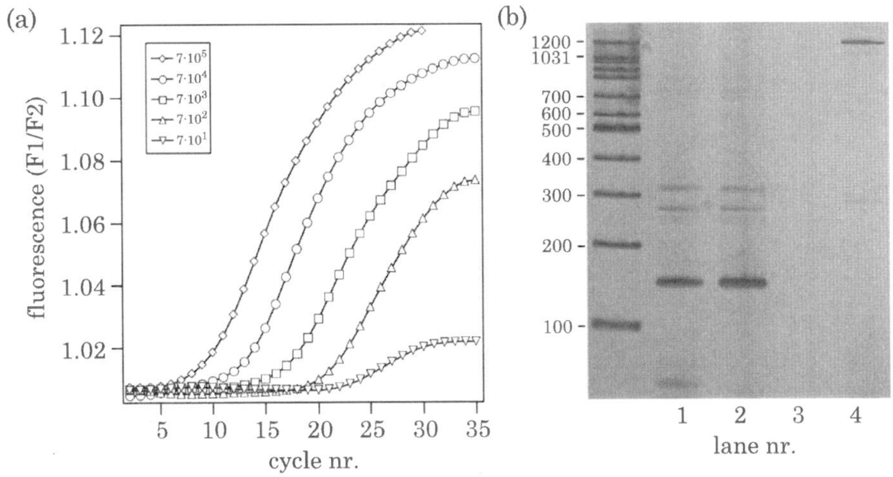

Figure 8.17 (a) Online PCR traces for 5 different template concentrations of plasmid: the threshold where the signal rises depends on the concentration. (b) Chip PCR (35 cycles) detected with electrophoresis gel. Lanes 1 and 2: 150 bp fragment of human genomic DNA cycled from 6000 genome equivalents (one genomic equivalent equals to 2 x 23 chromosomes); lane 3: no template; lane 4: 1382 bp fragment was amplified from 6 genome equivalents [48] (Courtesy Advalytix).

from a dye inside the oil covered PCR droplet has been incorporated. The unit is placed below the transparent chip, and the optical path is focused through the window in the heater into the aqueous solution. Fig. 8.17 (a) shows the optically monitored increase of fluorescence intensity for an intercalating dye versus the number of PCR cycles and for different concentrations of the initial copies [43]. The intensity is proportional to the actual number of copies after amplification. As can be seen from the figure, we observe a rapid increase in fluorescence for a few cycles, eventually saturating as the building blocks exhaust. Fig. 8.17 (b) shows a gel electrophorogram of the amplified DNA fragments. Lanes 1 and 2 represent a deliberate amplification from a given start copy concentration, lane 3 is a control experiment with no start copies at all, and lane 4 represents the result for a much lower start copy concentration as in lane 1. Within the experimental resolution, no cross contamination has been observed in this multi spot PCR.

The leftmost lane in Fig. 8.17.b is a marker lane for calibration of the molecular weight of the amplificants. The on-chip PCR presented here is able to operate with sub microliter volumes at a significantly increased amplification speed. Moreover, for fortunate but still unknown reasons, the on-chip PCR seems to be much more resistant against (airborne) contamination than conventional approaches. For a detailed description of the PCR processes employed here, see [43,48] and references therein.

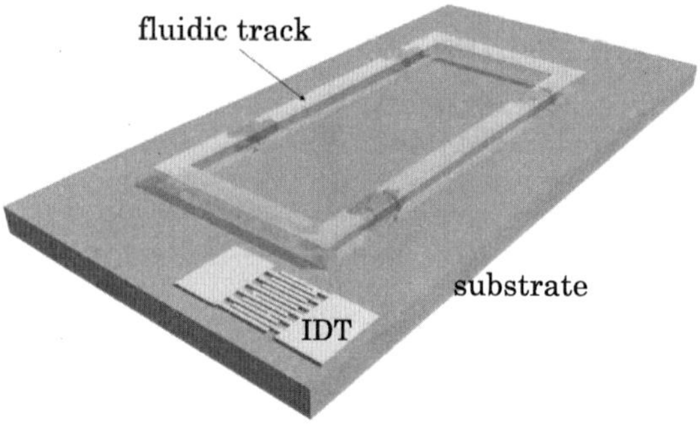

Figure 8.18 A SAW driven microfluidic chip for cell adhesion measurements. A SAW, excited electrically, interacts with the confined liquid at the solid/liquid interface and induces liquid circulation. The surface wave basically acts as a localized pump, as its mechanical energy is absorbed by the liquid over only a few micrometers. Due to the small scales of this microfluidic system, it creates a homogenous laminar flow along the channel. The total length of the track in this case is about $l = 40$ mm [50].

8.5.3 Cell Adhesion

8.5.3.1 Blood Flow on a Chip

Another powerful application for SAW driven microfluidic chips is the study of cell adhesion to either artificially fabricated surfaces or to biological surfaces. This study is very useful for the investigation of cell adhesion to medical implants. These can be coarsely divided into two different types: one type is meant to adhere firmly to the biological environment, like tooth implants. The other group needs to avoid cell adhesion—like artificial hip joints. Both types of implants possess surfaces that can be deliberately functionalized to be optimized into one or the other direction.

The combination of planar (semiconductor) technology for SAW driven fluidic chips is completely compatible with all kinds of thin film technologies. Hence, many different materials, textures, or functionalizations are easily investigated in terms of cell adhesion. For this purpose, closed loops of virtual fluidic tracks are used, where SAW induces acoustic streaming. As the SAW itself is attenuated over a propagation length of a few wavelengths, no real high frequency effects remain in an even microscopic closed loop. Instead, the pressure difference between the SAW entrance and the fluid volume drives a homogenous, easily controllable laminar flow whose profile covers the complete range of physiological flow and shear conditions from 0 to several 10 000 s^{-1} along the channel, which can be adjusted by means of the SAW amplitude. A simple version of such a "planar micro flow chamber" is sketched in Fig. 8.18. Employing a hydrophilic/hydrophobic surface functionalization, a closed loop virtual track is formed. In the case of Fig. 8.18, the total volume is about 8 μl.

Such acoustically driven planar micro flow chambers provide many advantages for cell adhesion investigations. Apart from the advantages of using SAW as a localized nano-pump, the free (functionalized) surface of the chip is easily accessible to exchange buffers or reagents, or to monitor the flow optically. Moreover, the flow chamber and hence the flow within, is electrically addressable, which makes it a versatile tool for investigations. As an example, we present here some recent investigations regarding blood flow and platelet adhesion. Studying cell or platelet adhesion (hemo-dynamics) in small capillaries and arteries is a particularly challenging topic, since it is difficult to be mimicked in an in-vitro system and hard to access by optical or mechanical means. As the diameters of capillaries and arteries become as small as one micrometer, the Reynolds number becomes very low, making the pumping mechanism extraordinary difficult. Using the chip design shown in Fig. 8.18, a blood vessel can be mimicked, allowing for the study of the blood flow and platelet adhesion. The flow chamber has no dead volume so that investigations of expensive or rare substrates—with a volume 100 to 1000 times less than usually required—are possible. The particularly small sample volume of only 8 μl is probably the most prominent advantage of the nano-pump driven planar flow chamber chip (FCC). Moreover, the chip components are inert and entirely compatible with biological systems and the system is free of all movable parts, making the handling extremely simple. The complete optical transparency of the $LiNbO_3$ substrate allows for individual cells to be tracked over long periods of time, when the μ-FCC is mounted directly onto a fluorescence microscope.

Lanes ranging from a few μm up to 10 mm both in width and length can be realized. The open structure on a free surface allows for direct access to the channel at all times (e.g., for the addition of antibodies or drugs). The technique also provides maximum freedom for mimicking all possible vessel architectures and conditions as they may exist in nature with curved, branched, or restricted vessels. Finally, no restrictions are put on the surface functionalization. Artificial lipid membranes, protein coats, and even confluent cell layers have been prepared successfully.

8.5.3.2 Proteins in a Microflow

Using the same micro-system, the effect of shear flow on a protein called von-Willebrand-Factor (vWF) has been investigated [50]. This protein is a glycoprotein being synthesized and stored in endothelial cells, and has been found to play an important role for blood coagulation by adhesion of the platelets, particularly in regions where the shear rates are high. At first glance, this adhesion and subsequent coagulation behavior seems to be counterintuitive, but it is the very special nature of this mechanically activated macromolecule that helps us to resolve this contradiction: under normal conditions, it assembles into multimers (biopolymers) which, when stretched, can reach sizes as long as 100 μm. The monomeric length is unusually large (~100 nm) and contains 2050 amino acid residues. Intuitively, one would expect that blood platelet adhesion is always decreasing with increasing shear force applied to the

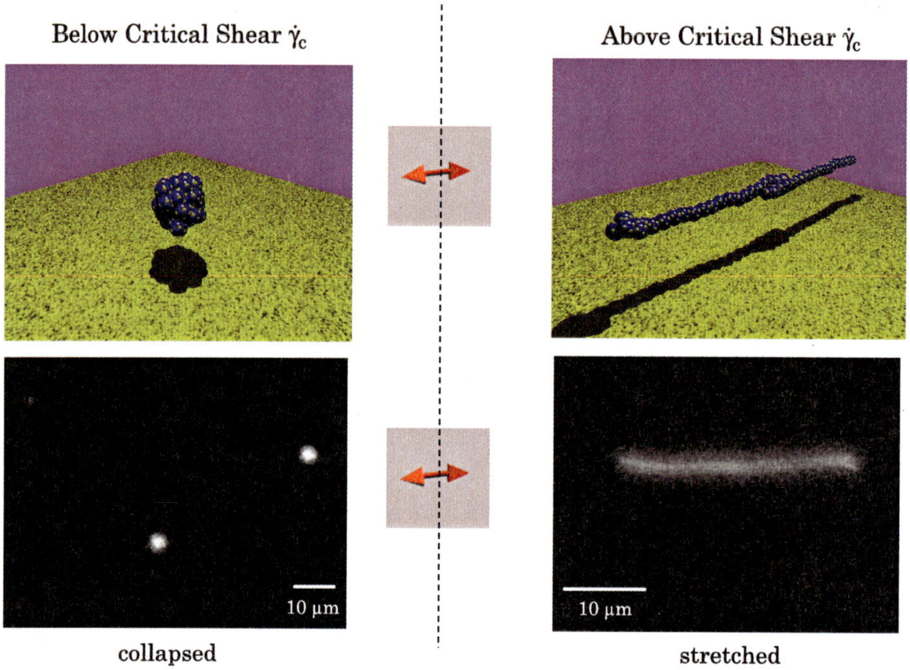

Figure 8.19 vWF proteins in a flow: the blood clotting protein vWF is exposed to various shear flow conditions. Only when a critical value is exceeded the protein elongates in a reversible fashion. The unusually high shear rate levels could not be explained by conventional hydrodynamic models; and a new theory was developed [52].

platelet. Surprisingly, vWF mediated adhesion is strongly enhanced under high shear-flow conditions [51]. vWF was spread over the hydrophilic track of the flow chamber chip and exposed to various shear flows. Images of vWF at concentrations typical for human blood ($c \approx 2$ $\mu g/ml$) are presented in the lower panel of Fig. 8.19 for two different SAW induced shear rates. At shear rates between $\dot{\gamma} \approx 10$ and $1000\ \mathrm{s}^{-1}$, the biopolymer exhibits a compact conformation (Fig. 8.19 left). This compact conformation remains unchanged as long as the shear rate is maintained below a certain (critical) value $\dot{\gamma}_{\mathrm{crit}} \sim 5000\ \mathrm{s}^{-1}$. Increasing the shear rate, a shape transformation of the vWF fibers from a collapsed to a stretched conformation of length $l \approx 15$ μm (Fig. 8.19 right) is observed. This transition is reversible, as we observe an immediate relaxation of the protein to its compact conformation when the flow is turned off.

This finding explains the apparent counterintuitive observation mentioned at the beginning of this section. Under high shear flow, vWF mediated blood platelet adhesion is more effective. In its collapsed state all binding sites are buried inside the coil. Once the critical shear is exceeded, vWF stretches into a long thin fiber exposing all its binding sites. If vWF touches the surface in this conformation, it will immediately be immobilized and serve as a "sticky" grid for free floating blood platelets.

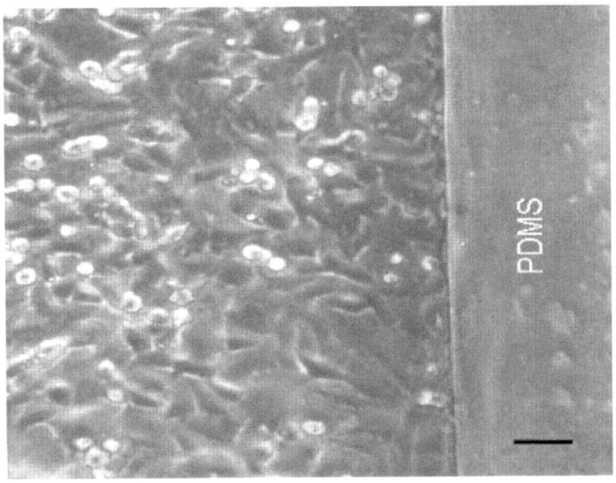

Figure 8.20 Cell-Cell Interactions on a Chip: cells are grown directly in the microfluidic channel delimited by PDMS walls. The cells form a confluent layer over the entire channel and do not grow on the boundaries (PDMS walls). The scale bar represents 50 μm.

8.5.3.3 Cell-cell Interactions on a Chip

Although the simple planar flow chamber sketched in Fig. 8.18 represents an elegant technique for examining many relevant topics, it does not perform satisfactorily for the direct culture of living cells on its surface: Initial attempts are hindered by the fact that the protein rich media influenced the channel architecture due to the protein adsorption on the hydrophobic surface. As a result, this led to an unwanted inhomogeneous flow profile. To avoid this, a more sophisticated 3D version of the system has been developed which easily allows for successful cell culture while retaining all of the advantages of the planar system previously described. This was achieved by a flexible yet stable 3D architecture by adding "walls" to the channel employing the fully transparent, biocompatible synthetic elastomer polydimethyl-siloxane (PDMS). This system was devised by examining the adhesion of human melanoma cells at various shear rates. The results are shown in Fig. 8.20, which shows the cell culture chip and the surrounding PDMS channel. Note that the cells form a confluent layer filling out the entire channel without growing outside of the boundaries. In addition, the cells appear healthy, divide readily, and are securely attached to the chip surface. This cultured layer remains completely unaffected by the SAW, which still induces a continuous flow along the channel. In such a bio-functionalized microfluidic chip, a true artificial blood vessel, for example, can be realized. We are presently investigating the possibilities of growing and culturing many different cell lines in such a channel, including endothelial cells.

8.5.3.4 Advanced Applications

Using SAW driven microfluidic chips, many other applications can be envisioned, some of which are presently under investigation. A representative example is SAW mediated alignment of carbon nanotubes (CNT). CNTs are a self assembled version of an ideal one-dimensional electronic system and hence have recently attracted considerable interest in terms of future applications as electronic devices. The preparation of CNT is quite advanced, and CNT suspensions are commercially available. However, to actually integrate CNT into an electronic environment like a microelectronic chip, a deliberate positioning with respect to the chip layout of the CNT is an important prerequisite.

Employing a flip-chip technique [53], an aqueous solution containing CNT in suspension is placed in a capillary gap in which a flow is induced by SAW technique. As CNTs are polarizable, and possess a streaming resistance, the combination of SAW induced piezoelectric fields and SAW induced acoustic streaming results in an alignment of CNTs with respect to the substrate. This alignment exhibits a clear preferential direction, as can be seen in Fig. 8.21.

Functionalization of the CNT, employing thiolization, makes the nano-tubes stick to the gold electrodes. Using this technique, the SAW induced alignment can be further improved and CNTs can be deliberately deposited between Au electrodes to investigate their electronic properties [54]. Recently, CNTs have been aligned between two Au electrodes even on a non-piezoelectric substrate like silicon. Here, the same flip-chip technique as described in [53] has been employed. However, the cover slip was replaced by a silicon substrate on which an array of microfabricated Au electrodes had been previously deposited.

Depending on the CNT concentration, even single CNT can be "spanned" between the Au electrodes on the silicon substrate [55]. Moreover, it could be shown that they can even be processed to be free standing between the Au "posts." These devices represent a truly one-dimensional electronic system with tunnel barrier contacts to manipulate them electrically. Using the very same SAW technique, acousto-electric currents can be induced in such one-dimensional electronic systems. The difference with a quasi two-dimensional system [15,17], however, is that the SAW represents a dynamical potential modulation acting like an escalator for single electrons. Hence, the SAW induced acousto-electric current measured between the Au contacts consists of single electrons, one by one driven through the CNT by the SAW, and resulting in a quantized current [56].

8.6 Summary

Surface acoustic waves induce an acoustically driven flow in fluids on the substrate surface on which they are propagating. SAWs are modes of elastic energy, propagating along the surface of a solid. They are easily excited and detected once a piezoelectric substrate is used. The SAWs used in our experiments are mostly of the Rayleigh type, having large surface displacements

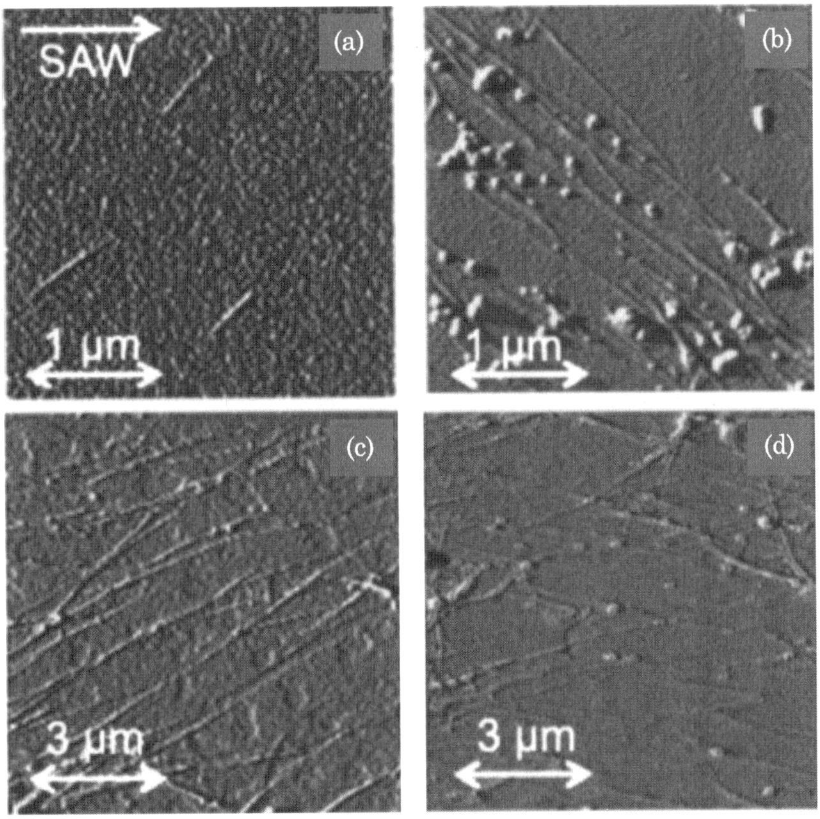

Figure 8.21 AFM images of aligned multi wall carbon nanotubes (MWNT) with different concentrations of the suspension. After experiencing the SAW induced shear flow in a capillary gap between the $LiNbO_3$ substrate and a cover slide, the MWNTs are aligned with an angle between $\pm 25°$ and $\pm 45°$ with respect to the SAW propagation direction. Reused with permission from [53]. © 2004 American Institute of Physics.

normal to the surface plane. Fluids on this surface interact with the SAW by means of viscous damping and react to the interaction by developing an internal streaming. This streaming is very similar to the well known acoustic streaming, originating from a sound induced pressure gradient along the direction of propagation. For small SAW amplitudes or in confined geometries, this acoustic streaming effect leads to complex flow line patterns in the liquid. In a microfluidic system, viscous forces by far dominate inertial forces, which in turn restrict the flow in such systems to laminar regime. Especially for biological and chemical applications, this represents a drawback when compared to macroscopic fluidic systems, as diffusion is the only source for mixing. At higher SAW amplitudes, fluids confined in the form of droplets on the surface on which the SAW is propagating can be actuated as a whole. This observation led to the idea of chemically patterning the surface of a SAW chip in order to generate areas of preferred wettability (hydrophilic areas), surrounded by hydrophobic regions of the chip. Very similar to the metal strip lines and circuit paths on a microelectronic chip, this way "virtual fluidic tracks" can be created on a chip, on which acoustically actuated droplets can be maneuvered at will. The SAW generators are electrically excited and addressable, opening the doors for the creation of programmable microfluidic processors. However, the liquids are virtually confined by the surface functionalization and all liquid processing takes place on the planar chip surface.

Typical applications of SAW digital systems are micro-array hybridization assays, and other microfluidic applications, where efficient mixing and stirring of sub microliter volumes is required. These include mixing in open droplets, capillary gaps, micro titer-plate wells, and the like. The droplet based fluidic systems have been shown to be able to perform very complex tasks, such as a complete on-chip polymerase chain reaction including micro-array analysis for DNA amplification purposes. SAW induced acoustic streaming in a planar microfluidic flow chamber has been used to investigate blood flow related phenomena. In a planar artificial blood vessel, specific macro-proteins turn out to represent mechanically activated molecules with a specific switching behavior being related to their adhesion properties. Finally, cell culture and proliferation on such chips is possible; besides, SAW induced acoustic streaming has found its way into nano-electronics. Here, we used the technique to deliberately align carbon nano-tubes with respect to contacts on a silicon chip. These structures exhibit a quantized acousto-electronic current, a true indication for a quasi one-dimensional behavior.

Acknowledgements

This work represents the summary of many years of effort by an even larger number of colleagues and students in the research group at Augsburg University and at Advalytix. It is nearly impossible to acknowledge all of their contributions in an appropriate manner, which, however, by no means should be forgotten at this point. At least, many of these invaluable contributions are summarized in the references to this chapter, which we refer the reader to. Special thanks, however, to our team of skilled technicians and administrative support in Augsburg: Funda Cevik, Sidonie Lieber, Andreas Spörhase, and Alexander Hupfer.

It is much easier to account for the financial support which enabled our research and led to the development of SAW actuated chips. In a series of different research programs, we have been supported in part by the Bayerische Forschungsstiftung BFS, the Deutsche Forschungsgemeinschaft DFG, the Fonds der Chemischen Industrie, the Bundesministerium für Bildung und Forschung BMBF, and the European Community. This support is gratefully acknowledged.

References

1. E.M. Purcell, *American Journal of Physics*, Vol. 45, p. 3, 1977.
2. J.G. Smits, *Sensors and Actuators A*, Vol. 21–23, p. 203, 1990.
3. A. Manz, C.S. Effenhauser, N. Burggraf, D.J. Harrison, K. Seiler and K. Fluri, *J. Micromech. Microeng.*, Vol. 4, p. 257, 1994.
4. G.P. Lopez, H.A. Biebuyck, C.D. Frisbie and G.M. Whitesides, *Science*, Vol. 260, p. 647, 1993.
5. G. Möller, M. Harke and H. Motschmann, *Langmuir*, Vol. 14, p. 4955, 1998.
6. H. Gau, S. Herminghaus, P. Lenz and R. Lipowski, *Science*, Vol. 283, p. 46, 1999.
7. S. Daniel, M.K. Chaudhury and J.C. Chen, *Science*, Vol. 291, p. 633, 2001.
8. K. Ichimura, S.K. Oh and M. Nakagawa, *Science*, Vol. 288, p. 1624, 2000.
9. R.M. Moroney, R.M. White and R.T. Howe, *Appl. Phys. Lett.*, Vol. 59, p. 774, 1991.
10. M.A. Unger, H.P. Chou, T. Thorsen, A. Scherer and S.R. Quake, *Science*, Vol. 288, p. 113, 2000.
11. W.L. Nyborg, *Acoustic Streaming, in Physical Acoustics 2B*, W.P. Mason, Ed., Academic Press, p. 265, 1965.
12. C. Kittel, *Introduction to Solid State Physics*, 6th edition, John Wiley & Sons Inc., 1986.
13. Lord Rayleigh, *Proc. London Math. Soc.*, Vol. 17, p. 4, 1885.
14. B.A. Auld, *Acoustic Fields and Waves in Solids*, John Wiley & Sons, New York, 1973.
15. A. Wixforth, J.P. Kotthaus and G. Weimann, *Phys. Rev. Lett.*, Vol. 56, p. 2104, 1986.
16. C. Rocke, S. Zimmermann, A. Wixforth, J.P. Kotthaus, G. Böhm and G. Weimann, *Phys. Rev. Lett.*, Vol. 78, p. 4099, 1997.
17. M. Rotter, A.V. Kalameisev, A.O. Govorov, W. Ruile and A. Wixforth, *Phys. Rev. Lett.*, Vol. 82, p. 2171, 1999. See also:
    M. Rotter, A. Wixforth, A.O. Govorov, W. Ruile, D. Bernklau and H. Riechert, *Appl. Phys. Lett.*, Vol. 75, p. 965, 1999.
18. G.W. Farnell, "Elastic Surface Waves," H. Matthews, Ed., *Surface Wave Filters*, Wiley, New York, 1977.
19. Thomas Frommelt, Dissertation University of Augsburg, 2007.
20. C. Eckart, *Physical Review Letters*, Vol. 73, p. 68, 1948.
21. M. Schindler, P. Talkner and P. Hänggi, *Physics of Fluids*, Vol. 18, p. 103303, 2006.
22. R. Hui Liu, R. Lenigk, R.L. Druyor-Sanchez, J. Yang and P. Grodzinsky, *Analytical Chemistry*, Vol. 75, no. 8, 2003.
23. A. Wixforth, *Superlattices and Microstructures*, Vol. 33, pp. 389–396, 2003.
24. D. Gershon, *Nature*, Vol. 416, pp. 885–891, 2002.
25. Z. Wu and N.T. Nguyen, *J. Micromech. Microeng.*, Vol. 15, pp. R1–R16, 2005.
26. J. Knight, A. Vishwanath, J. Brody and R. Austin, *Phys. Rev. Lett.*, Vol. 80, p. 3863, 1998.
27. Z. Wu and N.T. Nguyen, *Biomedical Microdevices*, Vol. 7, no. 1, p. 13, 2005.
28. V. Hessel, S. Hardt, H. Löwe and F. Schönfeld, *AIChE Journal*, Vol. 49, no. 3, p. 556, 2003.
29. S. Hardt and F. Schönfeld, *AIChE Journal*, Vol. 49, no. 3, p. 578, 2003.
30. H. Chou, M.A. Unger and S.R. Quake, *Biomedical Microdevices*, Vol. 3, no. 4, p. 323, 2001.
31. A. Scherer and S.R. Quake, *Science*, Vol. 290, p. 1536, 2000.
32. A.D. et al. Stroock, *Science*, Vol. 295, p. 647, 2002.
33. S. Hardt and F. Schönfeld, *AIChE Journal*, Vol. 50, no. 4, p. 771, 2004.
34. A. Groissman and V. Steinberg, *Nature*, Vol. 410, p. 905, 2001.
35. A. Vikhansky, *Physics of Fluids*, Vol. 15, p. 108, 2003.
36. P. Tabeling, M. Chabert, A. Dodge, C. Julien and F. Okkles, *Philos. Trans R. Soc. London Ser.*, Vol. A 362, p. 987, 2004.

37. Z. Guttenberg, A. Rathgeber, S. Keller, J.O. Rädler, A. Wixforth, M. Kostur, M. Schindler and P. Talkner, *Physical Review E*, Vol. 70, p. 056311, 2004.
38. K. Sritharan, C.J. Strobl, M.F. Schneider, A. Wixforth and Z. Guttenberg, *Appl. Phys. Lett.*, Vol. 88, p. 054102, 2006.
39. M. Streibl, A. Wixforth, J.P. Kotthaus, A.O. Govorov, C. Kadow and A.C. Gossard, *Appl. Phys. Lett.*, Vol. 75, p. 4139, 1999.
40. P. Hegde, R. Qi and K. Abernathy, Biotechniques, Vol. 29, pp. 548–550, 552–554, 556, 2000.
41. V. Chan, D.J. Graves and S.E. McKenzie, *Biophys Journal*, Vol. 69, pp. 2243–2255, 1995.
42. S.A. Allison, S.H. Northrup and J.A. McCammon, *Biophys Journal*, vol. 49, pp. 167–175, 1986.
43. http://www.Advalytix.de.
44. A. Rathgeber, M. Wassermeier and A. Wixforth, *Journal of ASTM International*, Vol. 2, p. 1, 2005.
45. A. Toegl, R. Kirchner, C. Gauer and A. Wixforth, *Journal of Biomolecular Techniques*, Vol. 14, pp. 197–204, 2003.
46. K.B. Mullis and F.A. Faloona, *Methods in enzymology*, Vol. 155, pp. 335–350, 1987.
47. http://www.wikipedia.com.
48. Z. Guttenberg, H. Müller, H. Habermüller, A. Geisbauer, J. Pipper, J. Felbel, M. Kielpinski, J. Scriba and A. Wixforth, *Lab-on-a-Chip*, Vol. 5, pp. 308–331, 2005.
49. S. Dong, E. Wang, L. Hsie, Y. Cao, X. Chen and T.R. Gingeras, *Genome Res.*, Vol. 11, pp. 1418–1424, 2001.
50. S.W. Schneider, S. Nuschele, A. Wixforth, C. Gorzelannzy, A. Alexander-Katz, R.R. Netz and M.F. Schneider, *PNAS*, Vol. 104, no. 19, pp. 7899–7903, 2007.
51. Z.M. Ruggeri, *Journ. Thromb. Haemost.*, Vol. 1, pp. 1335–1342, 2003.
52. A.A. Katz, M.F. Schneider, S.W. Schneider, A. Wixforth and R.R. Netz, *Phys. Rev. Lett.*, Vol. 97, p. 138101, 2006.
53. C.J. Strobl, C. Schäflein, U. Beierlein, J. Ebbecke and A. Wixforth, *Appl. Phys. Lett.*, Vol. 85, p. 1427, 2004.
54. T. Smorodin, U. Beierlein, J. Ebbecke and A. Wixforth, *Small*, Vol. 1, p. 1188, 2005.
55. K.M. Seemann, J. Ebbecke and A. Wixforth, *Nanotechnology*, Vol. 17, p. 4529, 2006.
56. J. Ebbecke, C.J. Strobl and A. Wixforth, *Phys.Rev. B.*, Vol. 70, p. 233401, 2004.

9 Introduction to Droplet Microfluidics and Multiphase Microflows

9.1 Introduction

In the preceding chapters, we have studied the behavior of droplets placed on a planar surface. However, this is not the only way to manipulate droplets. Another approach, pioneered in the late 1990s, is usually called droplet microfluidics. In such an approach, immiscible discrete droplets are formed and transported by a continuous microflow (Fig. 9.1).

It has been shown that these types of microflows, where droplets are continuously created and transported, can be of great interest for biological and chemical applications [2,3]. We emphasize here that digital microfluidics (planar microfluidics) and droplet microfluidics are not in opposition, and, in fact, are complementary. They just do not address the same problems. For example, digital microfluidics is well adapted to biorecognition and biodiagnostics because only an extremely small volume of liquid is required and droplets can be carefully monitored until the detection step. On the other hand, droplet microfluidics is well adapted to continuous processes, as for example the production of a large number of encapsulated biological objects. From a general standpoint, digital (planar) microfluidics can treat very small volumes of liquid (stored in reservoirs) in parallel, whereas droplet microfluidics treats small liquid volumes in series (continuous on line process).

More generally, droplet microfluidics can be considered as a subset of the more general two-phase or multiphase microflows. The physics of droplet microfluidics and multiphase-flow is a vast domain and this chapter presents some introductory notions. First, we focus on the flow of liquid plugs in capillary tubes, then we analyze two-phase flows in parallel channels, and finally we present some aspects and applications of droplet microfluidics.

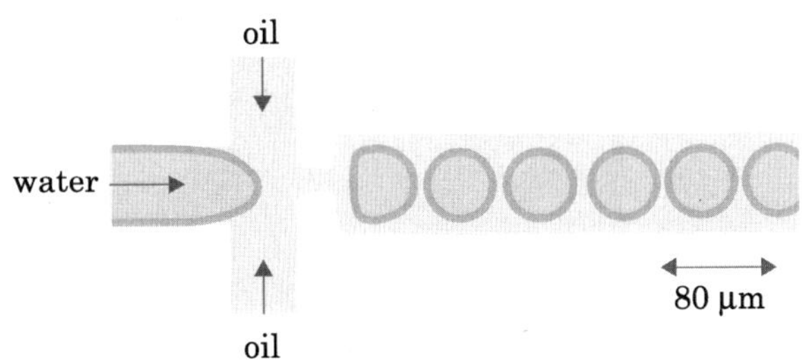

Figure 9.1 A continuous flow can be broken into discrete droplets [1].

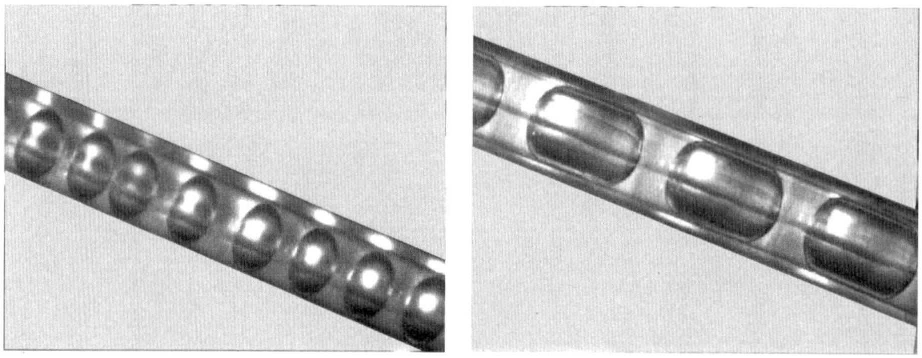

Figure 9.2 Microdrops and plugs in a capillary tube (Courtesy Y. Fouillet, CEA-LETI).

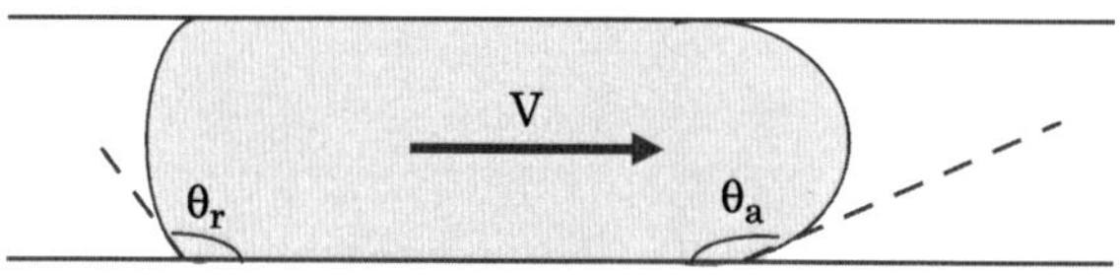

Figure 9.3 Schematic view of a liquid plug in a capillary tube with the advancing and receding contact angles.

9.2 Two-phase Flows: Plugs in Microchannels

Microflows are not always constituted by a single liquid. Frequently, one deals with small quantities of liquid (buffer liquid containing the molecules of interest) under the form of liquid plugs moving inside capillary tubes, and separated by inert, bio-compatible, non-miscible plugs of another liquid (Fig. 9.2).

Conceptually, liquid plugs may be seen as a transitional state between microflows and digital microfluidics.

9.2.1 Interface and Meniscus

The shape of a liquid plug in a capillary tube depends on the capillary forces. A liquid plug moving inside a capillary tube (or between two parallel plates) is limited by two menisci, one corresponding to the advancing front (index a), the other one corresponding to the receding front (index r) as shown in the Fig. 9.3. In micro-capillaries, menisci have spherical shapes. Note that receding, advancing, and static contact angles are not identical.

9.2.2 Dynamic Contact Angle

The contact angle formed between a flowing liquid front (advancing or receding) and a solid surface is not constant but reflects the balance between capillary forces and viscous forces. The relative importance of these forces is often expressed by the non-dimensional capillary number Ca defined by

$$\mathrm{Ca} = \frac{\mu U}{\gamma} \tag{9.1}$$

where μ is the dynamic viscosity of the moving fluid (unit kg/m/s), U its velocity (m/s), and γ its surface tension (N/m). The capillary number is a scale of the ratio between the drag force of the flow on a plug and the capillary forces. In a cylindrical tube of radius R, the friction pressure drop for a plug of length L is given by the Washburn law

$$\Delta P = \frac{8\mu U L}{R^2}. \tag{9.2}$$

We deduce an order of magnitude of the drag force (force necessary to push the plug in the tube)

$$F_{\mathrm{drag}} \approx \Delta P \pi R^2 \approx \mu U L. \tag{9.3}$$

On the other hand, the capillary/wetting force is given by

$$F_{\mathrm{cap}} \approx \gamma R. \tag{9.4}$$

From (9.3) and (9.4) we deduce

$$\frac{F_{\mathrm{drag}}}{F_{\mathrm{cap}}} \approx \frac{\mu U}{\gamma}\frac{L}{R} \approx \mathrm{Ca}\frac{L}{R}. \tag{9.5}$$

Hoffman first proposed an expression for the dynamic contact angle based on experimental observations [4], however, this correlation is rather complicated and Voinov and Tanner [5] have established the more workable correlation

$$\theta_d^3 - \theta_s^3 = A\mathrm{Ca} \tag{9.6}$$

where θ_d and θ_s are the dynamic and static contact angles. The value of the coefficient A is $A \sim 94$ when θ is expressed in radians. Tanner's law is plotted in Fig. 9.4.

For microflows, using the approximate values $\mu \sim 10^{-3}$ kg/m/s, $U \sim 10$ µm/s to 1 cm/s, and $\gamma \sim 50 \times 10^{-3}$ N/m, we find that typical values of the capillary number are in the range 2×10^{-7} to 2×10^{-4}. The capillary number is then small and corresponds to the linear part of the Tanner law. Linearization of

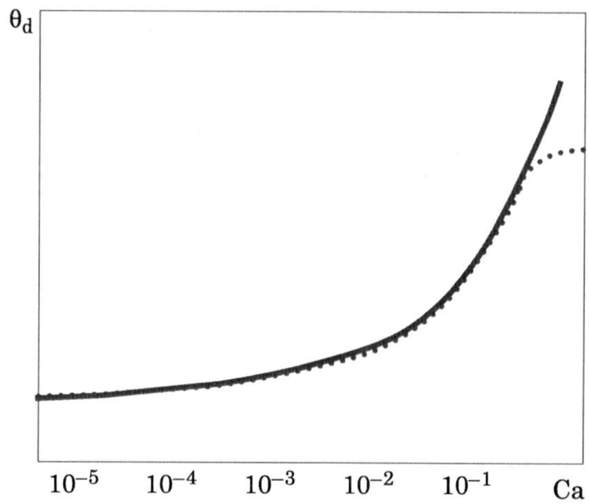

Figure 9.4 Experimental results for the dynamic contact angle versus the capillary number (dots) and Tanner relation (continuous line).

(9.6) yields [6,7]

$$\theta_d = \left(\theta_s^3 + A\text{Ca}\right)^{\frac{1}{3}} \approx \theta_s \left(1 + \frac{1}{3}\frac{A\text{Ca}}{\theta_s^3}\right) \tag{9.7}$$

or

$$\theta_d - \theta_s \approx \frac{1}{3}\frac{A\,\text{Ca}}{\theta_s^2}. \tag{9.8}$$

Note that the capillary number is signed. The relation (9.8) shows that $\theta_d - \theta_s$ is of the sign of Ca; the values of the advancing and receding contact angles are then given by

$$\begin{aligned} \theta_a &\approx \theta_s + \frac{1}{3}\frac{A\,|\text{Ca}|}{\theta_s^2} \\ \theta_r &\approx \theta_s - \frac{1}{3}\frac{A\,|\text{Ca}|}{\theta_s^2} \end{aligned} \tag{9.9}$$

where θ_a and θ_r are, respectively, the advancing and receding contact angles. Relations (9.9) confirm the experimental observation that an advancing contact angle is larger than the static contact angle and a receding contact angle is smaller than the static contact angle (Fig. 9.5).

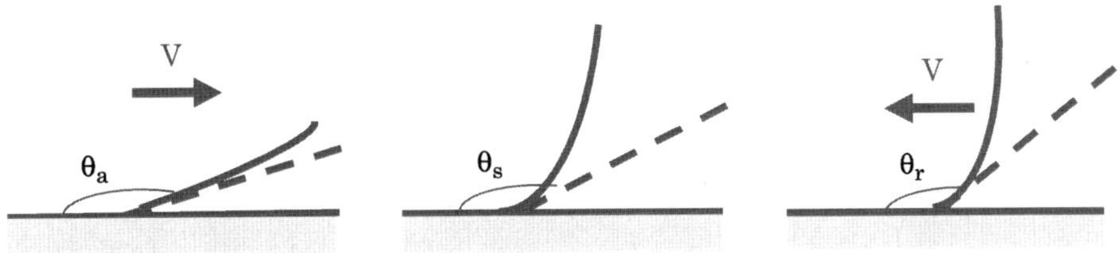

Figure 9.5 Sketch of an advancing, static and receding contact angle. The advancing contact angle is larger than the static contact angle, which is in turn larger than the receding contact angle.

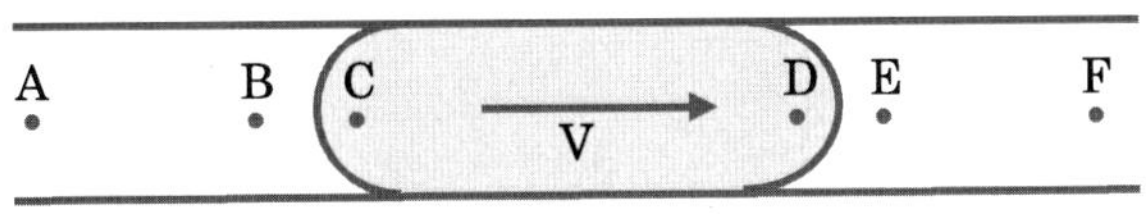

Figure 9.6 Decomposition of a two - phase flow in lumped elements. Between points A and B, C and D, E and F, the pressure drop is due to friction; between B and C, D and E the pressure drop results from capillary forces.

9.2.3 Plugs Moving Inside a Capillary

9.2.3.1 Cylindrical Tube

In this section, we analyze the motion of one or more liquid plugs inside a cylindrical capillary tube. We use a lumped model and we show that Bernoulli's equation combined with Tanner's law explains the main features of the behavior of liquid plugs moving inside capillary tubes [6,7]. Flow regions may be decomposed in two (Fig. 9.6): first, the regions where a fluid moves inside the capillary, inducing a friction pressure drop, and second, the interfaces that induce a capillary pressure drop.

The total pressure drop in the capillary is then

$$\Delta P_{\text{channel}} = \Delta P_{\text{cap}} + \Delta P_{\text{drag}}. \tag{9.10}$$

The pressure drop due to friction on the solid walls is given by the Washburn law [9]

$$\Delta P_{\text{drag}} = \frac{8U}{R^2}\left(\mu_1 L_1 + \mu_2 L_2\right) \tag{9.11}$$

where indices 1 and 2 address liquid 1 (liquid plug) and liquid 2 (surrounding carrier fluid). R is the radius of the capillary, U the average liquid velocity and L_1, L_2 the total length of contact of liquid 1, 2 with the solid wall ($L_1 + L_2 = L$, total length of the tube). Each interface—advancing and receding—contributes (positively or negatively in function of the contact angles) to the capillary

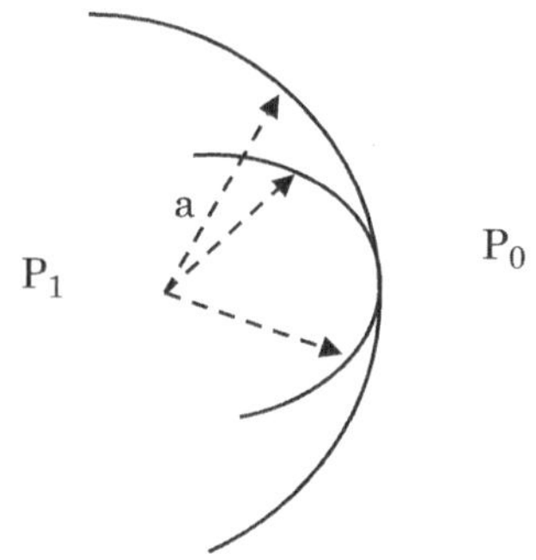

Figure 9.7 Sketch of a spherical interface separating two fluids. The pressure difference between the convex region and the outside region is $P_1 - P_0$ and is given by Laplace law (see Chapter 1).

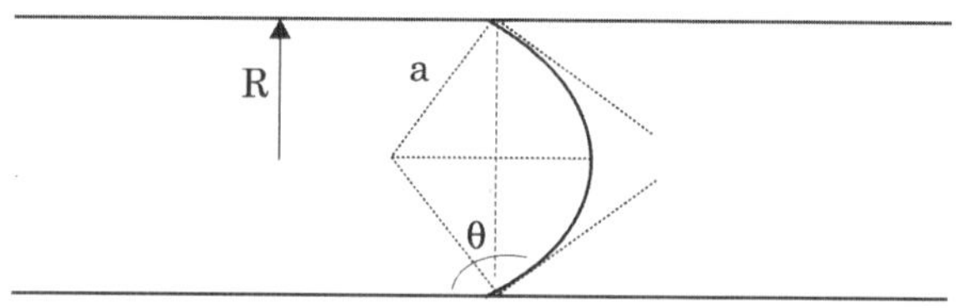

Figure 9.8 Schematic view of the meniscus in a cylindrical capillary tube.

pressure drop. We have seen in Chapter 2 that the capillary pressure drop derives directly from the Laplace law which relates the pressure difference at a spherical interface of curvature radius a (Fig. 9.7) by

$$\Delta P = \frac{2\gamma}{a}. \tag{9.12}$$

The meniscus has a spherical shape (if the capillary is small enough), and Fig. 9.8 shows that the contact angle is related to the tube radius R and the curvature radius a by

$$\cos\theta = -\frac{R}{a}.$$

Substitution of this relation into (9.12) yields

$$\Delta P_a = -\frac{2\gamma}{R}\cos\theta_a. \tag{9.13}$$

Similarly, the receding front contribution is given by the relation

$$\Delta P_r = \frac{2\gamma}{R}\cos\theta_r. \tag{9.14}$$

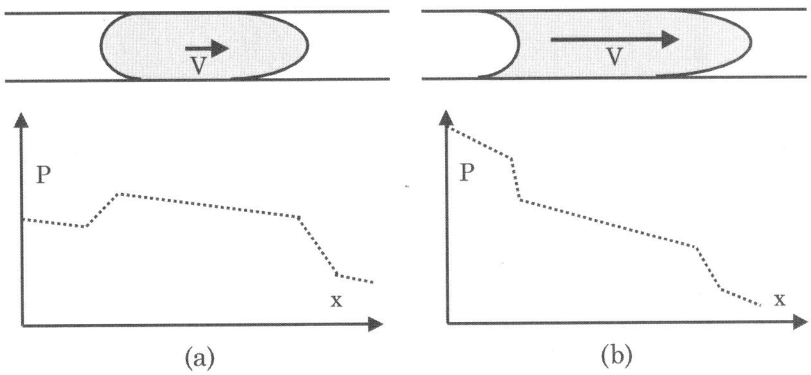

Figure 9.9 Two possible configurations for a plug moving inside a capillary tube: (a) at low velocity, the receding angle is larger than $\pi/2$ and the contribution to the pressure drop is negative, (b) at high velocity, θ_r is smaller than $\pi/2$ and the contribution to the pressure drop is positive. The slope of the pressure drop inside the different liquid is due to the friction pressure drop.

In our convention, the pressure drop is always taken following the fluid flow. Consider the two configurations of Fig. 9.9. If θ_a is larger than $\pi/2$, there is a positive pressure drop associated with the advancing interface. If θ_r is smaller than $\pi/2$, the receding front contributes positively to the pressure drop (right scheme in Fig. 9.9), and negatively in the opposite case (left scheme in Fig. 9.9).

The capillary pressure drop is due to the difference of the capillary forces between advancing and receding fronts because of the two different contact angles (advancing and receding) θ_a and θ_r

$$\Delta P_{\text{cap}} = \frac{2\gamma}{R}\left(-\cos\theta_a + \cos\theta_r\right). \tag{9.15}$$

Equation (9.15) shows that too many plugs in the capillary may rapidly block the flow. For N plugs flow, we will have for some N

$$\Delta P_{\text{cap}} = \frac{2\gamma}{R}N\left(-\cos\theta_a + \cos\theta_r\right) > P_i - P_o \tag{9.16}$$

and the flow will come to a stop (P_i and P_o are the inlet and outlet pressures). Let us introduce now the linearized Hoffman-Tanner law—equation (9.7)—to find a more workable expression of the capillary pressure drop

$$\theta_a = \theta_{s,a}\left(1 + \frac{1}{3}\frac{A\,\text{Ca}}{\theta_{s,a}^3}\right) \tag{9.17}$$

and

$$\theta_r = \theta_{s,r}\left(1 - \frac{1}{3}\frac{A\,\mathrm{Ca}}{\theta_{s,r}^3}\right) \tag{9.18}$$

with

$$\mathrm{Ca} = \frac{U\mu_1}{\gamma} \tag{9.19}$$

where the index s stands for the static contact angle, and $\theta_{s,r}$ and $\theta_{s,a}$ are the two static contact angles. They are equal if there is no static hysteresis, i.e., if the surface is perfectly smooth. The minus sign in (9.18) derives from the fact that we consider a positive Ca. After substitution of (9.18) and (9.17) into (9.16), and using some algebra, and keeping the higher order terms only, the capillary pressure drop can be cast under the form

$$\Delta P_{\text{cap}} \cong \frac{2\gamma}{R}N\left(-\cos\theta_{s,a} + \cos\theta_{s,r}\right) + \frac{2ANU\mu_1}{3R}\left(\frac{\sin\theta_{s,a}}{\theta_{a,s}^2} + \frac{\sin\theta_{s,r}}{\theta_{r,s}^2}\right). \tag{9.20}$$

9.2.3.1.1 No static hysteresis

If there is no static hysteresis of contact angle, the first term to the right hand side of (9.20) vanishes and the capillary pressure drop is proportional to the average velocity U of the flow. As the friction pressure drop is also proportional to U, so is the total pressure drop

$$\Delta P = U\left[\frac{8}{R^2}\left(\mu_1 L_1 + \mu_2 L_2\right) + \frac{2AN\mu_1}{3R}\left(\frac{\sin\theta_{s,a}}{\theta_{a,s}^2} + \frac{\sin\theta_{s,r}}{\theta_{r,s}^2}\right)\right]. \tag{9.21}$$

Let us write the ratio between the friction and capillary pressure drop

$$\frac{\Delta P_{\text{drag}}}{\Delta P_{\text{cap}}} = \frac{12\left(L_1 + \frac{\mu_2}{\mu_1}L_2\right)}{ANR\left(\frac{\sin\theta_{s,a}}{\theta_{a,s}^2} + \frac{\sin\theta_{s,r}}{\theta_{r,s}^2}\right)} \approx \frac{L}{12RN}\frac{\theta_s^2}{\sin\theta_s} \tag{9.22}$$

where θ_s is the median value of the static contact angle. In (9.22) we have supposed that the viscosities of the two fluids were similar, and $L = L_1 + L_2$. Suppose that θ_s is of the order of $\pi/2$, then the ratio of the two types of pressure drop becomes $\Delta P_{\text{friction}}/\Delta P_{\text{capillary}} \approx (L/R)(1/5N)$. Depending on the aspect ratio L/R and on the number of plugs N, the friction pressure drop or the interface pressure drop governs the flow. In the example of Fig. 9.2, the capillary radius is $R = 100$ μm, and the length $L = 2$ mm; the number of plugs

N is larger than 40, hence the capillary pressure drop is 10 times larger than the friction pressure drop.

9.2.3.1.2 Static hysteresis

In case of hysteresis of static contact angle, at very low velocities, the pressure drop has an asymptotic value which is the first term of the right hand side of (9.20).

$$\Delta P_{\text{cap}} \approx \frac{2\gamma}{R} N \left(-\cos\theta_{s,a} + \cos\theta_{s,r}\right). \tag{9.23}$$

In this case, the ratio of the friction to the capillary pressure drop is approximately

$$\frac{\Delta P_{\text{drag}}}{\Delta P_{\text{cap}}} \approx \frac{L}{R}\frac{1}{N}\frac{1}{\left(-\cos\theta_{s,a} + \cos\theta_{s,r}\right)}\text{Ca}. \tag{9.24}$$

Relation (9.24) for the ratio of the two pressure drops is comparable to relation (9.5) for the ratio of the two forces inducing the pressure drops. Both ratios are proportional to the capillary number.

9.2.3.2 Rectangular Channel

The same reasoning can be done for rectangular channels, with the precaution of substituting a capillary and a friction pressure drop adapted for rectangular micro-channels. The expression of the capillary pressure loss (9.13) becomes

$$\Delta P_a = -2\gamma\cos\theta_a\left(\frac{1}{h} + \frac{1}{w}\right)$$

where h and w are the height and width of the channel. On the other hand, the friction factor$f$ is given by the Shah and London approximated expression [7]

$$\begin{aligned} f\,\text{Re} &= 96\left(1 - 1.3553\alpha + 1.9467\alpha^2 - 1.7012\alpha^3 + 0.9564\alpha^4 - 0.2537\alpha^5\right) \\ &= 96g\left(\alpha\right) \end{aligned}$$

where Re is the Reynolds number, α is the ratio between the two dimensions of the rectangular channel $\alpha = \min(\text{h/w}, \text{w/h})$. Note that α is smaller than 1. After substituting this relation in the Darcy-Weisbach equation, we obtain

$$\Delta P_{\text{drag}} = f\frac{\rho}{2}U^2\frac{L}{D_h} = \frac{8\eta U L}{w^2}\frac{3}{2}\left(1+\alpha\right)^2 g\left(\alpha\right).$$

A more compact form has been recently obtained by Bahrami et al. [8] (including many other geometries of cross section)

$$\Delta P_{\text{drag}} = \frac{8\eta UL}{w^2} \frac{1}{2\left(\frac{1}{3} - \frac{64\alpha}{\pi^5} \tanh \frac{\pi}{2\alpha}\right)} = \frac{8\eta UL}{w^2} h(\alpha).$$

From hereon the reasoning is the same as that detailed in the previous section.

9.2.4 Hysteresis of Static Contact Angle

In Chapter 2, Young's law for contact angle was derived. Young's law predicts the value of the static contact angle as a function of the surface energy of the different materials (liquid plug, surrounding liquid, and solid substrate). Apparently, it should result in a unique value of the static contact angle. However, it happens frequently that the static contact angle is not uniquely defined. It can be comprised between two values, a first value obtained by slowing down to a stop an advancing front $\theta_{s,a}$, and another value (smaller) obtained by slowing down to a stop a receding front $\theta_{s,r}$ as shown in Fig. 9.10.

In fact, Young's law applies only for a perfectly homogeneous solid substrate. If the substrate is disordered, the static contact angle is not uniquely defined. Hysteresis of contact angle is linked to the smoothness and homogeneity of the substrate [10–13].

9.2.5 Example of Three-phase Flow in a Microchannel

A promising way to perform biological and chemical reactions in microvolumes is to use plugs transported by an immiscible carrier fluid. Fig. 9.11 shows the principle of such reactions. Different reagents transported by independent plugs are successively mixed with a solution containing a chemical or biochemical species. A condition for a proper functioning of such plug flow reactions is that the plugs do not coalesce. Coalescence would bring contamination between the liquids. Coalescence may occur when the plugs are not moving at the same velocity, due to differences of viscosity and surface tensions, according to equations (9.22) and (9.24). In order to keep the plugs separated, Chen et al. [14] use spacer plugs constituted by a third immiscible liquid.

We place ourselves in the case of slow moving liquids (small capillary and Weber numbers) so that surface tension forces dominate inertia and viscous forces. We assume also that the carrier fluid is the only fluid wetting the solid walls.

Obviously, a first condition for the efficiency of spacer plugs is that the liquid plugs do not engulf each other. Fig. 9.12 shows the satisfactory arrangement

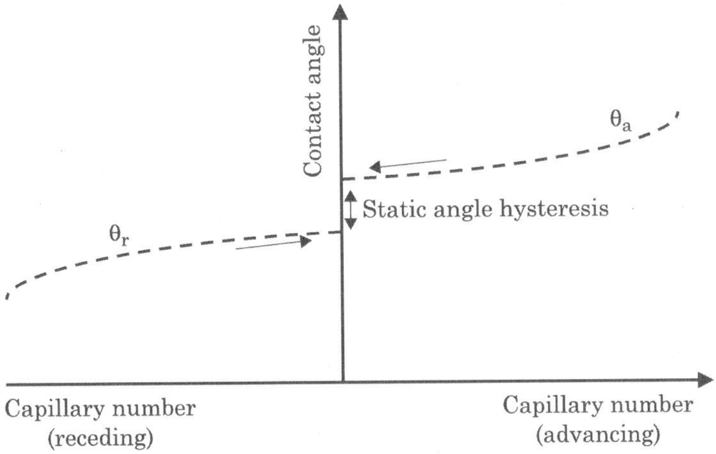

Figure 9.10 Hysteresis of static contact angle and Hoffman–Tanner law for advancing and receding contact angles versus Capillary number: the advancing contact angle is larger than the receding contact angle and there is a static hysteresis at zero velocity.

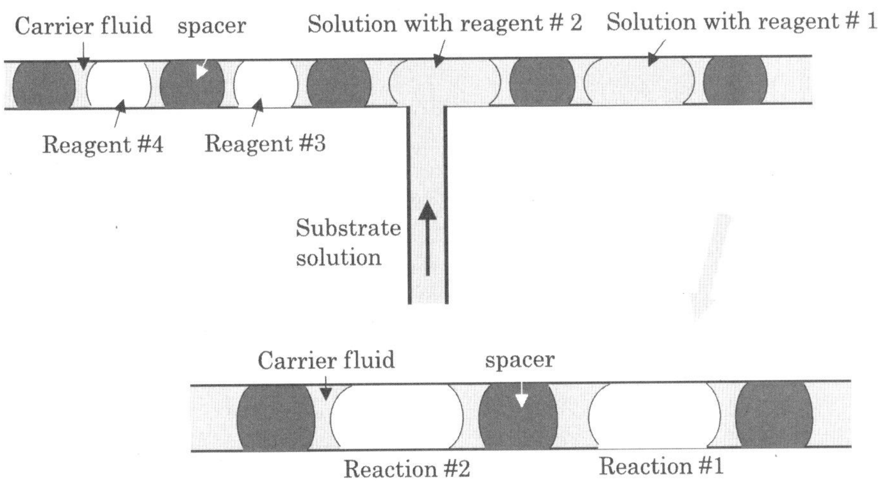

Figure 9.11 Principle of three-phase flow reactions: spacer plugs of immiscible liquid prevent coalescence of droplets.

of the plugs (a), and engulfment (b). The condition for stability of plugs in contact is given by the balance of the surface tension forces at the triple line (Neumann's construction)

$$\vec{\gamma}_{ct} + \vec{\gamma}_{tr} + \vec{\gamma}_{rc} = 0. \tag{9.25}$$

Relation (9.25) can be satisfied only if the magnitude of every force is smaller than the sum of the magnitudes of the other two forces. This statement can

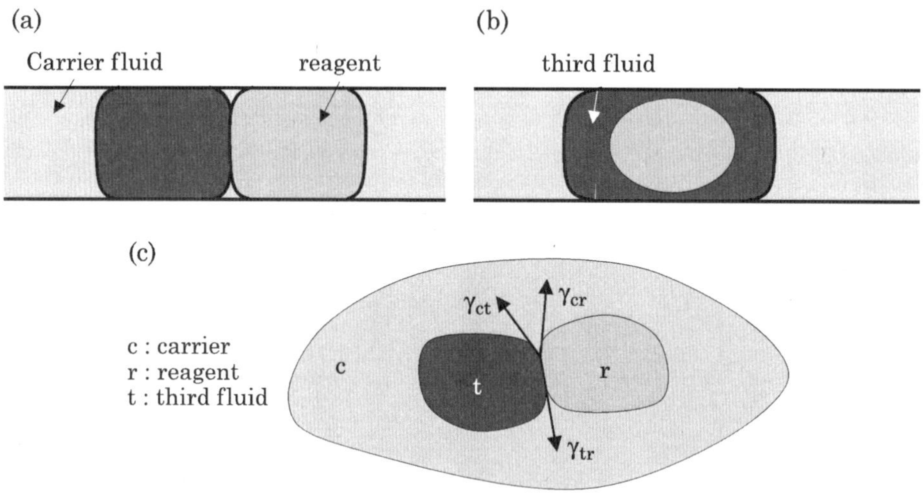

Figure 9.12 Sketch of two plugs in contact. (a) Plugs stay distinct; (b) spacer liquid plug engulfing reagent liquid; (c) Neumann's construction for the equilibrium of the interfaces.

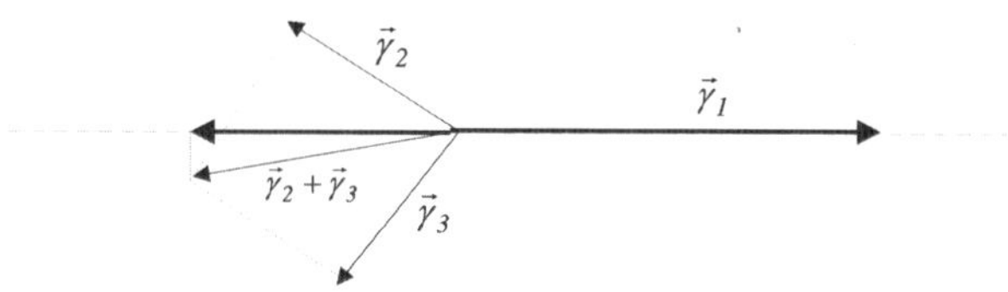

Figure 9.13 Assuming that γ_1 is larger than $\gamma_2 + \gamma_3$, the resultant of the forces $\vec{\gamma}_2 + \vec{\gamma}_3$ projected on the direction of γ_1 cannot equilibrate $\vec{\gamma}_1$.

be easily verified by remarking that if the magnitude of a force is larger than the sum of the magnitudes of the two others, equilibrium cannot be reached (Fig. 9.13).

Hence, it can be shown that the condition for non-engulfment is

$$\begin{gathered}\gamma_{rc} < \gamma_{ct} + \gamma_{tr} \\ \text{and} \\ \gamma_{ct} < \gamma_{tr} + \gamma_{rc}.\end{gathered} \tag{9.26}$$

Gas bubbles usually satisfy relation (9.26). However, gas bubbles can induce pressure fluctuation in the system. Chen et al. [14] have found organic spacer liquids (SID, DTFS) adapted to fluorinated carrier fluids and aqueous reagent plugs.

9.3 Two-phase Flows: Flowing Fluids Separated by an Interface

In biotechnology and micro-chemistry, one frequently proceeds the transfer of a species contained in a liquid A (primary fluid) to another immiscible liquid B (secondary fluid). Mass transfer is facilitated when the interfacial area between the two immiscible flowing fluids is important. This is, for example, the case of microfluidic systems aimed at the detection of viruses or of traces of heavy metals in large volumes of liquids.

In this section, we focus on the extraction of target molecules from a primary liquid and concentration of these molecules in a secondary liquid. We present a system constituted of two adjacent microchannels geometrically separated by vertical micro-pillars [15]. The primary and secondary liquids are chosen immiscible, and vertical interfaces attached to the pillars separate the two fluids. The system must be designed to withstand the largest possible interfacial area to favor molecule transfer from primary to secondary fluids. Stability of the interface is then the key for the design of the system.

9.3.1 Introduction

The transfer and concentration of macromolecules from a flowing primary carrier fluid to a secondary storage liquid is the first step of any recognition process. In the literature such types of concentration micro-devices have already been proposed [16,17]. We analyze the case sketched in Fig. 9.14 where the carrier fluid and the storage liquid are flowing side by side, separated by interfaces sustained by micro-pillars. The two immiscible fluids should not mix to form an emulsion, for two reasons: the system should use the economy of a de-mixing system, and on-line detection is facilitated. The best efficiency of such devices is obtained for a largest possible interfacial area. However, it is observed that, for a given geometry of the pillars, there are only a limited number of interfaces that can be stable. We investigate here the stability of the interfaces in function of the geometry (channel length and width, morphology of the pillars) and flow parameters (flow rates in the channels).

9.3.2 Model

The two fluids flow in parallel with different velocities, and their pressure evolves nearly independently—as long as the interfaces remain stable. Because of the differences of flow rate and geometry, a pressure difference between the two channels builds up. Hence the interfaces are submitted to two constraints: a normal stress due to the pressure difference between the two channels, and a tangential stress exerted by the friction of the flowing fluids.

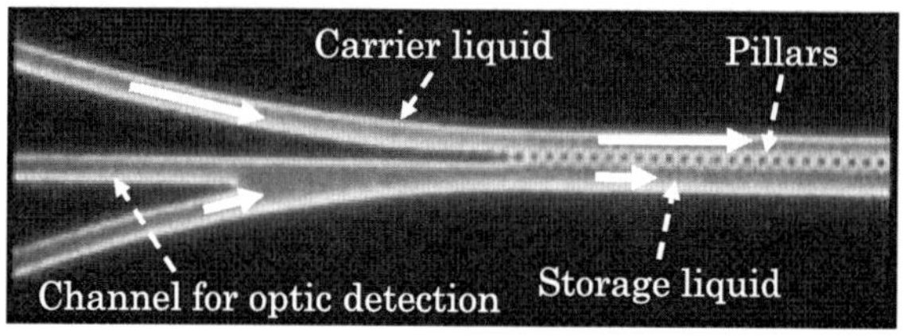

Figure 9.14 Schematic of the microsystem. From [14]. © 2007 NSTI; reprinted with permission.

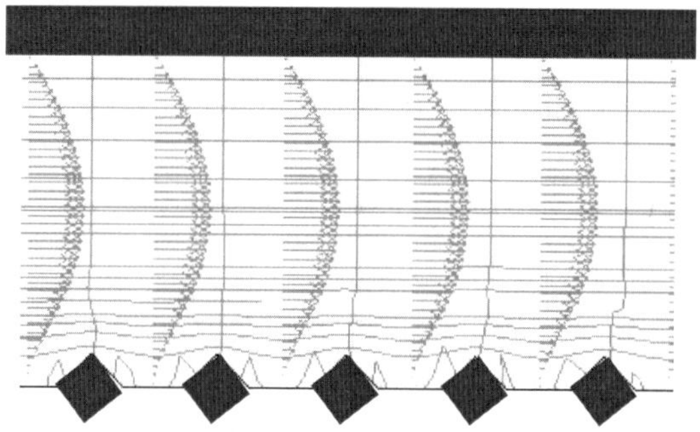

Figure 9.15 Velocity profile, pressure contour-lines and streamlines in a channel limited by lozenge pillars and a stable interface. From [14]. © 2007 NSTI; reprinted with permission.

A calculation of the flow with the Comsol numerical software shows that the tangential stress on the interface is much less than the normal stress (Fig. 9.15). Hence, we consider only the normal stress in the model.

Taking into account the outlet channels and using the Washburn [18] equation for the pressure drop, modified by the Purday, Shah and London expressions for rectangular channels [7,19], the pressure is a function of the distance to the channel entrance, noted x, according to

$$P(x) - P_0 = cf(\mu, Q, a, b)\, x + cg(\mu, Q, D) \tag{9.27}$$

with the notations of Fig. 9.16, and where μ is the liquid dynamic viscosity. The functions f and g are respectively the pressure drop functions for the "transfer" channel and the outlet channel. The pressure difference across the interface in a cross section of the transfer region is then

$$\begin{aligned} P_1(x) - P_2(x) = {} & c_1 \left[f(\eta_1, Q_1, a_1, b_1) - f(\eta_2, Q_2, a_2, b_2) \right] x \\ & + c_2 \left[g(\eta_1, Q_1, D_1) - g(\eta_2, Q_2, D_2) \right]. \end{aligned} \tag{9.28}$$

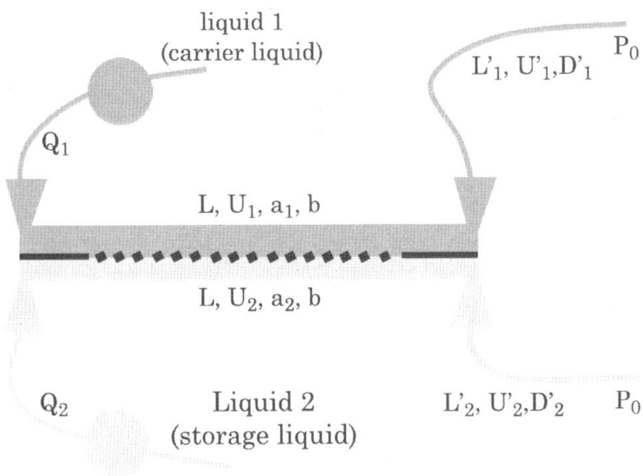

Figure 9.16 Schematic of the flow parameters: a is the channel width, b the channel depth, L stands for length, U for velocity, Q for flow rate, and D for hydraulic diameter.

If we specify the functions f and g, equation (9.28) yields

$$P_1(x) - P_2(x) \cong 8(L-x)\left[\frac{\eta_1 Q_1}{\frac{a_1^3 b^3}{(a_1+b)^2}} - \frac{\eta_2 Q_2}{\frac{a_2^3 b^3}{(a_2+b)^2}}\right] + 32\left(\frac{\eta_1 Q_1 L_1}{\pi D_1^4} - \frac{\eta_2 Q_2 L_2}{\pi D_2^4}\right). \qquad (9.29)$$

The maximum pressure difference $P_1 - P_2$ for which an interface remains stable is related to the maximum theoretical stability length, sketched in Fig. 9.17. We analyze here the maximum pressure that an interface can withstand for lozenge and cylindrical pillars (Fig. 9.18).

9.3.2.1 Lozenge pillars

The stability approach for lozenge pillars is identical to Tsori's approach for wedged pipettes [20], and we will not reproduce the equations here. It suffices to say that the interface is always anchored to the pillar edges (because the surfaces of the two adjacent pillars make sufficiently large angles). Hence, the maximum pressure difference that the interface can sustain (if the pillar edges are perfectly straight) is given by the Laplace equation for a minimum curvature radius

$$\Delta P_{\max} = \frac{\gamma}{R} = \frac{2\gamma}{\delta} \qquad (9.30)$$

where R is the curvature radius and $\delta = 2R$ is the distance between two pillar edges.

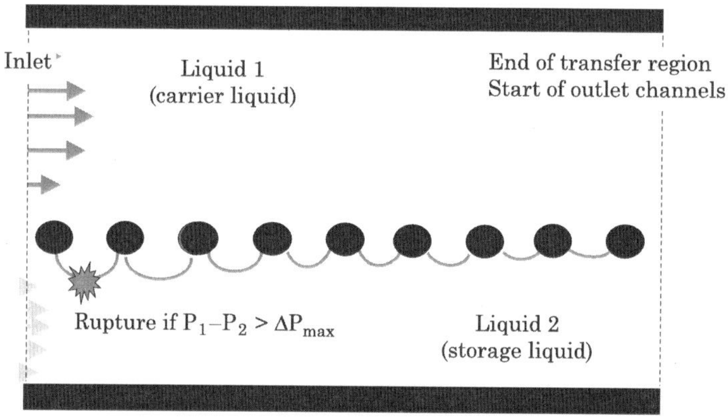

Figure 9.17 Sketch of the stability of the interfaces. The rupture occurs always at the beginning of the channel (under the condition of perfectly microfabricated pillars).

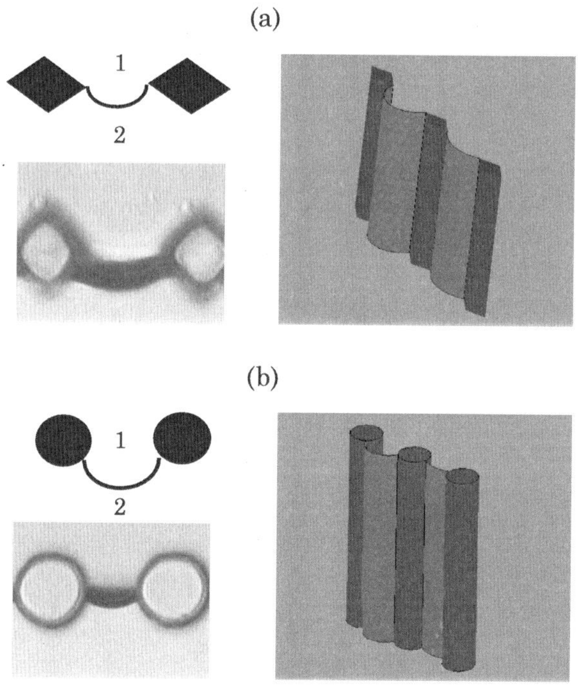

Figure 9.18 Interface position: (a) the interface is anchored to the pillar edge, (b) the interface adjusts to respect the Young constraint. Experimental photos (left) and Evolver calculation (right) [20]. From [14]. © 2007 NSTI; reprinted with permission.

9.3.2.2 Circular pillars

In this case there is no pinning of the interface; the location of the triple contact line is determined by Young's constraints. Using the notations of

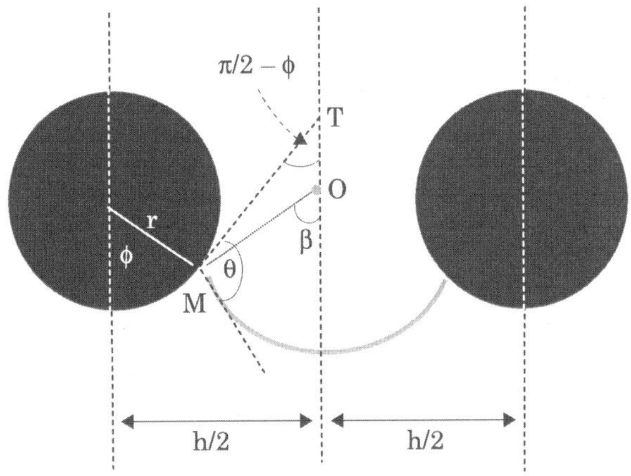

Figure 9.19 Schematic of the location of the interface. O is the curvature center, T the intersection of the tangent with the median axis, and M the contact point.

Fig. 9.19, we locate the position of the contact point M by its angle ϕ, and we have the following relation

$$\frac{h}{2} = r\sin\varphi + R\sin\beta. \tag{9.31}$$

The angle β can be expressed in function of the two angles ϕ and θ by

$$\left(\frac{\pi}{2} - \varphi\right) + (\pi - \beta) + \left(\theta - \frac{\pi}{2}\right) = \pi. \tag{9.32}$$

And we find

$$\beta = \theta - \varphi. \tag{9.33}$$

Substitution of (9.33) in (9.31) produces the curvature radius

$$R = \frac{\frac{h}{2} - r\sin\varphi}{\sin(\theta - \varphi)}. \tag{9.34}$$

The pressure difference across the interface is then

$$\Delta P = \frac{\gamma}{R} = \frac{\gamma\sin(\theta - \varphi)}{\frac{h}{2} - r\sin\varphi}. \tag{9.35}$$

The important point now is that the function $\Delta P(\phi)$ has a maximum; in other words, there is a position of the interface corresponding to a minimum curvature

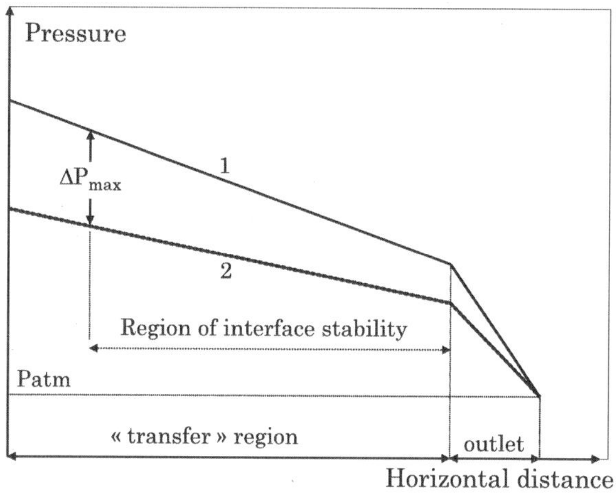

Figure 9.20 Pressure profiles and determination of the length of interface stability.

radius, i.e., to a maximum pressure difference. This position is determined by

$$\frac{\partial(\Delta P)}{\partial \varphi} = 0.$$

After some algebra, we find that the solution is given by

$$\varphi_{\max} = \theta - \arccos\left(\frac{2r\sin\theta}{h}\right). \tag{9.36}$$

And the maximum sustainable pressure is then

$$\Delta P_{\max} = \frac{\gamma \sin(\theta - \varphi_{\max})}{\frac{h}{2} - r\sin\varphi_{\max}}. \tag{9.37}$$

After examination of equations (9.36) and (9.37), we deduce that the maximum pressure difference is a function of the Young contact angle θ, the distance between two pillars centers h, and the radius of the pillar r.

9.3.2.3 Maximum stability length

On a general standpoint, the horizontal pressure profiles P_1 and P_2 along the channels are shown in Fig. 9.20. According to equation (9.29) the profile is piecewise linear, with a jump of the slope between the "transfer" region and the outlet channels.

Examining the figure, we deduce that there is a maximum stability length corresponding to a pressure difference of $\Delta P_{\max}$. Depending on the lozenge or

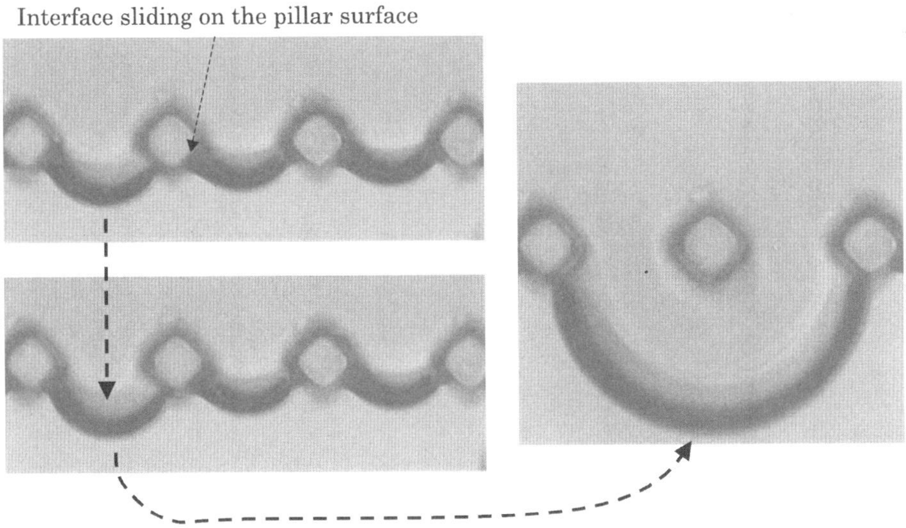

Figure 9.21 Once the triple contact line has detached from the edge—under the effect of the pressure difference or of a defect in the edge of the pillar—the interface inevitably invades the neighboring channel. From [14]. © 2007 NSTI; reprinted with permission.

circular geometry, we use equations (9.30) or (9.37). Substitution in equation (9.29) leads to the maximum value of the length of the transfer channel

$$\Lambda = \frac{\Delta P_{\max} - 32\left(\frac{\eta_1 Q_1 L_1}{\pi D_1^4} - \frac{\eta_2 Q_2 L_2}{\pi D_2^4}\right)}{8\left(\frac{\eta_1 Q_1}{\frac{a_1^3 b^3}{(a_1+b)^2}} - \frac{\eta_2 Q_2}{\frac{a_2^3 b^3}{(a_2+b)^2}}\right)}. \tag{9.38}$$

9.3.3 Experimental Results

Experiments have been conducted using microfabricated channels of silanized Ordyl. Lozenge and circular pillars (150 μm high) have been tested. Stability has been investigated for water-cyclohexane interfaces.

9.3.3.1 Lozenge pillars

Interfaces attach to the edges of the pillars, as predicted by the theory. When the pressure increases above the rupture threshold or if the edge is not perfectly sharp, the interface slides along the sides of the pillars and carrier liquid invades the storage channel (Fig. 9.21).

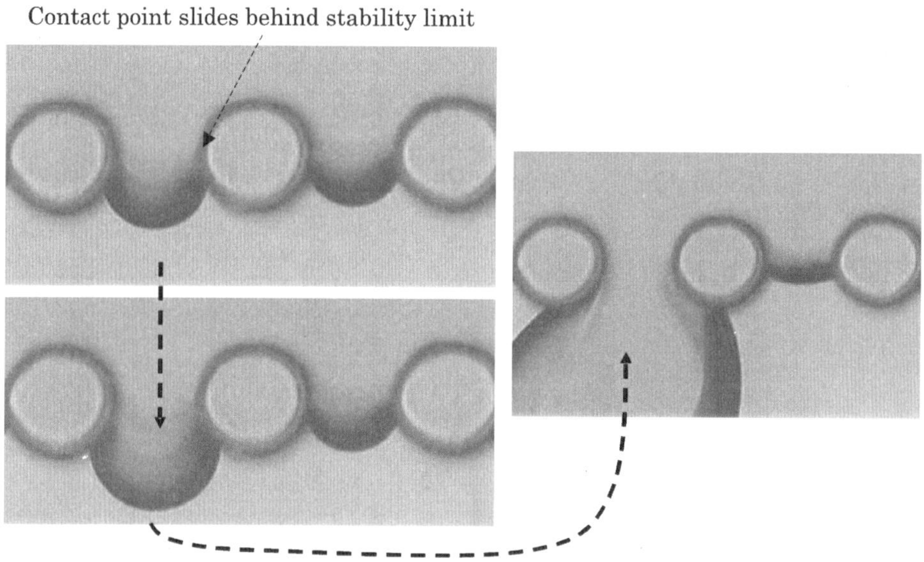

Figure 9.22 When the pressure difference becomes larger than ΔP_{max}, the contact slides beyond the limit position, the interface invades the neighboring channel. From [14]. © 2007 NSTI; reprinted with permission.

9.3.3.2 Circular pillars

We observe that interfaces take the position predicted by the theory, as shown in Fig. 9.22. Beyond the limit angle ϕ_{max}, the interface invades the other channel.

9.3.4 Discussion and Conclusions

Relation (9.38) shows that, in order to maximize the stability length L_{max}, it is necessary (i) to reduce as much as possible the difference of pressure drop in the outlet channels, (ii) to minimize the pressure drop difference in the transfer region, and (iii) to have the larger possible rupture threshold ΔP_{max}. The rupture threshold ΔP_{max} depends essentially on the free distance between two neighboring pillars, on the interface tension between the two liquids, and on the shape of the pillars. Fig. 9.23 shows a comparison of the rupture pressure threshold between lozenge and circular pillars. Comparison has been performed by considering same exchange surfaces and same free distances between two pillars. The pinning of the interface on the lozenge arêtes stabilizes the interface. However, the model assumes perfectly microfabricated pillars. In reality, defects on edges deteriorate the pinning of the interface and may lead to a more instable situation.

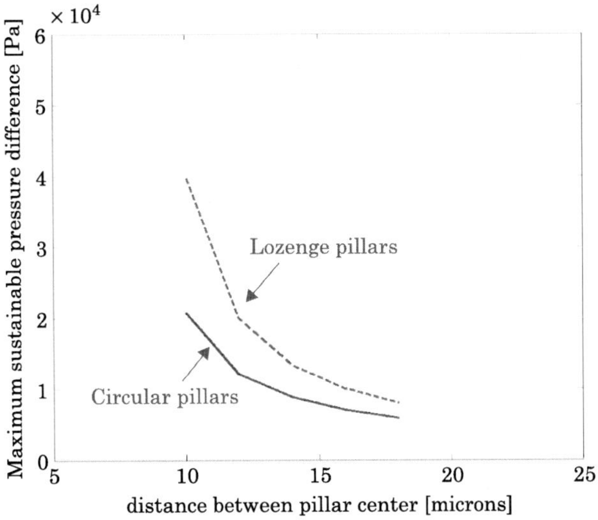

Figure 9.23 Comparison of stability between lozenge and circular pillars: maximum pressure difference ΔP_{max} versus the distance between pillar centers.

In conclusion, we have shown here that the outlet capillary tubes must be carefully designed (length and diameter adapted to the velocities) in order that their specific pressure drop is similar. It is advantageous to have the interfaces attached (pinned) to edges of the vertical pillars. However, if the quality of the edges is not satisfactory, the attachment is poor and sudden breakup of the interfaces can take place. Finally, equations (9.30) and (9.37) show that the smaller the size of the interfaces, the larger the pressure difference sustained by the interface. Assuming that h denotes the distance between pillar center and r the radius or the 1/2 axial dimension of the pillars, let us analyze what happens when the dimensions are homotetically decreased, keeping the number r/h constant. Equation (9.30) can be re-written as

$$\Delta P_{\text{lozenge,max}} = \frac{2\gamma}{h\left(1 - 2\frac{r}{h}\right)} \propto \frac{1}{h}$$

and equation (9.37) can be cast under the form

$$\Delta P_{\text{cylinder,max}} = \frac{\gamma \sin\left(\theta - \varphi_{\text{max}}\right)}{h\left(\frac{1}{2} - \frac{r}{h}\sin\varphi_{\text{max}}\right)} \propto \frac{1}{h}.$$

Thus the maximum pressure that the interface can withstand is inversely proportional to h. At the same time, the interfacial area remains unchanged by the homothetic transform, because

$$\frac{S_i}{S} = \frac{n\,(h-2r)}{nh} = 1 - 2\frac{r}{h}$$

where n is the number of intervals between pillars. Hence, it is advantageous to reduce the size of the interfaces together with the size of the pillars, the limit being the constraints of the microfabrication process.

9.4 Droplets in Two-phase Flows

In this section, we focus on the formation and behavior of droplets in a microflow, and especially on the emission of mono-dispersed droplets. It is an extremely important challenge in biotechnology to be able to produce mono-dispersed droplets in a continuous flow. Such droplets can be produced either in T-junctions of in flow focusing devices (FFD) [22,23]. In this section we investigate the mechanisms of droplet break-up; we show that the mechanisms are not the same for T-junctions and FFD. Finally, we present applications of such devices in biology and biotechnology.

9.4.1 Flow Focusing Devices (FFD) and T-junctions

Two different systems corresponding to two types of instabilities are used to create micro-drops in a microflow. The first type of instability occurs in T-junctions at low velocity—small capillary and Weber numbers—where the detachment of a droplet is governed by the pressure drop created by the merging droplet (Fig. 9.24 (a)). The second type is obtained in flow focusing devices and in T-junctions with higher flow velocity, where a flowing liquid is reduced to a filament under the action of another surrounding fluid (Fig. 9.24 (b)); because of the surface tension forces, the filament cannot be indefinitely stable; it breaks down in droplets at some distance of the channel entrance.

In the following sections, we present these two types of micro-devices and show their applications for the encapsulation of liquids and particles. An important observation at this point is that multi-step devices using T-junctions and FFDs can be realized, as shown in Fig. 9.25. This principle is the key to multi layering encapsulation.

9.4.1.1 Droplet Detachment in T-junctions

T-junctions that we consider are typical of microfluidics: due to microfabrication constraints, they are formed by two rectangular channels of the same depth b, usually merging at an angle of 90° (Fig. 9.26). It has been observed that

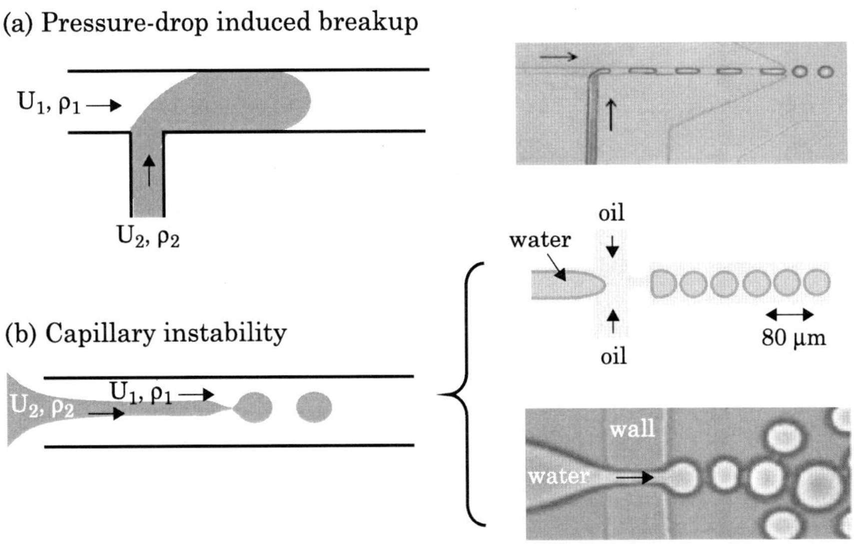

Figure 9.24 Two different types of instabilities leading to droplet break-up: (a) in a T-junction, (b) in a FFD. Figure (a) from [23]; reprinted with permission; figures (b) reused with permission from [22]. © 2003 American Institute of Physics.

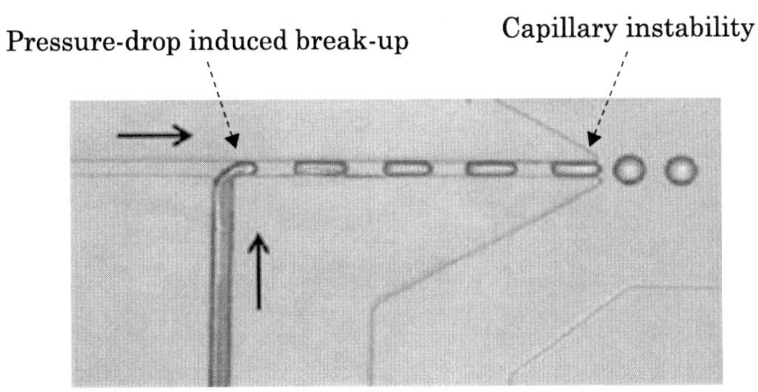

Figure 9.25 The two types of instabilities can be used successively. From [23], reproduced by permission of The Royal Society of Chemistry.

droplet detachment in T-junction depends on the flow velocity: the instability leading to droplet detachment is different whether $\mathrm{Ca} < 10^{-2}$ or $\mathrm{Ca} > 10^{-2}$. Microsystems for biotechnology usually function with small flow rates, so that we assume here that the carrier fluid (continuous phase) flows at low speed ($\mathrm{Ca} < 10^{-2}$). The case of large velocities has been treated by Nisisako et al. [25] and Thorsen et al. [26]. We assume also that the carrier fluid wets the walls.

9.4.1.1.1 Principle of fluid segments (plugs) formation

T-junctions are one of the most frequently used microfluidic geometries to produce immiscible fluid segments (plugs) and droplets. The droplet formation

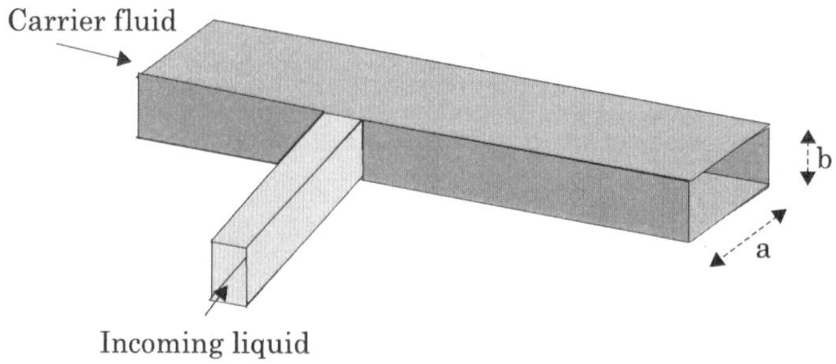

Figure 9.26 Sketch of a T-junction composed of rectangular microfabricated channels.

proceeds in several steps: the liquid penetrates the main channel, forms a blob, and develops a neck. The neck elongates and becomes thinner as the blob advances downstream. It eventually breaks up and the droplet detaches.

At low capillary and Weber numbers, interfacial forces dominate shear stress, and breakup is triggered by the pressure drop across the droplet (or the bubble). In such a flow regime, the size of the droplets is determined solely by the ratio of the volumetric rates of flow of the two immiscible fluids. For rectangular cross sections, if L is the length of the fluid segment, a the width of the channel, Q_{disp} and Q_{cont} the flow rates of the discontinuous and continuous phases respectively, it has been observed that the relation

$$\frac{L}{a} = 1 + \alpha \frac{Q_{\text{dis}}}{Q_{\text{cont}}} \tag{9.39}$$

links the length L to the flow rates [27]. In relation (9.39), the constant α is positive and of the order of 1. Hence the length of the droplet L is always larger than a, and the droplet is in reality a fluid segment. Note that relation (9.39) is not valid for the entire domain of variation of the ratio $Q_{\text{disp}}/Q_{\text{cont}}$. For small values of this ratio, L is constant, as indicated in Fig. 9.27. Relation (9.39) should be corrected by

$$\frac{L}{a} = 1 + \alpha \frac{Q_{\text{dis}}}{Q_{\text{cont}}} H\left(Q_{\text{dis}} - Q_{\text{cont}}\right) \tag{9.40}$$

where H is the Heaviside function: $H(x) = 1$ if $x > 0$, else $H(x) = 0$ [28].

The physics behind equation (9.39) or (9.40) is complex. The process can be decomposed into four steps (Fig. 9.28). In a first phase, the stream of discontinuous fluid enters the main channel. In a second phase, it forms a blob with approximately the size of main channel width ($L \sim a$). If the flow rate of "discontinuous" liquid Q_{dis} is sufficiently large compared to the flow rate of the "continuous" liquid Q_{cont} ($Q_{\text{dis}} > Q_{\text{cont}}$), the droplet elongates in the main channel. This third phase does not take place in the opposite case. Finally, the droplet detaches.

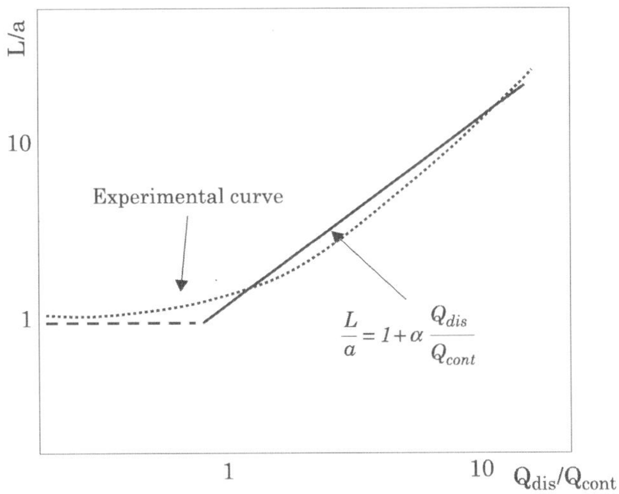

Figure 9.27 Graph of L/a versus $Q_{\mathrm{dis}}/Q_{\mathrm{cont}}$: the relation is independent of the dynamic viscosity μ showing that the shear stress has no influence if the Capillary number is smaller than a critical value $\mathrm{Ca}_{\mathrm{crit}} \sim 10^{-2}$.

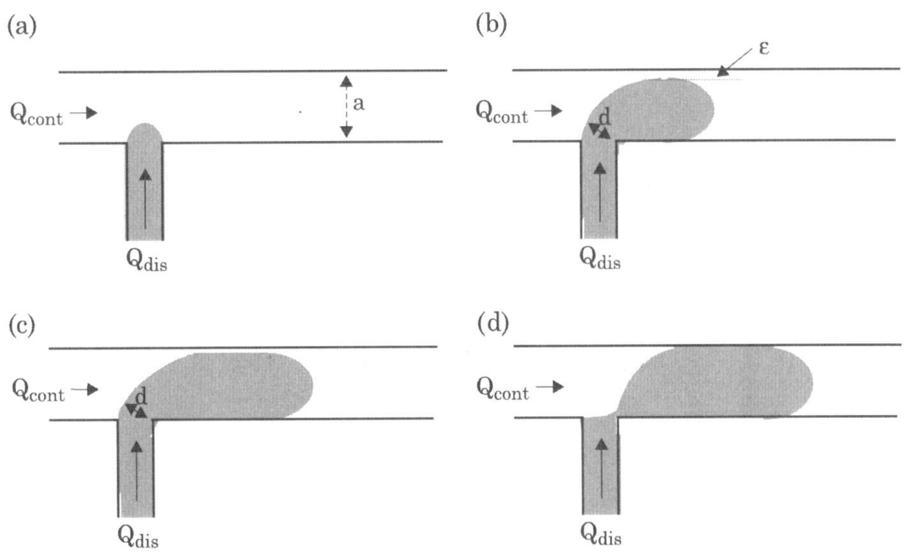

Figure 9.28 The four phases of droplet formation in a T-junction: (a) the stream of discontinuous fluid enters the main channel; (b) the stream blocks the main channel (nearly totally, except for a very small gap ε); (c) the droplet elongates downstream; (d) the droplet separates from the inlet.

During the two first phases, the droplet reaches a length $L \sim a$. If $Q_{\mathrm{dis}} < Q_{\mathrm{cont}}$ the droplet does not have the time to elongate and separation occurs immediately. Hence, $L \sim a$ when $Q_{\mathrm{dis}} < Q_{\mathrm{cont}}$. Conversely, if $Q_{\mathrm{dis}} > Q_{\mathrm{cont}}$, the droplet elongates. Let us calculate the elongation length. During the elongation phase, the droplet growth rate is approximately given by

$$u_{\text{growth}} \approx \frac{Q_{\text{dis}}}{ab}. \tag{9.41}$$

Remember that b is the height of the channel. On the other hand, the neck shrinks at a rate

$$u_{\text{squeeze}} \approx \frac{Q_{\text{cont}}}{ab}. \tag{9.42}$$

If d denotes the width of the neck, the time needed to achieve the squeezing is approximately

$$\tau_{\text{squeeze}} \approx \frac{d}{u_{\text{squeeze}}} \approx \frac{dab}{Q_{\text{cont}}}.$$

The total length of the droplet when it detaches is then

$$L \approx a + u_{\text{growth}}\tau_{\text{squeeze}} \approx a + d\frac{Q_{\text{dis}}}{Q_{\text{cont}}}.$$

If we note $\alpha = d/a$, and scale by a, we nearly recover equation (9.39)

$$\frac{L}{a} \approx 1 + \frac{d}{a}\frac{Q_{\text{dis}}}{Q_{\text{cont}}} = 1 + \alpha\frac{Q_{\text{dis}}}{Q_{\text{cont}}}. \tag{9.43}$$

However, at this point, α is not a constant ($\alpha = d/a$) whereas in (9.40) α is a constant, with a value close either to 0 or to 1. In the case where $Q_{\text{dis}} < Q_{\text{cont}}$, we have seen that $L \sim a$, which is equivalent to $d = 0$ in (9.43). In the case where $Q_{\text{dis}} > Q_{\text{cont}}$, $u_{\text{growth}} > u_{\text{squeeze}}$, which means that the growth velocity is larger than the squeeze velocity. It is observed that the width of the neck d does not vary quickly during the elongation phase; it suddenly goes to zero at breakup. This is due to a little gap between the blob and the wall (ε in Fig. 9.28) that vanishes suddenly at breakup. Hence, the squeezing velocity is somewhat smaller than its value from equation (9.42). These observations explain why the ratio d/a can be approximated by a constant α, of the order of 1. Finally, we can approximate

$$\frac{d}{a} \approx \alpha H\left(Q_{\text{dis}} - Q_{\text{cont}}\right)$$

and relation (9.43) collapses to relation (9.40).

9.4.1.1.2 Droplet formation: frequency control of the droplet size

As we have seen above, according to equation (9.39) or (9.40), the size of the droplet is of the order of the channel width, (always larger than the

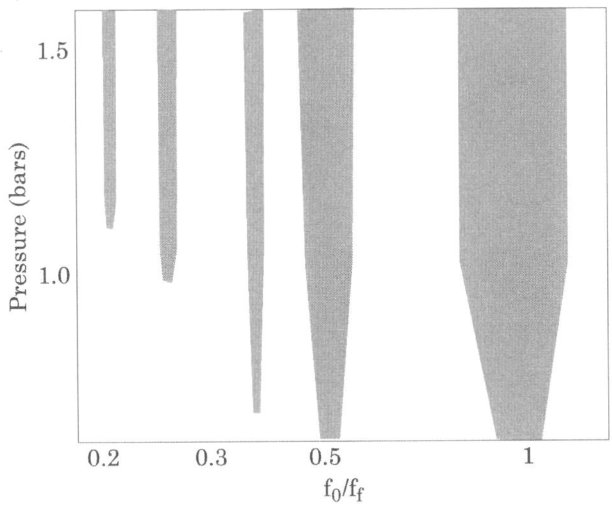

Figure 9.29 Arnold tongues: the grey areas correspond to the synchronized regimes. f_f is the forcing frequency.

channel width). It has been found [29,30] that smaller droplet sizes can be produced if the incoming rate of liquid was modulated in frequency. Without frequency modulation, there is a natural frequency f_0 of droplet detachment. By superposing a tunable forcing frequency, resonances can be found leading to the formation of mono-disperse droplets whose sizes differ notably from that of equation (9.39). Such regimes are called synchronized regimes. In such regimes, the system delivers regular drops at regular time intervals. One must be careful to avoid quasi-periodic regimes where droplets are emitted irregularly and have irregular sizes. The physics behind the frequency actuated droplet emission is not yet completely understood. It involves complex nonlinear fluid dynamics. However, it has been observed [29,30] that, depending on the forcing frequency, there were domains of synchronized regimes. Such domains are shown in Fig. 9.29, and are called "Arnold tongues."

A very important experimental observation is that droplet volumes vary as the inverse of the emission frequency (Fig. 9.30). Hence droplet volumes can be considerably reduced, approximately by an order of 10, and the range of droplet size is extended to the interval $[a/3, a]$; moreover, the size of the droplet can be adjusted in line by varying the frequency.

9.4.1.1.3 Mixing in T-junctions

As we have mentioned above, T-junctions are particularly well suited for biochemical and chemical reactions. However, the crux for obtaining a high efficiency of such reactions is that the constituents that react are well mixed in a very short time. For example, in the case of rapid polymerization, the components should be rapidly mixed in order to obtain a homogeneous

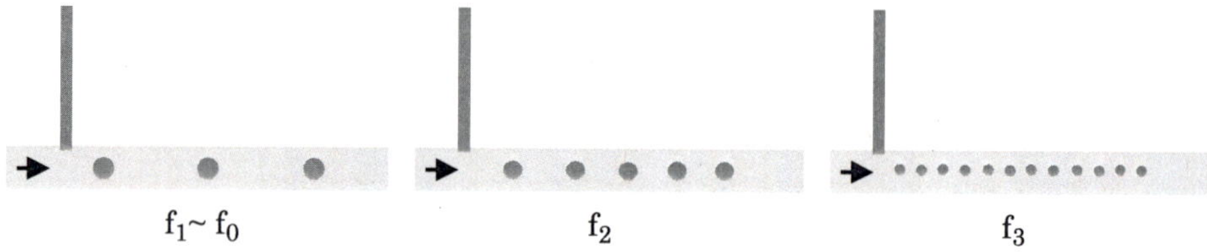

Figure 9.30 Droplet size is inversely proportional to the frequency.

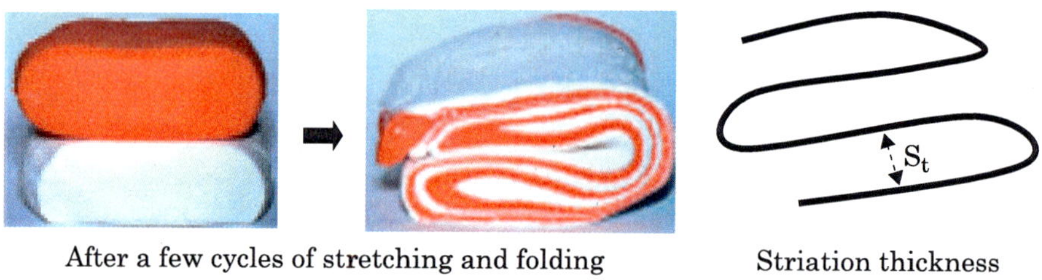

Figure 9.31 Left: stretching and folding a highly viscous fluid (or a very laminar flow). Right: striation thickness. Photograph by J. Tice; from [46]. © 2006 Wiley; reprinted with permission.

polymerization. Mixing in liquid plugs has been thoroughly studied by Handique et al. [30], Song et al. [31], Tice et al. [32] and Bringer et al. [33].

The principle is the same that we developed in Chapter 5. In the process of folding and stretching, the striation thickness is the distance between the filets of the diffusing species (Fig. 9.31); the diffusion time is then given by Fourier's law

$$t \approx \frac{s_t^2}{2D}. \tag{9.44}$$

Ottino [35] has shown that the striation thickness is reduced after each folding of the filet according to

$$s_t(n) = s_t(0)\sigma^{-n} \tag{9.45}$$

where σ is called the Lyapunov coefficient. We analyze here two types of channels: straight channels and channels with turns (Fig. 9.32). We show that the mixing process is much more efficient in channels with turns.

Suppose first that the microchannel is straight. Because the walls are fixed and the plug is moving, two recirculation patterns form inside the plug (Fig. 9.33.a). Striation thickness in the plug is given by the relation [31,34]

$$s_t = s_t(0)\frac{L}{d} \tag{9.46}$$

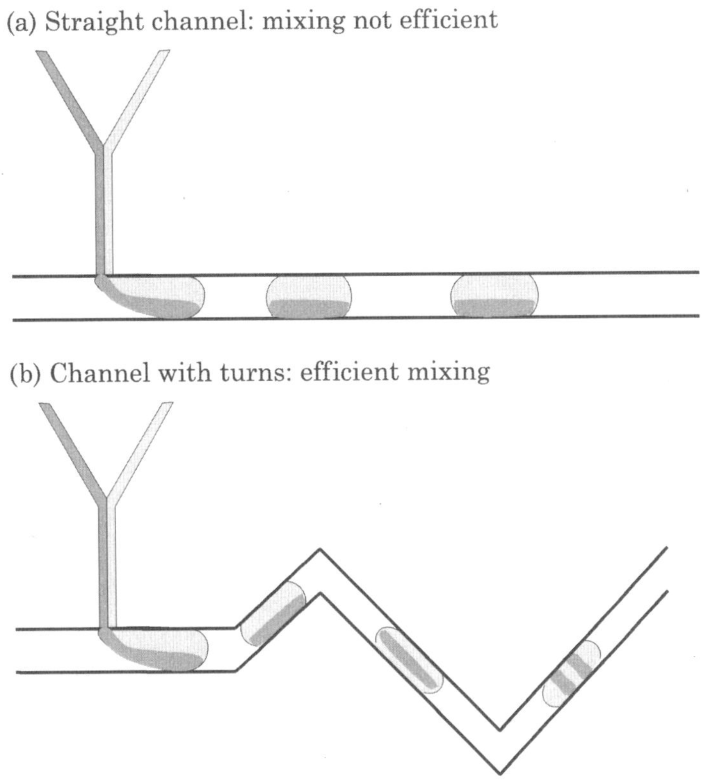

Figure 9.32 Two different types of microchannels: the channel with turns is much more efficient for mixing the components inside the droplets.

where L is the length of the plug and d the distance traveled by the droplet. Each time the plug has traveled a distance $d = 2L$, the striation thickness is divided by two. Striation patterns are schematically shown in Fig. 9.33.b. The two halves homogenize first due to the reduction of the striation. Taking into account that $s_t(0) \approx a$ and using equation (9.44), the time for homogenization of each 1/2 plug is approximately

$$t \approx \frac{a^2}{2D}\frac{L^2}{d^2}. \tag{9.47}$$

However at this time, the concentration in the plug is not uniform and the situation is schematized in Fig. 9.33.c.

It has been observed that winding microchannels reduce homogenization time [34]. We analyze here the role of the turns of the capillary tube. Suppose we have a capillary tube constituted of n linear segments of length $d \sim 2L$, the segments being individualized by sufficiently pronounced turns. First, the recirculation flow inside the plug is modified by the turns, as shown in Fig. 9.34.

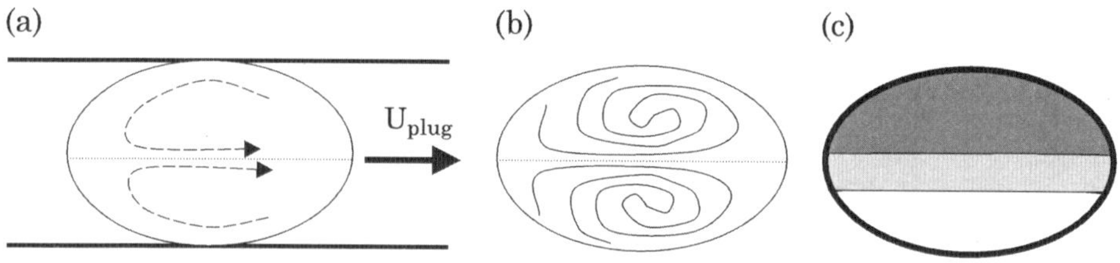

Figure 9.33 (a) Internal motion in a moving plug; (b) striation pattern in a moving plug; (c) once diffusion has homogenized the two halves of the plug, the mixing is not complete.

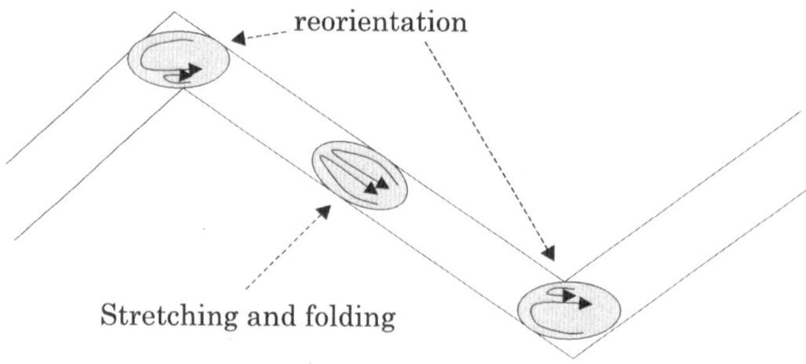

Figure 9.34 Dissymmetry of internal recirculation flow is induced by turns of the capillary tube.

Second, the dissymmetry of the recirculation flow in the turns induces a re-orientation of the fluid domains as shown in Fig. 9.35. This re-orientation is essential for the mixing of liquids in the plug. This phenomenon is called Baker's transformation and is schematized in Fig. 9.36. Re-orientation is necessary to increase the number of striations. Using Ottino's formula (9.45), the striation thickness is

$$s_t(n) = a\sigma^{-n}.$$

Using (9.46) with $d \sim 2L$, we find $\sigma = 2$. The diffusion time at step n is then derived from equation (9.44)

$$t_{\text{diffusion}}(n) = \frac{a^2\sigma^{-2n}}{2D}. \tag{9.48}$$

On the other hand, because $d(1)$ is of the order of L, the convection time is given by

$$t_{\text{convection}}(n) \approx \frac{d(n)}{U} = n\frac{L}{U}. \tag{9.49}$$

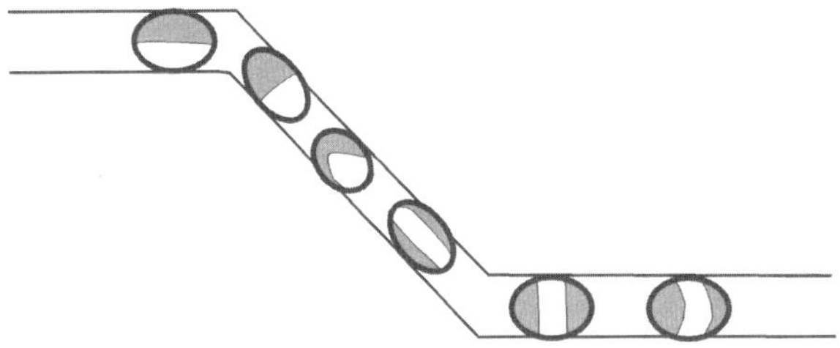

Figure 9.35 Sketch of the effect of stretching and folding in the straight parts and re-orientation in the turns.

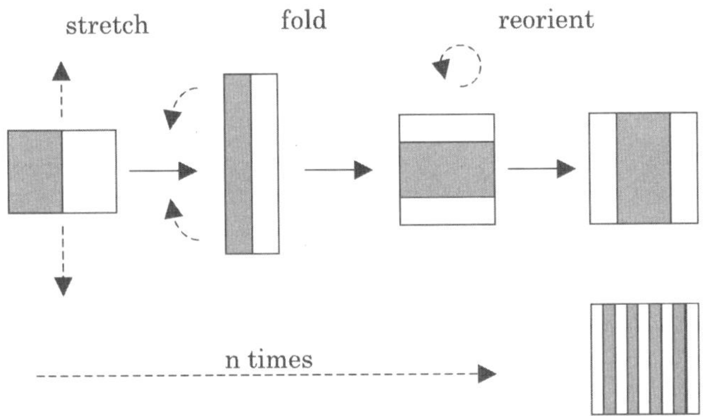

Figure 9.36 Baker's transformation and reduction of the striation length.

Following Bringer et al. [34] and Stroock et al. [36], the mixing time is obtained by equating the diffusion and convection time, so that

$$t_{\text{convection}}(n) \approx n\frac{L}{U} \approx \frac{a^2\sigma^{-2n}}{2D} = t_{\text{diffusion}}(n). \tag{9.50}$$

After rearrangement, we find

$$2n\sigma^{2n} \approx \frac{aU}{D}\frac{a}{L} \approx \mathrm{Pe}\frac{a}{L} \tag{9.51}$$

where $\mathrm{Pe} = aU/D$ is the Peclet number. Equation (9.51) produces the value of n (number of segments) necessary to obtain mixing. Note that the solution of the equation $xe^x = q$ is given by the Lambert-W function [28]

$$xe^x = q \Rightarrow x = W(q).$$

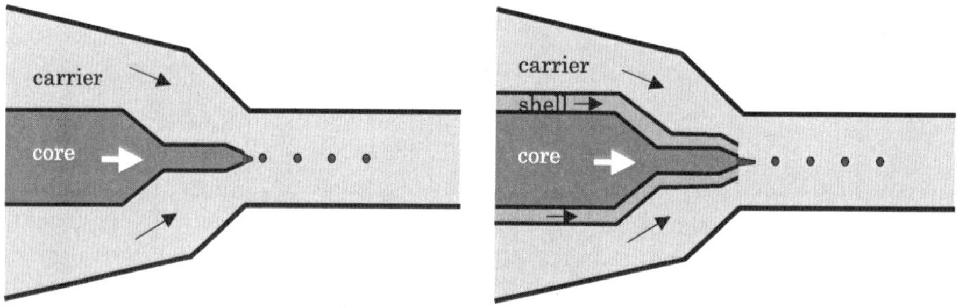

Figure 9.37 Principle of a FFD. Left: mono-dispersed core liquid droplets; right: encapsulated droplets.

The Lambert-W function is programmed in most numerical software. Then n is given by the relation

$$n \approx \frac{W\left(\ln \sigma \, \mathrm{Pe}\frac{a}{L}\right)}{2 \ln \sigma}. \tag{9.52}$$

Now we introduce the value $\sigma = 2$ of the Lyapunov coefficient. Suppose that the flow rate $Q_{\mathrm{dis}} < Q_{\mathrm{cont}}$, equation (9.40) gives $L \sim a$ and the preceding equation (9.52) becomes

$$n \approx \frac{W(0.7\,\mathrm{Pe})}{1.4}.$$

Using typical values of $D \sim 10^{-9}$ m^2/s, $U \sim 1$ mm/s and $a = 100$ μm, the value of the Peclet number is 100 and $n \sim 3$. In the case where $D \sim 10^{-10}$ m^2/s, the number of segments required to obtain mixing is five. Hence, a small number of sharp turns of the capillary tube is very effective to completely mix the liquid.

9.4.1.2 Flow Focusing Devices (FFD)

The principle of a flow focusing device is shown in Fig. 9.37. A filament of core fluid forms at the nozzle and finally breaks up into a droplet. So far, the functioning of such devices has been mostly approached experimentally [23–25,37–39].

Contrary to T-junctions, little is known about the detailed physics of the droplet formation. The first theoretical investigations on the stability of microflows in FFD have been recently done by Guillot et al. [40]. It seems that there are two regimes depending on the velocities of the core flow: a dripping regime at low velocity, and a jetting regime at large velocity. Here we focus on the low velocity case (dripping regime). Apparently, many parameters influence

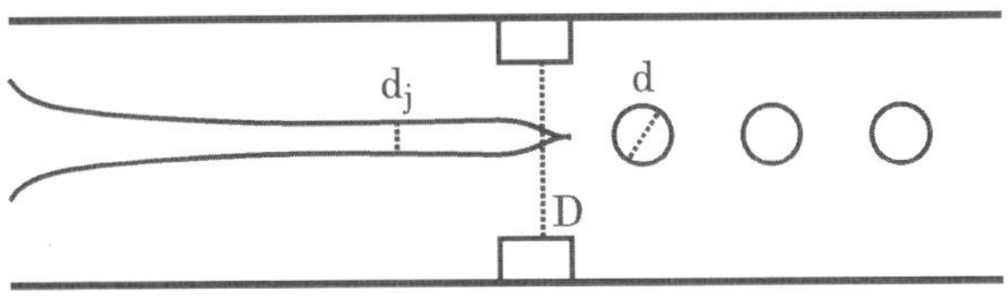

Figure 9.38 Sketch of a FFD.

the droplet size: the ratio of the flow rates of the two fluids Q_{in}/Q_{ext}, the nozzle diameter D, the Reynolds numbers Re, the ratio of the densities of the two fluids ρ_{in}/ρ_{ext}, and the surface tension γ. For the geometry of Fig. 9.38, and a gaseous core, Ganan-Calvo et al. [21,38] have proposed the following correlation for the bubble diameter d

$$d = \left(\frac{Q_{in}}{Q_{ext}}\right)^{\frac{1}{3}} D^{\frac{2}{3}} d_j^{\frac{1}{3}} f(\mathrm{Re}_{in}) \tag{9.53}$$

where the filament diameter d_j is given by

$$d_j \approx \left(\frac{Q_{in}}{Q_{ext}}\right)^{\frac{1}{2}} D G(\rho_{ext}, \rho_{in}, D, \gamma). \tag{9.54}$$

A comparison with experimental results, shows that the flow rates are the dominant terms in the correlation, and the correlation can be simplified by

$$\frac{d}{D} \approx \left(\frac{Q_{in}}{Q_{ext}}\right)^{0.37}. \tag{9.55}$$

The interesting thing here is that relation (9.55) shows that the ratio of the flow rates is controlling the droplet diameter. This observation confirms the results of Anna et al. [22]. The situation is then similar to that of the T-junction, with the exception that the diameter of the droplet is smaller that that of the orifice or nozzle.

Let us mention an interesting numerical approach has been performed by Davidson et al. [41] for an axisymmetric nozzle, using a VOF (volume of fluid) numerical method. Their numerical results show that the flow rate ratio is not the only parameter controlling the size of the produced droplets, and that the Reynolds and Weber numbers are also important parameters. Fig. 9.39 shows the results for a ratio $Q_{in}/Q_{ext} = 0.025$, a Reynolds Re_{in} number equal to 0.045, and a Weber number $\mathrm{We}_{in} = 0.23 \times 10^{-5}$.

In conclusion, the detailed physics of FFD is not yet completely understood; however, the experimental approach has shown that FFD seems to be a very

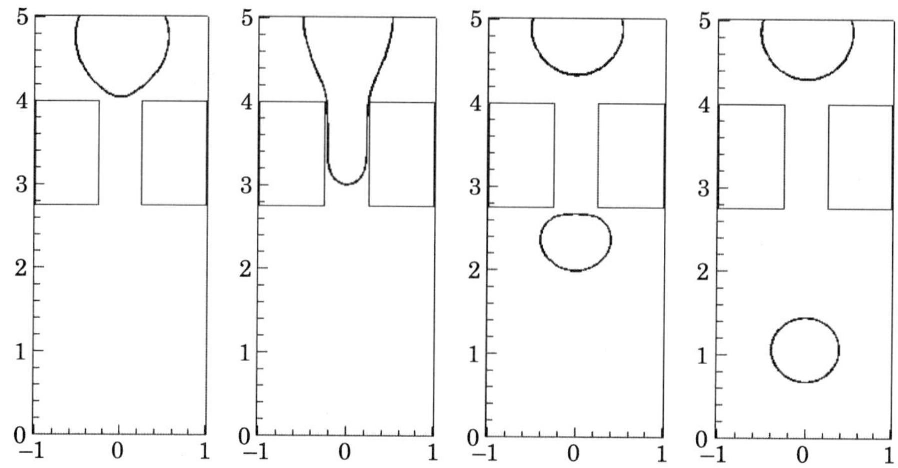

Figure 9.39 Numerical modeling of droplet break-up in a FFD [41].

promising technique to realize on-line, mono-dispersed droplets, and especially to encapsulate liquids for biotechnological and pharmaceutical applications.

9.4.1.3 Droplets or Separated Two-phase Flow?

We have seen in the preceding sections the formation of droplets in a microflow using T-junctions or flow focusing devices. Let us remark here that the formation of droplets at the merging of two immiscible flows is not automatic. Guillot and Colin [40] have investigated the stability of parallel flows in a microchannel after a T-junction (Fig. 9.40). They have observed different regimes depending on the flow rate of the two liquids (and more generally on the capillary numbers of the two regimes). Note that in their case the bottom plate is hydrophilic (glass) and the other walls are hydrophobic (PDMS).

Based on similar considerations as those of Garsteki et al. [27], the authors have derived a semi-empirical expression for the limit of the parallel flow domain.

However, not much is known on the stability of co-flowing immiscible fluids far from the inlet. In Section 9.3 we have investigated the case of co-flowing microflows when micro-pillars are present to increase the stability of the interface. The picture is largely different than that of macroscopic channels. In particular, the flows are not stratified because gravity is negligible at this scale (the Bond number $Bo = \rho g D_h^2/\gamma$ is much smaller than 1), and the non-wetting liquid forms a cylindrical filament inside the channel, as shown in Fig. 9.41. It appears that, if the filament touches the wall by capillary contact, the stability may be ensured on a long distance, but if it detaches from the wall, Rayleigh-Plateau type instabilities contribute to break the filament into droplets.

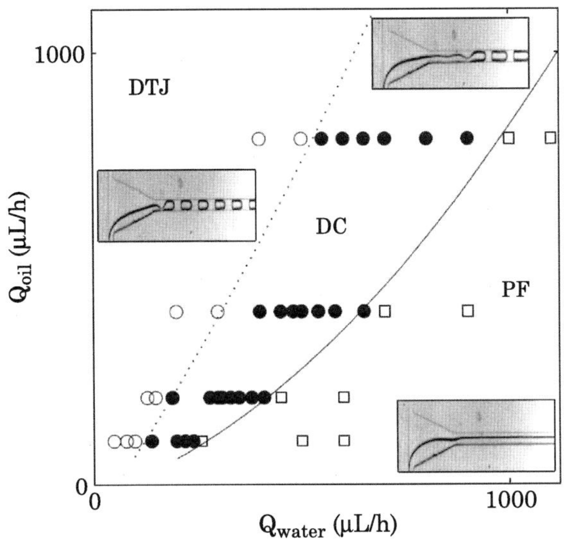

Figure 9.40 Merging of water and oil microflows: three regimes are observed: parallel flows (PF) for small values of oil flow rate and large values of water flow rate, droplets in the channel (DC) at the inlet in the opposite case, and an intermediate regime with a short parallel flow that breaks up into droplets. From [39]. © 2006 APS; reprinted with permission.

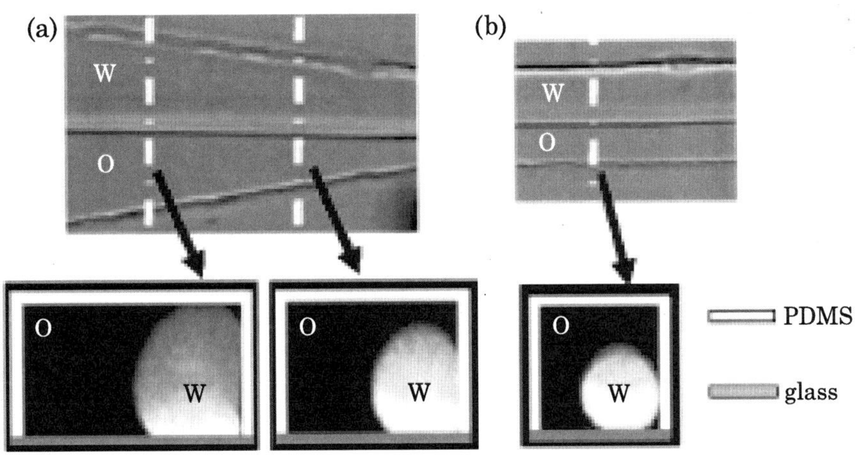

Figure 9.41 (a) Microchannel inlet: the oil/water interface forms a spherical segment. Progressively the water flow forms a filament attached to the hydrophilic glass plate; (b) at the outlet the two-phase are still flowing in parallel and the water filament is nearly cylindrical. From [39]. © 2006 APS; reprinted with permission.

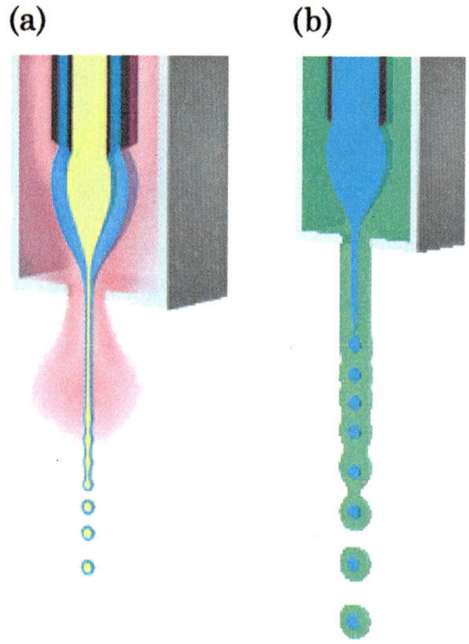

Figure 9.42 Two different FFDs for the encapsulation of droplets: (a) concentric nozzles can be used to create complex particles from two liquids or from a liquid and a gas resulting in the production of hollow spheres; (b) A funnel shaped lens of liquid can be used to create microencapsulated spheres. From [41], courtesy Flow Pharma Inc. of California, USA.

9.4.2 Encapsulation

Flow focusing technique, and to a lesser extent T-junctions, are used in laboratories and in the industry to encapsulate liquids, bubbles, particles, and biological objects like proteins and cells. T-junctions can be used in series with FFD as shown in Fig. 9.25. Fig. 9.42 shows two types of FFDs aimed at producing encapsulated micro-drops [42].

Encapsulation is an essential process in many applications like drug delivery, protein crystallization [2,43], complex particles, cell preservation [44], etc. According to the application, the coating may be programmed to become solid by polymerization, or the interior to crystallize when in contact with adequate chemical species. Fig. 9.43 schematically shows some applications of encapsulation: (b) micro-drops of two liquids (of different nature) encapsulated in a droplet of a third liquid obtained in a T-junction [45,46]; (c) a droplet of liquid ink encapsulated by a polymerized coating of resin [42]; (d) encapsulated live liver cells after 2 days storage in a non-aseptic water [3].

9.5 Summary

In this chapter we have introduced the physics of two-phase flows in microchannels, under the form of droplets and plugs, and the stability of an interface between two immiscible fluids flowing in parallel. The perspectives of

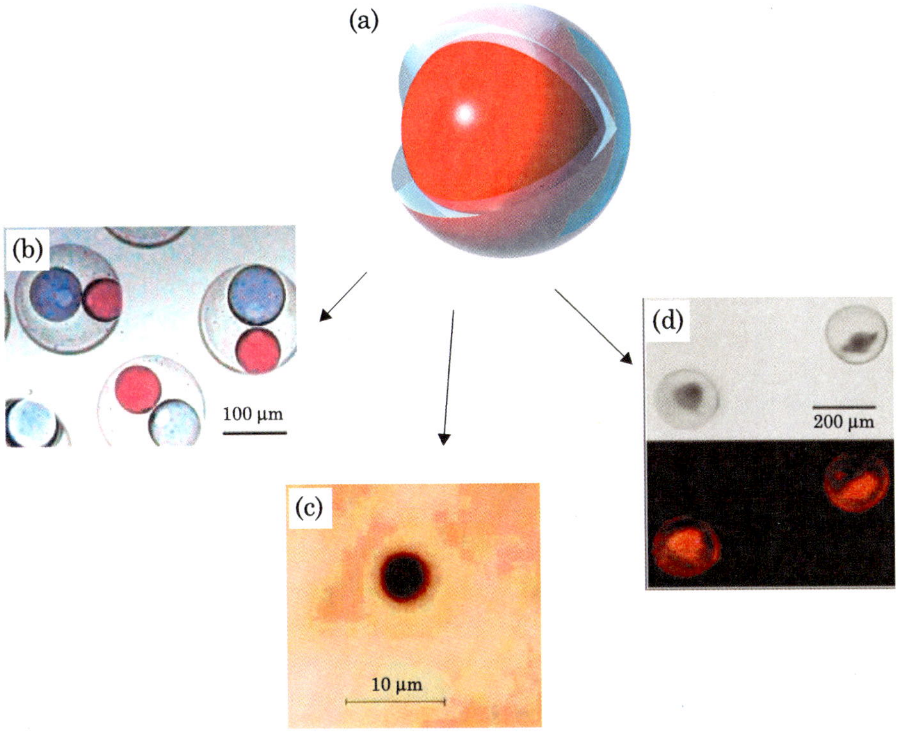

Figure 9.43 Encapsulation has many applications: (b) two immiscible liquids encapsulated by a third liquid; (c) liquid ink micro-drop encapsulated by a polymer resin; (d) live liver cells encapsulated in a protective gel, the fluorescent image below shows that the cells are still alive. Fig. 9.43 (b) reprinted with permission from [44]. © 2004 American Chemical Society. Fig. 9.43 (c) from [41], courtesy Flow Pharma Inc. of California, USA. Fig. 9.43 (d) from [3]. © 2006 IOP; reprinted with permission.

droplet microfluidics are especially promising. Two-phase—and multi-phase—flows are already used in T-junctions and flow focusing devices (FFD) to generate mono-dispersed droplets and to encapsulate them in a liquid or a gel and even a solid (polymerizable) coating, to generate applications in biology, biotechnology, and microchemistry.

To this point, it may be useful to compare digital microfluidics, which we have presented in Chapters 4 through 8, and droplet microfluidics. In reality, the two techniques have different scopes of application; they address different problems. Digital microfluidics can handle extremely small volumes of liquids and realize operations and manipulation in parallel, and with great accuracy, whereas droplet microfluidics is well adapted to perform operations in series. These properties are illustrated in Fig. 9.44.

Volumes and forces are generally smaller in digital microfluidics, and very small objects can be handled easily. Besides, the precision of the manipulations is very high. As a very coarse rule, digital microfluidics is particularly dedicated

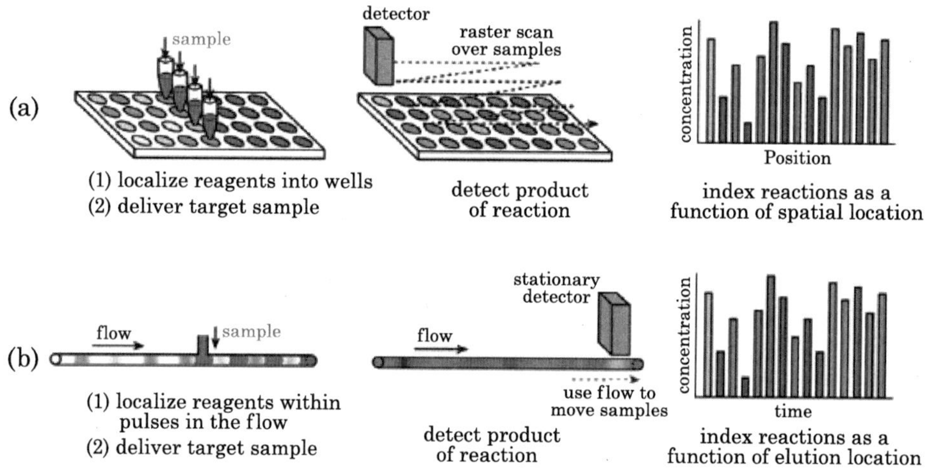

Figure 9.44 Comparison between digital and droplet microfluidics. From [46]. © 2006 Wiley; reprinted with permission.

to precise biorecognition and bioanalysis, T-junctions are very efficient for on-line reactions, FFDs are essential for encapsulation processes, and separated two-phase microflows are particularly adapted as mass exchanger micro-devices.

References

1. M. Joanicot and A. Ajdari, "Droplet Control for Microfluidics," *Science*, Vol. 5, pp. 887–888, 2005.
2. Bo Zheng, L.S. Roach and R.F. Ismagilov, "Screening of protein crystallization conditions on a microfluidic chip using nanoliter-size droplets," *JACS*, Vol. 125, pp. 11170–11171, 2003.
3. Hyun-Jik Oh, So-Hyun Kim, Ju-Yeoul Baek, Gi-hun Seong and Sang-Hoon Lee, "Hydrodynamic micro-encapsulation of aqueous fluids and cells via 'on the fly' photo polymerization," *J. Micromech. Microeng.*, Vol. 16, pp. 285–291, 2006.
4. R.L. Hoffman, "A study of advancing interface," *Journal of Colloid Interface Science*, Vol. 50, pp. 228–241, 1975.
5. M. Fermigier and P. Jenffer, "An experimental investigation of the dynamic contact angle in liquid–liquid systems," *Journal of Colloid and Interface Science*, Vol. 146, pp. 226–241, 1990.
6. J. Berthier and F. Ricoul, Numerical modeling of ferrofluid flow instabilities in a capillary tube at the vicinity of a magnet, Proceedings of the 2002 MSM conference, 22–25 April 2002, San Juan, Puerto Rico, 2002.
7. J. Berthier and P. Silberzan, *Microfluidics for Biotechnology*, Artech House Publishers, 2006.
8. M. Bahrami, M.M. Yovanovich and J.R. Culham, "Pressure drop of laminar, fully developed flow in microchannels of arbitrary cross-section," *ASME J. of Fluid Engineering*, Vol. 126, pp. 1036–1044, 2006.
9. P. Collet, J. de Coninck and F. Dunlop, "Dynamics of wetting with a disordered substrate: the contact angle hysteresis," *Europhys. Lett.*, Vol. 22, no. 9, pp. 645–650, 1993.

10. E. Chibowski, A. Ontiveros-Ortega and R. Perea-Carpio, "On the interpretation of contact angle hysteresis," *J. Adherence Sci. Technol.*, Vol. 16, no. 10, pp. 1367–1404, 2002.
11. S.M.M. Ramos, E. Charlaix and A. Benyagoub, "Contact angle hysteresis on nano-structured surfaces," *Surface Science*, Vol. 540, pp. 355–362, 2003.
12. Bo He, Junghoon Lee and N.A. Patankar, "Contact angle hysteresis on rough hydrophobic surfaces," *Colloids and Surfaces A: Physicchem. Eng. Aspects*, Vol. 48, pp. 101–104, 2004.
13. D.L. Chen, Liang Li, S. Reyes, D.N. Adamson and R.F. Ismagilov, "Using three-phase flow of immiscible liquids to prevent coalescence of droplets in microfluidic channels: criteria to identify the third liquid and validation with protein crystallization," *Langmuir*, Vol. 23, pp. 2255–2260, 2007.
14. J. Berthier, Van Man Tran, H. Jeanson and N. Sarrut, On the stability length of fluid/liquid interfaces sustained by microfabricated pillars, Proceedings of the 2007 Nanotech Conference, 21–24 May, Santa Clara, Ca, 2007.
15. A. Aota, M. Nonaka, A. Hibara and T. Kitamori, Micro counter-current flow system for high efficiency extraction, 7th International Conference on Miniaturized Chemical and Biochemical Analysis Systems, Squaw Valley, Ca, USA, 5–9 October, 2003.
16. W. van der Wijngaart, T. Frisk and G. Stemme, A micromachined interface for transfer of liquid or vapor sample to a liquid solution, Proceedings of the 13th International Conference on Solid-State Sensors, Actuators and Microsystems, Seoul, 5–9 June, 2005.
17. E.W. Washburn, "The dynamics of capillary flow," *Phys. Rev.*, pp. 273–283, 1921.
18. I. Papautsky, B.K. Gale, S. Mohanti, T.A. Ameel and A.B. Frazier, Effects of rectangular microchannel aspect ratio on laminar friction constant, Proc. SPIE, Vol. 3877, pp. 147–158, Microfluidic Devices and Systems II, 1999.
19. Y. Tsori, "Discontinuous liquid rise in capillaries with varying cross-sections," *Langmuir*, 22, pp. 8860–8863, 2006.
20. K. Brakke, "The Surface Evolver," *Exp. Math.*, Vol. 1, p. 141, 1992.
21. A.M. Gañán-Calvo and J.M. Gordillo, "Perfectly monodisperse microbubbling by capillary flow focusing," *Phys. Rev. Lett.*, Vol. 87, no. 27, p. 274501-1-4, 2001.
22. S.L. Anna, N. Bontoux and H.A. Stone, "Formation dispersions using flow focusing in microchannels," *Appl. Phys. Lett.*, Vol. 82, no. 3, pp. 364–366, 2003.
23. J.T. Cabral and S.D. Hudson, "Microfluidic approach for rapid multicomponent interfacial tensiometry," *Lab Chip*, Vol. 6, pp. 427–436, 2006.
24. T. Nisisako, T. Torii and T. Higuchi, "Novel microreactors for functional polymer beads," *Chemical Engineering Journal*, Vol. 101, pp. 23–29, 2004.
25. T. Thorsen, R.W. Roberts, F.H. Arnold and S.R. Quake, "Dynamic pattern formation in a vesicle-generating mirofluidic device," *Phys. Rev. Lett.*, Vol. 86, pp. 4163–4166, 2001.
26. P. Garstecki, M.J. Fuerstman, H.A. Ston and G.M. Whitesides, "Formation of droplets and bubbles in a microfluidic T-junction: scaling and mechanism of break-up," *Lab. Chip.*, Vol. 6, pp. 437–446, 2006.
27. http://mathworld.wolfram.com/HeavisideStepFunction.html.
28. P. Tabeling, V. Barbier, J. Goulpeau, B. Lonetti and H. Willaime, Playing with actuators in microfluidic systems, Proceedings of the NSTI-Nanotech Conference 2006, Vol. 2, 2006.
29. H. Willaime, V. Barbier, L. Kloul, S. Maine and P. Tabeling, "Arnold tongues in a microfluidic drop emitter," *Phys. Rev. Let.*, Vol. 96, p. 054501-1-4, 2006.
30. H. Handique and M.A. Burns, "Mathematical modeling of drop mixing in a slit-type microchannel," *J. Micromech. Microeng.*, Vol. 11, pp. 548–554, 2001.
31. H. Song, J.D. Tice and R.F. Ismagilov, "A microfluidic system for controlling reaction networks in time," *Angew. Chem.*, Vol. 42, pp. 767–771, 2003.
32. J.D. Tice, H. Song, A.D. Lyon and R.F. Ismagilov, "Formation of droplets and mixing in multiphase microfluidics at low values of the Reynolds and the Capillary numbers," *Langmuir*, Vol. 19, pp. 9127–9133, 2003.

33. M.R. Bringer, C.J. Gerdts, H. Song, J.D. Tice and R.F. Ismagilov, "Microfluidic systems for chemical kinetics that rely on chaotic mixing in droplets," *Phil. Trans. R. Soc. Lond. A*, Vol. 362, pp. 1087–1104, 2004.
34. J.M. Ottino, *The Kinematics of Mixing: Stretching, Chaos, and Transport*, Cambridge University Press, 1988.
35. A.D. Stroock, S.K.W. Dertinger, A. Ajdari, I. Mezic, H.A. Stone and G.M. Whitesides, "Chaotic Mixer for Microchannels," *Science*, Vol. 295, pp. 647–650, 2002.
36. A. Günther and K.F. Jensen, "Multiphase microfluidics: from flow characteristics to chemical and material synthesis," *Lab. Chip.*, Vol. 6, pp. 1487–1503, 2006.
37. P. Garstecki, H.A. Stone and George M. Whitesides, "Mechanism for flow-rate controlled breakup in confined geometries: a route to monodisperse emulsions," *Phys. Rev. Lett.*, Vol. 94, p. 164501-1-4, 2005.
38. A.M. Ganan-Calvo, L. Martin-Banderas, R. Gonzalez-Prieto, A. Rodriguez-Gil, T. Berdun-Alvarez, A. Cebolla, S. Chavez and M. Flores-Mosquera, "Straightforward production of encoded microbeads by Flow Focusing: Potential applications for biomolecule detection," *International Journal of Pharmaceutics*, Vol. 324, pp. 19–26, 2006.
39. Pierre Guillot and Annie Colin, "Stability of parallel flows in a microchannel after a T-junction," *Physical Review E*, Vol. 72, p. 066301, 2005.
40. M.R. Davidson, D.J.E. Harvie and J.J. Cooper-White, "Flow focusing in microchannels," *ANZIAM J.*, Vol. 46, pp. C47–C58, 2005.
41. http://www.flowfocusing.com/Home_Page_Frameset_Source/ff_Home_Content.htm.
42. B. Zheng, J.D. Tice and R.F. Ismagilov, "Formation and arrayed droplets by soft lithography and two-phase fluid flow and application in protein crystallization," *Advanced materials*, Vol. 16, no. 15, pp. 1365–1368, 2004.
43. Kan Liu, Hui-Jiang Ding, Jing Liu, Yong Chen and Xing-Zhong Zhao, "Shape-Controlled Production of Biodegradable Calcium Alginate Gel Microparticles Using a Novel Microfluidic Device," *Langmuir*, Vol. 22, pp. 9453–9457, 2006.
44. S. Okushima, T. Nisisako, T. Torii and T. Higuchi, "Controlled production of monodisperse double emulsions by two-step droplet breakup in microfluidic devices," *Langmuir*, Vol. 20, pp. 9905–9908, 2004.
45. J. Atencia and D.J. Beebe, "Controlled microfluidic interfaces," *Nature*, Vol. 437, pp. 648–655, 2005.
46. H. Song, D.L. Chen and R.F. Ismagilov, "Reactions in droplets in microfluidics channels," *Angewandte Chemie*, Vol. 45, no. 44, pp. 7336–7356, 2006.

10 Epilog

The place and importance of digital microfluidics in the development of microsystems have been pointed out through this book. It has been seen that the advantages of such systems are related to the reduction of the liquid volumes to be handled; besides, their high potential for reconfigurability has been shown by Chakrabarty [1]. With the background acquired through the different chapters, a reflection on digital microfluidics and on its most needed and promising developments can be started. Questions such as "can miniaturization be pushed further?" and "what are the next developments that will extend the field of applications of digital microfluidics?" can reasonably be raised.

Of course, the answer to these questions is not straightforward and necessitates a long development. This Epilog chapter only presents a few perspectives.

10.1 Increasing the Domain of EWOD Applicability

In Chapter 4, three limitations to the EWOD technology were pointed out: first, a minimum potential is required to move a droplet, due to contact angle hysteresis; second, the electrowetting force is limited by a maximum actuation potential linked to the saturation phenomenon; and third, the electrowetting force is also limited by a maximum potential linked to dielectric breakdown. Decreasing hysteresis and increasing the dielectric breakdown level require improving the electric characteristic and the surface quality of the materials separating the electrodes from the liquid: high capacity layers supporting large critical electric fields with highly hydrophobic surface are needed. This line of work is presently the subject of active research [2–5], and experimentation on new nano-textured surfaces is in progress.

10.2 Interconnecting the Different Microfluidic Toolboxes

A second remark stems from the observation of Fig. 1.1, redrawn slightly differently in Fig. 10.1: is it possible to switch from one type of microfluidics to another one? Is it useful?

In Fig. 10.1 the interconnections between the different toolboxes have been materialized by arrows. A few connections have been developed (continuous arrows); for example, two microflows merging in a T-junction or in a flow focusing device (FFD) generate droplets or a two-phase microflow. It is also the case of inkjet printing, or spotting robots, transforming a microflow into

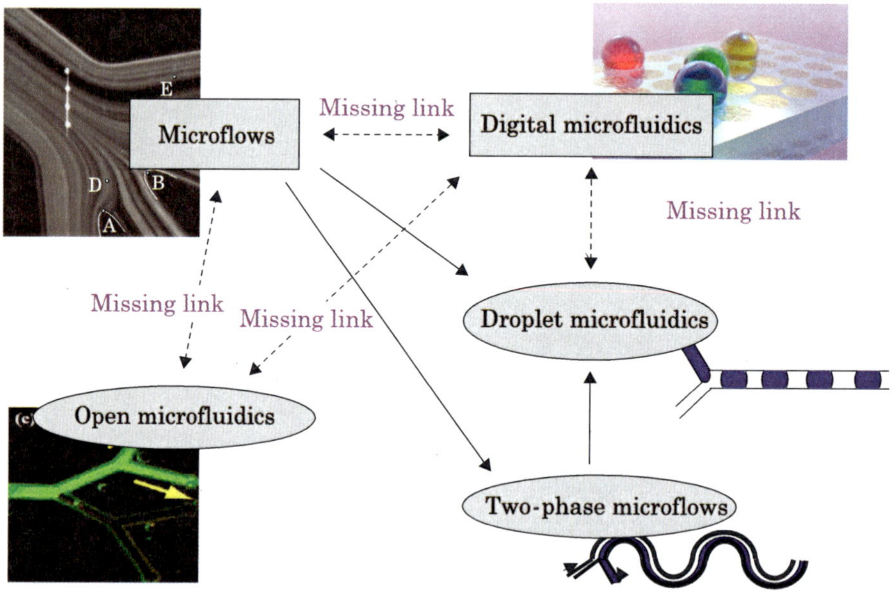

Figure 10.1 Sketch showing that links between the different types of microfluidics are still largely undeveloped. Photo bottom left reprinted with permission from Satoh et al., Anal. Chem., Vol. 77, pp. 6857–6863, 2005. © 2005 American Chemical Society. Photo top right courtesy Advalytix.

droplets. Conversely, Günther et al. [6] have shown it was possible to separate a plug flow into two single fluid microflows. However, many links are still missing.

The establishment of links between the different toolboxes is currently the object of active study: this is the case of the transformation of a two-phase microflow to droplet microflow [7,8]; it has been shown by Guillot et al. [9] that this transformation depends heavily on the stability or the instability of the flow (Fig. 10.2). This observation implies that the reverse path—from a droplet microflow (or an emulsion) to a separated two-phase flow will be difficult.

Another missing link is presently under scrutiny: digital microfluidics would benefit from being connected to microflows. Indeed, digital microfluidic systems are currently manually loaded in sample liquids by using micro-pipettes, or in some cases by robots. Automatic loading of digital microsystems using a microflow would be a considerable improvement. The same reasoning applies for "open microfluidic" systems. It would also be useful to extract micro-drops from a continuous flow and direct them onto a digital system. That would, for instance, improve substantially the detection of pathogens in flowing liquids. Conversely, it could be convenient to be able to convert a droplet or a digital microflow into a continuous flow. In particular, this would be an advantage for the manipulation and the study of cells. A first system of this kind has recently been studied by E. Berthier and D.J. Beebe [10] where droplets deposed on

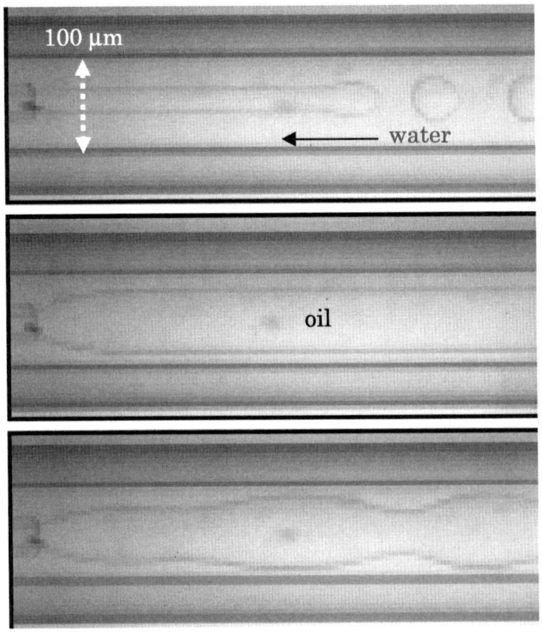

Figure 10.2 Two phase flow degenerating into a droplet flow. From [9]. © 2007 APS; reprinted with permission.

an inlet port give birth to a continuous, almost constant microflow, under the condition that a calibrated refill is implemented (Fig. 10.3).

These few examples show that interconnections between the different types of microfluidics are essential. However, at the present time, they are still largely underdeveloped.

10.3 Miniaturization: Nano-EWOD

It is obvious that the trend towards miniaturization has been the engine behind microsystems development. A second question that arises at this stage concerns miniaturization of digital microfluidics: can it be pursued further and what will it brings? The word nano-EWOD is even evoked to denote the scale down of EWOD-based microsystems. The idea behind this is the manipulation of droplets of less than 5-10 nl. One of the potential applications would be the manipulation of single cells, or at least small collections of cells, or bacteria, etc.

The first idea of a rule for down scaling is to homothetically reduce the dimensions of the usual EWOD-based systems described in Chapter 5. However, such a scale down raises some complicated questions, like: would the electrowetting force on the droplet be sufficient to perform all the operations required for the process? Remember that this force is proportional to the width

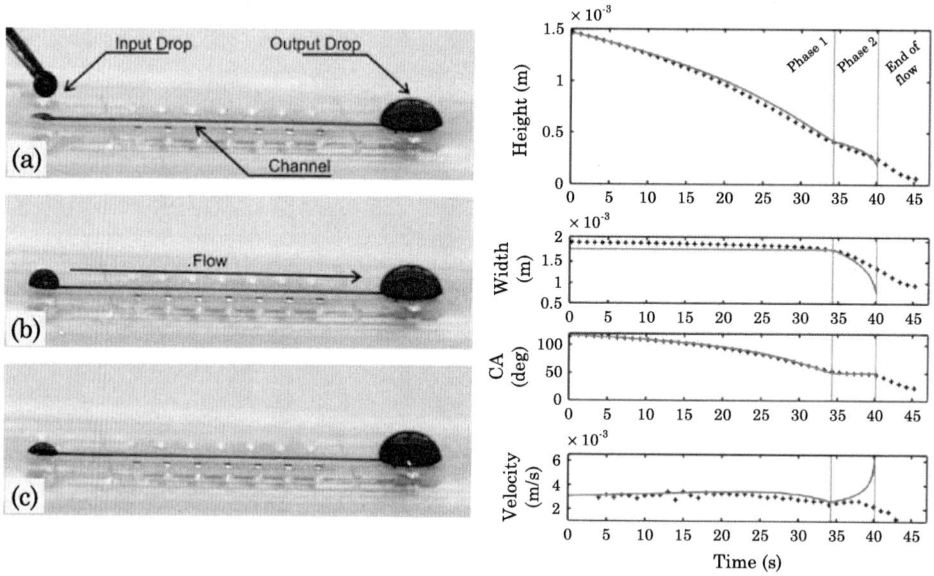

Figure 10.3 Droplet deposed on an inlet port gives birth to a continuous microflow (left). In a first phase the contact area remains constant while the contact angle decreases; during this first phase, the velocity remains constant. Reprinted with permission from [10], © of The Royal Society of Chemistry (2007).

of the electrode. And how can the hysteresis level be lowered sufficiently to palliate this loss of electrowetting force? Can the thickness of the dielectric layer be sufficiently reduced to compensate for the loss of electrowetting force? These two last questions refer to the remarks of Section 10.1. Another difficulty is the reduction of the size of the horizontal gap between the electrodes [11]: can it be sufficiently reduced to avoid pinning? And finally, what are the effects of the increase in micro-drop pressure associated with the reduction of the vertical gap?

All these questions make nano-EWOD a difficult but interesting challenge.

10.4 Other Applications Specific to Digital Microfluidics

The focus of this book was mostly on biotechnological, biological, and chemical applications of digital microfluidics. However, digital microfluidics is not limited to biotechnology. New promising applications have recently appeared in optics, with the development of a "liquid" screen based on electrowetting display technology that becomes brighter in direct sunlight [12] and with the development of adjustable "liquid" lenses [13] (Fig. 10.4). The principle derives directly from the electrowetting effect: in the case of the Liquavista^{TM}, each "pixel-electrode" of the screen is occupied by an oil droplet, and when

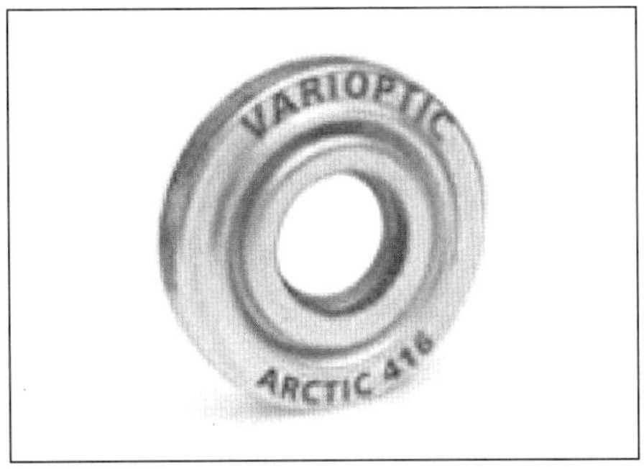

Figure 10.4 Left: display screen based on electrowetting by LiquaVista™: ColorMatch prototype display manufactured by Liquavista. The display diagonal is 2,5" [12]. Right: adjustable optical lens by Varioptic™: Varioptic liquid lens Artic 416 [13].

the actuation is switched off, the droplet shrinks and is pushed towards a small corner of the pixel-electrode. In the case of the Varioptic™lens, the convexity of the liquid lens is controlled by the intensity of the electric actuation.

10.5 Summary

This book has presented the background for the study of the behavior of micro-drops and its applications to digital and droplet microfluidics. Considering the whole domain of today's microfluidics, schematized in Fig. 10.1, the amount of progress done since the beginning of that science is considerable; however, a considerable amount of work remains to be done to cover all the applications that microtechnologies require. Coincidentally, the same conclusion can be made, based on totally different grounds, by comparison with the developments of microelectronics. The parallel is done principally because the first microfluidic systems were realized with devices obtained by microfabrication techniques used in microelectronics. The comparison is based on the following observation: it has taken approximately 50 years for microelectronics to become mature—approximately between the years 1950 and 2000. If the same is to be expected in the field of biotechnology, we are just at midway of the development of microfluidic systems. And this leaves a considerable field for new developments in the years to come.

References

1. Krishnendhu Chakrabarty and Fei Su, *Digital Microfluidic Biochips: Synthesis, Testing, and Reconfiguration Techniques*, Taylor and Francis, 2007.
2. Nicolas Verplanck, Elisabeth Galopin, Jean-Christophe Camart, Vincent Thomy, Yannick Coffinier and Rabah Boukherroub, "Reversible Electrowetting on Superhydrophobic Silicon Nanowires," *Nano Lett*, Vol. 7, no. 3, pp. 813–817, 2007.

3. T.N. Krupenkin, J.A. Taylor, T.M. Schneider and Shu Yang, "From rolling ball to complete wetting: the dynamic tuning of liquids on nanostructured surfaces," *Langmuir*, Vol. 20, pp. 3824–2827, 2004.
4. M.S. Dhindsa, N.R. Smith, J. Heikenfeld, P.D. Rack, J.D. Fowlkes, M.J. Doktycz, A.V. Melechko and M.L. Simpson, "Reversible electrowetting of vertically aligned superhydrophobic carbon nanofibers," *Langmuir*, Vol. 22, pp. 9030–9034, 2006.
5. G. McHale, D.L. Hebertson, S.J. Elliott, N.J. Shirtcliffe and M.I. Newton, "Electrowetting of nonwetting liquids and liquid marbles," *Langmuir*, Vol. 23, pp. 918–924, 2007.
6. A. Günther, M. Jhunjhunwala, M. Thalmann, M.A. Schmidt and K.F. Jensen, "Micromixing of miscible liquids in segmented gas-liquid flow," *Langmuir*, Vol. 21, pp. 1547–1555, 2005.
7. Todd Thorsen, Richard W. Roberts, Frances H. Arnold and Stephen R. Quake, "Dynamic pattern formation in a vesicle-generating microfluidic device," *Phys. Review Letters*, Vol. 86, no. 18, pp. 4163–4166, 2001.
8. Joshua D. Tice, Helen Song, Adam D. Lyon and Rustem F. Ismagilov, "Formation of droplets and mixing in multiphase microfluidics at low values of the Reynolds and the Capillary numbers," *Langmuir*, Vol. 19, no. 19, pp. 9127–9133, 2003.
9. Pierre Guillot, Annie Colin, Andrew S. Utada and Armand Ajdari, "Stability of a Jet in Confined Pressure-Driven Biphasic Flows at Low Reynolds Numbers," *Phys. Review Letters*, Vol. 99, p. 104502, 2007.
10. Erwin Berthier and David J. Beebe, "Flow rate profile of a surface tension driven passive micropump," *Lab-on-a-Chip*, Vol. 11, pp. 1475–1478, 2007.
11. Jean Berthier and Christine Peponnet, "A model for the determination of the dimensions of dents for jagged electrodes in electrowetting on dielectric microsystems," *BioMicrofluidics*, Vol. 1, no. 1, p. 014104, 2007.
12. Liquavista^{TM}: http://www.liquavista.com/files/LQV060828XYR-15.pdf.
13. Variioptic^{TM}: http://www.varioptic.com/en/tech/technology01.php.

Index

MICRO & NANO TECHNOLOGIES

PUBLISHED 2008

1. *Microdrops and Digital Microfluidics* · Jean Berthier · 978-0-8155-1544-9
2. *Micromixers: Fundamentals, Design and Fabrication* · Nam-Trung Nguyen · 978-0-8155-1543-2

FORTHCOMING 2008

Fabrication and Design of Resonant Microdevices · Behraad Bahreyni · 978-0-8155-1577-7

Hot Embossing: Theory and Technology of Microreplication · Matthias Worgull · 978-0-8155-1579-1

Introduction to Quantum Information Processing (QIP) · Timothy P. Spiller and William J. Munro · 978-0-8155-1575-3

Nanotechnology Applications: Solutions for Improving Water Quality · Edited by Mamadou Diallo, Jeremiah Duncan, Nora Savage, Anita Street, Richard Sustich · 978-0-8155-1578-4

Micro-machining Using Electrochemical Discharge Phenomenon: Fundamentals and Application of Spark Assisted Chemical Engraving · Rolf Wüthrich · 978-0-8155-1587-6

Industrial Micro and Nano Fabrication · J.G.E. Gardeniers and R. Luttge · 978-0-8155-1582-1

Physics of Carbon Nanotube Devices · François Léonard · 978-0-8155-1573-9

Small Scale Mechanics: Principles and Applications · David Mendels · 978-0-8155-1590-6

Emerging Nanotechnologies for Manufacturing · Edited by Waqar Ahmed and M.J. Jackson · 978-0-8155-1583-8

Risk Governance of Nanotechnology: Environmental, Health and Safety Concerns About Nanotechnology and Their Implication for the Nanotechnology Industry · Edited by Steffi Friedrichs · 978-0-8155-1586-9

Handbook of MEMS Materials and Technologies · Edited by Veikko Lindroos, Markku Tilli, Ari Lehto, and Teruaki Motooka · 978-0-8155-1594-4

Micromanufacturing Engineering and Technology · Edited by Yi Qin · 978-0-8155-1545-6

OF RELATED INTEREST

MEMS: A Practical Guide to Design, Analysis and Applications · Edited by Jan Korvink and Oliver Paul · 978-0-8155-1497-8 · 2006

Nanostructured Materials: Processing, Properties and Applications, 2nd Edition · Edited by Carl C. Koch · 978-0-8155-1534-0 · 2007

Ultrananocrystalline Diamond: Synthesis, Properties, and Applications · Edited by Olga A. Shenderova and Dieter M. Gruen · 978-0-8155-1524-1 · 2006

For the latest information on related titles visit **www.williamandrew.com/MNT/**